The Proceedings of the 2nd International Conference
October 10-12, 2002, Shenyang, P. R. China
on

New Development in Rock Mechanics & Rock Engineering

Chief Editor:

Yunmei Lin   Chun'an Tang

Vice Chief Editor:

Xiating Feng   Shuhong Wang

Rinton Press

Library of Congress Control Number: 2002109138

12 Castleton Road
Princeton, New Jersey 08540
USA
rinton@rinton.com
http://www.rinton.com

Published by Rinton Press, Inc.

ISBN: 1-58949-021-5

Foreword

It is a great honor of our Northeastern University to host the Second International Conference on New Development in Rock Mechanics and Rock Engineering (NDRM'2002) in the coming golden October. The purpose of the conference is to exchange new ideas and advanced development in problem solving in rock mechanics theory and practice, to provide an opportunity to strengthen academic relation among experts from home and abroad, thereby to promote the development of rock mechanics science. Participants of the conference are from Australia, Canada, China, France, India, Japan, Korea, Portugal, Russia, Sweden, United Kingdom, and United States of America. Many of them are internationally well known experts and have been actively working at the forefront of rock mechanics field for many years.

May I express the hope that this international conference contributes to making strong ties among the scholars from homeland and abroad, raises the level of research and practice in rock mechanics science, and promotes the friendships among scientists from the World.

The Northeastern University, established in 1923, is one of the leading universities in China. In the days to celebrate the coming 80th anniversary of this University, I wish to express my sincerely appreciation to every author joining with this conference. Meanwhile, we hope to further develop relation with local and foreign universities, institutes and enterprises. We also expect to have more experts and scholars at home and abroad come and visit, give lectures, and have cooperation in research with our university as well.

Professor Jicheng He

President

Northeastern University

June, 2002

Preface

The Second International Conference on New Development in Rock Mechanics and Rock Engineering is held in The Northeastern University from 10th to 12th of October 2002. It provides a forum for these engaged in research and development of rock mechanics and engineering. There are presentations and discussions on original researches, new developments, and case studies in rock mechanics and rock engineering for topics as wide as possible throughout the industry. Both scientists and engineers from civil, mining, geological, hydro-electrical and related fields participate in the conference.

This volume of proceedings comprises of 113 papers selected from these submitted to NDRM'2002. According to different aspects of rock mechanics science, these papers have been divided into 6 parts: **General Problem and Fundamental Research; Experimental Techniques and Field Measurements; Numerical Analysis and Computer Simulation; Artificial Intelligence and System Science; Rock Mass Engineering Stability Analysis, Design, Support and Construction; Mining Practice, Rock Blasting and Rock Burst.** The day to day job for the conference is planned by the local organizing committee.

The full support by colleagues from **The China National Natural Science Foundation Committee; The Shenyang Scientific Technical Committee; Institute of Rock and Soil Mechanics, Chinese Academy of Sciences; College of Resources and Civil Engineering, Northeastern University; Chinese Rock Mechanic Northeast Society; Rinton Press (USA)** makes it possible for the proceedings being published before the conference. We appreciate very much the hard works by authors and all colleagues.

The members of international advisory committee have supported the conference in the whole process of preparation. We are indebted to them for their guidance and experiences in setting up this conference.

A final thank is owed to the authors and conference's participants for their interests in the conference and their scientific contributions.

Chief Editor:
Yunmei Lin, Chun'an Tang
Vice Chief Editor:
Xiating Feng, Shuhong Wang

The Second International Conference on

New Development in Rock Mechanics and Rock Engineering

October 10 to 12, 2002 Shenyang, P. R. China

Organized by

Northeastern University (NEU)
Chinese Rock Mechanic Northeast Society

Sponsored by

The China National Natural Science Foundation Committee
Shenyang Scientific Technical Committee
Institute of Rock and Soil Mechanics, Chinese Academy of Sciences
College of Resources and Civil Engineering, Northeastern University
Chinese Rock Mechanic Northeast Society
Rinton Press (USA)

CONFERENCE CHAIRS

Prof. Yunmei Lin (China)
Prof. Chun'an Tang (China) Prof. Xiating Feng (China)
Prof. Jean-Claude Roegiers (USA) Prof. Chung-In Lee (Korea R)

PROGRAM COMMITTEE

Prof. Sijing Wang (Honorary President) Prof. Xiaohe Xu (President)
Prof.Yongjia Wang Prof. Baozhi Chen
Prof. Li Liang Prof. Qing Wang

LOCAL ORGANIZING COMMITTEE

Prof. Yunmei Lin (Chair)
Prof. Shiwen Zhu Prof. Wen Zhao
Dr. Shuhong Wang Dr. Wei Chen

INTERNATIONAL ADVISORY COMMITTEE

Prof. Sijing Wang
Institute of Geology & Geophysics, Chinese Academy of Sciences
President, Chinese Society for Rock Mechanics & Engineering
Member, Chinese Academy of Engineering, (China)

Prof. Jun Sun
Tong-Ji University
Member, Chinese Academy of Science (China)

Prof. Chung-In Lee
School of Civil, Urban and Geo-system Engineering, Seoul National University
Vice-president for Asia, ISRM (Korea R)

Dr. Satoshi Hibino
Central Research Institute of Electric Power Industry
Vice-president at Large, ISRM (Japan)

Prof. Pekka S. Sarkka
Helsinki University of Technology
Vice-president for Europe, ISRM (Finland)

Prof. Euripedes do A. Vargas Jr.
Dept. of Civil Engineering , Catholic University of Rio de Janeiro
Vice-president for South America, ISRM (Brazil)

Dr. Jose Delgado Rodrigues
International Society for Rock Mechanics
Secretary General, ISRM (Portugal)

Prof. Jean-Claude Roegiers
Rock mechanics Institute, University of Oklahoma
McCasland Chair (USA)

Prof. Barry H. Brady
The University of Western Australia
Dean, Dept. of Engineering and Mathematical Sciences (Australia)

Prof. Minh Phong Luong
Centre National de la Recherche Scientifique (France)

Prof. Alexander M. Linkov
Institute for Problems of Mechanical Engineering
Russian Academy of Sciences (Russia)

Prof. Alexander B. Fadeev

St. Petersburg University of Architecture and Civil Engineering (Russia)

Prof. John A. Hudson
Imperial College and Rock Engineering Consultants (United Kingdom)

Prof. Zhifa Yang
Institute of Geology & Geophysics, Chinese Academy of Sciences
Secretary General, Chinese Society for Rock Mechanics & Engineering (China)

Prof. Bing Jun Fu
Institute of Geology & Geophysics, Chinese Academy of Sciences (China)

Dr. Victor V. Zoubkov
State Scientific Research Institute of Rock Mechanics and Mining Surveying
Corporative Scientific Center VNIMI. (Russia)

Prof. Yoshiaki Mizuta
Yamaguchi University,
President, The Mining and Materials Processing Institute of Japan (Japan)

Prof. P. A. Lindqvist
Lulea University of Technology, Civil and Mining Engineering
Member, the Royal Institute of Engineering (Sweden)

Dr. Daniel Francois Malan
CSIR Miningtek (South Africa)

Prof. Nielen van der Merwe
University of Pretoria (South Africa)

Prof. Wangxi Zhu
Dean of Metallurgy & Mining Subject
National Natural Science Foundation of China (China)

Prof. Li Li
Vice-chairman of Shenyang Scientific Technical Committee (China)

Prof. John Phillip Carter
Dept. Of Civil Engineering, University of Syndey (Australia)

CONTENTS

2. *EXPERIMENTAL TECHNIQUE AND FIELD MEASUREMENT*

3. *NUMERICAL ANALYSIS AND COMPUTER SIMULATION*

4. ARTIFICIAL INTELLIGENCE AND SYSTEM SCIENCE

5. ROCK MASS ENGINEERING STABILITY ANALYSIS, DESIGN, SUPPORT AND CONSTRUCTION

6. MINING PRACTICE, ROCK BLASTING AND ROCK BURST

1. General Problem and Fundamental Research

PRINCIPLES OF THE STRESS PATH AND ITS VISUALISATION FOR ROCK MECHANICS AND ROCK ENGINEERING

J.A. HUDSON and J.P. HARRISON
Department of Earth Science and Engineering, Imperial College, London, UK
jah@rockeng.co.uk and j.p.harrison@ic.ac.uk

A set of stress path principles is presented for rock mechanics modelling and rock engineering design. Methods for visualizing changes in the stress tensor with time are also discussed.

1 Introduction

The basic procedure in considering the rock stress aspect of rock mechanics design is to consider the pre-existing stress-state in the rock mass and calculate the disturbance to this stress state as a result of engineering excavation. Failure criteria are then applied to the stress states as excavation proceeds to establish whether failure will occur. Recently, there has been research activity in this area: Hakala[1] studying stress changes during overcoring; Eberhart[2] evaluating the 3-D stress changes in the proximity of an advancing tunnel face; and Kaiser et al.[3] assessing the damage that can occur during mining of complex excavation geometries. If some portion of the stress path intersects and then follows the failure locus (considered as either initiation or propagation of damage), see Figure 1, both the strength and the stiffness of the rock around an excavation can be altered from the pre-construction values.

An overview of the stress path principles during generic excavation in rock masses does not seem to have been presented. Thus, in this paper we present a set of stress path principles and offer suggestions for stress path visualisation.

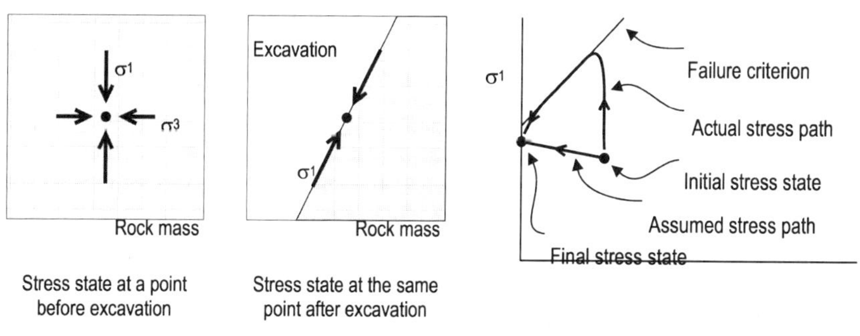

Figure 1. Illustration in $\sigma_1 - \sigma_3$ space of the changes in magnitudes and orientations of principal stresses as induced by excavation.

2 Principles of stress paths in the context of excavation for rock engineering

In order to consider the stress path, it is first necessary to establish a set of stress path principles. These are listed below.

1. The stress path* is defined as the variations in magnitudes and orientations of the three principal stresses in the stress tensor as a result of natural or engineering-induced changes.
2. Stress paths are geometrically constrained by two factors. One is the convention that $\sigma_1 \geq \sigma_2 \geq \sigma_3$, which has the effect that only part of the entire $\sigma_1 - \sigma_2 - \sigma_3$ space can be traversed by the stress path. The other constraint is that certain states are not sustainable in an engineering context, due to the fact that rock follows a failure criterion, and hence the traversable region of $\sigma_1 - \sigma_2 - \sigma_3$ space is limited.
3. In an elastic material, by definition failure does not occur, and so the stress path is unconstrained in terms of failure and of no practical significance.
4. In an inelastic material, the stress path does have practical significance, because if it reaches the failure initiation locus, not only will the properties of the material change, but the final stress state will be different from that based on elasticity.
5. Time is considered to be the independent variable for the stress path, although this can often be correlated directly to a spatial variable, such as tunnel advance.
6. The stress path in terms of changes in the principal stress magnitudes can be plotted directly in $\sigma_1 - \sigma_2 - \sigma_3$ space; the stress path in terms of changes in principal stress orientations can be plotted directly on the hemispherical projection. To plot both on the same diagram requires a hybrid presentation.
7. Because rock is an inelastic material, the stress path should be tracked in all analyses of stress redistribution, to account properly for those circumstances when the stress path reaches the failure locus.
8. Different excavation sequences will give rise to different stress paths, some of which may reach the failure locus; it may be possible to ameliorate (or exacerbate) this effect for engineering purposes by considerations of the excavation sequence.

With reference to these principles, in Figure 2 we illustrate unconstrained and constrained stress paths for the process of excavation shown in Figure 1. Case (a) is for an elastic material. Such materials are defined as those that undergo no permanent deformations, and all strain energy can be recovered. This means that the material has no failure locus (Principle 3), and so the stress path is unconstrained (except for the convention that $\sigma_1 \geq \sigma_2 \geq \sigma_3$, Principle 2). In case (b), which is for an inelastic material having a failure locus, the stress path is constrained by the presence

---

* The term 'stress path' has been used (e.g. Lambe and Whitman, 1979 [4]) to refer to the sequence of loading in a laboratory test. Here we use the term in a more general sense to refer to the changes in the stress tensor with time.

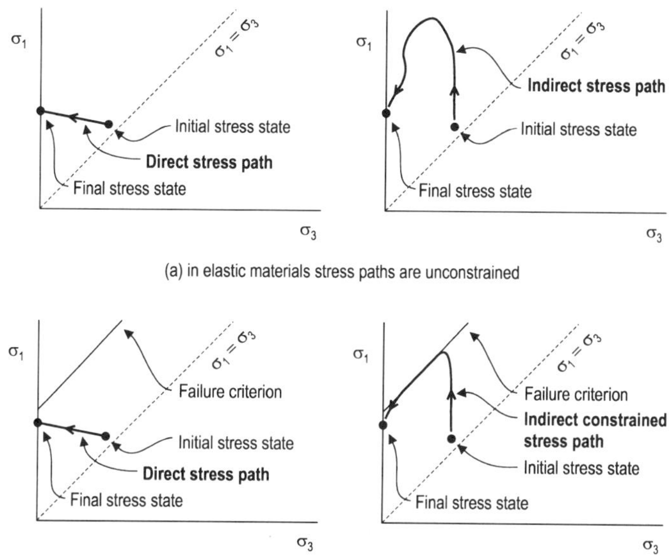

(a) in elastic materials stress paths are unconstrained

(b) in inelastic materials, stress paths are constrained by the failure locus

Figure 2. Unconstrained and constrained stress paths.

of the failure locus (Principle 4). Moreover, attainment of the failure locus means that the rock will become damaged, and this may affect the final stress state. On this basis, it is possible that the two final stress states, a) and b) in Figure 2, will be different — which has ramifications for practical rock engineering.

3 Visualising changes in the six-component stress tensor

In Figure 3a), we show a generic stress path in $\sigma_1 - \sigma_2 - \sigma_3$ space, which illustrates the changes in the stress magnitudes. This diagram also shows the constraints noted in Principle 2: i.e. the convention of naming the principal stresses and how the stress path must be wholly contained beneath the locus of a failure criterion. Note here that we have drawn the failure criterion as being independent of the intermediate principal stress. Another excavation constraint is that all unsupported excavation surfaces must be principal stress planes (because no shear stresses act on them), and the normal stress on these surfaces is zero (by Newton's Third Law, neglecting atmospheric pressure). Therefore, the stress path for a point in the rock mass that will eventually become part of an unsupported excavation surface is additionally constrained to terminate on the $\sigma_3 = 0$ plane.

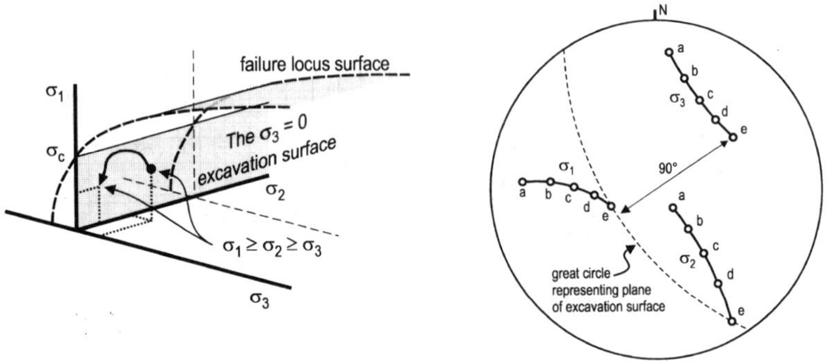

Figure 3. a) Changes in principal stress magnitudes. b) Changes in principal stress orientations

In Figure 3b) we show a generic stress path in *xyz* space, plotted on a hemispherical projection, illustrating the changes in the stress orientations. This diagram shows the fundamental constraint of the orthogonality of principal stresses. Corresponding to the constraint introduced by the excavation plane in Figure 3, the orientational locus for the minor principal stress in Figure 3b) terminates at the orientation which is normal to the excavation surface. This surface is represented by a great circle on a hemispherical projection, as indicated in Figure 3b), and shows how the locus of σ_3 terminates at its pole.

The two options for a composite presentation are either to include orientational information on the stress magnitude plot, Figure 3a), or stress magnitudes on the orientational plot, Figure 3b) — both of which are difficult to construct. The authors are currently working on alternative methods of presentation that will provide both the technical rigour and the presentational clarity required.

4 Conclusions

In this paper we have noted the importance of understanding the stress path in rock mechanics modeling and for rock engineering design. We presented eight stress path principles as the necessary precursor to considering methods for the visualisation of changes in the stress tensor with time.

5 References

1. M. Hakala, *Personal communication* (2002)
2. E. Eberhardt, *Int. J. Rock Mech. Min. Sci.*, **38**,4, 499-518 (2001)
3. P.K. Kaiser et al., *Int. J. Rock Mech. Min. Sci.*, **38**, 2, 167-180 (2001)
4. T.W.Lambe et al, *Soil Mech., SI version*. John Wiley, New York, p.553 (1979).

DEVELOPMENT OF A ROCK MECHANICS STRATEGY FOR SITING THE SWEDISH REPOSITORY FOR SPENT FUEL

R. CHRISTIANSSON
Swedish Nuclear Fuel and Waste Management Co., Stockholm, Sweden.
rolf.christiansson@skb.se

J.A. HUDSON
Rock Engineering Consultants and Imperial College, UK
jah@rockeng.co.uk

A series of site investigations is planned as part of the Swedish nuclear waste disposal programme. In order to provide the necessary rock mechanics support for these site investigations, a strategy has been developed which includes the follwing items:

- quality control of rock mechanics data, including techniques to reduce uncertainties;
- a descriptive rock mechanics methodology (which has been verified by a Test Case using protocols and an auditing procedure) enabling the crystalline rock mass to be characterized in terms of constructability and performance assessment; and
- the use of the rock mechanics information for designing the repository.

The development of the rock mechanics strategy and these items are explained in the paper.

1 Introduction

The Swedish Nuclear Fuel and Waste Management Co. (SKB) is responsible for the handling and final disposal of the nuclear waste produced in Sweden. In 2002, SKB plans to start site investigations using deep boreholes at two different sites with crystalline rock[1]. The site investigation phase will provide the broad knowledge base that is required to evaluate the suitability of investigated sites for a deep repository. The Site Investigation is planned to continue for a number of years and the field investigations will be carried out in increments. The interpretation of the measured data is made in terms of a *Site Descriptive Model* covering geology, rock mechanics, thermal properties, hydrogeology, hydrogeochemistry, transport properties and surface ecosystems. The Site Descriptive Model is a cornerstone for the understanding of the investigated site and forms a basis for subsequent planning of the repository layout, as well as for Safety Assessment studies. Work on the Design and Safety Assessment, as well as studies of the Environmental Impact Assessment, will be carried out in parallel and will therefore allow feedback to the characterization programme[2] which defines the different characterization stages and describes the interaction needed between different disciplines.

As the reference planning case, SKB is preparing for disposal of 4,500 canisters of spent fuel using the KBS-3 method. This may require a ground surface for the repository of 2 – 4 km^2. The repository location is planned to be between 400 m and

700 m depth in granitic rock. The selection of information to be considered during the site characterization program will depend on several assessments of what is required for use in the safety assessment and design. This paper describes how SKB has developed the full Site Investigation Programme, with special emphasis on Rock Mechanics, as being one of the important diciplines for design of the deep repository.

2 The Identification of Parameters of Importance

The Swedish programme for disposal of nuclear waste is based on a relatively long experience, including geology studies, the Stripa mine work, underground cavern studies, the CLAB facility close to the Oskarshamn nuclear power station, the low and intermediate waste repository at Forsmark, and the AEspoe Hard Rock Laboratory[3] excavated from 1990 to 1995. The research is presented in many reports, see www.skb.se.

The parameters of importance in geoscientific site investigations for geological disposal were first summarised by Andersson[4] and a deeper analysis of the importance of these parameters was presented subsequently[5]. Within the rock mechanics subject, there is the risk of extensive spalling — identified as a criterion that could be disadvantageous, for costs and workers safety, but most likely not for the long term safety. However, more extensive and reliable information is required to enable understandiing of the possible rock mass response to the construction, operation, thermal load from the spent fuel and the possible impact of scenarios considering future development of the geological environment, for example by climate changes (e.g. glaciations).

3 Development of a Program for Data Collection

In 2001, SKB presented a complete generic Site Investigation Program[2]. The key issues are:

- design the programme so that the information needed is actually collected,
- be able to describe the properties of the site in space and be able to judge the uncertainty in the description based on measurement results from the ground surface and from a limited number of boreholes,
- make sure that different disciplines are coordinated and build up a consistent description of geosphere and biosphere,

- identify whether there are measurements that must be carried out in a certain sequence, since certain measurement methods have a short-term impact on the site, and
- make sure that the programme focuses on the different information needs at different points in the decision-making process.

The generic program also describes with what methods data will be collected. The input to the rock mechanics is also given[2]. Detailed descriptions on how each method should be applied have been collated. These documents form one of the cornerstones in the SKB Quality Assurance strategy for the Site Investigation Program. The documents are structured to cover the purpose of the method, the reference for the method, technical equipment needed, preparational work, testing/measuring procedures, analysis and interpretation (if required), data management, the review process of results and management of documents.

By developing further specifications, SKB has set up a number of Tender Dossiers for the various data collection contracts during the Site Investigations. Analysis of risks for completion of the Programme has identified for example the risk of lack of resources (manpower, equipment) and the risk of systematic errors during data collection. Long term agreements have been signed, based on technical and finacial offers, to secure resources for the Programme. Overall quality aspects, for example the need to conduct comparative tests between laboratories that carry out rock mechanics testing, have also been considered.

For the current Site Investigation Programme efforts been put into increasing the understanding of the accurancies and limitations for methods used for *in situ* stress measurements. Various types of equipment have been tested in two orthogonal boreholes under controlled conditions in a limited volume of rock at 455 m depth at the AEspoe Hard Rock Laboratory[6]. The limitations in methods have also been studied, and a scheme for Technical and Quality Auditing has been developed[7].

4 Development of a Strategy for Site Characterisation

As a part of the planning work, SKB has developed strategies for Site Descriptive Modelling. The Site Descriptive model is established on regional and local scales. The local scale model will cover the possible location for the repository and its immediate surrounding. Interpretations made in different disciplines should be mutually consistent. Spatial variability as well as uncertainty due to sparse data, irregular distribution of data and interpretative issues should be handled and illustrated. The approach to the Rock Mechanics Site Descriptive Model[7] must consider the analyses and parameters required in the repository design, see Fig. 1.

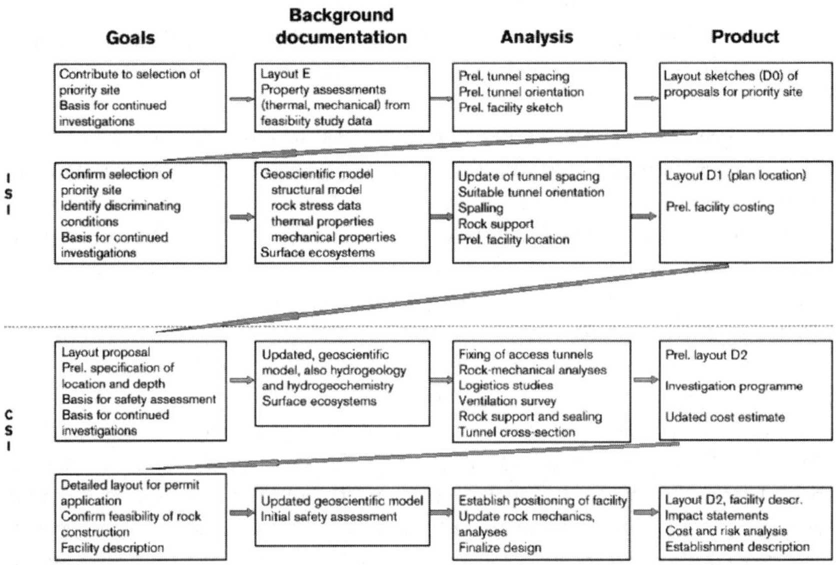

Figure 1. Steps in the initial site investigation (ISI) and complete site investigation (CSI)[2]. The third column indicates the types of rock mechanics and rock engineering design studies that will take place.

Thus, the Rock Mechanics Site Descriptive Model should describe the initial stresses and the variability of rock mechanics properties, such as the deformability and strength properties of the intact rock, of fractures and fracture zones, and of the rock mass viewed as a unit consisting of intact rock and fractures. The parameters for the rock mass should primarily be provided on the tunnel scale, capturing the mechanical processes that may be expected around underground openings of the sizes planned for the KBS-3 repository in the Swedish system.

5 Testing of the Strategy for Site Characterisation

The approach to the Rock Mechanics Site Descriptive Model was tested by predicting the rock mechanics properties of a 600 m x 180 m x 120 m rock volume at the AEspoe Hard Rock Laboratory (HRL)[8], see Fig. 2, using limited borehole data of the type typically obtained during a site investigation. These predicted properties were then compared with 'best estimate' properties obtained from a study of the test rock volume using additional tunnel and tunnel borehole data. This exercise was known as the Test Case.

Three modelling techniques were used to predict the rock properties:

- the 'empirical approach' — the rock properties were estimated using rock mass classification schemes and empirical correlation formulae[9];
- the 'theoretical approach' — the rock properties were estimated using numerical modelling techniques[10]; and
- the 'stress approach' — the rock stress state was estimated using primary data and numerical modelling[11].

A report on the 'best estimates' of the actual rock conditions is also available[12].

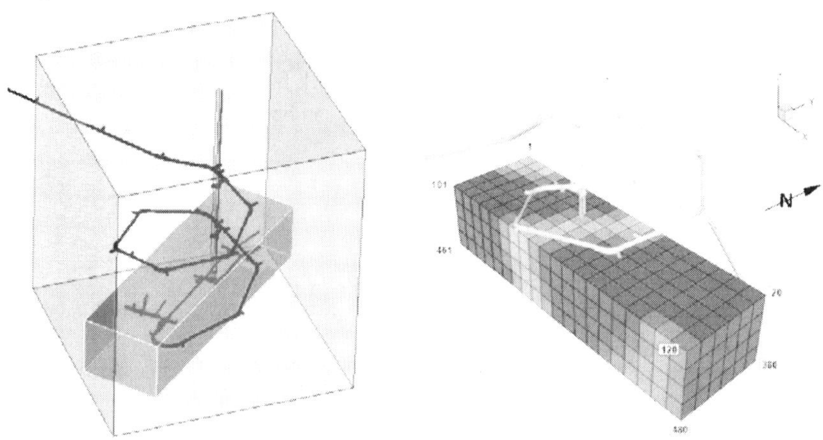

a) b)

Figure 2. a) The rock volume used for the Test Case, showing the access tunnel and shaft.
b) Numbering of the blocks in the Test Case 600m x 180m x 120m volume, from Staub et al. (2002).

The Test Case exercise was invaluable in forcing the generic approach to the Rock Mechanics Site Descriptive Strategy to become a specific approach for the AEspoe HRL rock volume being considered. The predictions of the rock mass modulus between fracture zones made by the empirical and theoretical teams were similar and close to the 'best estimate' values. The predictions of the rock mass modulus in fracture zones were more varied, because few data were available and hence there was greater choice in the assumptions necessary to make the predictions. The empirical and theoretical teams' predictions of the rock strength were also more varied, mainly because few data were available and different interpretations of rock mass strength were used. The rock stress predictions, based on database information and numerical modelling, were satisfactory and helped to focus attention on the key factors involved. In particular, the use of numerical modelling assists in evaluating potential stress changes in the vicinity of fracture zones. Methods of dealing with conceptual uncertainty and spatial variability of stress were successfully introduced into the rock stress characterization and predictive methodology

Aspects of the Test Case work that, with hindsight, could have been improved were: better definition of the rock properties to be predicted; more attention paid to difficulties in 3-D extrapolation, i.e. how to assign values to points distant from known information; more rigorous application of the Quality Control instruments; more encouragement of inter-team discussions; and earlier implementation of external reviewing of the work.

Thus, the Test Case provided a opportunity to test the methodology and smooth out problems that are highlighted by having to apply a general approach to a specific site. Also, the Äspö HRL can be used as a benchmark for crystalline rock conditions against which further Test Cases in different areas can be compared.

6 Overview of the Design Process

When the site investigations are finished, the design shall have:

- presented one site-adapted deep repository facility among several analyzed and proven its feasibility,
- identified facility-specific technical risks, and
- developed detailed design premises for the detailed characterization phase.

The facility description is developed stepwise, as illustrated schematically in Fig. 1. The non-site-specific facility description, called layout E (second column in Fig. 1), is available before the site investigations are begun. It is based on a generic coordinated proposal for the geometric design of the deep repository and serves as a basis for the continued site-specific design process. The final development step during the site investigation phase will be based on the information from the complete site investigation. In each step, various analyses are performed whose results are used as a basis for the detailed steering of the continued site investigation programme, as a basis for the safety assessment work and for the continued design work. The flow of background documentation, analyses and results is in Fig. 1.

The facility description comprises one of the prerequisites for the preparation of draft documents for the rock and construction work, as well as the preparation of programmes for layout-governing systems in the underground facility, such as ventilation, rock drainage and electricity supply. The results of the site investigations are used in the design process to propose locations for the various facility sections, whereby the site-adapted layout of the deep repository is obtained.

The design work is updated as more geoscientific descriptions becomes available With the guidance of the layout and the geological-structural model, critical passages of tunnels through fracture zones are identified. The need for rock support is assessed. With the guidance of the hydrogeological model, an assessment is made of the sealing needs, particularly at passages of water-bearing fracture zones. Sealing needs are influenced by technical and environmental requirements.

7 Rock Mechanics Analyses for Designing the Repository

A number of rock mechanics analyses will carried out, such as:

- detailed thermo-mechanical sensitivity analyses for deposition holes and the rock between deposition tunnels, based on measured ranges of variation and estimated uncertainties in mechanical properties and boundary conditions,
- updating (choice of different alternatives) of suitable tunnel orientation based on stresses (levels and directions) and directions of local minor fracture zones and fractures within the rock blocks,
- updated stability and rock support forecasts based on updated tunnel geometry, updated stresses, stress directions and rock mechanics properties,
- stability forecasts for deposition holes and deposition tunnels based on tunnel geometry, fracture geometry, rock stresses and the mechanical properties of the fractures, and
- optimization of cross-sections and directions for operating and disposal chambers in the deep repository.

The analyses influence the main layout for the deep repository based on the number of deposition holes and the space requirements for other facility sections.

8 Conclusions

Based on SKB experience and the identification of important parameters for long term safety, as well as consideration of the construction and operation of the repository, programmes for data collection and interpretation have been developed. The verification of methods and strategies have been tested and the AEspoe HRL has been an important facility for these purposes. The Site Investigation programme has now been approved by the Swedish Government, athourities, and locally by communities.

References

1. SKB, Integrated account of method, site selection and programme prior to the site investigation phase. SKB TR-01-03. Swedish Nuclear Fuel and Waste Management Co., Stockholm (2001).
2. SKB, *Site investigations: investigation methods and general execution programme.* SKB TR-01-29, Swedish Nuclear Fuel and Waste Management Co., Stockholm, (2001).
3. SKB, *Äspö Hard Rock Laboratory***.** SKB TR 94-11. Swedish Nuclear Fuel and Waste Management Co., Stockholm, Sweden (1994).
4. J.Andersson, K.-E. Almén, L.-O. Ericsson, A. Fredriksson, F. Karlsson, R. Stanfors, A. Ström, *Parameters of importance to determine during geoscientific site investigation.* SKB TR 98-02, Swedish Nuclear Fuel and Waste Management Co., Stockholm, Sweden (1998).
5. J. Andersson, A. Ström, C. Svemar, K.-E. Almén, L.-O. Ericsson, *What requirements does the KBS-3 repository make on the host rock? Geoscientific suitability indicators and criteria for siting and site evaluation.* SKB TR-00-12, Swedish Nuclear Fuel and Waste Management Co., Stockholm, Sweden (2000).
6. R. Christiansson and T. Janson, Test with three different stress measurement methods in two orthogonal bore holes. *Proceedings of the NARMS2002-TAC Conference* held in Toronto (2002).
7. J. Andersson, R. Christiansson and J.A. Hudson. *Site investigations strategy for development of a rock mechanics site descriptive model.* SKB TR 02-01, Swedish Nuclear Fuel and Waste Management Co., Stockholm (2002).
8. J A Hudson (ed.), *Strategy for a rock mechanics site descriptive model: A Test Case based on data from the AEspoe HRL.* SKB Report R 02-04 (2002).
9. K. Röshoff, F. Lanaro and L. Jing, *Strategy for a Rock Mechanics Site Descriptive Model: Development and testing of the Empirical Approach.* SKB report R 02-01, (2002).
10. I. Staub, A. Fredriksson and N. Outters, *Strategy for a Rock Mechanics Site Descriptive Model: Development and testing of the Theoretical Approach.* SKB report R 02-02, (2002).
11. E. Hakami, H. Hakami and J. Cosgrove, *Strategy for a Rock Mechanics Site Descriptive Model: Development and testing of an approach to modelling the state of stress.* SKB report R 02-03, (2002).
12. A. Makurat, F. Løset, A.W. Hagen, L. Tunbridge, V. Kveldsvik and E. Grimstad. *Äspö HRL: A Descriptive Rock Mechanics Model for the 380-500 m level.* SKB report R 02-11, (2002).

SUSTAINABLE DEVELOPMENT OF MINING IN CHINA AND COAL MINING ENVIRONMENTAL PROBLEMS

JIAN-GUO YAO
Beijing Mining Research Institute, Central Coal Research Institute,
Beijing, 100013, China
E-mail:yaojg@sina.com

This paper indicates the potential of resources development, such as great potentials of traditional mineral resources on the land, considerable potentials of mineral reserves in the sea area, considerable potentials of natural gas, coal-bed methane and thermal heat resources. Second, this paper points out the rock mechanic's trouble problems, which must to be solved for sustainable development of mining in China. Meanwhile, the present paper discusses coal mining environmental problems, including coal mining damages to land, damages to the buildings in the villages, influence of coal mining to water resource, pollution of water resources, influence of coal extraction and utilization on atmospheric environment etc, which have great influence on ecological balance. We put forward a lot of measures to against environment pollution in the mine areas, such as popularization of mining techniques with less subsidence, protection and re-use of mine water resources, as well as clean mining method.

1 Introduction

The general strategic framework for achieving sustainable development was depicted by white paper entitled "China Agenda of 21st Century" published in March of 1993 by the State Council. Entering into the 21st century, China is facing more opportunities with further development of economy. The Chinese government formulated a strategic plan for developing vigorously the western area. The 10th National Five-year Plan also started in year of 2001 in China. The rock mechanics and rock engineering are closely linked with development of national economy. Rock mechanics and rock engineering, as important pillars of national economic constructions, also face new opportunities and severe challenges when entering the 21st century.

2 Sustainable Development of Mining in China

2.1 Resources Potentials of Sustainable Development of Mining in China

Currently, the product of mining in China accounts for 6.27% of the gross national product. If this ratio is to be kept by 2010, the mining product value will reach 1,060 billion Yuan RMB. According to experts' prediction, production of mining products shall increase to a great extent in order to meet the demand in 2010 and in 2020.

In the past 50 years, large-scale geological surveys of the mineral deposits were made in China. A great number of mineral deposits that can be exploited and used were found. Considerable mineral reserves were proven. Although, large quantities of mineral resources were consumed in the development of construction in half a century, and the production capacity of some mines declined, China still has a considerable potential of mineral resources to meet the need of sustainable development. The potential of resources is mainly embodied in the following aspects:

2.1.1 Great potentials of traditional mineral resources on the land

The proven reserves of some important minerals such as oil, coal, copper and gold are only 20-25% of the predicted reserves. A group of new mineral deposits can be found in the West China, and the potentials of some old mine areas in East China can be further tapped. For example, it is predicted that there exist 5,000 billion tons of coal reserves in the extent from surface to 2,000m below the surface. However, the present proven reserves are only 1,000 billion tons.

2.1.2 Considerable potentials of mineral reserves in the sea area

There are considerable potential mineral deposits to be tapped in littoral placer, under sea coal deposits, ocean oil and gas. For example, the ocean oil and gas resources, accounting for 30-40% of the national oil and gas reserves, remain to be developed.

2.1.3 Considerable potentials of natural gas, coalbed methane and thermal heat resources

The experts predict that China has 35 trillion m^3 of natural gas and coal bed methane each. Considerable terrestrial heat resources can be developed.

2.2 Rock Mechanics Face difficulties for Sustainable Development of Mining in China

For sustainable development of mining, the circles of rock mechanics face a lot of difficulties. For example, the coal mine areas in East China are characterized by thick overburden of Cainozoic era, occurrence of coal seam at great depths, basement of seam being confined water-bearing strata of Ordovician System. The mining depth in the most of mines has reached 500-1,000m. In view of high water pressure, high ground stresses, high gas and high ground temperature conditions of deep mines, research should be concentrated on the following subjects. They are:

- movement law and dynamics of Karst water in the depth of coal mine;
- forecast of surface subsidence and prevention of surface subsidence under the common effect of mining and underground drainage;
- characteristics of ground temperature and characteristics of stress field at deep coal mines in East China ;
- basic theory for high resolution 3-D seismic exploration for prospecting geological conditions at depth of the coal field.

3 Coal Mining Environmental Problems

- *Coal mining damages to land.*

Based on incomplete statistics, up to now, the cumulative figure of the collapsed farmland exceeds 400,000ha in China. In average, mining of 10,000 tons of coal brings about collapsed farmland of 0.2ha. In average 20,000ha of farmland collapses each year, which represents the average cultivable land for a population of 30,000 population in China. Farmland collapse induced by mining causes loss of farmland that the farmers relay on for their existence, which in turn leads serious social problems. In the mountain areas, mining induces increasing slope failure of the rock mass and debris flow.

- *Coal mining damages to the buildings in the villages*

According to the statistics the coal reserves occurred under the buildings exceeds 4.8 billion tons, of which 2/3 under the villages. In the plain area, namely in five provinces of Shandong, Hebei, Henan, Anhui and Jiangsu, there are 1,100 villages with a population over one million underlain by coal reserves. Immigration of residents for coal mining is much more difficult than water conservancy projects. First of all, the expenditure for moving is increasingly high, and coal enterprises can not bear the cost. For example in the south western area of Shandong, the cost for moving one household reaches100,000 RMB to over 200,000 Yuan RMB; Secondly, the distance is long. It is very difficult to choose the site. Long distance of moving will bring about inconveniences in farming and in transport. It is very difficult to reach an agreement with the farmers. This will affect normal replacement of production in mining area.

- *Influence of coal mining to water resource.*

Mine water drainage, as one of the safety techniques in coal mining, results in fall of underground water level, which in turn will bring some difficulties to farmland irrigation and water for living. In some mine areas, water drainage made the surface subsidence factor (the ratio of maximum surface subsidence to the mining height of coal seam) exceed 1.2. Fall of water level in surface alluvium caused damage of shafts in some of mine areas in East China.

In average for mining one ton of run-of-mine coal it is necessary to discharge 2t of polluted water. Pollution of water sources in the mine area due to mine water, waste water from the coal preparation plant and seepage water from the rejects bring about great danger to the environment. In some areas the residents are in water shortage because of serious pollution of the water sources, the rivers, lakes and sea.

- *The Influence of coal extraction and utilization on atmospheric environment.*

The waste gasses discharged from ventilation system to the surface contain harmful compositions, such as, carbon monoxide, carbon dioxide, methane, dust etc. The quantity of methane drained from coal mining in China accounts to 1/4 to 1/3 of the total from other countries . Methane is extremely harmful to the ozone layer.

4 The counter-measures of rock mechanics against environment pollution

4.1 Clean mining method

In aspect of reducing wastes, it is necessary to change entry-layout and support-layout. The purpose on the whole is to eliminate or reduce the quantity of wastes. The techniques for underground disposal of wastes (e. g. driving of wide entry, leaving entry along the goaf, wastes stowing, and so on) should be introduced.

In aspect of reducing harmful gases, the clean mining techniques are, for example, pre-exploitation of coalbed methane, underground gas drainage, etc.

Underground coal gasification will bring about a complete change and big leap forward of the coal mining technique. China has developed an underground coal gasification technology, that is big cross-section, long gallery, two-stage underground gasification process. Trials have been made successively in Majiazhuang Mine in Xuzhou, Liujiazhuang Mine in Tangshan, and in Xinwen Mine area.

4.2 Popularization of mining techniques with less subsidence

There are two approaches for controlling surface subsidence. One of the techniques is to leave coal pillars as support; the other concentrates on filling body for support.

The main principle of the technique of leaving coal pillars is to extract coal resources to the greatest extent within the limit of allowable surface subsidence by choosing the optimal design of coal pillars. In China there are more than 40 mines adopted the strip mining method. Coal recovery rate varies from 40% to 60%. The surface subsidence factor is in the range of 0.1 to 0.3. The problems with this proposal are low mining productivity and difficulties with fully mechanized mining. If modern equipment, like continuous miners, etc., are used, the productivity will be high. Room and pillar mining method is worthy of popularizing when conditions are suitable.

The principle of controlling strata movement by fillings is to inject the filling materials into the stowing space that might occur in the movement of overlying strata. The purpose is to form a stable and balancing structure by the interaction of fillings and the rock strata, thus controlling further movement of strata towards the surface. Based on positions of the filling materials, the techniques fall into two categories: stowing of the mined-out–area and cement injection into a space created by bed separation of rock strata. The aim of stowing of goaf is to replace coal by stowing materials. Prompt and compact stowing and selection of suitable technology can ensure that there is no caving zone in the overlying strata after mining. Thus, the surface subsidence can be greatly reduced. Grouting into bed separation space is a new technique for reducing subsidence. Grout is injected through a surface borehole into bed separation cavity formed by mining. Its advantages are no interference to underground production, and low cost of grouting. This technique is applied to coal mines in Fushun, Yanzhou, Xinwen, Datun, etc. It has a broad prospect in controlling surface subsidence when used with other techniques.

4.3 Protection and re-use of mine water resources

In order to reduce the effect of mining to water resources, first of all it is necessary to reduce the total quantity of water flowing into coal mines. Leaving water-resistant coal pillars and rock-pillars, building of water-resistant curtain by grouting, and so on, are all effective approaches for reducing the total water quantity. Secondly, to control disposal of polluted water, it is necessary to take measures such as, control pollution of water source and innovation of technology. At the same time, physical, chemical and biological methods for treatment of polluted water. Mining techniques that can conserve mine water should be adopted. The direction of development is to regard mine water as one of the resources.

4 Conclusions

At present, the newly found problems in rock mechanics are beyond the scope of current theories and knowledge of rock mechanics. New growth points and new domains of research in front of us are: the interaction of multi-media coupling, performance of rock engineering projects in an ultra-long period under the influence of environmental factors, interaction of water and rock in environmental protection and hazards prevention, etc. The circles of rock mechanics are required to make preparations. Crossing and combination of multiple disciplines are necessary.

CHARACTERIZATION OF THE ANISOTROPY OF ROCK JOINTS SURFACE MORPHOLOGY

TIKOU BELEM
Department of Applied Sciences, University of Quebec (UQAT)
445, Boulevard de l'Université, Rouyn-Noranda, PQ, J9X 5E4, Canada
E-mail: tikou.belem@uqat.ca

FRANÇOISE HOMAND
LAEGO, École de Géologie de Nancy
Rue du Doyen M. Roubault, BP 40, 54501 Vandoeuvre-lès-Nancy Cedex, France

The proposed approach is based on the *primary* and *secondary* asperities concept where roughness is subdivided into two components: the secondary or second-order roughness and the primary or first-order roughness. The 2nd-order asperities correspond to the *sensu stricto* roughness while the 1^{rst}-order asperities describe anisotropy of surface morphology. The 1^{rst}-order roughness was characterized in terms of apparent structural anisotropy of surface morphology (k_a) with the aid of profile angularity parameter (θ_p).

1 Introduction

It is well known that the presence of discontinuities strongly affects the stability of rock engineering structures such as tunnels, underground mine excavations and open pits. As an example, the knowledge of the shear strength of rock joints is necessary when determining the stability of jointed rock masses. The dominant factor influencing the mechanical behavior of rock joints is their morphology and roughness, and the most difficult thing is their characterization and modeling. The role of the 1^{rst}-order roughness on dilatancy development is explicitly taken into account in the formulation of shear strength criteria starting from the results of the first works done by Patton[1]. In the literature, most of the experimental studies for the purpose of modeling mechanical behavior of rock joints are often carried out on various types of artificial mortar joints. These artificial joints exhibit anisotropic surface morphology such as sawteeth, irregular or regular triangles, combination of different triangles, undulations, etc.

The aim of this paper is to describe apparent structural anisotropy of rock joint surface morphology using profile angularity parameter calculated along both horizontal (x-direction) and transversal (y-direction) directions.

2 Joint surface roughness profile angularity parameter

Numerous new parameters have been proposed by Belem *et al.*[2] to supplement and overcome the inefficiency or the subjectivity of some of the already existing linear parameters (*e.g.* $Z2$, R_L, P_s and JRC). One of these parameter is the surface linear angularity parameter (θ_p). For a given single topographic profile, the arithmetic mean angle of its inclinations, θ^*_p, is calculated along x- or y-axis by:

$$\left(\theta^*_p\right)_{k=x,y} = \operatorname{atan}\left(\frac{1}{N_k - 1}\sum_{i=1}^{N_k - 1}\left|\frac{z_{i+1} - z_i}{\Delta k}\right|\right) \tag{1}$$

where k denotes x-axis or y-axis; z_i, the discrete values of profile heights; (N_k-1), the number of intervals used for slopes calculation along k-axis; Δk, the sampling step along k-axis. For the whole joint surface involving non-identical profile lengths (e.g. circular sections), the directional surface mean angle (θ_p), defined as the weighted mean of the parameter θ^*_p, is given as follow:

$$\left(\theta_p\right)_{k=x,y} = \operatorname{atan}\left(\frac{\sum_{j=1}^{M_k} L_k^j\left(\frac{1}{N_k^j - 1}\sum_{i=1}^{N_k^j - 1}\left|\frac{z_{i+1} - z_i}{\Delta k}\right|\right)}{\sum_{j=1}^{M_k} L_k^j}\right) \tag{2}$$

where M_k is the total number of profiles in k-direction; N_k^j the number of discrete points corresponding to the jth profile along k-axis; L_k^j the nominal length of the jth profile along k-axis. When all the surface profiles have the same nominal length ($L_k^j = L_k$; i.e. identical L_k^j in k-direction), Eq. (2) will be reduced to an arithmetical mean.

3 Degree of apparent anisotropy of surface morphology

The surface linear mean angle (θ_p), calculated along both x- and y-direction can well describe the apparent structural anisotropy of surface morphology with regard to either x- and y-axis or (x, y) plane. The degree of apparent anisotropy of surface morphology, k_a, is defined based on the ratio of the anisotropy ellipse small half-axis b to its big half-axis a by the following relationships :

$$k_a = \frac{b}{a} = \frac{\min\left\{\theta_{p(x)}, \theta_{p(y)}\right\}}{\max\left\{\theta_{p(x)}, \theta_{p(y)}\right\}} \quad \text{with } 0 \le k_a \le 1 \tag{3}$$

Subscript x or y denotes respectively the direction in which the weighted mean angle (θ_p) is calculated. Table 1 summarized the range of variation of the degree of apparent anisotropy of surface morphology. When $k_a = 0$ corresponds to a perfectly anisotropic morphology (e.g. surface with saw teeth, undulations, etc.); when $k_a = 1$ corresponds to a perfectly isotropic surface morphology.

Table 1. Classification of the anisotropy of surface morphology

Range of variation of k_a	Shape of surface morphology
$0 \leq k_a < 0.25$	surface morphology is anisotropic
$0.25 \leq k_a < 0.5$	surface is more anisotropic than isotropic
$0.5 \leq k_a < 0.75$	surface is more isotropic than anisotropic
$0.75 \leq k_a \leq 1$	surface morphology is considered isotropic

4 Anisotropy function for linear parameters

From the linear parameters calculated along x-axis and y-axis, we propose an anisotropy function derived from the one proposed by Wong[3] to predict a directional parameter by the relationship:

$$P(\varphi) = P_0 + (P_{90} - P_0)\sin^2\varphi \tag{4}$$

where φ is the measurement direction; P_0 is the parameter value at $\varphi = 0°$ (i.e parallel to x-axis); P_{90} is the parameter value at $\varphi = 90°$ (i.e. perpendicular to x-axis). Even if Eq. (4) is different from the anisotropy function proposed by Aydan *et al.*[4], they give the same results (see Figure 2).

5 Application examples

Table 2 contains the results of calculated parameters (θ_p, k_a) for a granite rock joint with hammered surface, a man made mortar joint with a regularly undulated surface (Figure 1a) and a natural schist joint mortar replica with rough surface (Figure 1b).

Table 2. Calculated parameters for three different types of rock joint surface morphology

Type of joint surface	θ_p			k_a	Classification
	x-axis (0°)	y-axis (90°)	(x, y) plane		
Hammered	8.73°	8.78°	8.76°	0.99	isotropic
Undulated	10.27°	2.16°	6.17°	0.21	anisotropic
Rough	10.18°	4.38°	7.16°	0.43	anisotropic

(a) Man made mortar joint with regular undulations

(b) Rough and undulated natural schist joint mortar replica

Figure 1. Test samples: (a) undulated surface; (b) rough surface.

Figure 2 presents the results of calculation of θ_p in three directions (0°, 45°, 90°) for the two anisotropic surface morphology (Figure 1) and compared to the proposed anisotropy function $P(\varphi)$ given by Eq. (4). A very good agreement can be observed.

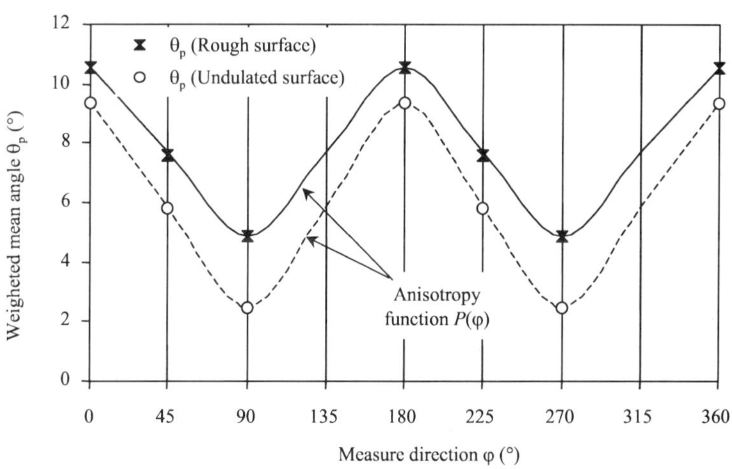

Figure 2. Comparison between calculated and predicted θ_p using the proposed function $P(\varphi)$

References

1. F.D. Patton, *Proc. 1rst. Congr. Int. Soc. Rock Mech.*, Lisbon, 509-513 (1966).
2. T. Belem et al., *Rock Mech. Rock Engng.* **33**(4): 217–242 (2000).
3. T.F. Wong, *Mech. Mater.* **4**, 261-276 (1985).
4. Ö. Aydan et al., *Rock Mech. Rock Engng* **29**(1), 47-59 (1996).

EVALUATION OF DEFORMATION AND FRACTURING CHARACTERISTICS OF ROCK BY MOMENT TENSOR ANALYSIS AND BONDED-PARTICLE MODELING

SOO-HO CHANG, KYUNG-JIN YOUN, CHUNG-IN LEE

School of Civil, Urban and Geosystem Engineering, Seoul National University, Korea

In this study, laboratory experiments as well as PFC^{2D} modeling were carried out to study characteristics of deformation and fracturing of rocks under uniaxial and triaxial compressive conditions. Laboratory experiments include stress-strain measurements to carry out the moving point regression analysis, and acoustic emission measurements for the moment tensor analysis. Damage thresholds showed linear relationships with confining pressures. From the moment tensor analysis, most of crack motion vectors were parallel with a failure surface, and it was shown that shear failure was a major failure mechanism of rocks under uniaxial and triaxial compression. In addition, the feasibility of the application of the bonded-particle model to reproduce mechanical behaviors of rocks was explored by comparing model behaviors to results from experiments with different confining pressures. The model could exhibit macroscopic behaviors such as strain softening, dilation and fracture that arise from extensive microcracking throughout a bonded assembly. However, axial stress increments with increasing confining pressures were much smaller than those from experiments.

1 Introduction

The recent studies to investigate damaged rocks include numerical analyses, experimentally quantifying damage characteristics in rocks and the moment tensor analysis. Especially, the moment tensor analysis has been recently applied to estimate failure mechanisms and damage progression in rocks.

In this study, damage thresholds of rocks under uniaxial and triaxial compressive conditions were determined by the moving point regression analysis (Eberhardt *et al*, 1999) and acoustic emission (AE) measurements. In each damage threshold, crack types of AE sources were estimated by the moment tensor analysis in order to investigate cracking mechanisms of rocks. In addition, the feasibility of the application of the bonded-particle model (PFC^{2D} model) to reproduce mechanical behaviors of rocks was explored by comparing model behaviors to results from experiments with different confining pressures.

2 Theories

Damage thresholds include the crack closure stress (σ_{cc}), the crack initiation stress (σ_{ci}), the secondary cracking stress (σ_{ci2}), the crack coalescence stress (σ_{cs}), the crack damage stress (σ_{cd}), and peak strength (σ_{peak}). They were determined by the moving point regression analysis and AE measurements (Eberhardt *et al*, 1999).

A moment tensor inversion procedure named SiGMA (Simplified Green's function for moment tensor analysis) was developed by Ohtsu (1991). It uses only P wave amplitudes. Amplitude $A(\boldsymbol{x})$ of the first motion is represented as:

$$A(x) = \frac{C_S \mathrm{Ref}(\mathbf{t},\mathbf{r})}{R} \begin{pmatrix} r_1 & r_2 & r_3 \end{pmatrix} \begin{pmatrix} m_{11} & m_{12} & m_{13} \\ m_{12} & n_{22} & m_{23} \\ m_{13} & m_{23} & m_{33} \end{pmatrix} \begin{pmatrix} r_1 \\ r_2 \\ r_3 \end{pmatrix} \tag{1}$$

where C_s is the calibration coefficient, **t** is the direction of the sensor sensitivity, Ref(**t**, **r**) is the reflection coefficient at the observation location **x**.

Since the moment tensor is symmetric, the number of independent unknowns m_{pq} to be solved is six. Thus, multi-channel observation of the first motions at more than six channels is required to determine the moment tensor components. From eigenvalues and eigenvectors of a moment tensor, a crack type, unit crack-motion vector $\boldsymbol{l}$ and unit normal vector $\boldsymbol{n}$ on a crack surface can be estimated (Ohtsu, 1991).

The *PFC* (Particle Flow Code) is a distinct element geomechanical modeling program in which rock is represented by an assembly of thousands of round particles bonded together at points of contact (Itasca Consulting Group, 1999). Each particle in a *PFC* model of rock is assigned a normal and shear stiffness. These micro-parameters influence Young's modulus and Poisson's ratio of the rock being modeled. The input parameters of a PFC^{2D} model include contact modulus (E_c), ratio of particle normal to shear stiffness (k_n/k_s), particle friction coefficient (□), average normal bond strength ($□_c$), standard deviation of normal bond strength, average shear bond strength ($□_c$) and standard deviation of shear bond strength. Based on Young's modulus, Poisson's ratio and peak strength obtained from uniaxial compressive tests, the input parameters were determined in this study. The post-failure behaviors were also considered. After models for granite and marble under uniaxial compression were determined, they were applied to triaxial compressive conditions.

3 Experiments

Hwangdeung granite and Yeosan marble produced in Korea were the main rock types used for testing. The servo-controlled hydraulic testing machine, INTERLAKEN 2000 was used, and the test was displacement controlled at the rate of 1mm/min. MISTRAS 2001 system and DA100 system were also used to measure AE and strains respectively. For the moment tensor analysis, six AE sensors were used. The AE sensor of 3.6mm diameter has 150~1,000 kHz of operating frequency range. In order to obtain the same sensitivity of sensors, they were tightly attached to a specimen by electron wax. The signals from AE sensors were amplified 40 dB by the pre-amplifiers, then sent to the AE monitoring system and further amplified 40dB. Considering background noise level, the trigger level was set to 45dB. Sample rate was set to 8MHz. In addition, a triaxial compression cell comprised of 8 BNC connectors for AE sensors and 14 plugs for strain gauges was newly developed and used to reduce electric noise. During tests, confining pressures of 4.9, 9.8, 14.7 and

19.6 MPa were regulated by the SBEL Model AP-1000 pump.

4 Deformation and fracturing characteristics of rock

4.1 Stress-strain curves

Under uniaxial compressive condition, the stress-strain curves obtained from PFC^{2D} modeling could approximately simulate deformation behaviors of granite and marble. Especially, young's modulus, poisson's ratio and peak strength from the PFC^{2D} model were very close to those from experiments. The reason can be deduced from the fact that input parameters of a PFC^{2D} model were determined based on experimental results.

In Figure 1, it can be seen that as confining stress increased, differential axial stresses increased in both cases of experiments and PFC^{2D} modeling. However, in *PFC* modeling, axial stress increments with increasing confining stress were much smaller than experiments. In addition, it was also difficult to exactly represent plastic deformation and strain-hardening characteristics shown after a yielding stress, since the linear contact-stiffness model based on a linear relationship between contact force and relative displacement was used in *PFC* modeling.

4.2 Damage thresholds

In Figure 2, it can be seen that as confining pressure increased, the crack closure stresses (σ_{cc}) were nearly constant or slightly decreased. Such a reason may be found in the fact that existing cracks or pores might be closed by application of confining pressure prior to triaxial tests. However, because confinement inhibits the initiation and propagation of new cracks, other damage thresholds represented linear relationship with confining pressures.

For granite, the damage thresholds from the PFC^{2D} model were close to those from experiments. However, for marble, in order to more realistically represent such a deformation behavior of marble, standard deviation of contact strengths was chosen greater than that of a granite model. From such a reason, bond breakages occurred even at low stress levels, and the crack initiation thresholds of marble was determined smaller than that of experiments. In addition, since plastic deformation and strain-hardening characteristics were not simulated well, the crack damage thresholds of marble were determined greater than experiments.

4.3 Crack types

Crack types in AE sources can be determined from eigenvalues of a moment tensor. In beginning stage of loading, ratios of shear, tensile and mixed-mode cracks were similar. However, as axial differential stress increased, ratio of shear cracks became bigger and occupied about 60% until failure. Such a reason may be due to the fact that propagation and coalescence of micro-cracks might be induced by shear cracks.

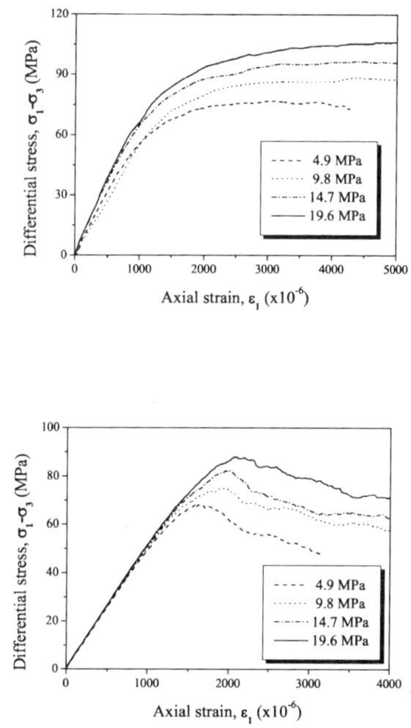

(a) Experiments (b) *PFC^{2D}* modeling

Figure 1. Stress-stain curves of marble under triaxial compression.

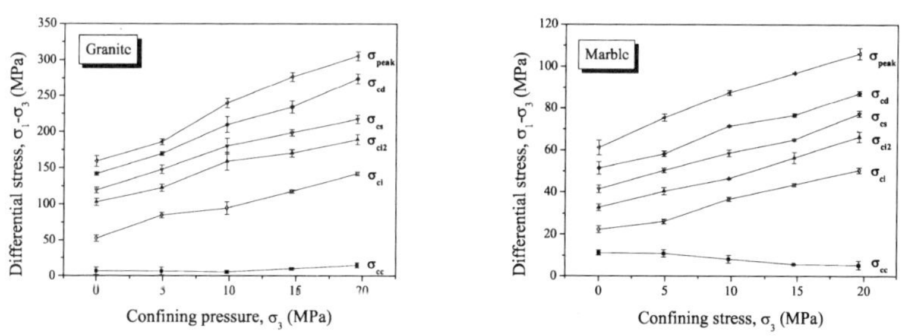

Figure 2. Damage thresholds obtained from experiments at different confining pressures.

At different confining pressures, the occupied ratio by each crack type among all cracks that had occurred until failure was almost constant (Figure 3). However, as confining pressure increased, percent of cumulative relative crack volumes (D_r) by shear cracks also increased (Figure 4). The relative crack volume obtained from the moment tensor analysis can be considered as a damage index (Shigeishi & Ohtsu, 2001). Therefore, it is concluded that as confining pressure increase, damage accumulated by shear cracks becomes greater. From such results, cracking by shear cracks may be considered to be a major failure mechanism of rock under uniaxial and triaxial compression. In addition, orientation of a major failure plane can be

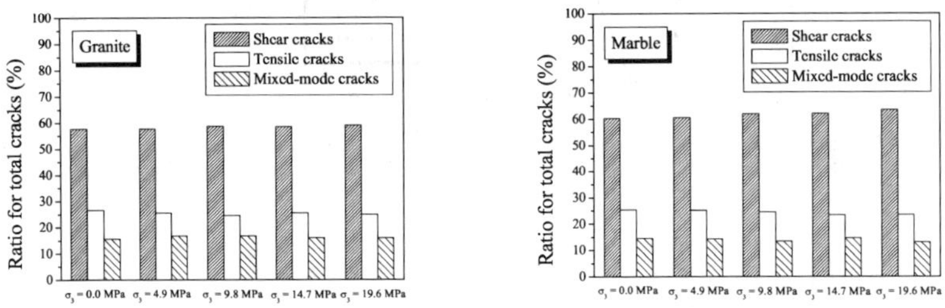

Figure 3. Ratios of each crack type for total cracks at different confining pressures.

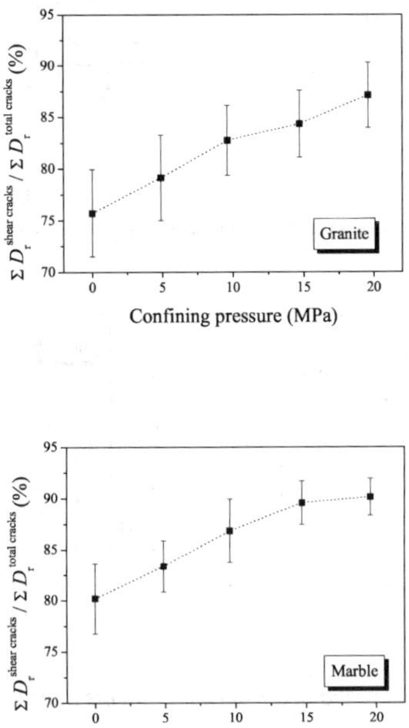

Figure 4. Variation of cumulative damage induced by shear cracks at different confining pressures.

estimated by distributions of crack motion vector ***l*** and crack normal vector ***n***.

The *PFC* model considers only shear and tensile cracks. However, percentages of shear and tensile cracks were generally close to those from the moment tensor analysis. Although many cracks occurred in the beginning stage of loading even for marble, shear cracks became much more dominant than tensile cracks with increasing axial stress. It exactly agrees with results from the moment tensor analysis.

5 Conclusions

Because confinement inhibits the initiation and propagation of new cracks, damage thresholds except the crack closure stress represented linear relationship with confining pressures. From the moment tensor analysis, it was shown that shear cracks occupied approximately 60% of all cracks until failure, and as confining pressure increased, damage accumulated by shear cracks became greater. Therefore, cracking by shear cracks may be considered to be a major failure mechanism of rock under uniaxial and triaxial compression. In addition, orientation of a major failure plane can be estimated by distributions of crack motion vector ***l*** and crack normal vector ***n***. The PFC^{2D} model could exhibit macroscopic behaviors under uniaxial compression such as strain softening and fracture that arise from extensive micro-cracking throughout a bonded assembly. In addition, damage thresholds and percentages of crack types were close to those from experiments. However, axial stress increments with increasing confining pressures could not be simulated. It was also difficult to exactly represent plastic deformation and strain-hardening characteristics shown after a yielding stress.

Acknowledgments

This work was supported by grant No. R01-2000-00375 from the Korea Science & Engineering Foundation.

References

1. E. Eberhardt *et al*, *Int. J. Rock Mech. Min. Sci.*, **36**, 361 (1999).
2. Itasca Consulting Group, PFC2D Version 2.0 (Minneapolis, Minnesota, 1999)
3. M. Ohtsu, *J. Geophys. Res.*, **96**, 6211 (1991)
4. M. Shigeishi and M. Ohtsu, *Construction and Building Materials*, **15**, 311 (2001)

STABILITY THRESHOLD OF ROCK MATERIALS

MINH-PHONG LUONG
CNRS UMR 7649, LMS, Ecole Polytechnique, 91128 Palaiseau Cedex France
Email : luong@lms.polytechnique.fr

A non-destructive testing technique is used to detect the occurrence of material instability subsequently causing damage and a specific data reduction procedure to assess damage accumulation. The evolution of the increase in non-linear energy is assumed to correspond to the extent of rock damage. The damaging process was detected by analysing the signal evolution of ultrasonic pulses propagating through a deep argillaceous rock specimen subjected to increasing shear loads up to failure. A non-linear analyser based on a multidimensional Fourier transform permits to separate linear and non-linear parts of the mechanical response. Experimental results readily evidence the stability threshold of brittle geomaterials.

1 Introduction

This paper introduces a non-destructive technique, based on ultrasound propagation characteristics, to detect the stability limit, thus providing a very useful *early warning* for the security of underground work at high risk for environment. An input-output non-parametric based on ultrasonic pulse propagation[25] and a non-linear analyser for data reduction procedure[10] were chosen to portray the non-linear behaviour of a deep argillaceous rock subjected to increasing static shear loads. Based on multidimensional Fourier transform, the non-linear analyser permits to separate linear and non-linear parts. It reveals to be very useful for a highly sensitive detection of the material stability threshold beyond which the micro cracking process occurs and subsequently leads to a propagating damage.

2 Existing techniques

Common experimental methods used strain gauges, photoelasticity, moiré, ultrasound and radiography, as well as acoustic emission and thermographic methods to obtain information concerning deformations, strains, structural integrity and failure mechanisms in rock infrastructure. According to Kotonski and Malecki[6], granite for instance depending on its grain structure could reach values from 1,700 to 5,000 m/s. This invites the use of non-destructive testing (NDT), directly applied to the part being examined rather than a representative sample. Several authors[5-6-24] used sonic and ultrasonic methods to study fracture of concrete. A "discontinuity stress" has been proposed to describe the end of the quasi-elastic behaviour of concrete[20]. The first attempts to measure the discontinuity stress were made using ultrasonic pulse velocity[22-26], unfortunately without giving details on the method of such acoustic measurement. Acoustic emission AE testing is relatively simple, needing only a sensor and basic signal-analysis equipment. A wide variety of mechanisms are assumed to be responsible for AE generation, ranging from dislocation motion to crack propagation. Basic AE monitoring techniques are therefore applied to a wide class of materials and structures[21]. But it provides a late warning because it results from damage manifestation.

3 Theoretical background

Techniques using ultrasonic waves are appealing because of the direct connection between the characteristics of wave propagation and the damage states of a solid[1]. Many factors affect wave attenuation and dispersion, such spreading of a wave beam, scattering, absorption due to various mechanisms and mode conversion, resulting in partitioning of the energy among two or more wave modes, each travelling at its own velocity. This analysis considers, in the framework of small perturbation, the propagation mechanism of a wave through a solid presenting locally an unstable behaviour. Within the theory of plasticity, several postulates that guarantee mechanical stability have been suggested based either on energy or wave propagation considerations. Mechanical stability is taken as the capability to sustain a given stress state. Instability can also occur in saturated loose soil prior attaining the failure stress state, without formation of shear bands. This type of instability, referred to as prefailure flow instability, characterised by a runoff decrease in stress-carrying capacity[7]. Note the instability of a material can occur due to a lack of ability to propagate a small perturbation in the form of stress wave in a certain direction[17]. Based on the hypothesis of a rigid plastic material, Hill[4] established a condition that guarantees stability for an aggregate of infinitesimal volume. This condition is referred to as the maximum work principle. Its application to a small stress increment $d\sigma$ requires the second increment of work to be positive.

$$d^2W = \mathbf{d\sigma}_{ij} \cdot \mathbf{d\varepsilon}_{ij} \geq 0 \tag{1}$$

Since Hill considered the material to be inelastic in his derivation, the strains in Eq. (1) are entirely plastic. Hence, this inequality can be rewritten as

$$d^2W = \mathbf{d\sigma}_{ij} \cdot \mathbf{d\varepsilon}^{p}_{ij} \geq 0 \tag{2}$$

Using a completely different approach based on thermodynamical considerations, Drucker[2-3] proposed Eq. (2) as the necessary condition for material stability. Drucker's quasi-thermodynamical postulate is based on the assumption that energy is dissipated during a cycle of application and removal of a small load increment. Hence, energy must be supplied to the material for plastic deformations to take place. Drucker's postulate and Hill's maximum work principle are based on different assumptions, but they provide the same stability equation expressed by Eq. (2). If contrary to Hill's hypotheses, the elastic strains are included in Eq. (1); this inequality includes the term Eq. (3) which is always positive. This implies that Eq. (1) guarantees mechanical stability a little beyond Drucker's condition.

$$d^2W^e = \mathbf{d\sigma}_{ij} \cdot \mathbf{d\varepsilon}_{ij}^e \tag{3}$$

Although these two postulates evolved from different assumptions, the only difference between them is the use of the total strain increment or the plastic strain increment. If either Drucker's or Hill's stability postulate was necessary and sufficient for stability, then the violation of one of the criteria could be used as a condition of instability. These criteria state that the second-order work, produced by loads and free displacements, must be positive for arbitrary kinematically admissible variations of the displacement field.

The stability condition contained in Eq. (2) was analysed and discussed by Mandel[18]. He showed that Drucker's postulate was a sufficient but not a necessary condition for a material to be stable. Additionally, one of the hypotheses behind the maximum work principle is that sliding between grains obeys Schmid's law, which assumes a constant sliding resistance independent of the intergranular forces, such as those found in metals. The frictional nature of sliding in geomaterials does not obey this law. The stresses and strains in Hill's postulate, defined for an aggregate, are therefore applicable to an assembly of grains. Therefore discrepancies between Eq. (2) and the actual behaviour of geomaterials should be expected.

Based on the assumption that a stable material is able to propagate a small perturbation in the form of waves, Mandel proposed a necessary condition for stability. He showed[17] that a wave could propagate in a material with an elastic plastic matrix **A**, along the direction α, if and only if all the eigenvalues λ of the matrix **B** are real and positive, where

$$\mathbf{d\sigma}_{ij} = \mathbf{A}_{ijkl} \cdot \mathbf{d\varepsilon}_{kl} \tag{4}$$

$$\mathbf{B}_{ik} = \mathbf{A}_{ijkl} \cdot \alpha_j \cdot \alpha_l \tag{5}$$

$$k = 1, 2, 3 \qquad \lambda_k > 0 \tag{6}$$

If one of the eigenvalues λ is negative or null, one of the components of the perturbation cannot propagate. This phenomenon is used as a highly sensitive manifestation of geomaterial instability[12-13-14-16].

4 Proposed method

Volterra analysis is used here to deal with unknown non-linear systems. Although attempts have been made to identify the governing equation of SDOF systems[19], it is more usual to identify high order transfer functions characterising the non-linear behaviour of the mechanical system. Volterra has described a series decomposition to express the system dynamic response y(t) as a function of excitation x(t). The response y(t) can be expanded in a Volterra series in the following way

$$y(t)=\sum_{i=1}^{\infty} y_i(t)=\sum_{i=1}^{\infty}\int_{-\infty}^{+\infty}\cdots\int_{-\infty}^{+\infty} h_i(\tau_1,\ldots,\tau_i)\prod_{k=1}^{i} x(t-\tau_k)d\tau_k \quad (7)$$

These series result from the summation of multidimensional convolution products between excitation and Volterra kernels $h_i(\tau_1, \ldots \tau_i)$. The term y_1 of the series corresponds to the linear component of the response, whereas the other terms correspond to the non-linear components. A more complete description of the Volterra series is given in references[9-23-27]. Working out the Volterra kernels allows the characterisation of the system by separating linear and non-linear parts of the dynamic response. It should be noted that the kernels are entirely dependent on the governing differential equation, i.e. the system itself, and thus independent on excitation. If the system is linear, high order kernels are constant and equal to zero. Activation of non-linear phenomena gives rise to non-zero values in these high order kernels. For a non-linear system, appearance of supplementary non-linearities (such as unstability, microcracking, damage, etc.) increases the amplitude of original high order kernels. The proposed method aims to evaluate the significance of non-linear terms in the global response. If these non-linear aspects exhibit a sudden increase, it will be associated to the occurrence of material unstability causing additional non-linear behaviour.

5 Non-linearity ratio and non-linearity contrast

A simple quantitative parameter on which a criterion could be applied is necessary to characterise the system response. The determination of kernels gives access to the different terms of the series, and to their energies. The energy ${}_lE_m^i$ of the ith component of ${}_ly(t)$ for the mth excitation level assumes the form:

$${}_lE_m^i=\int_0^{\infty} {}_ly_i^{(m)}(t)^2\,dt \quad (8)$$

The non-linear phenomenon significance could be evaluated by comparing these energy values. To quantify the non-linear contributions in the global response, the non-linearity ratio ${}_lr_m$ at point l for the mth excitation is defined in the following way:

$$ {}_{l}r_m = \frac{\sum_{i=2}^{N} {}_{l}E_m^i}{\sum_{i=1}^{N} {}_{l}E_m^i} = \frac{\text{non-linear energy}}{\text{total energy}} \qquad (9) $$

According to Eq. (8), Volterra kernels can be introduced into the former equation. The non-linearity ratio is then written as:

$$ {}_{l}r_m = \frac{a_m^4 \, {}_{l}\eta_2 + \ldots + a_m^{2N} \, {}_{l}\eta_N}{a_m^2 \, {}_{l}\eta_1 + a_m^4 \, {}_{l}\eta_2 + \ldots + a_m^{2N} \, {}_{l}\eta_N} \qquad (10) $$

with ${}_{l}\eta_i = \int_0^{\infty} {}_{l}h_i(t)^2 \, dt$

For N=3, the non-linearity ratio ${}_{l}r_m$ can be expressed as a function of a, so-called ${}_{l}\rho(a)$, assuming the form :

$$ {}_{l}\rho(a) = \frac{a^2 \, {}_{l}\eta_2 + a^4 \, {}_{l}\eta_3}{{}_{l}\eta_1 + a^2 \, {}_{l}\eta_2 + a^4 \, {}_{l}\eta_3} \qquad (11) $$

This function expresses the energetic part of non-linear components in the system response to an impulse excitation of magnitude a. To compare two measurements A and B, it is necessary to calculate η_i and to build functions ρ_A and ρ_B. The difference between the two curves reveals the range of impulse magnitude in which the function $\rho(a)$ varies. In our example, the absolute sensitivity thus defined is maximal in the range $3 \leq a \leq 4$. The area of the absolute sensitivity curve could be used to compare results obtained from two tests. It is then necessary to fix integration boundaries. Minimum and maximum values a_{min} and a_{max} can be arbitrarily chosen so that:

$$ \rho(a_{min}) = 0.01 \quad \text{et} \quad \rho(a_{max}) = 0.99 \qquad (12) $$

These boundary values should be useful in characterising in a global way the non-linearity as a function of impulse magnitude a. Nevertheless a_{max} values, thus defined, may have nothing to do with test values. It would be better to choose boundary conditions in relation with experimental values. The absolute sensitivity takes the form:

$$ \delta_B^A(a) = \rho_B(a) - \rho_A(a) \qquad (13) $$

A two tests comparison parameter, so-called "non-linearity contrast" is then suggested:

$$ \Delta_B^A = \int_{a_{min}}^{a_{max}} \left| \delta_B^A(a) \right| da \qquad (14) $$

The main interest of the ρ function is that it gives a global appreciation of non-linearity. And, as revealed by numerical examples, it clearly appears that the difference $\rho_B(a) - \rho_A(a)$ may change sign with the variation of excitation magnitude

a. Usually, results are given for a unit magnitude, but their extrapolation to test values may not be justified.

6 Experimental validation

This technique has been applied in laboratory on a specimen of deep argillaceous rock subject to increasing levels of static mode II direct shearing (Figure 1). The pulse transmission technique, using scan piezoelectric transducer V150 (0.10 MHz), has been applied as non-destructive testing. The pulser section produces an electrical pulse to excite a piezoelectric transducer, which emits an ultrasonic pulse. The pulse travels through the specimen that is subjected to a given static shearing, to a second transducer acting as a receiver. The transducer converts the pulse into an electric signal available for non-linear analysis. Three input pulses were used at three different amplitude levels. Response signal records at different static stress levels have indicated the evolution of the ultrasonic pulse travelling through the specimen that is loaded by increasing static shear loads. Ultrasonic pulse responses showed very slight changes, even when the shear stress levels τ/τ_f approaches 100% at failure. The data processing technique in use provides the variation of non-linear energy normalised by the total energy or non-linearity ratio (Figure 2) and non-linearity contrast (Figure 3) as a function of stress levels. They have evidenced the occurrence of material instability leading subsequently to microcracking and failure. The slope change of both non-linearity ratio and non-linearity contrast suggests that the quasi-brittle rock specimen presents two quite different behaviours: the former is quasi elastic and stable, the latter may lead to a sudden failure caused by the extension of the unstable zone that generates non-linear effects affecting the ultrasonic wave propagation mechanisms.

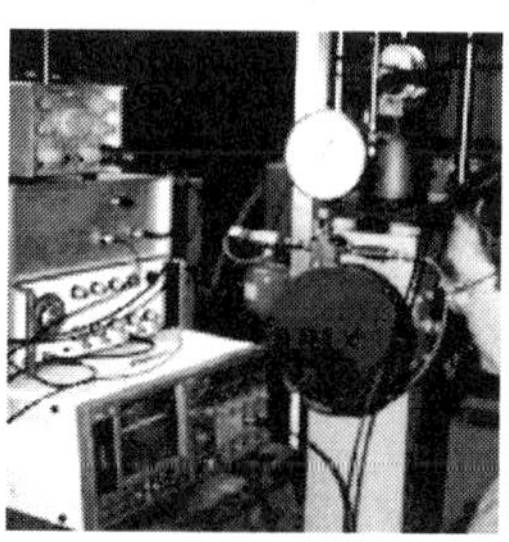
1a.

1b.

Figure 1. Deep argillaceous rock specimen under direct shear loading between the 3 parts up to failure.

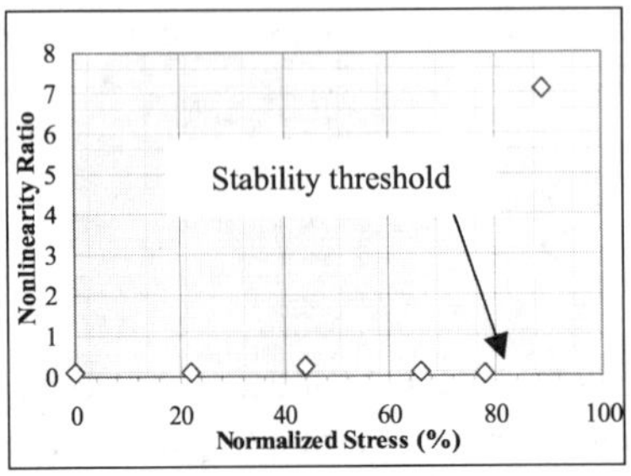

Figure 2. Non-linearity ratio versus normalised shearing stress up to failure.

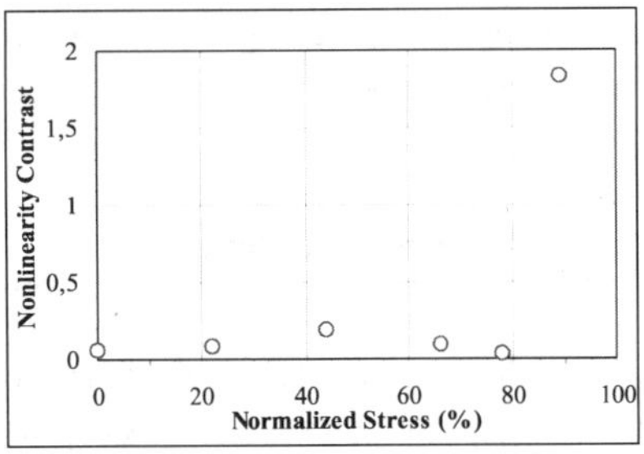

Figure 3. Non-linearity contrast versus normalised shearing stress up to failure.

7 Concluding remarks

This paper aims to demonstrate that the highly sensitive detection of material instability occurred in deep argillaceous rock could provide a very useful early warning (compared to traditional techniques such as acoustic emission, increase of wave attenuation, etc. that occur after crack initiation) for the security of work at high risk for the environment. The underground excavation environment is often very complex. A non-destructive evaluation of material instability could offer an efficient field monitoring before failure initiation in order to reduce risk of imminent failures by giving advanced and sufficient warning for remedial measures to be designed. The application of ultrasound scanning to inspection and monitoring of underground geotechnical structures relies on the fact that preceding the process of microcracking, the deep argillaceous rock may be locally unstable and hence modifies and partially obstructs the propagation characteristics of ultrasound pulses.

References

1. J.D. Achenbach in *Elastic Waves and Ultrasonic NDE*, ed. by S.K. Datta *et al.*, North-Holland, Amsterdam, (1990).
2. D.C. Drucker in *Proc. 1st U.S. Nat. Congress of Appl. Mech.*, USA, 487 (1951).
3. D.C. Drucker, *J. Appl. Mech.*, ASME, USA, **26**, 101 (1959).
4. R. Hill, *J. Mech. Phys. Solids.*, Pergamon Press, London, **6**(3), 236 (1958).
5. R. Jones, *British J. Applied Physics*, UK, V **3**(7), 229 (1952).
6. W. Koltonski and I. Malecki, *Acoustica*, **8**, 307 (1958).
7. P.V. Lade and D. Pradel, *J. Engrg. Mech.*, ASCE, USA, **16**, 2532 (1990).
8. R. L'Hermite, *Bull.* **18**, RILEM, Paris, France, 27 (1954).

9. H. Liu, *Thesis* ISMCM, St Ouen, France (1989).
10. H. Liu, and T. Vinh, *Mechanical Systems and Signals Processing*, **5**(1), 61 (1991).
11. M.P. Luong in *Comprehensive Rock Engineering*, ed. J. A. Hudson, **4**(26), 715 (1992).
12. M.P. Luong in *Micromech. Concrete & Cement. Comp.*, ed. C. Huet, 105 (1993a).
13. M.P. Luong in *Int. Conf. NDT of Concrete in the Infrastructure*, ed. SEM, 199 (1993b).
14. M.P. Luong in *Rock Foundation*, ed. Yoshinaka & Kikuchi, 155 (1995).
15. M.P. Luong in *Size-Scale Effects in the Failure Mechanisms of Materials and Structures*, ed. Alberto Carpinteri, 157 (1996).
16. M.P. Luong in *Advances in Rock Mechanics*, ed. Yunmei Lin, 3, (World Scientific, Singapore, 1998).
17. J. Mandel in *Int. Symp. Stress Waves in Anelastic Solids*, Brown University (1963).
18. J. Mandel in *Rheology and Soil Mech.*, ed J. Kravtchenko & P.M. Sirieys, 58 (1966).
19. V.Z. Marmarelis, *Mathematical and Computer Modelling*, **12**(8), 991 (1989).
20. K. Newman in *Proc. Int. Conf. Structure of* Concrete, CCA, UK (1968).
21. C. Ouyang, E. Landis and S.P. Shah, *J. Eng. Mech.*, **117**(11), 2681 (1991).
22. G.S. Robinson, *J. Structural Div.,* ASCE, **93**(ST1), 71 (1967).
23. W.J. Rugh in *Nonlinear system theory – The Volterra-Wiener Approach.*, UK (1981).
24. R. Sell, *Bull.* **5**, RILEM, Paris, 5 (1959).
25. J.J. Thomas, *Thesis* Ecole polytechnique, France (1995).
26. G.W.D. Vile in *Proc. Int. Conf. Structure of Concrete,* CCA UK, 131 (1968).
27. T. Vinh and G. Tomlinson in *Dynamique des systèmes non linéaires.* St Ouen (1990).

IMPROVED PROCEDURES FOR EVALUATING PHYSICAL DETERIORATION OF ARGILLACEOUS ROCKS

IMAM A. SADISUN
Dept. of Geology, Faculty of Earth Sciences and Mineral Technology, Bandung Institute of Technology, Jl. Ganesha No.10, Bandung, Indonesia
E-mail: imam@mine.kyushu-u.ac.jp

HIDEKI SHIMADA, MASATOMO ICHINOSE, KIKUO MATSUI
Dept. of Earth Resources and Mining Engineering, Graduate School of Engineering, Kyushu University, 6-10-1 Hakozaki, Higashi-ku, Fukuoka, Japan

Numerous foundation and slope stability problems worldwide have been contributed to the presence of argillaceous rocks. The main difficulties with argillaceous rocks are generally attributed to their properties that possess tendency to physically deteriorate within relatively short time frames. This paper deals with improved procedures used to evaluate the susceptibility of argillaceous rocks for physical deterioration, known as slaking index test, which it can provide valuable complementary data when the classical slake durability test is not allowed.

1 Introduction

Evaluation of argillaceous rocks to physical deterioration requires testing which generally consists of two types, dynamic and static slaking tests [1]. The slake durability test is a widely known dynamic slaking test proposed by Franklin and Chandra [2], and has been standardized by ISRM in the Suggested Method of Rock Characterization Monitoring and Testing [3]. However, there are some infirmities of this test because of its insensitiveness to those rocks that preferably broken off in small pieces with average size coarser than the 2 mm opening sieve. Moreover, equidimensional roughly spherical in shape of rock lumps with rounded corners needed in rotation process during the test is frequently difficult to obtain, particularly for the samples having peculiar behavior of easily split along lamination or fissile present. In addition, the test also employs a tumbling

Table 1. Observation criteria for classifying of argillaceous rocks to slaking.

Class*	Criteria
0	No visible change or change can be allowable as only very minor hairline cracking.
1	Severe hairline cracking without spalling, and/or mud-sand particles starting to breakaway from the rock surface.
2	Crack opening with minor spalling and/or little suspended mud-sand particles.
3	Further process of crack opening with major spalling and/or suspended mud-sand particles.
4	Complete disintegration to a mound of chips and/or remarkable portion of suspended mud-sand particles.

*Classes correspondingly denote the degree of disintegration.

factor, which generally does not exist in the field.

To overcome the aforementioned infirmities, through this study static slaking tests was examined in differences with the standard slake durability test. This study was also aimed to improve procedures of static slaking test and to evaluate its reliability. As a case study, the test was performed on some argillaceous rocks obtained from an industrial estate development of the southern Karawang area of West Java Province, Indonesia (Subang claystone) and from an underground mining activity of the Ikeshima coal mine, Nagasaki Prefecture, Japan (Ikeshima shale).

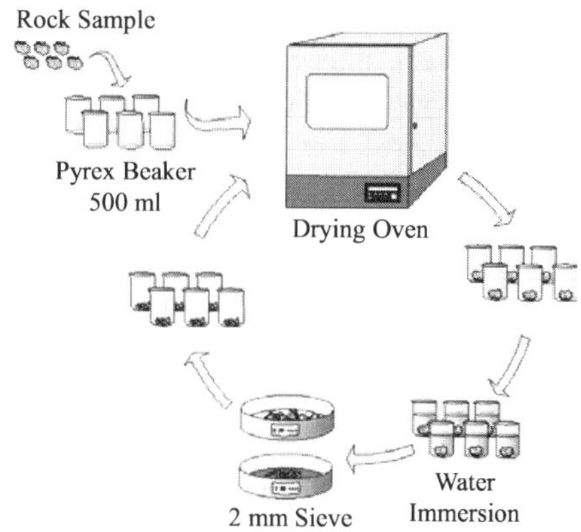

Figure 1. Schematic diagram of slaking index test.

2 Testing for Susceptibility to Physical Deterioration

The basis of slaking tests is to analyze rock susceptibility to short-term weakening and breakdown when subjected to simulated rapid physical disintegration processes.

In the slake durability test, the essence of the test is to accelerate physical disintegration to a maximum by soaking, rotating and sieving. The test results indicate that the Subang claystone are characterized by low to very low durability having a maximum slake durability index (I_{d2}) of 31.7% and a mean I_{d2} of only 17.6%, which indicate low resistance of a rock specimen to the breakdown by water and mechanical agitation. The Ikeshima shale is generally more durable than the claystone having a mean I_{d2} of 46.9%. Meanwhile, the main procedure of the static slaking test performed in this study is subjected to five dry-wet cycles of the intact rock in a 500 ml beaker, namely slaking index test. About this procedure, the detailed explanation can be found in Sadisun et al. [4] as schematically shown in Figure 1. At a more fundamental level to understand the detail mechanism of physical deterioration, the development of incipient discontinuities into open fractures is also investigated by visual tests including slaking exposure and slaking immersion tests. All samples are described in terms of crack development and any disintegration that might occur as listed in Table 1. Similar results are shown when the specimens are assessed by being subjected to slaking index test although the dynamic slake durability test always possesses a stronger effect on the deterioration process. All results of static slaking tests are listed in Table 2.

Table 2. Summary results of static slaking tests.

Sample#	Time to initially slake (h:m:s)		Slaking Index (I_{s1})
	Exposure	Immersion	
SC-1	24:12:00	00:05:17	35.1
SC-2	04:30:00	00:03:08	57.4
SC-3	19:58:00	00:04:31	45.7
IKS-1	55 days	00:07:45	8.5
IKS-2	92 days	00:10:55	2.7
IKS-3	42 days	00:05:42	11.3

To further evaluate the stability index of physical deterioration in the slaking index test, the samples are rated based on class value of the observation index during the first 30 minutes of immersion. Then the stability indexes are obtained by adding between the class value of observation index and slaking index. The stability indexes of the claystone resulted from further evaluation of this test are ranging from 4 to 8, which meant medium to high acceleration in physical deterioration after their exposure and moisture level; whereas, the stability indexes of the shale are ranging from 2 to 4 having more sturdy to physical deterioration.

3 Reliability of Slaking Index Test

In order to assess the reliability of the slaking index test, samples used in this study were also modified by using some different shapes ranged from spherical to disk forms. Plots are drawn of slaking index versus rock shape for four sets of rock samples, and are linked graphically (Figure 2) showing a general linearity and relatively unchangeable values, which give the evidence that there is no influence of rock shape in the slaking indexes (I_{s1} and I_{s2}).

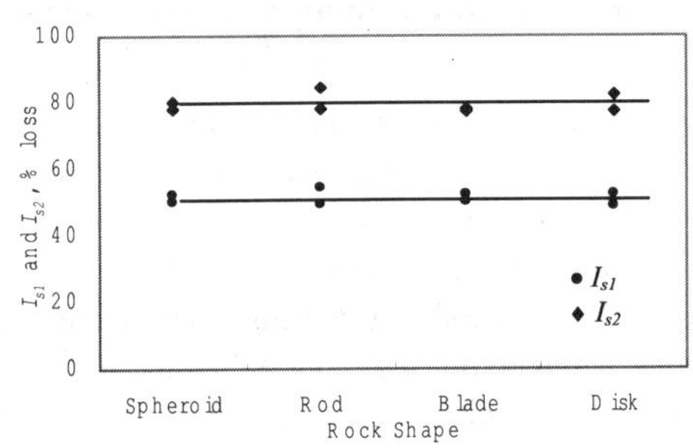

Figure 2. Relatively unchanged slaking indexes vs. rock shapes.

The relationship between the slake durability index and the slaking index are determined by regression analysis indicating that the best fit to experimental data is a negative exponential fit through successive cycles of the treatment (Figure 3). The relationship derived from analysis give a general trend equation $I_{d2} = 65e^{-0.035\,Is1}$ having a statistically significant correlation of 0.93 (r^2=0.86) at the 95% confidence level. The similar result is also obtained from the relationship among the first cycle of both tests ($I_{d1} = 100\ e^{-0.02\,Is1}$; $r^2 = 0.88$).

The lack of tumbling factor in the slaking index test consequentially causes the increasing index values to move quite smoothly and allow the use of samples in different shape. In addition, it can be recognized that there is a close relationship between I_s and I_d to the set of samples. Moreover, concerning the rate to disintegrate, the utilizing of observation procedure in the slaking index test offer any distinction in rate of disintegration, which is fundamental in the evaluation of stability index among various rock types with different slaking behavior.

These all-aforementioned results can assure that a simple static slaking index test is meaningful which can be inexpensive, repeatable, easy to assemble with readily available material, having minimal tools, and further can provide valuable complementary data when the classical slake durability test is not allowed.

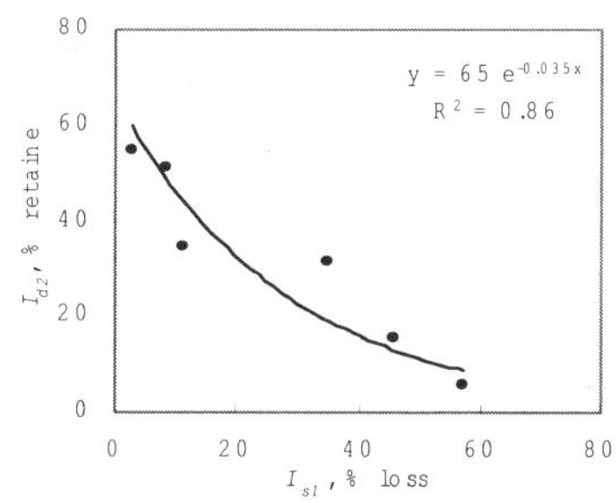

Figure 3. Slaking vs. slake durability indexes.

4 Conclusions

Some testing procedures of physical deterioration, in particular the susceptibility for slaking subjected to a static slaking index test was examined in detail. An improved procedure used to evaluate the susceptibility for physical deterioration was introduced based on a combination of observation and experimental techniques.

References

1. I.A. Sadisun, H. Shimada, M. Ichinose, and K. Matsui, *Proc. Intl. Workshop on Intelligent Mining Systems*, Japan, pp. 65-74 (2002).
2. J. A. Franklin and R. Chandra, *Int. J. Rock Mech. Min. Sci.*, Vol. 9, pp. 325-341 (1972).
3. ISRM, *Rock Characterization Testing & Monitoring, ISRM Suggested Methods*. Ed. E.T. Brown, Pergamon Press, Oxford, 211 pp. (1981).
4. I.A. Sadisun, H. Shimada, M. Ichinose, and K. Matsui, *Proc. 11th Japan Symp. on Rock Mechanics (CD-ROM)*, Paper No. E-01 (2002).

THE MATHEMATICAL PRINCIPLES OF PERMEABILITY MEASUREMENT BY TRANSIENT METHOD FOR CYLINDRICAL SPECIMEN MADE OF LOW-PERMEABILITY ROCK

KEN KURIYAMA YOSHIAKI MIZUTA
Professor, Dr., Faculty of Engineering, Yamaguchi University, Tokiwadai, Ube755-8611, Japan

Strict formulation of mathematical equations for permeability determinations by transient method were established and those are extended to the case by use of cylindrical specimen, in which fluid leakage along specimen boundary can be easily prevented. Intrinsic permeability k was given by the following formula: $k = \alpha\kappa\mu\log(R_0/R_i)V_1V_2/(2\pi L(V_1+V_2))$ where α is the value to be determined from test data, κ is fluid compressibility, μ is fluid viscosity, R_0, R_i and L are the outer radius, the inner radius and the height of the cylindrical specimen, respectively. V_1 and V_2 are capacity of the spaces communicating with the inside and the outside of the cylindrical specimen in the pressure cell. Authors developed a simple apparatus for permeability measurement. Pneumatic permeability tests using the apparatus were carried out and the permeability of a specimen made of limestone was determined. Permeability test takes a few hours even for a specimen made of low-permeability rock because of low viscosity of air.

1 Derivation of a fundamental equation

A method to obtain the value of permeability of rock by observing the change of pressure of fluid in a vessel enclosing a sample is called a transient method. W.F.Brace et al studied transient method and established the equation for permeability determinations by use of specimen of rectangular prism by solving differential equations. But they did not show how to derive differential equations in detail. First we shall derive a fundamental equation and secondly find differential equations for the case of cylindrical specimen.

A specimen is contained in a vessel and air flows into and out of the specimen through the interface of the vessel. Let V be the volume of air in the vessel, P the air pressure, N the number of particles of air, T the absolute temperature of air, S the boundary between the specimen and the vessel, n outward unit normal vector on S, k_B Boltzmann's constant, μ the viscosity of air, v the velocity of air in the specimen and k the permeability of the specimen.

Assuming that the equation of state of ideal gas $PV=N\,k_B\,T$ is satisfied in the vessel, Darcy's law $v=-(k/\mu)\cdot P$ holds in the specimen and compressibility of ideal gas $\kappa=1/P$ does not change so much during an experiment, we can get a fundamental differential equation in the following way.
From the following equality

$$\frac{dP}{dt}=\frac{dP}{dN}\frac{dN}{dt}=\frac{k_BT}{V}\left(\iint_S vndS\Big/V\right)N=P\frac{1}{V}\iint_S\left(-\frac{k}{\mu}\nabla P\right)ndS=-P\frac{k}{\mu}\frac{1}{V}\iint_S\frac{\partial P}{\partial n}dS$$

we obtain the equation (1).

$$\frac{dP}{dt} = -\frac{k}{\kappa\mu}\frac{1}{V}\iint_S \frac{\partial P}{\partial n} dS \tag{1}$$

2 Differential equations for cylindrical specimen

We shall derive differential equations for a cylindrical specimen of which outer radius is R_o, inner radius is R_i and height is L. Let P_1 be the pressure of air in the pressure cell on the outside of the specimen, V_1 the volume of the outside cell, P_2 the pressure of air in the cell on the inside of the specimen and V_2 the volume of the inside cell. From Eq. (1) the following equations are obtained.

$$\left.\begin{aligned} \frac{dP_1}{dt} &= -\frac{k}{\kappa\mu}\frac{1}{V_1}\iint_{S_1} \frac{\partial P_1}{\partial n} dS \\ \frac{dP_2}{dt} &= -\frac{k}{\kappa\mu}\frac{1}{V_2}\iint_{S_2} \frac{\partial P_2}{\partial n} dS \end{aligned}\right\}$$

In order to get differential equations from these equations, the surface integrals should be transformed. Since Darcy's law holds in the specimen, it is easily shown that $\Delta P = 0$ in the specimen. So we obtain

$$0 = \Delta P = \frac{\partial^2 P}{\partial r^2} + \frac{1}{r}\frac{\partial P}{\partial r} + \frac{1}{r^2}\frac{\partial^2 P}{\partial \theta^2} + \frac{\partial^2 P}{\partial z^2}$$

where (r,θ,z) is a cylindrical coordinate system. Since P depends on r only because of the axial symmetry, it follows that

$$0 = \Delta P = \frac{\partial^2 P}{\partial r^2} + \frac{1}{r}\frac{\partial P}{\partial r}.$$

The general solution of this ordinary differential equation is $P = a\log r + b$. Constants a and b must be determined to satisfy the boundary conditions, that is, $P = P_1$ at $r = R_o$ and $P = P_2$ at $r = R_i$. Hence $a = (P_1 - P_2)/\log(R_o/R_i)$.
Therefore we get

$$\frac{\partial P_1}{\partial n} = \left.\frac{\partial P_1}{\partial r}\right|_{r=R_o} = \frac{(P_1 - P_2)}{\log(R_o/R_i)}\frac{1}{R_o}$$

on the outer boundary and

$$\frac{\partial P_2}{\partial n} = -\frac{\partial P_2}{\partial r}\bigg|_{r=R_i} = -\frac{(P_1 - P_2)}{\log(R_o/R_i)}\frac{1}{R_i}$$

on the inner boundary.

Hence ordinary differential equations with respect to P_1 and P_2 are obtained.

$$\left.\begin{aligned}\frac{dP_1}{dt} &= -\frac{k}{\kappa\mu}\frac{2\pi L}{\log(R_0/R_i)}\frac{1}{V_1}(P_1 - P_2)\\ \frac{dP_2}{dt} &= -\frac{k}{\kappa\mu}\frac{2\pi L}{\log(R_0/R_i)}\frac{1}{V_2}(P_2 - P_1)\end{aligned}\right\} \quad (2)$$

The solutions of Eqs. (2) under the initial condition $P_1(0) = P_1^0$, $P_2(0) = P_2^0$ is

$$\left.\begin{aligned}P_1 &= \frac{V_1P_1^0 + V_2P_2^0}{V_1 + V_2} + \frac{V_2}{V_1 + V_2}(P_1^0 - P_2^0)\exp\left(-\frac{k}{\kappa\mu}\frac{2\pi L}{\log(R_0/R_i)}\frac{V_1 + V_2}{V_1V_2}t\right)\\ P_2 &= \frac{V_1P_1^0 + V_2P_2^0}{V_1 + V_2} - \frac{V_1}{V_1 + V_2}(P_1^0 - P_2^0)\exp\left(-\frac{k}{\kappa\mu}\frac{2\pi L}{\log(R_0/R_i)}\frac{V_1 + V_2}{V_1V_2}t\right)\end{aligned}\right\} \quad (3)$$

W. F. Brace established a model of flow of fluid in rock by considering porosity in rock. They assumed that there was a porosity in the center of a specimen of rectangular prism. A model for a cylindrical specimen can be established by their idea and the preceding method in this paper. The governing differential equations gained in such a way are simultaneous differential equations having three unknown functions; P_1, P_2 and another. It is shown that approximate solutions of P_1 and P_2 are equal to Eqs. (3).

3 Determination of permeability

Taking logarithm both sides of Eqs. (3), the following equality is given:

$$\log\left(P_1 - \frac{V_1P_1^0 + V_2P_2^0}{V_1 + V_2}\right) = \log\left(\frac{V_2}{V_1 + V_2}(P_1^0 - P_2^0)\right) - \frac{k}{\kappa\mu}\frac{2\pi L}{\log(R_0/R_i)}\frac{V_1 + V_2}{V_1V_2}t\,.$$

This equality means that the graph of $\log\left(P_1 - (V_1P_1^0 + V_2P_2^0)/(V_1 + V_2)\right)$ has a slope

$$-\frac{k}{\kappa\mu}\frac{2\pi L}{\log(R_0/R_i)}\frac{V_1 + V_2}{V_1V_2}\,.$$

Comparing this theoretical slope with an slope $-\alpha$ obtained from experimental data, the permeability k of the specimen from the following equality is determined:

$$k = \alpha\kappa\mu \frac{\log(R_0/R_i)}{2\pi L} \frac{V_1 V_2}{V_1 + V_2} \quad . \tag{4}$$

The authors developed a simple apparatus for permeability measurement. Pneumatic permeability tests were carried out and the permeability of a cylindrical specimen made of limestone was determined by using the data as shown in Fig.1 and Table 1.The permeability k of the limestone was $6.1 \times 10^{-17} (m^2)$ and the hydraulic conductivity was $6.7 \times 10^{-10} (m/s)$.

The permeability test takes only a few hours even for a specimen made of low-permeability rock. This method was shown effective.

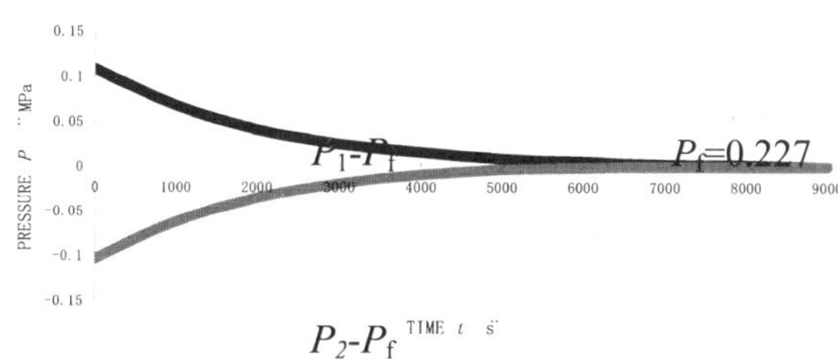

Figure 1. Pressure-time records obtained from the pneumatic permeability test for a limestone specimen.

Table 1. Specification of the permeability test using cylindrical specimen.

Outer radius of specimen R_0 (m)	4.45×10^{-2}
Inner radius of specimen R_i (m)	3.09×10^{-2}
Height of specimen L (m)	4.92×10^{-2}
Volume of outside cell V_1 (m^3)	2.90×10^{-3}
Volume of inside cell V_2 (m^3)	2.90×10^{-3}
Compressibility of air κ (Pa^{-1})	3.33×10^{-6}
Viscosity of air μ_a (Pa・s)	1.82×10^{-5}
Viscosity of water η (Pa・s)	8.90×10^{-4}

References

Brace, W.F., Walsh, J.B. & Frangos, W.T. (1986): J.Geophys.Res., Vol.73, p.2225-2236

AESPOE HARD ROCK LABORATORY, SWEDEN

C. SVEMAR R. CHRISTIANSSON
Swedish Nuclear Fuel and Waste Management Co., Stockholm, Sweden.
christer.svemar@skb.se rolf.christiansson@skb.se

The AEspoe Hard Rock Laboratory (AEHRL), which has been in operation since 1995, provides the opportunity to conduct research and development work in a rock environment similar to the one that will exist in a future repository at approx. 500 m depth. The The experimental work in the AEHRL is focused on two main areas: the retention properties of the rock against the transport of radionuclides, and development, testing and demonstration of methods and equipment for excavation and disposal. The results of research provide important references for results from the planned site investigations for siting the Swedish repository for spent fuel. The testing and demonstration is also an important task in parallel to these investigations. This paper presents the geo-scientific programme, including geology, geohydrology, geochemistry, rock mechanics and heat transport, but focuses on the programme for rock mechanics characterisation and investigation which has been undertaken using laboratory testing and *in situ* measurements in boreholes, large scale experiments covering both mechanical and acoustic monitoring, as well as studies of the performance of the entire tunnel system. A major verification experiment is being planned that consists of heating the pillar between two full scale deposition holes (1.75 m in diameter) until failure occurs. The objective is to demonstrate the capability for predicting the conditions when the risk for spalling becomes high.

1 Introduction

AEspoe Hard Rock Laboratory (AEHRL) consists of a 3 600 m long ramp down to 450 m level, close to the Oskarshamn nuclear power station. Figure 1 shows the surface facility on the island of AEspoe, the start of the TBM part of the ramp at 420 m level, and the location of the ramp under the island as well as. Besides the ramp there is one shaft with a hoist on the island and two ventilation shafts.

The activities in the AEHRL are separated into four phases.

- Verification of site investigation methods.
- Development of detailed investigation methodology.
- Testing of models for description of the rock's barrier properties — this focuses on the ability of the rock to retard transport of radionuclides from the repository level to ground surface in case a defect is developed in the canister, water has free access to the interior, radionuclides are dissolved and migrate out of the canister and further through the bentonite into the rock.
- Demonstration of technique for and function of important parts of the deep repository. This includes all practical application and development of

equipment and methodology for deposition and sealing of the deep repository.

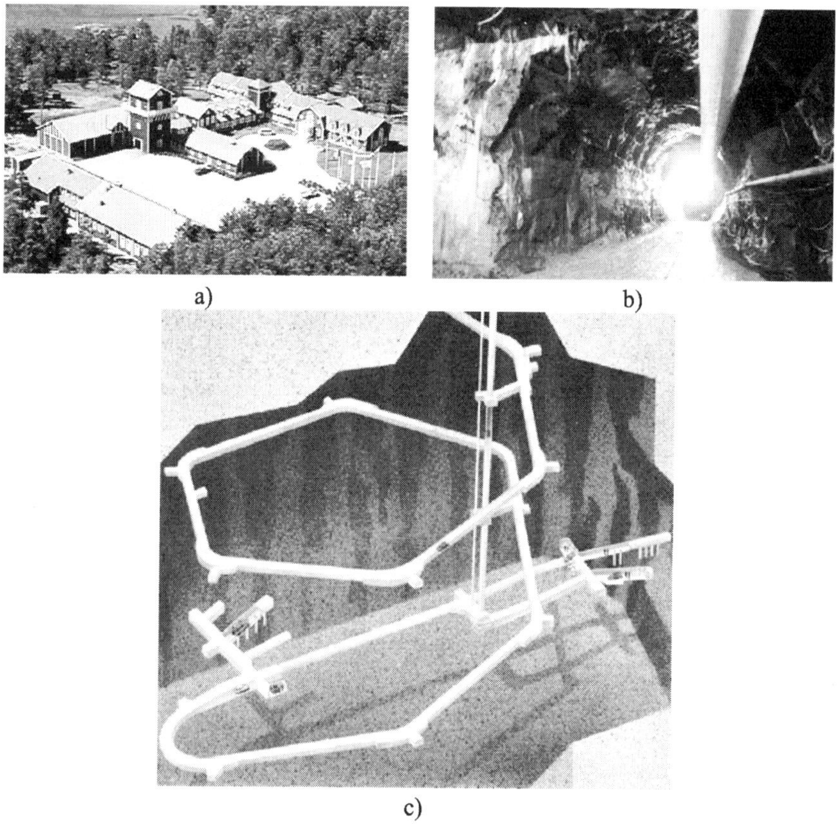

Figure 1. The AEspoe HRL. a) the surface buildings, b) the start of the TBM part of the ramp at the 420 m level, c) the layout of the shaft, ramp and other tunnels.

The two first phases were conducted during 1986-1995 and concerned pre-investigations and ramp and shaft excavations respectively down to the 450 m level, while the two latter phases have been conducted since 1995. Planned activities have been described[1].

2 Site Conditions

The rock types are dominated by AEspoe diorite and Smaoland (AEvroe) granite. These rock types are quite similar, but the diorite has a lower quartz content2. Greenstone and fine-grained granite occurs frequently as intrusions. The granites and the diorite may be of similar age. The uniaxial strength is in the range of 150 to 260 MPa, average 200 – 210 MPa. The Young's modulus is in the range of 70 to 78 GPa in the normally unaltered rock.

Fracturing characteristics has been described on the different scales[2]:

Regional scale: Major fracture zones

Site scale: Major fracture zones, minor fracture zones, fracture swarms, small scale fracturing

Block scale: Minor fracture zones, fracture swarms, and small scale fracturing

Detailed scale: Small scale fracturing.

The island is cut north of the underground laboratory by an east–west trending fracture zone that is estimated to be regional. South of the island is a northeast – southwest trending major fracture zone. In the area close to the experimental tunnels there exist some minor fracture zones and fracture swarms. The small scale fracturing, for example, as observed in the tunnels, form a systematic pattern, Figure 2.

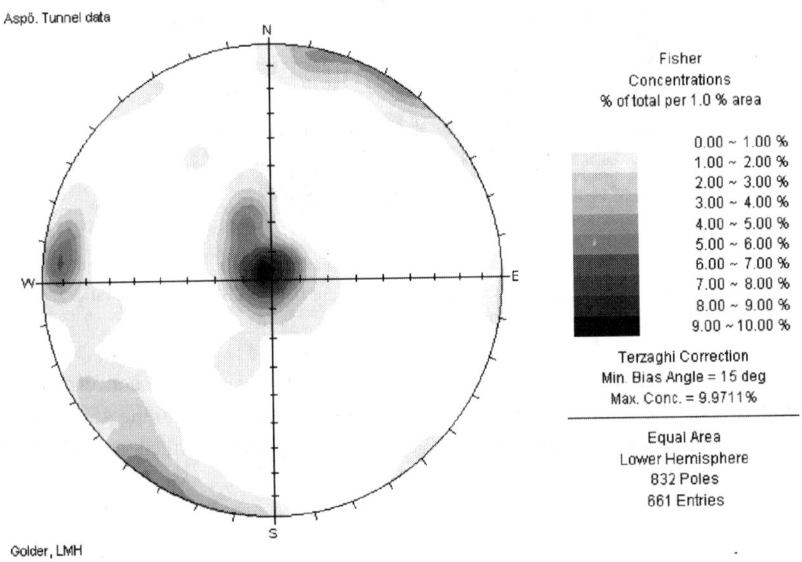

Figure 2. Fracture statistics from tunnels and shaft between the 350 – 450m levels, observations from three orthogonal directions[3].

3 Geoscience

The geoscientific activities concern the five geo-science areas, which SKB has chosen to describe separately and which integrated represent the properties determining the potential of a candidate site. The five areas are:

Geology — rock type and structure
Hydrogeology
Geochemistry — groundwater composition
Thermal properties of the rock
Rock mechanics

A further area is of importance for a site's potential is the migration properties of radionuclides. The properties are affected, in addition to factors relating to the five areas presented above, by geometrical, physical and chemical conditions in and along transport channels between the repository and the ground surface (the biosphere).

4 Monitoring

In the hydrogeological programme the piezometric pressure is measured. Besides long term variations during different seasons, it is also possible to register changes in the piezometric pressure caused by seismic events. The system is immediately affected when subject to a pulse of 1 kPa (0.1 m water pillar) or more. The earthquake in Turkey in August 1999 was such an event, and pressure increases of up to 25 kPa were observed by AEHRL's Hydro Monitoring System (HMS) about 8 minutes after the event.

5 Rock Stress Measurements

In the site investigations and the interpretation of site characteristics, the stress conditions in the rock are difficult to determine with the required accuracy for the rock mechanics site characterisation. Because of this, SKB has undertaken a number of projects in order to:

- compare different measurement equipment under controlled conditions
- develop control methods for quality assurance
- define conditions for which the different methods can be applicable
- study methods for evaluation, and

- analyse the accuracy with which stresses can be measured at large depths.

The single largest activity is connected with a field test in two mutually perpendicular drilled holes[4]. Measurements have been made by three different methods: 2-D overcoring, 3-D overcoring, and hydraulic fracturing.

Conventional methods for evaluation of rock stress measurement are based on different assumptions. The most important is that the rock is assumed to be continuous, homogeneous, isotropic and to behave elastically. These assumptions have been used in different rock mechanics laboratories. Results from investigations of the rock from the test area show that the diorite there can be considered to fulfill the assumptions of material properties on which the conventional evaluation methods are based upon. The basically elastic response of the rock at the 450 m level has also been confirmed by the analysis of tunnel maintenance[5].

The results of the field test in the diorite are as follows:

- All studied measurement methods show agreement for the direction of the principal stress (better than ± 10°) and the area has a significantly anisotropic stress field
- Overcoring measurements result in a rather large spread, especially if the content of feldspar is large, but the rule of thumb that 3-5 overcoring measurements, with good repeatability, provides a reasonable mean value, has been shown to be realistic
- A 3-D stress cell in a pilot hole which has been overcored can show somewhat too high stresses due to elastic expansion from induced cracks
- The 2-D stress cell systematically indicates high stress levels in the current geological environment due to induced micro-cracks in the machined bottom where the cell is glued to the rock
- Hydraulic fracturing is the best method for estimating the minor principal stress perpendicular to the borehole, providing the induced fracture is along and not perpendicular to the hole
- Hydraulic fracturing tend to underestimate the major principal stress perpendicular to the borehole in a strongly anisotropic stress field
- Measurement of the P-wave velocity in rock samples is a quick indicator method for judgement whether the overcoring method′s demand on homogeniety is fulfilled or not. If the rock responds elastically to the overcoring, the rock stress magnitudes can be determined with an accuracy of approx. ± 15 – 20%. It is more probable that the stresses are overestimated than underestimated.

6 Rock Mechanics Experiences through the Maintenance of Tunnels

A compilation of archived data of all the tunnel maintenance performed at AEHRL has been completed. Certain niches and tunnel crossing were chosen to be studied with numerical stress modelling. The major part of the tunnel maintenance in the ramp is confined to the areas where the tunnel area is widened from $25m^2$ to $42m^2$. Areas with larger amounts of fine-grained granite can also be associated with a relative higher need for maintenance, since especially the fine-grained granite has a higher fracture density than the rest of the rock types.

The angle between the tunnel and the major principal stress impacts the need for maintenance. Tunnel sections with a large angle to the principal stress direction are require more maintenance than tunnel sections that are almost parallel to it. The numerical stress modelling shows that the maintenance is required for tunnel sections with either high stress or low confinement. AEHRL is in general terms a stable facility seen from a rock mechanics point of view. Some parts of the tunnel system are subjected to stress concentrations due to the choice of geometry in relation to the *in situ* stresses. If the tunnel system was subjected to the *in situ* stresses at approx. 550m depth, we assess that the rock mass response would be brittle instead of elastic. Possible problems of spalling could however be controlled at least at that depth by aligning the tunnels more parallel to the major principal stress.

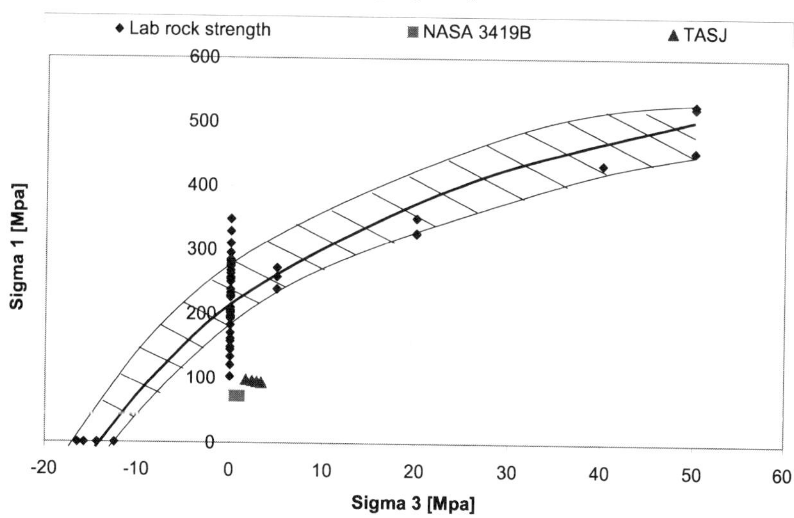

Figure 3. Compilation of laboratory rock strength measurements, and results from highly stressed areas, according to the numerical modelling

The results from the numerical modeling agree well with observations made in the tunnel system. The stress-modelled areas indicate quite high loads, but they are below the unconfined compressive strength, athough they are rather close to the limit where crack initiation occurs — 118 MPa according to laboratory tests on cores[6]. Results from different kinds of laboratory rock strength tests, together with two of the most stressed areas according to the numerical modeling, are presented in Figure 3. The hatched area is the assessed interval of the rock's compressive strength.

An empirical relation between the depth of failure and the maximum tangential stress at the boundary of an opening has been presented[7]. The linear relation is described by the equation

$$\frac{Rf}{a} = 0.49(\pm 0.1) + 1.25\frac{\sigma \max}{\sigma c}$$

where the depth of failure (Rf) is normalized to the effective tunnel radius (a) and the failure initiates when $\sigma_{max}/\sigma_c \approx 0.4 \pm 0.1$. The maximum stress obtained from numerical modeling is approximately 100 MPa and the unconfined compressive strength for AEHRL diorite is approx. 200 MPa. Using these values, the σ_{max}/σ_c ratio is 0.5. This is ery close to the limit to where spalling is initiated[7], e.g. supporting the results of observations and modelling discussed above.

7 A Rock Mechanics Descriptive Model

In connection with the work to develop a strategy for compiling a rock mechanics site descriptive model[8], studies been conducted with data from the AEHRL. The first model of the geo-scientific conditions was compiled as a prognosis prior the start of the ramping[9]. The result of all collected geo-scientific knowledge was compiled after the ramp had been completed and the rock laboratory was ready for operation in year 1995[10]. The different experiments which have been conducted since 1995 have been reported in sub-models.

As a first step to update all geo-scientific site characterisations over AEspoe, a detailed, site-specific rock mechanics model was integrated with a test of SKB's planned strategy for compilation of rock mechanics site specific models. The result from this is a characterisation of the rock's mean value deformation and strength properties as function of the depth below ground surface, as well as specific values in different locations close to the major deformation zones, which exist outside the AEHRL. The result has been reported[8], and is also discussed by Christiansson and Hudson in a paper at this conference.

8 Demonstration Experiments

8.1 Boring of Deposition Holes

Of a total of thirteen bored deposition holes with a diameter of 1.75 m and depth 8.5 m, geophones were installed around four: two in the Canister Retrieval Test; and two in the Prototype Repository. The equipment measured in a traditional way acoustic signals from stress induced displacements in the fractures, but could also transmit ultrasonic sound and measure its velocity. By the fact that the boring head could be located very precisely and bored the hole vertically, the ultrasonic ray path came to pass through the disturbed zone. Figure 2 indicates that the properties of this zone are changed when the boring head passes the level where the ray touches the hole[11].

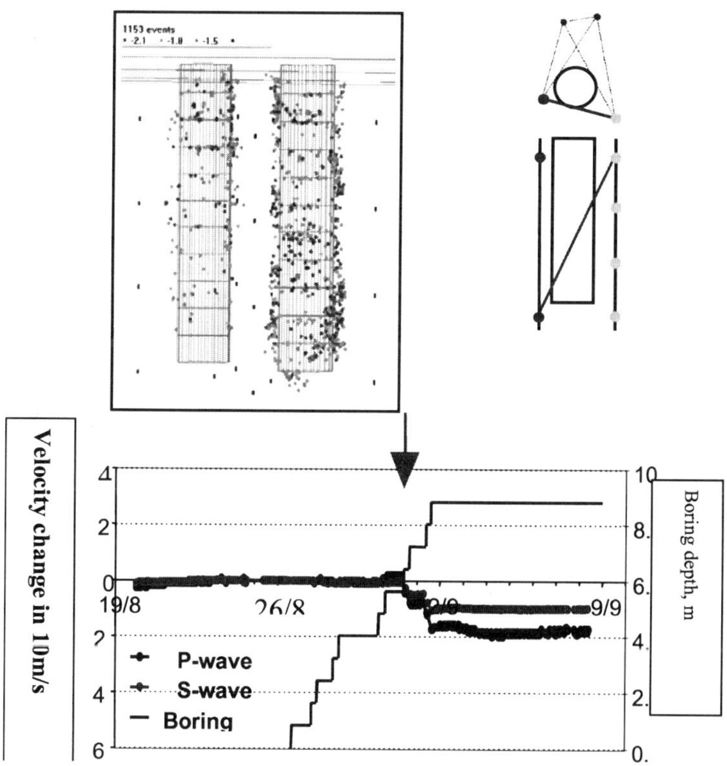

Figure 4. Acoustic emission and change in P and S wave ultrasonic velocity during boring. The P and S waves ray path is shown in the figure to the top right. The arrow above the lower diagram indicates when the boring head passes the ray path.

8.2 Disturbed Zone Around a Bored Deposition Hole

In order to be able to measure the force the cutter applies on the rock surface and correlate this with the 'damage' the tool creates in the deposition hole walls (hole diameter 1.75 m), two cutters were equipped with strain gauges on the cutter axis for measurement of strain in three directions. Samples were taken from the walls and hole bottom for analysis of the newly created cracks' lengths. The thrust was changed along the hole so the relation between the crack length and the cutter force could be established. It was then concluded that the theoretical length for a certain force is 2-3 times larger than the observed length. The difference is considered to depend on three factors, which the numerical model does not take into account today: a) rock confinement, b) roughness of rock surface, and c) rock debris on the surfaces. In quantitative terms, it was concluded that cracks in the cored samples varied between 10 and 15 mm[12]. As a comparison, the Posiva result on cores from the Olkiluoto Research Tunnel is approx 10 mm in the experimentally bored deposition holes there[11].

8.3 Verification Test

Extensive rock mechanics studies and numerical modelling of the AEHRL rock mass have been carried out during the last two years. The evaluation methods indicate that the rock mass, from 420 m level and downwards, is subject to such stresses that brittle failure may occur for marginally higher loads. A full scale failure test is being planned to demonstrate the capability to correctly measure the state of stress, and of predicting the rock mass response to loading in the zone where elastic response turns into brittle failure, by conducting a test with a heated rock mass. There will be two adjacent full scale deposition holes (diameter 1.75 m). One of the holes is filled with water, while the pillar between the holes is heated. Acoustic emission will be monitored. At the peak load, the water table is gradually lowered in the hole in order to study the response for small changes in the confining pressure. The plan is to make necessary preparations in 2002 and carry through the experiment in 2003.

9 Conclusions

The AEHRL is locaded in crystalline rock with various fracture conditions. The site is suitable for a varity of experiments and demonstrations for nuclear waste geological disposal. An extensive characterization of the site has been undertaken since 1986. The experiences from characterization, testing of instruments and demonstration tests have strongly supported the development of the SKB programme for Site Investigations, as well as for justification of the KBS-3 concept for geological disposal. The AEHRL is planned to be in operation at least until the final repository has been taken in operation, say within 20 years.

References

1. SKB, AEspoe Hard Rock Laboratory Annual Report 2000. SKB Technical Report, TR 00-10, SKB (2000).
2. I. Rehn, G. Gustavsson, R. Stanfors, P. Wikberg, AEspoe HRL. Geoscientific eavaluation 1997/5. Models based on site characterization 1986 – 1995. SKB report TR-97-06 (1997).
3. L. Maersk Hansen and J. Hermansson, Local model of geological structures close to the TASF-tunnel. SKB report IPR-02-15 (2002).
4. J. Jansson and M. Stigsson, Test with three different stress measuring methods in two othogonal boreholes. SKB report R-02-26 (2002).
5. C. Andersson and J. Soederhaell, Rock mechanics conditions at the AEspoe HRL. A study of the correlation between geology, tunnel maintenance and tunnel shape. SKB report R-01-53 (2001).
6. E. Nordlund, C. Li, et al Prototype repository. Mech. properties of the diorite in the prototype repository of the AEspoe HRL. SKB report IPR-99-25 (1999).
7. C.D. Martin, R. Christiansson, J. Soederhaell, Rock stability considerations for siting and constructing a KBS-3 type repository. SKB report TR-01-38 (2001).
8. J. Andersson, R. Christiansson, J.A. Hudson, Site investigations. Strategy for a rock mechanics site descriptive model. SKB report TR-02-01 (2002).
9. G. Gustafson, M. Liedholm, I. Rhén, R. Stanfors, P. Wikberg, AEspoe HRL – Predictions prior to excavation and the process of their validation. SKB Technical Report TR 91-23. SKB (1991).
10. R. Stanfors, P. Olsson, H. Stille. AEspoe HRL – Geoscientific evaluation 1997/3. Results from pre-investigations and detailed site characterization. Comparison of predictions and observations. Geology and mechanical stability. SKB Technical Report TR 97-04. SKB (1997).
11. Autio, Characterization of the excavation disturbed zone caused by the boring of the full scale deposition holes in the research tunnel at Olkiluoto. Swedish Rock Mechanics Day (1996).
12. Z. Zhang, S. Kou, P.-A. Lindqvist, Measurements of cutter forces and cutter temperature of boring machine in AEspoe HRL. SKB Technical Report TR-01-34. SKB (2001).

ELASTO-PLASTIC BEHAVIOUR OF SINGLE ROCK JOINT

YOUN-KYOU LEE

Dept. of Ocean System Eng., Kunsan National Univ., Kunsan, Chonbuk, Korea

SEOKWON JEON

School of Civil, Urban and Geosystem Eng., Seoul National Univ., Seoul, Korea

The response of some empirical formula for rock joint behavior is investigated through numerical simulation of direct shear tests of single rock joint. The joint is modeled by a 2D interface finite element. After initiation of slip, the behavior of the interface element is governed by an elasto-plastic constitutive law. The yield and plastic potential functions are derived from empirical equations for joint shear strength and dilation, expressed in terms of JRC and JCS. The pre- and post-peak shear stresses are approximated by evoking the mobilized JRC concept to both the yield and potential functions

1 Introduction

During the shearing of rough rock joints, the displacements normal to the shearing direction take place with the increasing shear displacements. This phenomenon of dilation is due to the roughness of joints and commonly characterized by dilation angle or dilation rate. Suppressing dilation leads to the increment of normal stress, followed by the increment of shear resistance. Dilation is, therefore, one of the most important components in mobilizing the shear strength of in-situ rock blocks whose deformations may be constrained to some extent by surrounding rocks.

In the plastic formulation of rock joints, the non-associated plastic flow rule is used to avoid the overestimation of dilation given by taking the associated flow rule. The selection of both yield and potential functions, therefore, is of importance and may have much influence on the prediction of shear resistance rock joints. Although several empirical expressions for shear behaviour are available, their predictions seem to be quite different. In this study, an elasto-plastic constitutive law for rock joints is described and the response of some empirical plastic potential functions, which reflecting the characteristics of dilation, is investigated under different boundary conditions.

2 Elasto-plastic Constitutive Model

The dimensionless model for shear stress-displacement modeling, proposed by Barton *et al.*[1], is shown in Figure1 and the typical shape of shear stress-

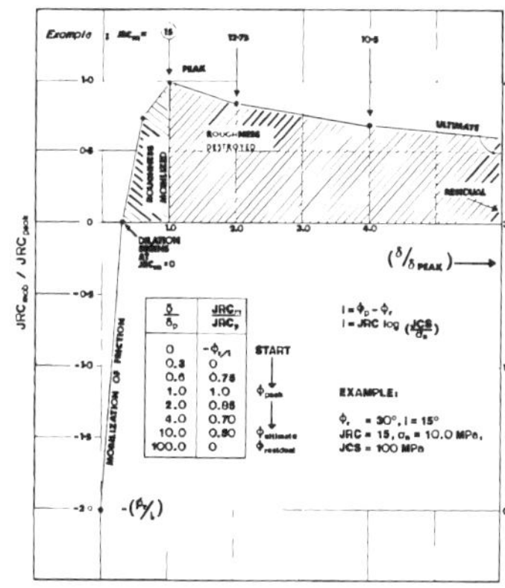

Figure 1. Dimensionless model of shear Behaviour of joints[1].

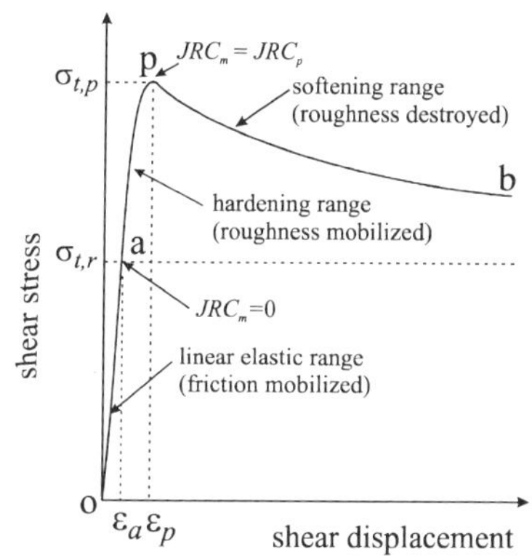

Figure 2. Elasto-plastic model of joints.

displacement relation is also given in Figure 2. Assuming that the intervals, o-a, a-p, and p-b characterize elastic, hardening, and softening behaviour respectively, the curve in Figure 2 suggests the possibility of making use of elasto-plastic formulation for numerical simulation of joint behaviour. Cuisiat *et al.*[2] tried this approach in their numerical investigation of joint behaviour using boundary element method. Similar approach is used in this study. While the mobilized JRC curve shown Figure 1 was fitted with piecewise continuous polynomial functions in Cuisiat *et al.*'s research, the variation of *JRC* is approximated by appropriate exponential functions of plastic shear work in this study.

The elasto-plastic formulation of joint behaviour yields the incremental stress-deformation relation,

$$d\sigma = \left(\mathbf{D} - \frac{\mathbf{D}(\partial Q/\partial\sigma)(\partial F/\partial\sigma)^T \mathbf{D}}{(\partial F/\partial\sigma)^T \mathbf{D}(\partial Q/\partial\sigma) - H} \right) d\varepsilon \qquad (1)$$

where $d\sigma$ is the vector of incremental stress acting on the joint surface and $d\varepsilon$ is the vector of the increment of relative deformation. F and Q are yield and plastic potential functions, respectively. The elastic matrix $\mathbf{D}$ consists of the shear and normal stiffness of joint. H is the hardening/softening parameter. If Q=F, so called associated formulation is obtained.

Eq.(1) shows that the numerical predictions of joint behaviour highly depend on the selection of the yield/potential functions and the plastic flow rule. Substitution of Barton's mobilized joint roughness coefficient, JRC_m, for peak *JRC*, JRC_p in Barton's empirical shear strength equation provides a yield function that enables to

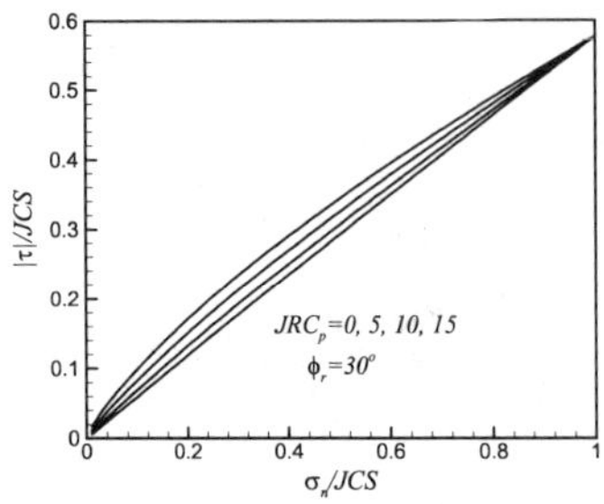

Figure 3. Normalized plot of yield functions.

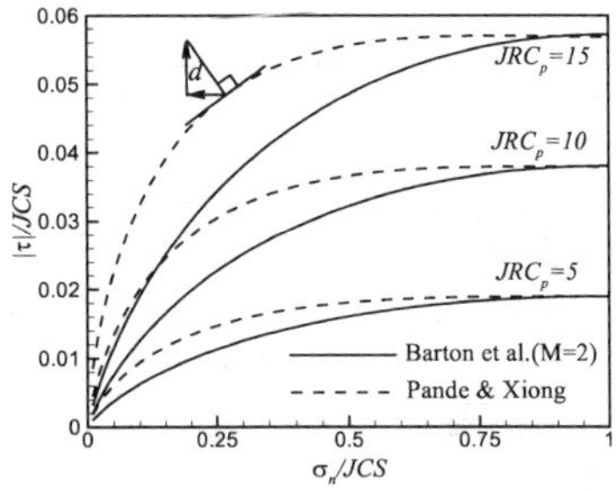

Figure 4. Normalized plot of potential functions

describe pre-peak or post-peak behaviour of rock joint. Thus, the following equation can be served as a yield function,

$$F = |\tau| - \sigma_n (JRC_m \log_{10}(JCS/\sigma_n) + \phi_r) \tag{2}$$

where τ and σ_n are the shear and normal stresses acting on the joint surface and ϕ_r is the residual friction angle. Figure 3 shows the function F for some values of JRC_p..

Following Lü & Brown[3], it is assumed that plastic potential functions have the form,

$$Q = |\tau| - G(\sigma_n) = \text{constant} \tag{3}$$

Then plastic potential function can be derived by integrating the tangent of dilation angle with respect to σ_n.

$$\tan(d) = \frac{d\delta_n}{d\delta_t} = \frac{\partial Q / \partial \sigma_n}{\partial Q / \partial \tau} = \frac{\partial G}{\partial \sigma_n} \tag{4}$$

where d is the dilation angle. Once the function G is determined from eq.(4), the corresponding potential function is given by Eq.(3).

By evoking the mobilized JRC concept, Barton & Choubey's empirical equation[4] for peak dilation angle, d_p, may be used in deriving a potential function that could account for pre-peak and post-peak shearing.

$$d_m = (1/M) JRC_m \log_{10}(JCS/\sigma_n) \tag{5}$$

where d_m is the mobilized dilation angle and M is a damage coefficient, given values of 1 or 2 for shearing under low or high normal stresses, respectively.

Another empirical equation by Pande & Xiong[5], which is also combining the indices of *JRC* and *JCS*, is considered. If JRC_p is replaced by JRC_m, their equation can be written as

$$\tan(d_m) = \frac{\tan i}{c_1} - \frac{JRC_m}{c_1^2 c_2}(1+\tan^2 i) - \frac{JRC_m \sigma_n}{c_2 JCS} \tag{6}$$

where $i = JRC_m \log_{10}(JCS/\sigma_n)$, $c_1 = (1 - \tan i \tan \phi_r)$, and $c_2 = (180/\pi)\log 10$. The plastic potential functions obtained from Eq.(5) and Eq.(6) are plotted in Figure 4. It displays the fact that dilation angle decreases with increase in normal stress, σ_n. The differences in two potential functions, however, are substantial in low normal stress range, and become larger with increase in JRC_p values.

The mobilization and destruction of *JRC* in pre- and post-peak shear stress regions are approximated by exponential functions of plastic work done by shear stress, $W_t = \int \sigma_t d\delta_t$.

$$JRC_m = JRC_p\left(1 - \exp\left(-\frac{c_1}{JRC_p}W_t\right)\right), \quad W_t \le W_{t,p} \tag{7}$$

$$JRC_m = JRC_p \exp\left(-\frac{c_1}{JRC_p}(W_t - W_{t,p})\right), \quad W_t > W_{t,p} \tag{8}$$

where $W_{t,p}$ is the value of W_t corresponding to JRC_p and can be determined by assuming that the JRC_p is achieved when JRC_m is sufficiently close to JRC_p. The constants c_1 and c_2 reflect how rapidly the roughness is mobilized and degraded. In order to implement the joint model mentioned above, therefore, requires seven input parameters, which are k_s, k_n, ϕ_r, JRC_p, JCS, c_1, and c_2.

3 Numerical Direct Shear Tests

In order to investigate the response of the plastic potential functions, obtained by use of Eq.(5) and Eq.(6), on the shear behavior of single rock joint, numerical direct shear tests are performed under both prescribed normal displacement and constant normal stress conditions. The damage coefficient, *M*, in Barton *et al.*'s equation, Eq.(5), is assumed to be 2.

The horizontal joint between lower and upper block is modeled by 2-D

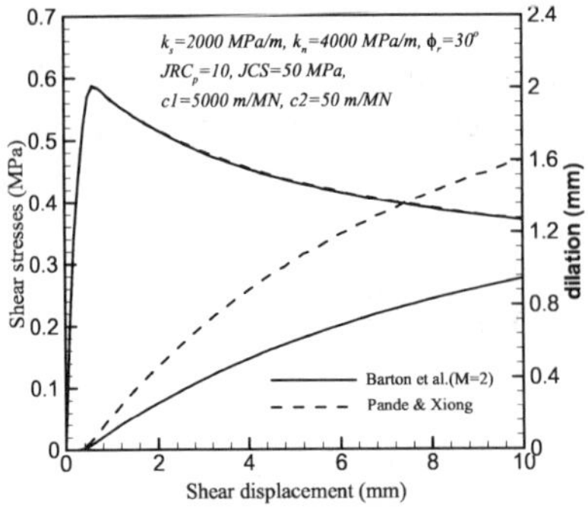

Figure 5. Shear stress and dilation curves under under constant normal displacement condition.

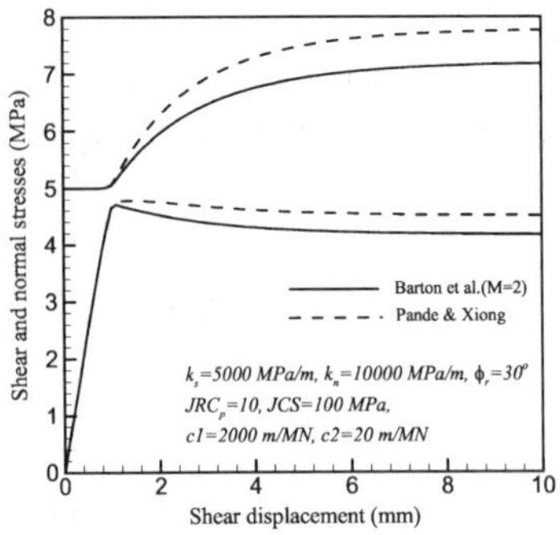

Figure 6. Shear and normal stress curves under constant normal stress condition.

maximum prescribed shear displacement of 10mm. While the upper block, compressed by uniform pressure, can move freely to the vertical direction under constant normal stress condition, no normal displacement is permitted on the joint surface after initiation of slip when the constant normal displacement condition is adopted.

History curves of shear stress and dilation during the slip, obtained when the initial normal stress value of 0.5MPa is maintained and the vertical displacements are not prescribed throughout the shearing, are shown in Figure 5. The shape of curves shows good qualitative agreement with experimental observations. It can be seen that the use of Pande & Xiong's function predicts larger dilation than the use of the function from Barton *et al.*'s formula. The figure also shows that the shear stress is not affected by the selection of plastic potential functions, because the upward movement of upper block is not restricted.

Figure 6 shows the shear and normal stress calculated with initial normal stress of 5 MPa under constant normal displacement condition. Because of the constrained dilation, the normal stress is increased after initiation of slip and the increase prevents the shear stress from decreasing rapidly after the peak. This phenomenon is due to the fact that the increase of normal stress contributes to developing the shear resistance of the joint. The more increase of normal stress is predicted and the larger shear stresses after the peak are calculated, when Pande & Xiong's potential function is considered. The differences in the predicted normal and shear stresses may become larger with increasing initial normal stress. The result demonstrates that the difference of estimated shear stresses of joint, resulting from taking different potential functions, could be substantial when the normal displacements of joints are constrained, so that care must be taken in selecting the plastic potential function for plastic formulation of rough rock joints.

4 Conclusions

In this study, an elasto-plastic model for rock joints, where both yield and plastic potential functions are based on Barton's empirical joint model, has been presented. Numerical direct shear tests were carried out in order to investigate the response of plastic potential functions, which were derived from empirical equations for dilation. Under constant normal displacement condition, the predicted shear resistance of rock joints was quite different, which suggested that the careful selection of plastic potential functions or dilation equations should be taken for elasto-plastic analysis of real rock joints which are constrained to some extent by surrounding rock masses.

Acknowledgements

This work was supported by grant No. R01-2000-00375 from Korea Science & Engineering Foundation.

References

1. N. Barton *et al.*, *Int. J. Rock Mech. Min. Sci. & Geomech. Abstr.* **22**, 121 (1985).
2. F.D.E. Cuisiat *et al.*, *Rock Joints*, eds. N. Barton and O. Stephansson, 611 (1990).
3. M. Lü and E.T. Brown, *Num. Meth. in Geomech.* ed. G. Swoboda, 1473 (1988).
4. N. Barton and V. Choubey, *Rock Mechanics* **10**, 1 (1977)
5. G.N. Pande and W. Xiong, *Proc. Int. Symp. on Num. Models in Geomech*, eds. R. Dungar *el al.* 218 (1982)

INFLUENCE OF STRESS-PATH ON THE STRESS-STRAIN RELATIONS OF JOINTED ROCKS

MING CAI, PETER KAISER
GRC/MIRARCO, Laurentian University, Sudbury, Ontario, Canada

HARUHIKO UNO, YOSHIAKI TASAKA
Tokyo Electric Power Services Co. Ltd., Tokyo, Japan

The actual stress path at a point in a rock mass near an underground excavation is complex. It is generally not in accordance with the loading path applied in a triaxial compression test (σ_3 = const.), the standard loading path used to determine the stress-strain relationship in laboratory tests. In most cases, a loading path of fixed mean principal stress ($(\sigma_1+\sigma_3)/2 = \sigma_m$ = const.) would be a better approximation of the stress path encountered in-situ near an underground excavation. In the present study, the effect of stress path on the stress-strain relationship of jointed rocks is studied using Distinct Element Method code UDEC. The simulation results demonstrate that the non-linearity in the stress-strain curve is stronger under a loading path of fixed mean principal stress than with a fixed confining stress, which is in agreement with laboratory and in-situ test results. It is recommended that for stability analyses of underground excavations, the constitutive relations of jointed rock mass be obtained from tests conducted in a constant mean principal stress loading condition.

1 Introduction

Stress changes near an underground excavation can drastically affect the stability of the opening. Stress changes not only influence the demand on the rock support, they also change the support capacity of frictional support components such as plain cablebolts [1]. Hence, for stability assessment as well as for support design, it is important to understand factors leading to detrimental stress changes [2].

Figure 1 presents the stress paths of points located at 0.5, 2.5, and 10 m from the cavern boundary. The cavern, which is 33 m wide and 52 m high, is excavated with 16 benches and the principal stresses are obtained from 2D linear elastic analysis. The maximum and minimum initial stresses are 12.48 MPa and 4.8 MPa, respectively, with the maximum stress dipping 64° to the right. It is seen that although σ_3 first decreases and then increases during the excavation process, the stress paths between the initial state and the final state can be approximated by the mean principal stress path. In a series of FEM analysis of large-scale underground excavations, we find that if stronger non-linearity in the deformation behavior of jointed rock mass is assumed, the predicted rock mass deformations agree better with the field measurement [3]. This implies that jointed rock masses under a fixed mean principal stress loading path might show stronger non-linear stress-strain relationships than that under a constant confinement stress path.

In terms of deformability of have shown that the stress-strain curves of intact rocks, laboratory test results rocks loaded under constant mean principal stress demonstrate stronger non-linearity than those under constant confining stress [4;5]. For jointed rock masses, very few test data exist because the difficulty in sample preparation and testing. Hence, in the present study, we applied distinct element modeling to investigate the response of jointed rocks under different stress paths and to verify the hypothesis of stronger non-linearity under fixed mean principal stress paths.

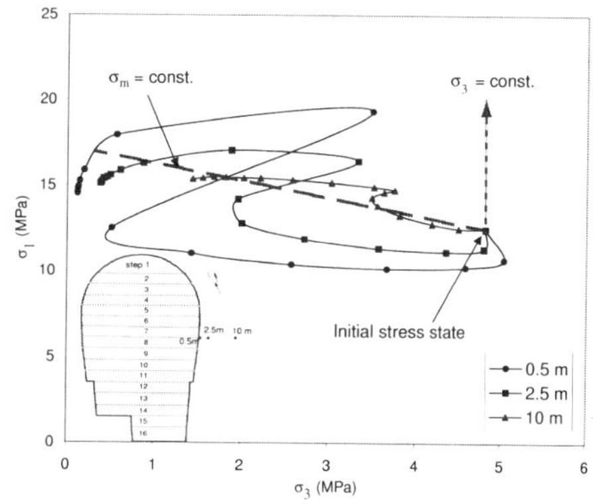

Figure 1. Stress paths of three points located on the right cavern sidewall.

2 Modeling

2.1 Loading Paths and Simulation Models

The Universal Distinct Element Code (UDEC) [6] is used in this investigation. The simulation models and loading paths are presented in Figure 2. In all models, an initial stress σ_1^0 and σ_3^0 is applied first and the models are brought to equilibrium. During the subsequent loading, only the load in the σ_1 direction is applied in the fixed confining stress path, but in the fixed mean principal stress path, σ_3 is decreased and σ_1 is increased, while $\sigma_m = (\sigma_1 + \sigma_3) / 2$ is kept constant. Two types of joints are considered in the models, i.e., uni-directionally distributed multiple joints and randomly oriented Voronoi joints.

2.2 Material Parameters

The material parameters are presented in Table 1. The mechanical property of the intact rock is linear elastic, and the joint strength properties are defined by the Mohr-Coulomb criterion and the joint deformation is governed by linear normal and shear stiffnesses. Other case studies with yielding intact rock and non-linear deformation properties of joints were also conducted. Due to space limitation, only the results with the material properties shown in Table 1 are presented in this paper.

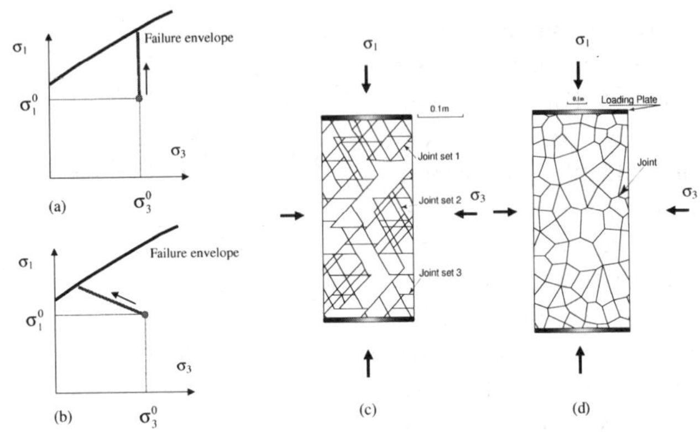

Figure 2. Stress paths for the loading and UDEC models: (a) fixed confining stress; (b) fixed mean principal stress; (c) 3 sets of laminated joints (model 1); (d) Voronoi joints (model 2).

Table 1. Input parameters.

	Model 1	Model 2
Bulk modulus (rock) GPa	16.7	16.7
Shear modulus(rock) GPa	10.0	10.0
Joint normal stiffness (K_n) GPa/m	120	120
Joint shear stiffness (K_s) GPa/m	120	40/120
Joint friction angle (ϕ)	40°	40°
Joint dilation angle (ϕ_d)	5°	5°
Joint cohesion (c)	0	0

3 Results and discussion

3.1 Laminated Joint Model

Figure 3 presents the differential stress versus axial strain curves[a] for laminated joint distribution (model 1) under different states of starting stress ($K_o = 1$). It is seen that nonlinear responses exist for both loading paths of fixed σ_3 and fixed σ_m. The sample shows gradual non-linearity but does not reach failure at σ_3 = constant (because the intact rock is not allowed to yield). On the other hand, for the loading path of fixed mean stress σ_m, the sample first responds in a linear fashion, followed by a distinct non-linearity near the peak strength. For the loading path with constant

[a] Because axial strain is plotted against the differential stress (σ_1-σ_3), the two loading paths (σ_m or σ_3 = constant) produce different slopes. This does not imply that the moduli differ at the initial loading stage.

confinement σ_3, nonlinear response is observed over a wide range of loading stage and the non-linearity is stronger when the initial stress is small.

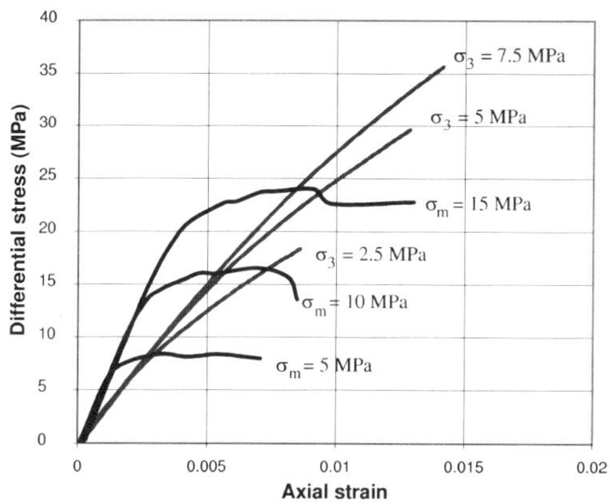

Figure 3. Differential stress-axial strain relations of a jointed rock mass with three sets joints (model 1).

3.2 Voronoi Joint Model

The differential stress and axial strain relationship for jointed rocks with Voronoi joints (K_s = 40 GPa/m) is presented in Figure 4. It is seen that the stress-strain curves for the loading path of fixed σ_3 is linear at the initial loading stage. Departure from linearity is more pronounced if the confining stress is low but, generally speaking, the nonlinear deformation is small for the stress path of fixed σ_3. On the contrary, the stress-strain relationship for the loading path of fixed σ_m shows very strong non-linearity near the peak strength (at about 80% of peak) due to slip of joints with low normal stress.

A view of the joints at the shear limit for the two loading paths (σ_3^0 = 7.5 MPa and σ_m^0 = 15 MPa) at peak loads (Points D and d in Figure 4.a) is shown in Figure 4. The slipped joints are plotted and viewed as if they were in an equivalent elastic material made up of the rock and joints. It can be seen that slipped joints remain isolated for the loading path of fixed σ_3 (Figure 4.b). The joint persistence for the slipped joints is small for the loading path of fixed σ_3, but large for the loading path of fixed σ_m. More persistent joints tend to deform more easily, resulting in strong non-linearity near the peak for the loading path of fixed σ_m.

For the loading path of fixed σ_3 (Figure 4.b), the number of joints reaching the shear limit increases earlier as the load is increased, but, because of the existence of confining stress, the slipping joints remain isolated. They act just like isolated cracks in an equivalent elastic solid and thus have relatively little effect on the deformation properties of the rock mass. For the loading path of fixed σ_m (Figure 4.c), there is almost no joint slip in the first half of the loading process because of the presence of relatively high confining stresses. As σ_3 decreases to low values, the number of joints reaching the shear limit increases rapidly and the slipping joints connect to the neighboring ones to form continuously sheared joints, resulting in strongly nonlinear deformation pattern.

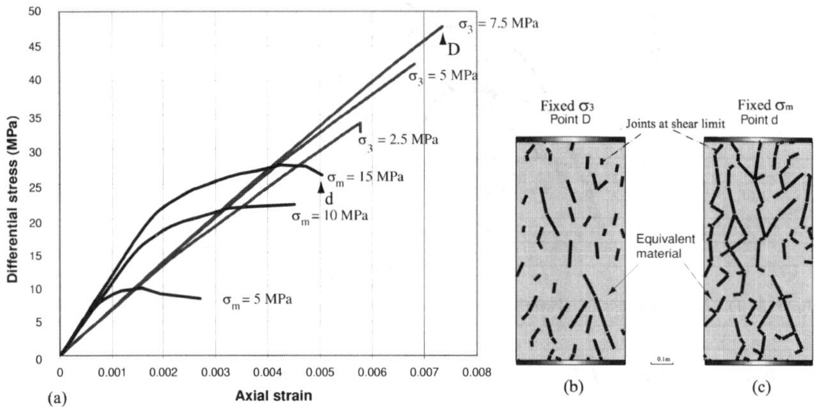

Figure 4. Simulation results of jointed rocks with Voronoi joints (model 2): (a) differential stress and axial strain relations; joints at shear limit at peak load for two loading paths (ioint pattern with shear slip): (b) fixed σ_3; (c) fixed σ_{m}.

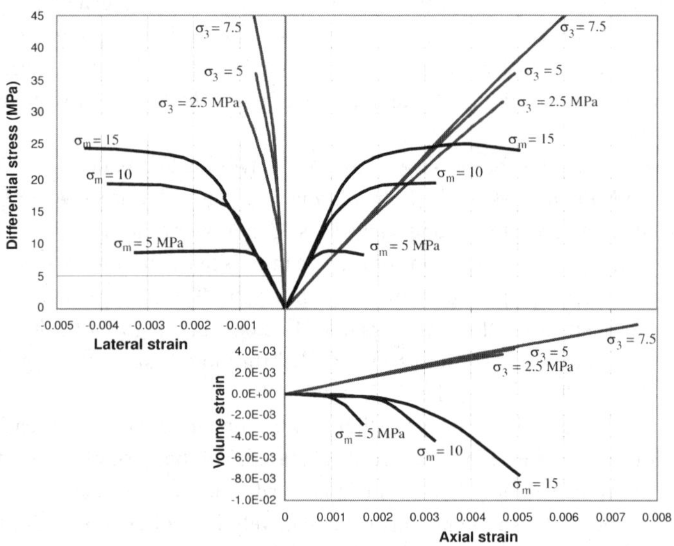

Figure 5. Stress-strain diagram obtained for jointed rocks with Voronoi joints ($K_n=K_s$).

Figure 5 presents the stress-strain diagrams for a case where the joint shear stiffness is equal to the joint normal stiffness ($K_n = K_s = 120$ GPa/m). The non-linearity in differential stress and axial strain relations in Figure 5 shows the same trend as that presented in Figure 4. Because the joints' shear stiffness is larger, the total strain of rock mass experiences at the same stress level is smaller. In this case, shear slip on joints occurs at a later stage. As is shown in Figure 5, the lateral strains for the loading path of fixed σ_m are larger, for the same differential stress level, than that for the loading path of fixed σ_3. It is interesting to note that the volume of the rock mass does not change during the early loading stage for the loading path of fixed σ_m. The rock mass suddenly starts to dilate when the slipping joints merge. The same trend was observed in laboratory tests [5;7].

4 Conclusions

This study confirms that certain stress paths cause strong nonlinear rock mass deformation behavior. Under the loading path of fixed confining stress, the stress strain curve shows an upward curvature at the initial loading stage, followed by a straight line response, ending with a downward curvature. Under the loading path of fixed mean principal stress, the stress strain relations are linear for most of the initial loading range. A distinct downward non-linearity is observed near the peak strength of the jointed rock mass, which is consistent with experimental results. This suggests that the departure from the elastic response of extensometer measurements near an underground opening can be directly related to the nonlinear rock mass behavior under the loading path of fixed mean principal stress, which is characterized by a decreasing tangential deformation modulus. It is recommended that for stability analysis of underground excavations, the constitutive relations of jointed rock mass be obtained from tests conducted with a constant mean principal stress loading condition.

References

1. P. K. Kaiser et al, *Int. J. Rock Mech. Min. Sci.* **29**, 293 (1992).
2. P. K. Kaiser et al, *Int. J. Rock Mech. Min. Sci.* **38**, 167 (2001).
3. Y. Tasaka et al, *Proc. of 10th National Symp. on Rock Mech.* 575 (1998).
4. M. Yasuda and T. Kataoka, *10th Conf. on Soil Mech. Found. Eng.*, 173 (1975).
5. M. Shen et al, *Proc. 3^{d} Rock Mech. System Eng.* 49 (1995).
6. UDEC-Universal Distinct Element Code, Itasca (1996).
7. Y. Osato et al, *33th Conf. on Soil Mech. and Foundation Eng.*, 1223 (1998).

ANALYTICAL SOLUTION OF PRESSURE TUNNEL WITH ELASTIC THEORY CONSIDERING DIFFERENT MODULUS IN COMPRESSION AND TENSION

ZHENDE ZHU AIJUN ZHANG WEIYA XU

Research Institute of Geotechnical Engineering, Hohai University, Nanjing, 210098

Rock mass is a heterogeneous media and contains fissures and joints, which show anisotropy mechanical behavior. If the classical elastic theory is applied to analyze the deformation of underground engineering, the calculated results can't fit the practical measured data well. Many laboratory tests indicate that the Young's modulus of rock in compression and tension is quite different. The elastic theory considering different modulus in compression and tension is applied to analyze the deformation and the unit elastic resistance K of a pressure tunnel. The results are quite different if they are compared to those from classical elastic theory. It has great practical potentialities with the elastic theory considering different modulus in compression and tension applied to design and stability analysis of underground engineering.

1 Introduction

С.А.Амбарцумян[1] (Soviet Union) first studied the variations of elastic constants of homogeneity materials in compression and tension, and developed the elastic theory considering different modulus of material in compression and tension. In the recent 2 decades, with the developments in new material, e.g. concrete, porcelain, glass steel, plastic, some metal, and especially the innovation of design in engineering and commonly used computers, new achievements have been obtained with the theory applied to analyze the strength, stiffness and stability of engineering. In rock engineering, rock mass is a heterogeneous media and contains fissures and joints, which makes it show anisotropy mechanical behavior. Some parts of surrounding rock mass is probably to be in tensile stress state. The deformation calculated by the classical theory is normally to be in small side if it is compared to the practical measured data. In order to obtain the optimized design in engineering and engender obviously economical benefit, it is very important to apply the elastic theory considering different modulus in compression and tension to analyze the mechanical properties and failure mechanism of rock mass.

2 The constitutive law of the elastic theory considering different modulus in compression and tension

*Science and technology Innovation Fellowship of Hohai University (2001401743)
Author: Zhende Zhu, 37 years old, acquired Doctor's degree of Tongji University in 1997, associate professor, chiefly engaging in teaching and researching of Rock Mechanics.

It's well known that the basic assumptions in elastic theory are that the material is continuum, homogeneous, perfect elastic, small deformation and isotropic. While the anisotropic elastic theory assume that physical parameters in each point is different from its direction. The constitutive law of anisotropic elastic theory is defined as :

$$\{\varepsilon\} = [a'_{ij}]\{\sigma\} \quad (1)$$

According to Long's theory, that is, $a'_{ij} = a'_{ji}$, even completely anisotropic material, a'_{ij} is composed of 21 independence elastic constant; for plane problem, if coordinate system xoy is middle plane, the number of independence constants will reduce to 6, if x y are principal direction, then can be expressed as:

$$\begin{aligned} \varepsilon_x &= a'_{11}\sigma_x + a'_{12}\sigma_y + a'_{13}\sigma_z \\ \varepsilon_y &= a'_{21}\sigma_x + a'_{22}\sigma_y + a'_{23}\sigma_z \\ \varepsilon_z &= a'_{31}\sigma_x + a'_{32}\sigma_y + a'_{33}\sigma_z \end{aligned} \quad (2)$$

For plane stress case ($\sigma_z = 0$), in engineering, $a'_{11} = 1/E_x$, $a'_{22} = 1/E_y$, $a'_{12} = a'_{21} = -\mu_x / E_x = -\mu_y / E_y$. Where: E_x, E_y, μ_x, μ_y are the elastic modulus and Poisson ratio, in isotropic case, $E_x = E_y = E$, $\mu_x = \mu_y = \mu$.

For elastic theory considering different modulus in compression and tension, the former four basic assumptions are the same as that of the classical theory. The fifth assumption should consider different modulus of material if it is in compression or tension stress state. It means that the deformation of material is composed of tension elastic modulus (E^+), compression elastic modulus(E^-)and Poisson ratio(μ^+ 、 μ^-). Their relationships can be written as:

$$\begin{aligned} \varepsilon_x &= a_{11}\sigma_x + a_{12}\sigma_y + a_{13}\sigma_z \\ \varepsilon_y &= a_{21}\sigma_x + a_{22}\sigma_y + a_{23}\sigma_z \\ \varepsilon_z &= a_{31}\sigma_x + a_{32}\sigma_y + a_{33}\sigma_z \end{aligned} \quad (3)$$

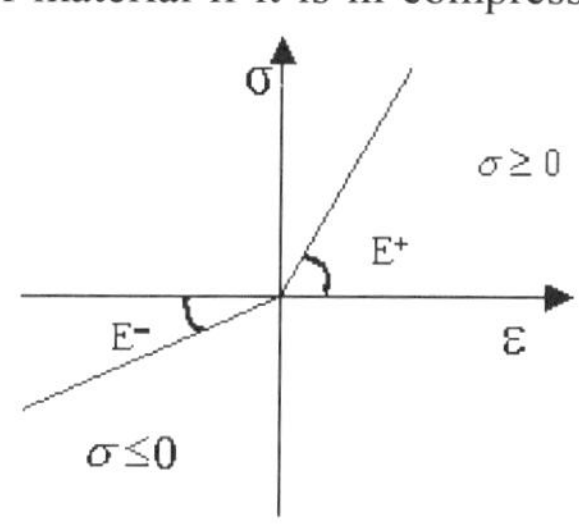

Figure1.The curve of constitutive relation with different modulus of compression and tension

Where: a_{ij} is related to the state of tension or compression stress. For isotropic material, $a_{11} = a_{22} = a_{33} = 1/E$, $a_{12} = a_{23} = a_{13} = -\mu / E$, $a_{ij} = a_{ji}$. Based on the results of most laboratory rock experiments, the curve of stress—strain is a double fold line (Fig.1), which has different E^+ 、 E^- along with the denotation change of main stress. Yet there is always an infinitesimal lubricious transition sector. On the basis of different tension and compression stress state of principal stress, table 1 lists the value of a_{ij} For plane strain, $\varepsilon_z = 0$ and the dominant function is that σ_x different with σ_y while σ_z can change its denotation.

It seems that equation (3) and equation (2) are quite similar, actually the elastic constants of anisotropic elastic theory are independent of the denotation of principal stress, while the elastic theory considering different modulus just as transverse isotropic material will use different modulus according to its principal stress. Therefore, with different calculating theory applied, the results will be quite different even for the same boundary conditions.

Table 1. The value of elastic constant on the state of tension or compression

a_{ij} \ σ_i	$\sigma_x,\sigma_y,\sigma_z>0$	$\sigma_x,\sigma_y,\sigma_z<0$	$\sigma_x>0,\sigma_y<0,\sigma_z>0$	$\sigma_y>0,\sigma_x,\sigma_z<0$
a_{11}	$1/E^+$	$1/E^-$	$1/E^+$	$1/E^-$
$a_{12}(a_{21})$	$-\mu^+/E^+$	$-\mu^-/E^-$	$-\mu^-/E^-$	$-\mu^-/E^-$
$a_{13}(a_{31})$	$-\mu^+/E^+$	$-\mu^-/E^-$	$-\mu^+/E^+$	$-\mu^-/E^-$
a_{22}	$1/E^+$	$1/E^-$	$1/E^-$	$1/E^+$
$a_{23}(a_{32})$	$-\mu^+/E^+$	$-\mu^-/E^-$	$-\mu^-/E^-$	$-\mu^+/E^+$
a_{33}	$1/E^+$	$1/E^-$	$1/E^+$	$1/E^-$

$*-\mu^+/E^+=-\mu^-/E^-$ That is $a_{ij}=a_{ji}$.

3 The elastic solution of pressure tunnel

Several representative rock specimens are selected and tests have been done according to experimentation criterion of rock mechanics, uniaxial compression and splitting tests on omnipotence experiment or rigidity servo testing system with standard specimen of ϕ 50mm×100m. Tensile and compressive strength of the rocks are shown in table 2.From the results shown in table 2, it is very clear that $E^+/E^->1$ and its range is from 1.1 to 1.8; If $E^+/E^-<1,\mu^+$ is so different with μ^-, in this case, the theory of $\mu^+/E^+=\mu^-/E^-$ normally can't be satisfied. For practical application, even lacking the data of μ^+ or μ^-, the equation stated above still can be used.

Considering a pressure tunnel with radius a and shown figure 2, the inlet and outlet pressure are p_1、p_2. It is an unlimited long thick-wall cylinder is plane strain problem, that is, $\tau_{rz}=\tau_{\theta z}=0$，$\varepsilon_z=0$.Therefore, σ_r，σ_θ，σ_z are principal stress. Except of bulk stress, from equation (2-3), the following is easy to obtain:

$$\sigma_z=\mu^+(\sigma_r+\sigma_\theta)\quad(\sigma_z>0)\tag{4}$$

$$\sigma_z=\mu^-(\sigma_r+\sigma_\theta)\quad(\sigma_z<0)\tag{5}$$

Because σ_r is different with σ_θ ,that is, σ_r 0，σ_θ 0, if equation (4) or equation (5) is introduced into equation (3), constitutive laws of plane strains problem of infinite long thick wall cylinder can be written as:

$$\varepsilon_r=b_{11}\sigma_r+b_{12}\sigma_\theta\tag{6}$$

$$\varepsilon_\theta=b_{21}\sigma_r+b_{22}\sigma_\theta\tag{7}$$

Though geometry equation $\varepsilon_r = \frac{du}{dr}$, $\varepsilon_\theta = \frac{u}{r}$;

Table 2 . .the value, E^+ 、E^- 、μ^+ 、μ^- , of several rocks

Rock type	Usually compressive test			Splitting test			μ^+/μ^-	E^+/E^-
	Rya(Mpa)	$E^- \times 10^3$(Mpa)	μ^-	Rla(Mpa)	$E^+ \times 10^3$(Mpa)	μ^+		
Calcite-marble	69.7	44	0.19	4.5	59	0.18	0.947	1.341
Skarn	83.1	73	0.21	6.2	69	0.07	0.333	0.945
Lampritite	52.4	110	0.23	8.7	149	0.17	0.739	1.355
Skarn-marble	77.8	60	0.19	4.9	65	0.20	1.053	1.083
Granodiorite	157.3	76	0.25	11.7	62	0.16	0.64	0.816
Skarn-slate	47.8	72	0.23	11.7	44	0.20	0.870	0.611
Orebreceia	73.6	---	---	3.0	65	0.11	---	---
Eruptive breccia	63.1	72	0.20	7.0	55	0.28	1.40	0.764
Limestone	---	68	0.20	4.6	---	0.15	0.75	---
Sandy shale	---	30	0.21	---	---	---	---	---
Marble	---	18	0.14	---	20	0.11	0.786	1.111
Quartzite	---	51	0.05	---	19	0.16	3.20	0.373

equilibrium equation $\frac{d\sigma_r}{dr} + \frac{\sigma_r - \sigma_\theta}{r} = 0$;

comparability equations $r\frac{d\varepsilon_\theta}{dr} + \varepsilon_\theta - \varepsilon_r = 0$,

supposing stress function φ and ψ , joining

$\sigma_r = \frac{1}{r}\frac{d\varphi}{dr} = \frac{\psi}{r}$, $\sigma_\theta = \frac{d^2\varphi}{dr^2} = \frac{d\psi}{dr}$ into

equilibrium equation , finite solution equation can be written as:

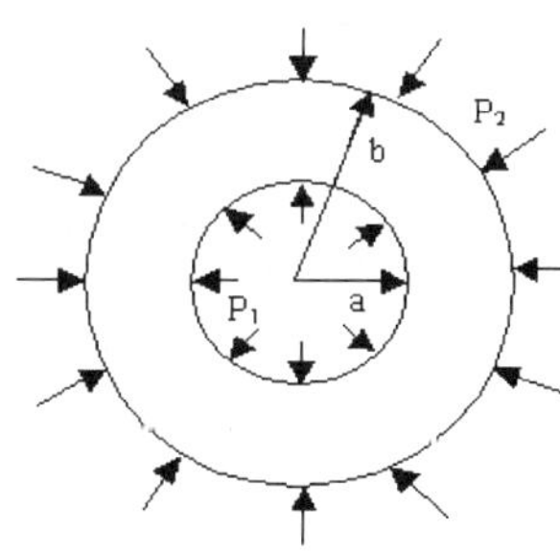

Fig. 2. Thick-wall cylinder with uniform inlet and outlet press P_1、P_2

$$r^2 \frac{d^2\psi}{dr^2} + r\frac{d\psi}{dr} - \xi^2\psi = 0 \tag{8}$$

where: $\xi = (b_{11} / b_{22})^{1/2} = \sqrt{(\alpha^2 - (\mu^+)^2)/(1-(\mu^+)^2)}$,

$\alpha = \left(E^+ / E^-\right)^{1/2}$ .Assuming $t = r/b$, from equation (8), we can get

$$\psi = Ct^{\xi} + Dt^{-\xi} \tag{9}$$

When $\xi = 1$, that is, $E^+ / E^- = 1$ and $\mu^+ = \mu^-$, equation (9) becomes $\psi = Ct + Dt^{-1}$ which is the classical elastic solutions.

When $\xi \neq 1$, assuming $\sigma_r\ 0$ and $\sigma_\theta\ 0$, then:

$$\sigma_r = Ct^{\xi-1} + Dt^{-(\xi+1)} \tag{10}$$

$$\sigma_\theta = C\xi t^{\xi-1} - D\xi t^{-(\xi+1)} \tag{11}$$

Base on boundary conditions $t = m = a/b$, $\sigma_r = -p_1$; $t = 1(r = b)$, $\sigma_r = 0$, we can yields C and D, finally:

$$\sigma_r = -p_1 \cdot \frac{m^{\xi+1}}{1-m^{2\xi}} \cdot \frac{1-t^{2\xi}}{t^{\xi+1}} \tag{12}$$

$$\sigma_\theta = p_1\xi \cdot \frac{m^{\xi+1}}{1-m^{2\xi}} \cdot \frac{1+t^{2\xi}}{t^{\xi+1}} \tag{13}$$

$$\sigma_z = \mu^+ \cdot \frac{p_1 m^{\xi+1}}{t^{\xi+1}(1-m^{2\xi})}\left[\xi(1+t^{2\xi}) - (1-t^{2\xi})\right] \tag{14}$$

When $\sigma_\theta\ 0$, $\sigma_r\ 0$, $\sigma_z > 0$, $m = 1/3$ and $\mu^+ = 1/6$, Table 3 list the value of σ_θ. From Table 3, $\alpha > 1$ has more influence than $\alpha < 1$ on the value of σ_θ.

Table 3. The affect of different α on σ_θ in pressure tunnel

α	0.50	0.64	0.81	1.00	1.44	1.69	2.00
σ_θ	1.082	1.130	1.187	1.000	1.388	1.464	1.555

Galerkin formulae $k = E[(1+\mu)a]^{-1}$, also named unit elastic resistance for pressure tunnel. If surrounding rock has variable elastic modulus, simply assuming $b \to \infty$, equation (12, 13) and equation (7) can been adopted. According to the relation of elasticity:

$$\varepsilon_\theta = u/r = \frac{\mu^+(1+\mu^+)}{E^+}\left[-\sigma_r + \frac{1-\mu^+}{\mu^+}\sigma_\theta\right] \tag{15}$$

equation (12) and equation (13) can deduce the following:

$$\frac{u}{r}=\frac{1+\mu^{+}}{E^{+}}\mu^{+}\frac{p_1 m^{\xi+1}}{(1-m^{2\xi})t^{(\xi+1)}}\cdot\left[(1-t^{2\xi})+\frac{1-\mu^{+}}{\mu^{+}}\xi(1+t^{2\xi})\right] .$$

When $r=a$ and $t=m=a/b\rightarrow 0$ (because of $b\rightarrow\infty$), the above equation can also been written: $\frac{u}{a}=\frac{1+\mu^{+}}{E^{+}}\mu^{+}p_1\left[1+\frac{1-\mu^{+}}{\mu^{+}}\cdot\xi\right]$ (16)

$$p_1=k_1\cdot u \quad (17)$$

Where: $k_1=\frac{E^{+}}{(1+\mu^{+})[\xi+\mu^{+}(1-\xi)]a}$ (18)

k_1 is the coefficient of unit elastic resistance when circle tunnel is in uniform inlet pressure. If $E^{+}>E^{-}$, then $\alpha>1$, $\xi>1$, $k_1<k$; If $E^{+}=E^{-}$, then $\alpha=1$, $\xi=1$, $k_1=k$; If $E^{+}<E^{-}$, then $\alpha<1$, $\xi<1$, $k_1>k$.

It is clear that the choosing of elastic resist coefficient must take the circumstances into consideration if rock has the variable elastic modulus. In order to reduce the cost of the testing tunnel, drilling hydraulic pressure sometimes substitute pressure cushion or hydraulic pressure test (such as the testing tunnel of Longtan hydropower station) in rock mass engineering. It is obviously significance for the design and calculation of press tunnel because the radius of bore is small, the solution is more exact and circularity is good.

4 Conclusions

(1) The constitutive relation of true rock, that is, the double fold line, is precise theoretically. When the denotation of principal stress changes, E^{+}、E^{-}、μ^{+}、μ^{-} are different and classical elastic mechanics solutions naturally is included.

(2) Adopting appropriate modulus radio of tension and compression ($\alpha=\sqrt{E^{+}/E^{-}}$), adjusting the denotation of σ_z (though loading inhibit stress) and sufficiently exerting the self-restraining capacity of surrounding rock, can achieve the most excellent design.

REFERENCES

1. СААмбарцумян Разнсмсдулъная Теосрия Упругостъ [M].Hayka, 1982:5-80
2. Baoshen Liu. The Introduce to Mine rock mechanics [M]. Changsha: Hunan science and technology publishing company, 1982：21—23
3. Zhiying Xu. Rock Mechanics[M]. Beijing: The public house of water conservancy, 1990：18—83
4. Ziying Cheng. Analytic Method in the Analysis of Surrounding Rock Mechanics [M]. Beijing: The public house of coal industry, 1994：291--300

GRADIENT-DEPENDENT CONSTITUTIVE FORMULATION FOR ELASTO-PLASTICITY COUPLED WITH ORTHOTROPIC DAMAGE

XINPU SHEN
Department of Architectural Engineering, Shenyang University of Technology, Shenyang, 110023, China; Email: xinpushen@yahoo.com

KHEMAIS SAANOUNI
University of Technology of Troyes – GSM/LASMIS, BP 2060, 10010 Troyes Cedex, France
URL: www.univ-troyes.fr e-mail: saanouni@univ-troyes.fr

Based on 'nonlocal energy dissipation', it is presented in this article the gradient dependent nonlocal constitutive model for elastoplasticity coupled with orthotropic damage for geo-materials and structures.

1. Introduction

For fracture/damage induced strain softening in geo-mechanical analysis, the loss of well-posedness of the boundary value problem results in the so-called mesh-dependency. Further, as the cracks being initiated in the material, the properties of the material become inhomogeneous and become the so-called complicate material, as that defined by Noll[1]. Noll[1] proposed that the mechanical properties of complicated materials are higher-order deformation gradient dependent.

This study aims to formulate a gradient-dependent nonlocal constitutive model. The formulation presented here is based on the local form of constitutive model presented by Shen, Mroz and Xu[2], and the nonlocal isotropic damage model presented by Shen and Saanouni[3].

2. Nonlocal Continuum Thermodynamics

The restriction of the formulation is made to infinitesimal deformation, which is classical for geo-mechanical analysis. The first principle of thermodynamics (energy balance) for the body occupying the domain Ω with boundary $\partial\Omega$ writes:

$$\frac{d}{dt}(E+K)=P+Q \tag{1}$$

and the kinetic energy principle writes:

$$\frac{dK}{dt}=P_{ext}+Q \tag{2}$$

where K and E are the internal and kinetic energy, respectively, of the body, whereas P and Q represent mechanical and thermal power supply.

On the other hand, the Second Law yields

$$\int_{\Omega}\left[\rho\dot{s}+\nabla\cdot\left(\frac{\mathbf{q}}{T}\right)-\frac{\xi}{T}\right]d\Omega\geq 0 \tag{3}$$

Eliminating ξ from and (3) yields the fundamental inequality:

$$\int_{\Omega}\left[\boldsymbol{\sigma}:\dot{\boldsymbol{\varepsilon}}+\rho T\dot{s}-\rho\dot{u}-\frac{\mathbf{q}}{T}\cdot\mathbf{g}\right]d\Omega\geq 0,\qquad \mathbf{g}=\nabla T \tag{4}$$

The following Helmholtz free energy is adopted

$$\Psi=u-Ts \tag{5}$$

The nonlocal form of *Clausius-Duhem* inequality is then obtained from Eq.(4) by using Helmholtz free energy in Eq. (5):

$$\int_{\Omega}\left[\boldsymbol{\sigma}:\dot{\boldsymbol{\varepsilon}}-\rho\dot{\Psi}-\rho s\dot{T}-\frac{\mathbf{q}}{T}\cdot\mathbf{g}\right]d\Omega\geq 0 \tag{6}$$

where u is the internal energy density (per unit mass), ε is the total strain tensor, ξ is the heat source and q is the heat flux, n is the outward normal at boundary $\partial\Omega$.

3. A gradient dependent nonlocal model for plasticity coupled with damage

It is assumed that the Helmholtz free energy function can be written in functional form as:

$$\Psi=\Psi\left(\boldsymbol{\varepsilon}^{e},r,\boldsymbol{\alpha},\mathbf{D},\nabla\mathbf{D}\right) \tag{7}$$

where $\boldsymbol{\varepsilon}^{e},r,\boldsymbol{\alpha},\mathbf{D},\nabla\mathbf{D}$ represent the elastic strain tensor, isotropic plastic hardening, kinetic plastic hardening tensor, orthotropic damage tensor, gradient of damage, respectively. Hereafter, the thermal influence is neglected for the purpose of brevity.

The total derivative of the Helmholtz free energy function can be written as

$$\dot{\Psi}=\frac{\partial\Psi}{\partial\boldsymbol{\varepsilon}^{e}}:\dot{\boldsymbol{\varepsilon}}^{e}+\frac{\partial\Psi}{\partial r}\dot{r}+\frac{\partial\Psi}{\partial\boldsymbol{\alpha}}:\dot{\boldsymbol{\alpha}}+\frac{\partial\Psi}{\partial\mathbf{D}}:\dot{\mathbf{D}}+\dot{\Psi}^{(\nabla\mathbf{D})},\ \ \dot{\Psi}^{(\nabla\mathbf{D})}=\frac{\partial\Psi}{\partial\left(\nabla_{i}D_{jk}\right)}\nabla_{i}\dot{D}_{jk} \tag{8}$$

Substituting Eq. (8) into (6), we get:

$$\int_{\Omega}\left[\left(\boldsymbol{\sigma}-\rho\frac{\partial\Psi}{\partial\boldsymbol{\varepsilon}^{e}}\right)\dot{\boldsymbol{\varepsilon}}^{e}+\boldsymbol{\sigma}:\dot{\boldsymbol{\varepsilon}}^{p}-\rho\frac{\partial\Psi}{\partial r}\dot{r}-\rho\frac{\partial\Psi}{\partial\boldsymbol{\alpha}}:\dot{\boldsymbol{\alpha}}-\rho\frac{\partial\Psi}{\partial\mathbf{D}}:\dot{\mathbf{D}}-\rho\dot{\Psi}^{(\nabla\mathbf{D})}\right]d\Omega\geq 0 \tag{9}$$

With the inequality (9), it is obtained the following state relations:

$$\boldsymbol{\sigma}=\rho\frac{\partial\Psi}{\partial\boldsymbol{\varepsilon}^{e}},\ R=\rho\frac{\partial\Psi}{\partial r},\ \mathbf{X}=\rho\frac{\partial\Psi}{\partial\boldsymbol{\alpha}},\ \mathbf{Y}=-\rho\frac{\partial\Psi}{\partial\mathbf{D}},\ \mathbf{Y}^{(\nabla\mathbf{D})}=\rho\frac{\partial\Psi}{\partial\nabla\mathbf{D}} \tag{10}$$

Substituting relations in Eq. (10) into (9) yields:

$$\int_{\Omega}\left[\boldsymbol{\sigma}:\dot{\boldsymbol{\varepsilon}}^{p}-R\dot{r}-\mathbf{X}:\dot{\boldsymbol{\alpha}}+\left(\mathbf{Y}+\nabla\cdot\mathbf{Y}^{(\nabla\mathbf{D})}\right)\dot{\mathbf{D}}\right]d\Omega-\oint_{\partial\Omega}\dot{D}\mathbf{n}\cdot\mathbf{Y}dS\geq 0 \tag{11}$$

The Eq. (11) is nothing other than the global power dissipated in the overall domain Ω by the various dissipative phenomena. The internal power dissipation in the damage phenomena indicated by D^{damage} can be expressed as:

$$\mathsf{D}^{\text{damage}} = \int_{\Omega} \left(\mathbf{Y} + \nabla \cdot \mathbf{Y}^{(\nabla\mathbf{D})}\right)\dot{\mathbf{D}} d\Omega - \oint_{\partial\Omega} \dot{\mathbf{D}} : \left[\mathbf{n} \cdot \mathbf{Y}^{(\nabla\mathbf{D})}\right] dS \geq 0 \tag{12}$$

From Eq. (11), it is obtained the nonlocal damage energy dissipation rate $\mathbf{Y}^{(\text{nonlocal})}$ and corresponding boundary condition as

$$\mathbf{Y}^{nonlocal} = \mathbf{Y} + \nabla \cdot \mathbf{Y}^{(\nabla\mathbf{D})} \quad in\ \Omega; \qquad \dot{\mathbf{D}} : \left[\mathbf{n} \cdot \mathbf{Y}^{(\nabla\mathbf{D})}\right] = 0, \quad on\ \partial\Omega \tag{13}$$

4. Damage Model and Nonlocal Damage Strain Energy Release Rate

The definition of the second order damage tensor is given as follows:

$$\mathbf{D} = \sum_{i=1}^{3} \omega_i \mathbf{n}_i \otimes \mathbf{n}_i \tag{14}$$

where $\mathbf{n}_i$ is the principal direction and ω_i the principal damage value of the second order damage tensor. The damaged elasticity tensor $\mathbf{E}$ can be expressed in terms of undamaged elasticity tensor $\mathbf{E}^0$ and the so-called active damage tensor $\widetilde{\mathbf{D}}$ as:

$$\mathbf{E} = (\mathbf{I} - \widetilde{\mathbf{D}}) \cdot \mathbf{E}^0 \cdot (\mathbf{I} - \widetilde{\mathbf{D}}) \tag{15}$$

The local damage conjugate force can then expressed as:

$$\mathbf{Y} = -\rho \frac{\partial \Psi}{\partial \mathbf{D}} = \left[\mathbf{E}^0 \cdot (\mathbf{I} - \widetilde{\mathbf{D}})\right] : \left(\boldsymbol{\varepsilon} - \boldsymbol{\varepsilon}^P\right) \tag{16}$$

Assuming the specific form of the free energy function Ψ for gradient dependent damage coupled with hardening plasticity for metal forming as

$$\Psi = \Psi^{local}(\mathbf{D}, \boldsymbol{\varepsilon}) + \Psi^{grad}(\nabla\mathbf{D}) \tag{17}$$

with specialization of the gradient form in a way similar to the one adopted by Liebe, Steinmann and Benallal[4] as

$$\rho\Psi^{grad}(\nabla\mathbf{D}) = \frac{1}{2}(\nabla\mathbf{D}) \cdot \mathbf{E}^D \cdot (\nabla\mathbf{D}), \ \mathbf{E}^D = c\mathbf{I} \tag{18}$$

where parameter c is used to make the term of the right-hand-side to be of unit of energy density, and is regarded as internal-length-related intrinsic parameter; $\mathbf{I}$ is the second order unit tensor. Finally we have:

$$\mathbf{Y}^{nonlocal} = \mathbf{Y} + \nabla \cdot \mathbf{Y}^{(\nabla\mathbf{D})} = \left[\mathbf{E}^0 \cdot (\mathbf{I} - \widetilde{\mathbf{D}})\right] : \left(\boldsymbol{\varepsilon} - \boldsymbol{\varepsilon}^P\right) + \nabla \cdot (c\nabla\mathbf{D}) \tag{19}$$

5. Coupling of Damage and Plasticity

The concept of effective stress is not adopted here. The way of coupling between orthotropic damage and plasticity as that described by Shen, Mroz and Xu[2] is adopted. The *homogenized Mohr- Coulomb criterion* is adopted as the plastic loading condition:

$$F^p = \tau_n - \left(c^* + \mu^* \sigma_n\right) \leq 0 \tag{20}$$

with

$$\mu^* = (1 - D_n^{ad})\left[\mu^0(1 - D_n^{ad} - D_n^{id}) + \mu^f D_n^{id}\right] \tag{21}$$

$$c^* = c^0(1 - D_n^{ad} - D_n^{id}) + c^f D_n^{id} \tag{22}$$

where $D_n^{ad} = \widetilde{D}_{ij} n_i n_j$, $D_n^{id} = (D_{ij} - \widetilde{D}_{ij}) n_i n_j$ are the components of active damage tensor and inactive damage tensor in the direction normal to the sliding surface respectively, μ^0, c^0 are internal frictional coefficient and cohesion in shear direction for intact material, and μ^f, c^f are values corresponding to final frictional sliding state.

The damage potential function used here can be expressed as:

$$F^D = \sqrt{\mathbf{Y}^{(nonlocal)} : \mathbf{P}^+ : \mathbf{Y}^{(nonlocal)}} - \phi_0 \leq 0 \tag{23}$$

where ϕ_0 is the threshold value for F^D, at which damage begin to propagate, P^+ is the 4th-order projection tensor.

6. Ending Remarks

Nonlocal aspects of orthotropic damage coupled with plasticity have been formulated in this paper. The evolution laws of internal variables will be discussed in another article.

References

1. W. Noll, *Arch Rational Mech. Anal.*, 27, 1-32 (1967).
2. X.P. Shen, Z. Mroz, B.Y. Xu, *Appl. Math. Mech., English edition*, 22(9): 1028-1034 (2001).
3. X.P. Shen, K. Saanouni, In: *Proc. ESAFORM'2002*, Pietrzyk M., Mitura Z., Kaczmar, (Eds), Publishing House Akapit, Krakow, Poland, 143-146 (2002).
4. T. Liebe, P. Steinmann, A. Benallal, *Comput. Methods Appl. Mech. Engng.*, 190, 6555-6576 (2001).

A FRACTURE-ENERGY-BASED ELASTO-SOFTENING-PLASTIC CONSTITUTIVE MODEL FOR JOINTS OF GEOMATERIALS

XINPU SHEN WENBO BAO YUXIU XU ZHIQIANG HUANG
Department of Architectural Engineering, Shenyang University of Technology, Shenyang, 110023, China

On the basis of plasticity and fracture mechanics for quasi-brittle materials, this article presented a constitutive model for gradual softening behaviour of joints of geomaterials. Corresponding numerical tests are carried out at the local level.

1 Introduction

For many years joint element has got wide applications in geo- engineering[1]. In recent years, as an extension of cohesive crack model[2], numerous researchers had proposed their fracture-mechanics-based new joint element constitutive models[3,4]. The principal novelties of these new joint elements are that interface behaviour was treated with advantage as a pseudo-material problem. The problem was then solved in the framework of conventional plasticity.

This study is aimed at the formulation of a fracture-energy-based piecewise linear elasto-plastic constitutive model for joints of geomaterials. Corresponding numerical tests are carried out.

2 Mathematical Formulation

Definitions of interface variables at a joint surface are given as follows: **p** represents the traction vector on the joint surface, $\mathbf{u}^+$ and $\mathbf{u}^-$ the displacement vector on two sides of the joint respectively, and **w** represents vector of joint opening, i.e., displacement discontinuity across the joint

$$\mathbf{p} = \{p_n \ \ p_t\}^T, \qquad \mathbf{w} = \mathbf{u}^+ - \mathbf{u}^- = \{w_n \ \ w_t\} \tag{1}$$

where p_n is normal traction component, p_t is tangential traction vector, w_n is normal component of joint opening, w_t is tangential vector of joint opening. In problems considered here, infinitesimal deformation is assumed, and it also is assumed the mechanical process is an isothermal process.

Two kinds of joint opening mechanism are assumed to be consisted in the failure process: mode I crack opening and mode II crack. The linear softening law is adopted in description of cohesive behaviour for both tensional and shear cracks. As shown in Figure1, the softening of strength parameters is defined on the space of displacement discontinuity across the joint surface and the traction, i.e., the spaces of $w_n - p_n$. Here it is defined that H_n denotes displacement softening modulus in direction normal to the joint surface; H_t denotes displacement softening modulus in tangential direction to the joint. H_n and H_t are related to fracture energy of mode I and mode II fractures respectively. The following relations can be obtained:

$$H_n = \begin{cases} -\dfrac{\chi_0^{\ 2}}{2G_f^I}, & if \quad w_n < w_{nc} \\ 0, \ if & w_n \geq w_{nc} \end{cases}, H_t = \begin{cases} -\dfrac{c_0^{\ 2}}{2G_f^{II}}, & if \quad w_t < w_{tc} \\ 0, \ if & w_t \geq w_{tc} \end{cases} \tag{2}$$

where w_{nc} is the Critical Opening Displacement value of mode I crack. w_{tc} is the Critical Opening Displacement value of mode II crack. The so-called *fracture-mechanics-based softening plasticity* means that, the cohesive law of the joint variables determines the softening properties of plasticity. For the case of *piecewise linear plasticity* adopted here, the softening behaviour related to strength vector **Y** is dependent on the internal variables $\lambda_i, i = 1{,}3$ at the joint surfaces. The model of piecewise linear plasticity is defined as Mohr-Coulomb type and two piecewise linear yielding surfaces are adopted, as shown in Figure 2. Then strength vector **Y** has 3 components: Y_1, Y_2 and Y_3. It can be proved that:

$$Y_1(\boldsymbol{\lambda}) = \chi_0 + H_n\lambda_1,\ Y_2(\boldsymbol{\lambda}) = Y_3(\boldsymbol{\lambda}) = c_0 + H_t(\lambda_2 + \lambda_3) + \mu H_n\lambda_1 \tag{3}$$

where $\mu = \tan\phi$ is the internal frictional coefficient.

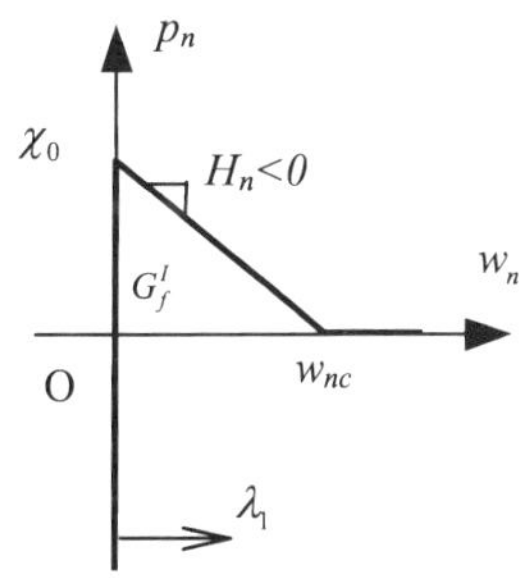

Figure 1. Cohesive law.

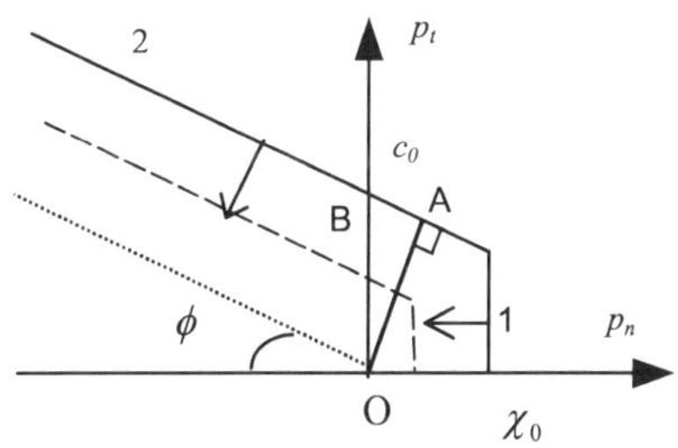

Figure 2. Softening linear yielding surface.

Based on the Eqs. (1) – (3), it is obtained the matrix form piece-wise linear Mohr-Coulomb-type plastic yielding functions as

$$\varphi = \mathbf{N}^T\mathbf{p} - \mathbf{H}\boldsymbol{\lambda} + \mathbf{Y} \leq \mathbf{0} \tag{4}$$

with the following expressions for vector of yielding surfaces φ, matrix of collection of normal directions of each yielding surface **N**, vector of plastic multipliers $\boldsymbol{\lambda}$, matrix of displacement-softening-modulus **H**, vector of initial strength parameters of both tension and shear behaviour **Y**, as:

$$\varphi^T = \{\varphi_1 \quad \varphi_2 \quad \varphi_3\}$$

$$\mathbf{N} = \begin{bmatrix} 1 & \mu & \mu \\ 0 & 1 & -1 \end{bmatrix}$$

$$\mathbf{H} = \begin{bmatrix} H_n & \frac{\chi_0}{c_0} H_t & \frac{\chi_0}{c_0} H_t \\ \mu H_t & H_t & H_t \\ \mu H_t & H_t & H_t \end{bmatrix} \tag{5}$$

$$\boldsymbol{\lambda}^T = \{\lambda_1 \quad \lambda_2 \quad \lambda_3\}$$

$$\mathbf{Y}^T = \{\chi_0 \quad c_0 \quad c_0\}$$

For the case of complicated external loading conditions the following incremental relationships have to be adopted,

$$\varphi = \mathbf{N}^T\left(\mathbf{p}^0 + \Delta\mathbf{p}\right) - \mathbf{H}\left(\boldsymbol{\lambda}^0 + \Delta\boldsymbol{\lambda}\right) - \mathbf{Y}^0 \leq \mathbf{0} \tag{6}$$

with

$$\Delta\mathbf{w} = \Delta\mathbf{w}^e + \Delta\mathbf{w}^p, \quad \Delta\mathbf{w}^e = \mathbf{K}^{-1}\Delta\mathbf{p}, \quad \Delta\mathbf{w}^p = \mathbf{V}\Delta\boldsymbol{\lambda} \tag{7}$$

The direction vector of the plastic flow related to the shear displacement is

$$\mathbf{V}_2 = \begin{cases} (\mu \quad 1)^T, \textit{if } Y_2(\boldsymbol{\lambda}) > 0 \\ (\mu \quad 0)^T, \textit{if } Y_2(\boldsymbol{\lambda}) = 0 \end{cases} \tag{8}$$

and that for $\mathbf{V}_3$ can be obtained by analogy to Eq. (8), see Figure .

As incremental displacement discontinuity vector $\Delta\mathbf{w}$ is given as input data, there is

$$\Delta\mathbf{p} = \mathbf{K}\left(\Delta\mathbf{w} - \Delta\mathbf{w}^p\right) = \mathbf{K}\left(\Delta\mathbf{w} - \mathbf{V}\Delta\boldsymbol{\lambda}\right) \tag{9}$$

The displacement-form of Eq. (6) is

$$\varphi = \mathbf{N}^T\left(\mathbf{w}^0 + \Delta\mathbf{w}\right) - \left(\mathbf{H} + \mathbf{N}^T\mathbf{K}\mathbf{V}\right)\left(\boldsymbol{\lambda}^0 + \Delta\boldsymbol{\lambda}\right) - \mathbf{Y}^0 \leq \mathbf{0} \tag{10}$$

Therefore the vector φ and $\Delta\boldsymbol{\lambda}$in (10) has the following complementary relationship:

$$\varphi \leq \mathbf{0}, \quad \Delta\boldsymbol{\lambda} \geq \mathbf{0}, \quad \varphi^T \cdot \Delta\boldsymbol{\lambda} = 0 \tag{11}$$

Solving above nonlinear complementary problems can then find the traction response for a given incremental displacement loading.

3 Numerical Tests

The models proposed here have been implemented in a set of subroutines for the purpose of constitutive verification. These problems are solved by using a mathematical programming solver. The calculation is carried out for mode II displacement loading, i.e., $w_n=0$, $w_t=w_t^0C(t)$, where C(t) is a loading factor. It is shown in Figure 3 that after the elastic part, dilatancy caused lateral compression continuous increase until the finish of the softening in shear strength. And after that point, only frictional sliding occurs on this interfacial point.Results are shown in Figure 3. The parameter values used in these calculations are: $K_n = K_t = 200MPa/mm$; $\mu = 0.9$; $\chi_0 = 2.8MPa$; $c_0 = 7.0MPa$; $G_f^I = 0.1N/mm$; $G_f^{II} = 1.0N/mm$

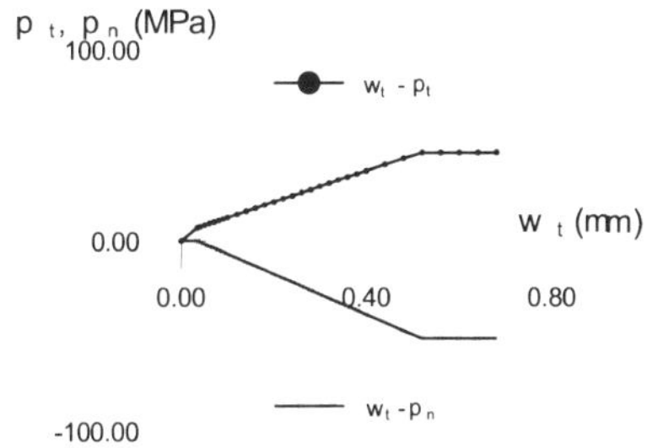

Figure 3. Mode II displacement - traction response.

4 Conclusions

This paper presents a piecewise linear elastoplastic constitutive model for the coupled failure process of decohesion and frictional sliding. The coupling between the cohesive fracture and corresponding softening of the macroscopic plasticity is modelled in a natural way by which no more material constant is required except those already adopted in geo-engineering. Numerical results indicate that this model can catch major phenomena existing in the failure process at the joint surface.

References

1. Goodman R. E., Taylor R. L., Brekke T. L., *Proc. ASCE*, 1968, **94**, SM3, 637-659.
2. Carol I., Prat P.C. and Lopez C.M., *J. Engng. Mech.*, ASCE, 1997, **123**(1), 1-9.
3. Hohnborg J. M., 1992, Birkhaeuser Publishers Basel, Berlin.
4. Fakharian K., Evgin E., *Int. J. Numer. Analyt. Meth. Geomech.*, 2000, **24**(1), 183-199.
5. Lofti H., Shing P., *J. Struc. Engng.*, ASCE, 1994, **120**(1), 63-80.

HAULING FORCES APPLIED ON FISSURE WALLS IN FRACTURED ROCK MASS BY FLUID FLOW

JUNRUI CHAI, HAOWU LIU
College of Hydroelectric Engineering, Sichuan University, Chengdu 610065, China
E-mail: jrchai@pridns.scu.edu.cn

YANQING WU
College of Hydroelectric Engineering, Xi'an University of Technology, Xi'an 710048, China

It is very important to investigate the forces applied on the rock mass by the fluid flow through fractured rock mass, which is one of the issues for analyzing the stability of rock mass. The forces applied on the fissure walls by the fluid flow include the hydrostatic seepage pressure and the dynamic seepage pressure, i.e., the tangent hauling force. Based on the cubic law of the single fissure flow, the equations of the hauling forces applied on the fissure walls by the single fissure flow are deduced by means of the momentum law of fluid mechanics. The equations are designated for the cases of no-filled fissure, filled fissure and the combined flow of fluid and the fillings, respectively. The equations have important values for analyzing the effect of fluid flow on the behavior of deformation and stability of fractured rock mass, and are the foundation of the interaction mechanism between stress and fluid flow.

1 Introduction

Couple analysis of seepage and stress fields in fractured rock mass is extensively concerned with by the researchers and engineers in the field of geotechnical engineering in the recent years[1, 2]. The key for couple analysis is the interaction mechanism between stress and the fluid flow. The principle of effective stress is extensively applied in the recent research for couple analysis, in which only the hydrostatic seepage pressure is considered and the effect of hydrodynamic seepage pressure on the deformation of fractures is neglected. The fluid flow through fractures exerts the hydrostatic seepage pressure and the hydrodynamic seepage pressure on the fissure walls. The hydrostatic seepage pressure makes the fracture aperture larger, and the hydrodynamic seepage pressure (i.e. the hauling force) exerts the tangent force on the fissure walls. The purpose of this paper is to deduce the equations for determining the hauling forces by means of the momentum law of fluid mechanics and the cubic law of the single fissure flow.

2 Basic Assumptions and the Cubic Law of Single Fissure Flow

The basic assumptions applied in this paper are the same as those for deducing the cubic law of the single fissure flow. Thus the fissure flow can be regarded as the narrow fissure flow between parallel plates (Figure 1).

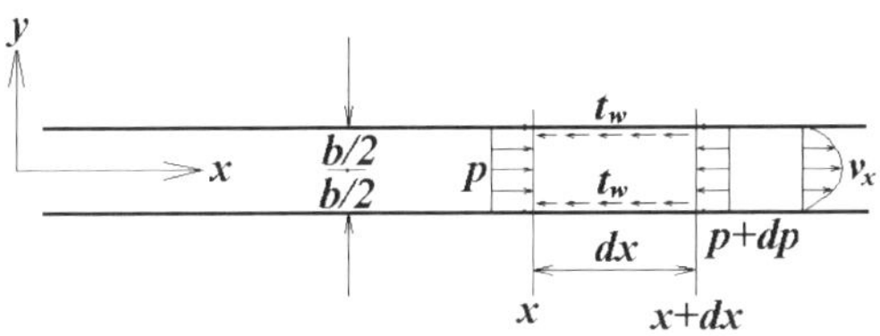

Figure 1. Flow in Single Fissure

Based on the above assumptions and the basic equations of fluid mechanics, the following cubic law of single fissure flow (Figure 1) can be obtained[3].

$$q = \frac{b^3\gamma}{12\mu} J_f ; \; v_x = \frac{\gamma}{8\mu}(b^2 - 4y^2) J_f ; \; \frac{\partial p}{\partial y} = 0 \tag{1}$$

where b is the aperture of the single fissure; q is the flow rate of the single fissure with the width of 1m; γ and μ are the specific weight and the coefficient of dynamic viscosity of the fluid, respectively; $J_f = -\frac{\partial H}{\partial x}$ is the hydraulic slope of fluid flow along the direction of x coordinate; $\frac{\partial H}{\partial x}$ is the hydraulic gradient of fluid flow along the direction of x coordinate; H is the hydraulic head of the fluid flow; v_x is the flow velocity of fluid; p is the fluid pressure; x 、 y are the coordinates, as shown in Figure 1.

3 Analysis of Hauling Forces Applied on Single Fissure Walls by Fluid Flow

It is noticed that if the fracture is not filled, the hauling force applied on the fissure wall by fluid flow is a kind of surface force, which is the reaction of the resistance exerted on the fluid flow by the fissure wall. The hauling force and the resistance must be equal in magnitude and opposite in direction.

It can be shown from equation (1) that the velocity of the single fissure flow

v_x varies in the parabolic form along y direction, and is independent on x coordination, i.e., the distribution of velocity v_x is the same within any x section (Figure 1).

It can be also shown from equation (1) that the fluid pressure p is independent on y coordination, it is the function of only x coordination, i.e., the pressure p is evenly distributed within any x section (see Fig. 1). For the fissure walls, the pressure p applies on the walls in the direction of the normal of the walls, which makes the fissure aperture larger.

Applying the momentum law for total flow of fluid mechanics along x direction to the fluid between section x and section $x+dx$ shown in Fig. 1, we can obtain[4]

$$\rho q(\beta_2 \overline{v_{x2}} - \beta_1 \overline{v_{x1}}) = \Sigma F_x \tag{2}$$

where ρ is the density of the fluid; q is the flow rate of single fissure with the width of 1m; $\overline{v_{x2}}$ and $\overline{v_{x1}}$ are the average velocity of fluid flow within section $x+dx$ and section x, respectively; β_2 and β_1 are the modification coefficient for momentum within section $x+dx$ and section x, respectively; $\sum F_x$ is the resultant force along x direction loaded on fluid of dx.

Because the distribution of velocity v_x is the same within any x section, the left term in equation (2) is equal to zero, i.e., $\rho q(\beta_2 \overline{v_{x2}} - \beta_1 \overline{v_{x1}}) = 0$. So we have

$$\sum F_x = 0 \tag{3}$$

According to the forces along x direction applied on the fluid between section x and section $x+dx$ shown in Fig. 1, we obtain

$$\sum F_x = pb - 2t_w dx - (p+dp)b \tag{4}$$

in which p and $p+dp$ are the fluid hydrostatic pressure on section x and section $x+dx$, respectively; t_w is the resistance exerted on the fluid flow by the fissure wall.

Substituting equation(4)into equation(3) we obtain the following equation.

$$t_w = -\frac{b}{2}\frac{dp}{dx} \tag{5}$$

4 Analysis of Hauling Forces Applied on Single Filled Fissure Walls

If the above single fissure is filled with some fillings, the fissure seepage exerts the dynamic seepage pressure on the fillings. Suppose the hydraulic slope of fissure seepage along x direction is J, the resultant force F_s of seepage body force applied on the fillings between section x and section $x+dx$ shown in Figure 1 can be determined as follows[5].

$$F_s = \int_{-\frac{b}{2}}^{\frac{b}{2}} \int_{x}^{x+dx} n\gamma J dx dy = nb\gamma J dx \tag{6}$$

where n is the porosity of the fillings.

So the resultant force of hauling force (the reaction of the resistance) is also F_s, and the pressure of hauling force t_w is determined as follows.

$$t_w = \frac{nb\gamma J dx}{2dx} = \frac{nb\gamma}{2} J \tag{7}$$

When piping occurs, both the fluid flow and the fillings move as a unit. The hauling force t_w applied on the fissure walls by the high-density fluid flow can be obtained from the following equation.

$$t_f = \frac{b}{2} \gamma_f J_f \tag{8}$$

in which γ_f is the specific weight of the high-density fluid.

References

1. M. Bai, Numerical modeling of coupled flow and deformation in fractured rock specimens, *Int. J. Num. Anal. Meth. Geomech.* **23,** 141-160 (1999).
2. H.R. Ghafour and R.W. Lewis, A finite element double porosity model for heterogeneous deformable porous media, *Int. J. Num. Anal. Meth. Geomech.* **20,** 831-844 (1996).
3. Wu Yanqing and Zhang Zhuoyuan, *Introduction to rock mass hydraulics*, (Southwest Jiaotong University Press, Chengdu, 1995).
4. Mao Chunpu, *Fluid mechanics* (Shanghai Jiaotong University Press, Shanghai, 1995).
5. J. Bear, *Dynamics of Fluids in Porous Media* (Elsevier, New York, 1979).

A NEW MODEL OF SEDIMENTARY ROCKS

TIANTANG YU
College of Civil Engineering, Hohai University, Nanjing , 210098,P.R.China
E-mail: Heaven_yu2000@21cn.com

ZHENBO WANG
College of Civil Engineering, Nanjing Technology University, Nanjing , 210009,P.R.China

J.F. SHAO
Laboratoire de Mechanique de Lille,Villeneuve d'Ascq Cedex, France

A new model of sedimentary rocks is presented. The inherent anisotropy of this kind of materials is described with a scalar anisotropy parameter. The time-dependent effect is considered with a damage factor. The uniaxial compressive strength is the function of time, the microstructure tensors and stresses. A complete plasticity model describing the deformation process in anisotropic sedimentary rocks is given. Several triaxial tests are simulated with the model presented in the paper, and the results show the inherent anisotropy and time-dependent behavior in sedimentary rocks can be efficiently described with the model.

1 Introduction

Sedimentary rocks exhibit strong inherent anisotropy. The anisotropy is strongly related to the microstructure, in particular the existence of bedding planes. The strength of anisotropy rocks is related to the direction of microstructure and the time.

An approach that retains the mathematical rigor and pragmatic is proposed. An anisotropy parameter expressed in terms of the stresses and structure-orientation tensors is introduced in the failure criterion. Assume that the uniaxial compressive strength is the function of time, the microstructure tensors and stresses. A complete plasticity model describing the deformation process in anisotropy sedimentary rocks is given.

2 A Constitutive Model For Anisotropic Rocks

2.1 *Anisotropy parameter* η

Assume that the failure criterion can be expressed

$$F = F(\sigma_{ij}, \alpha_{ij}) = F(I_1, I_2, I_3, \eta) \tag{1}$$

In the above equation, $\boldsymbol{\alpha}_{ij}$ is a microstructure tensor [1] , η is a scalar anisotropy parameter that represents the projection of the tensor on a suitably defined loading direction l, i.e. $\eta=\boldsymbol{\alpha}_{ij}\, l_i l_j$.

The generalized loading vector l is defined as

$$l_i = L_i (L_k L_k)^{-0.5} \quad , \quad L_i = L_1 e^{(1)} + L_2 e^{(2)} + L_3 e^{(3)} \tag{2}$$

where $e^{(i)}$ (i=1,2,3) is the principal triad of the microstructure tensor $\boldsymbol{\alpha}_{ij}$, $L_1=(\sigma_{11}^2+\sigma_{12}^2+\sigma_{13}^2)^{0.5}$, $L_2=(\sigma_{12}^2+\sigma_{22}^2+\sigma_{23}^2)^{0.5}$, $L_3=(\sigma_{13}^2+\sigma_{23}^2+\sigma_{33}^2)^{0.5}$.

η can be also expressed with the deviatoric part $\boldsymbol{A}$ of $\boldsymbol{\alpha}_{ij}$ [1], i.e.

$$\eta=\hat{\eta}\left(1+A_{ij}l_il_j\right) \quad , \quad A_{ij}=(\alpha_{ij}-\hat{\eta}\delta_{ij})\hat{\eta}^{-1} \quad , \quad \hat{\eta}=3^{-1}\alpha_{kk} \tag{3}$$

A more general expression for η is

$$\eta=\hat{\eta}\left(1+A_{ij}l_il_j+b\left(A_{ij}l_il_j\right)^2+c\left(A_{ij}l_il_j\right)^3+\ldots\right). \tag{4}$$

2.2 *Yield function*

The condition at failure for brittle-plastic materials can be described with a quadratic form [2]

$$F=a_1\bar{\sigma}(g(\theta)f_c)^{-1}+a_2\bar{\sigma}^2(g(\theta)f_c)^{-2}-(a_3+If_c^{-1})=0 \tag{5}$$

In the above equation, $\bar{\sigma}=J_2^{1/2}$, $I=-I_1$, $\theta=1/3\sin^{-1}(1.5\sqrt{3}J_3/\bar{\sigma}^3)$ (-30°≤θ≤30°). The function $g(\theta)$ defines the shape of the yield surface on the π plane, and it satisfies $g(30°)=1$, $g(-30°)=K$, where K is a constant. f_c is the uniaxial compressive strength. For an isotropic material f_c is a constant; for an anisotropic material, f_c varies with the orientation of the microstructure and time. The variation can be expressed by a distribution function similar to eq .(4)

$$f_c=\hat{f}(t)\left[1+A_{ij}l_il_j+b\left(A_{ij}l_il_j\right)^2+c\left(A_{ij}l_il_j\right)^3+\ldots\right] \tag{6}$$

In the above equation, $\hat{f}(t)$ is the average uniaxial compressive strength.

$$t=0,\quad \hat{f}(t)=\hat{f}_0,\ t=\infty,\quad \hat{f}(t)=\hat{f}_\infty \tag{7}$$

Define the damage factor $\xi(t)$ as

$$\xi(t)=(\hat{f}(t)-\hat{f}_0)(\hat{f}_\infty-\hat{f}_0)^{-1}\quad (t-0,\quad \xi(t)=0,\ t-\infty,\quad \xi(t)=1) \tag{8}$$

$\xi(t)$ is the function about time, and it represents the damage condition of the average uniaxial compressive strength. A large number of experimental data show that $\xi(t)$ can be expressed by a hyperbolic form..

$$\xi(t)=1-\left(1+\omega t\right)^{-\alpha} \tag{9}$$

where ω and α are constant that is related with the characteristics of the material.

The following expression can be obtained with eq.(8) and (9).

$$d\hat{f}(t) = \left(\hat{f}_\infty - \hat{f}_0\right)\left(1+\omega t\right)^{-\alpha-1}\alpha\omega dt \tag{10}$$

The condition at failure may be expressed in the form

$$F = \bar{\sigma} - g(\theta)\bar{\sigma}_c = 0 \tag{11}$$

where $\bar{\sigma}_c = (-a_1 + (a_1^2 + 4a_2(a_3 + If_c^{-1}))^{0.5})f_c(2a_2)^{-1}$

The yield function may be assumed

$$f = \bar{\sigma} - \beta(\varsigma)g(\theta)\bar{\sigma}_c = 0 \tag{12}$$

$\beta(\varsigma)$ is a hardening function, and it can be expressed with a hyperbolic form, i.e.

$$\beta(\varsigma) = \varsigma(A+\varsigma)^{-1}, \quad \varsigma = \int d\varsigma, \quad d\varsigma = \left(de_{ij}^p de_{ij}^p\right)^{0.5} \tag{13}$$

where e is the strain deviator.

The non-association flow rule may be written as

$$d\varepsilon_{ij}^p = d\lambda(\partial\varphi / \partial\sigma_{ij}) \quad \varphi = \varphi(I_1, I_2, I_3, \eta) \tag{14}$$

where φ is the plastic potential.

The plastic potential function for brittle-plastic materials can be written as [2]

$$\varphi = \bar{\sigma} + \eta_c g(\theta)\bar{I}\ln(\bar{I}\bar{I}_0^{-1}) = 0 \tag{15}$$

where $\bar{I} = a_3 f_c + I$ and η_c represents the value of $\bar{\sigma}(g(\theta)\bar{I})^{-1}$ at which the transition from compaction to dilatancy occurs. η_c can be defined as $\eta_c = \upsilon e f_c (d - a_1)(2a_2\bar{I}_0)^{-1}$, $d = (a_1^2 + 4a_2\bar{I}_0(ef_c)^{-1})^{0.5}$, and υ is a material constant.

Thus, the direction of plastic flow is defined by

$$\frac{\partial\varphi}{\partial\sigma_{ij}} = \frac{\partial\varphi}{\partial I}\frac{\partial I}{\partial\sigma_{ij}} + \frac{\partial\varphi}{\partial\bar{\sigma}}\frac{\partial\bar{\sigma}}{\partial\sigma_{ij}} + \frac{\partial\varphi}{\partial\theta}\frac{\partial\theta}{\partial\sigma_{ij}} + \frac{\partial\varphi}{\partial f_c}\frac{\partial f_c}{\partial\sigma_{ij}} \tag{16}$$

From the above equation, it is known that the effect of anisotropy is embedded in the last term.

2.3 *Constitutive model*

The consistency condition is

$$df = \frac{\partial f}{\partial \sigma_{ij}} d\sigma_{ij} + \frac{\partial f}{\partial \beta}\frac{d\beta}{d\varsigma} d\varsigma + \frac{\partial f}{\partial f_c}\frac{\partial f_c}{\partial \hat{f}}\frac{d\hat{f}}{dt} dt = 0 \tag{17}$$

where

$$d\sigma_{ij} = D^e_{ijkl}\left(d\varepsilon_{kl} - d\varepsilon^p_{kl}\right) , \quad d\varsigma = d\lambda\sqrt{dev\frac{\partial \varphi}{\partial \sigma_{ij}} dev\frac{\partial \varphi}{\partial \sigma_{ij}}} \tag{18}$$

Substituting eq.(18) into eq.(17), the plastic multiplier $d\lambda$ is obtained.

$$d\lambda = H^{-1}\left(\frac{\partial f}{\partial \sigma_{ij}} D^e_{ijkl} d\varepsilon_{kl} + \frac{\partial f}{\partial f_c}\frac{\partial f_c}{\partial \hat{f}}\frac{d\hat{f}}{dt} dt\right) \tag{19}$$

$$H = \frac{\partial f}{\partial \sigma_{ij}} D^e_{ijkl} \frac{\partial \varphi}{\partial \sigma_{kl}} - \frac{\partial f}{\partial \beta}\frac{d\beta}{d\varsigma}\sqrt{dev\frac{\partial \varphi}{\partial \sigma_{ij}} dev\frac{\partial \varphi}{\partial \sigma_{ij}}} \tag{20}$$

Employing eq.(16)、eq.(18) and eq.(19), the constitutive model can be obtained.

$$d\sigma_{ij} = \left(D^e_{ijkl} - \frac{D^e_{ijpq}\frac{\partial f}{\partial \sigma_{pq}}\frac{\partial \varphi}{\partial \sigma_{mn}} D^e_{mnkl}}{H} \right) d\varepsilon_{kl} - \frac{D^e_{ijkl}\frac{\partial f}{\partial f_c}\frac{\partial f_c}{\partial \hat{f}}\frac{d\hat{f}}{dt}\frac{\partial \varphi}{\partial \sigma_{kl}}}{H} dt \tag{21}$$

From the above equation, it is known that the effect of time is embedded in the last term.

3 Numerical Simulations

In order to test the correction of the anisotropy in the model, some triaxial tests are simulated with the model. Fig.1 shows the results of numerical simulation and experiment for different inclined angles at the confining pressure of 40Mpa. At this time, the following equation is satisfied: $\hat{f}(t) = \hat{f}_0$, i.e. the effect of time is neglected. From Fig.1, it is known that the anisotropy can be better embodied in the model.

Simulation of relaxation of the stress is done with the model. Assume that the inclined angle equals degree zero. From Fig.2, it is known that the effect of time can be better embodied in the model.

4 Conclusions

A new model of sedimentary rocks is presented in the paper. In the model, the uniaxial compressive strength is the function of time, the microstructure tensors and stresses. Numerical simulations show the inherent anisotropy and time-dependent behavior in sedimentary rocks can be efficiently described with the model.

References

1. Pietruszczak S., Mroz Z. Formulation of anisotropic failure criteria incorporating a microstructure tensor. Comp. and Geotechnics, 2000, 26(2): 105-112
2. S. Pietruszczak., S. J.Jiang and F.A.Mirza. An elasto-plastic constitutive model for concrete. Int. J. Solids Structures,1988,24(7):705-722

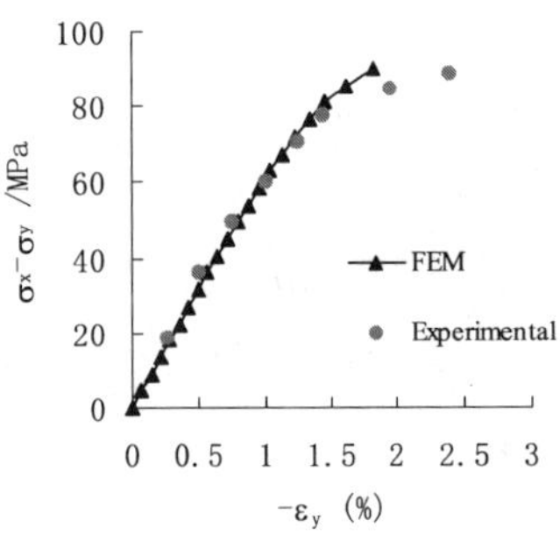

(a) α=0°

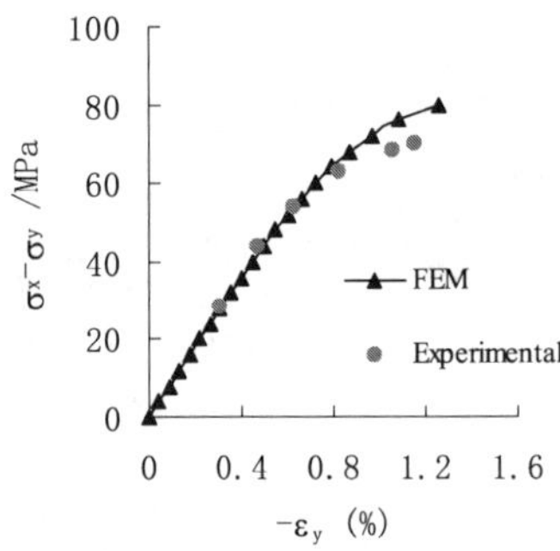

(b) α=45°

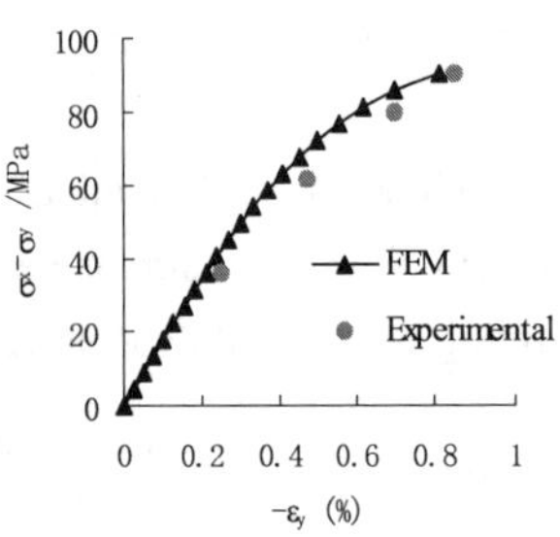

(c) α=90°

Figure 1. Numerical simulations of triaxial tests for different inclined angles

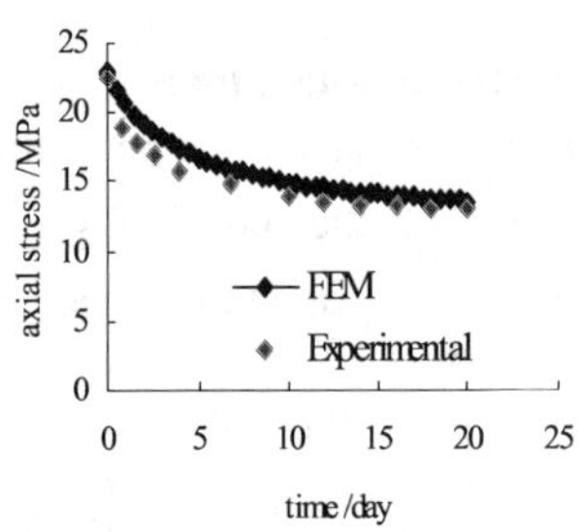

Figure 2. Relaxation of the stress

SIZE EFFECT AND SNAP BACK BASED ON CONSERVATION OF ENERGY AND GRADIENT-DEPENDENT PLASTICITY

XUE-BIN WANG YI-SHAN PAN

Dept. of Mechanics and Engineering Sciences, Liaoning Technical University, Fuxin 123000, China

The paper analyzed size effect and snapback for quasi-brittle materials due to shear strain localization. The traditional idea that localization zone was simplified as a band in a direction with respect to the direction of axial stress could not be accepted and it was modified by inclined shear band. The second-order shear strain gradient was introduced to obtained distributed plastic shear strain in shear band. By resorting to the conservation of energy that energy consumed by formation of shear band was equal to the dissipated energy, the size effect law was proposed. It is shown that the post peak response becomes steeper and even exhibits snapback with increasing length, as consists with experimental results. Shear strain localization is the reason for size effect and snapback behavior.

1 Introduction

For rock and concrete specimen under uniaxial compression, many experimental results (Hudson, 1971; Desai,1990; van Mier,1997; Jansen, 1997) showed that the post-failure portion of stress-strain curve becomes less steep as the specimen length is decreased and severe snapback can occur as the specimen length is very large. To investigate the size effect or snapback, a few models shown in Fig.1, have been put forward by Hudson (1971), Schreyer (1986), Bazent (1991), de Borst (1992) and Pan (1999). In fact, the axial cracks and inclined shear plane can be observed in many uniaxial compressive experiments. So, after the localization initiates, the assumption that the localized or plastic zone is simplified as a band in a direction with respect to the direction of axial stress is not an acceptable approximation. Moreover, according to Bazent (1997) the axial splitting cracks do not change the macroscopic continuum stress state due to uniaxial compression and cause no energy release. Similarly, the axial splitting cracks do not change axial displacement. So, the factor is not involved in our model.

Based on gradient-dependent plasticity, Wang[9, 10] performed theoretical analysis on the problem of size effect and snapback. Adopting the same theory, Wang[12] obtained an analytical solution on the inclination angle of the shear band dependent on specimen length, as can be observed in experiments or numerical simulation[11].

In the present contribution we shall utilize conservation of energy and shear strain gradient plasticity theory to discuss the problem of size effect and snapback.

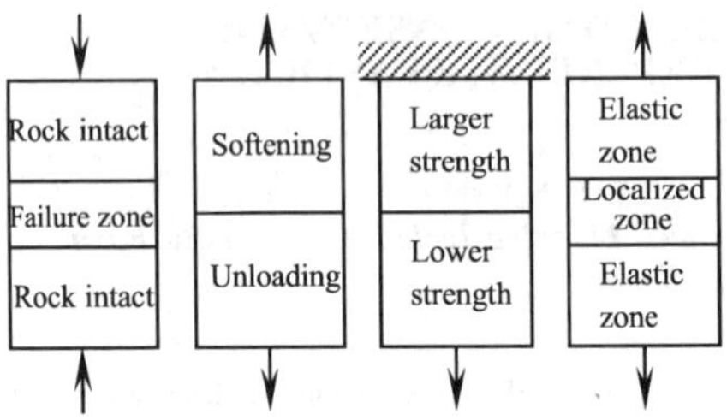

Fig.1 a few earlier models of size effect

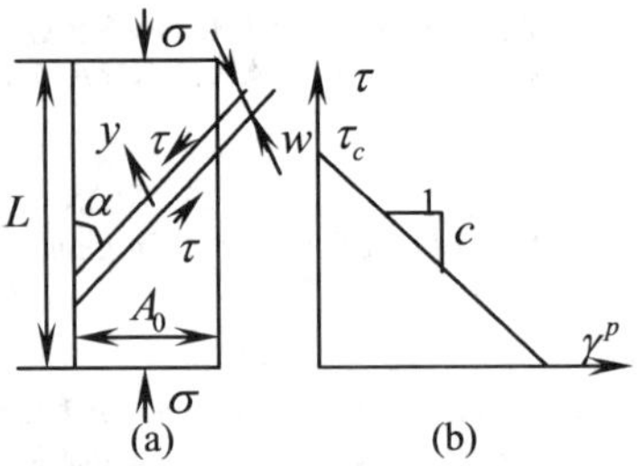

Fig.2 specimen and constitutive relation

2 Theoretical analysis

As shown in Fig.2(a), a specimen subjected to uniform compressive stress σ with length L and section area A_0. Localization initiates at the peak stress. The stable orientation of shear strain localization zone is α of being parallel to the axial direction and the width of the zone is w. Supposing that failure in shear band is caused by shear stress τ parallel to the tangential direction of the band. After localization initiates, the rock outside of the shear band is still in elastic state. The shear stress τ and its maximum τ_c can be expressed as

$$\tau = 0.5\sigma \sin(2\alpha), \tau_c = 0.5\sigma_c \sin(2\alpha) \qquad (1)$$

Where σ_c is compressive strength. In the elastic regime Yong's modulus governs the relation between stress and strain: $\sigma = E\varepsilon$. When σ_c is attained, softening occurs according to the linear diagram of Fig.2(b). For simplicity, the shear band is treated as a one-dimensional problem in the y-direction. For the situation a second-order gradient plasticity theory can be summarized by

$$\tau = \tau_c - c(\gamma^p + l^2 d^2\gamma^p dy^2) \qquad (2)$$

Where $c = G\lambda/(G+\lambda)$, G is shear modulus and E is elastic modulus; λ is called shear descending modulus and l is internal length parameter; γ^p is local plastic shear strain. The boundary condition is

$$\gamma^p = 0, \text{ at } y = \pm w/2 \qquad (3)$$

Application of the boundary condition (3) then results in

$$\gamma^p = (\tau_c - \tau)[1 - \cos(y/l)/\cos(w/2l)]/c \qquad (4)$$

The solution of the maximum plastic shear strain could be obtained by

$$d\gamma^p / dw = 0 \qquad (5)$$

The smallest non-trivial argument that satisfies Eq.(5) is π , so that $w = 2\pi l$

$$\gamma^p = (\tau_c - \tau)[1 + \cos(y/l)]/c \tag{6}$$

The expression shows that the plastic strain distributions alone the y-axis of the shear band is non-uniform. Equivalent plastic shear strain $\bar{\gamma}^p$ can be defined as

$$\bar{\gamma}^p = \frac{2}{w}\int_0^{w/2} \gamma^p dy = \frac{\tau_c - \tau}{c} \tag{7}$$

The energy consumed by formation of shear band may be written as

$$V = \frac{A_0 w}{\sin\alpha}\int \tau d\bar{\gamma}^p = \frac{A_0 w}{\sin\alpha}(\frac{1}{G} + \frac{1}{\lambda})(\tau_c^2 - \tau^2) \tag{8}$$

Conservation of energy requires that the sum of internal elastic strain energy and dissipated energy in specimen must equal to the work done by the external force on the specimen. In post-peak regime, the area under the complete stress-strain curve represents the sum of two parts of strain energy per unit volume. Elastic strain energy stores in entire specimen and this part of energy is recoverable, while rest of the energy is dissipated in failure zone. The latter energy causes increase of plastic shear strain in shear band. So, assume that dissipated energy is equal to the energy consumed by formation of shear band regardless of energy loss due to other reasons.

Typical complete stress-strain curve is simplified as bi-linear form shown in Fig.3. The negative value of slope λ' in strain softening region of the curve is called

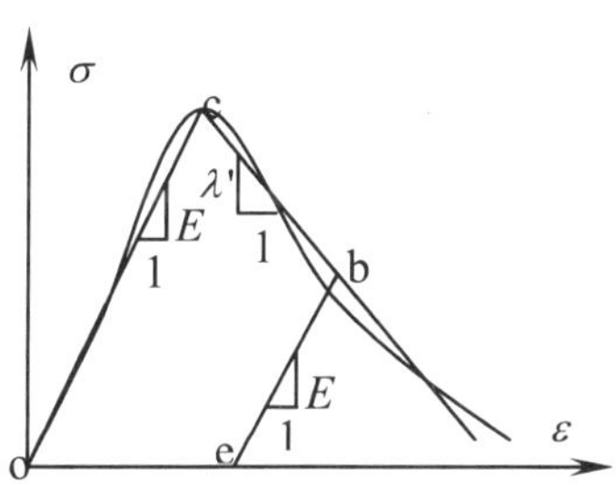

Fig.3 complete stress-strain curve and simplification

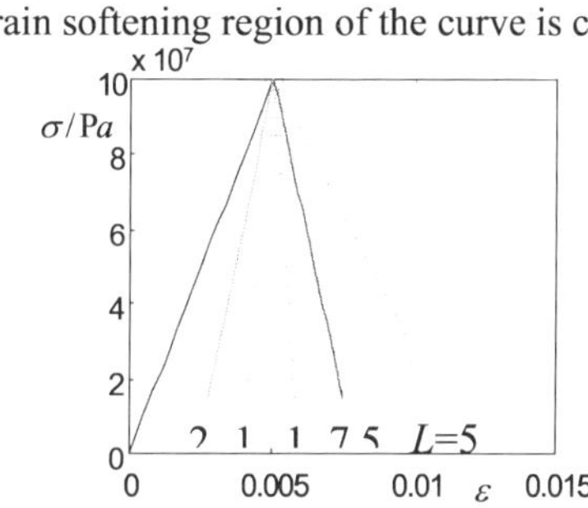

Fig.4 axial stress vs. axial strain for different specimen length

compressive descending modulus. The area described by stress-strain path ocbe is the dissipated energy per unit volume. In postpeak region, the specimen would unload alone a line parallel to the elastic modulus. With Eq.(1) the dissipated energy W can be expressed as

$$W = A_0 L\int \sigma d\varepsilon^p \tag{9}$$

Where ε^p is plastic axial compressive strain, $\varepsilon^p = \varepsilon - \sigma/E$, $\sigma = \sigma_c - \varepsilon^p E\lambda'/(E + \lambda')$. So, conservation of energy leads to

$$(G^{-1} + \lambda^{-1}) = L(E^{-1} + \lambda'^{-1})(\sin\alpha \cos^2\alpha)/w \tag{10}$$

Note that the expression above is consistent with the analytical solution obtained by force method [13]. Solving the equation above for $d\sigma / d\varepsilon$, we finally get the result

$$d\sigma / d\varepsilon = -\lambda' = (E^{-1} - w\sin\alpha\cos^2\alpha /(cL))^{-1} \tag{11}$$

This equation represents the size effect law. As an example the specimen with $E = 20\text{GP}a$, $\lambda = 5\text{GP}a$, $\sigma_c = 100\text{MP}a$, $\alpha = 30°$, $l = 0.009\text{m}$ and the Poisson's ratio $v = 0.25$. The relation between E and G is $E = 2G(1+v)$ .The complete stress-strain curves for L=5cm,7.5cm,10cm,15cm and 25cm are given in Fig.4. It is shown that the postpeak curves are highly dependent on the specimen length. For shorter specimens that exhibit Class I behavior tend to be somewhat ductile in nature. Otherwise, for longer specimens that exhibit Class II behavior, tend to respond in a brittle fashion. So, the analytical solution is consistent with many experimental results. Because there is localization, the post peak portion of the stress-strain curve is not only a material property but is dependent on specimen or measuring length.

3 Conclusions

A size effect law for quasi-brittle material specimen subjected to axial compressive stress is put forward considering conservation of energy. The post peak response becomes steeper and even exhibits snapback with increasing length. Shear strain localization is the reason for size effect and snapback behavior.

References

1. Z. P. Bazant et al. J. Engrg Mech., ASCE, D 122, 255 (1996).
2. R. de Borst. Int. J. solids and struct,D 35,521 (1992).
3. C. S. Desai et al. Int. J. Numer. Anal. Meth. Geomech., D 14 (1990).
4. J. A. Hudson et al. Engin. Geol., D 6,155(1972)
5. D. C. Jansen et al. J. Engrg. Mech., ASCE, D 123, 25 (1997).
6. H. L. Schreyer et al. J. Appl. Mech.,D 53, 791(1986).
7. Y. S. Pan et al. Chinese J. Geotech. Engrg, D 21, 472 (1999).
8. J. G. M. van Mier et al. Mater. Struct, D 30, 195 (1997).
9. X. B. Wang et al. Chinese J. Geomech.,D 7, 265
10. X. B. Wang et al. Chinese J. Geotech. Engrg, D 23,711
11. X. B. Wang et al. Chinese J. Rock Soil Mech., D 22,323
12. X. B. Wang et al. Chinese J. Rock Soil Mech. (will be published in 2002).

EFFECTIVE STRESS LAW OF MULTI-POROSITY MEDIUM SATURATED MULTI-PHASE FLUID

YAN JIN MIAN CHEN
University of Petroleum, China
E-mail: jiny@www.bjpeu.edu.cn

With the development of the theory and technology of petroleum extraction, the effective stress law of multi-porosity medium saturated multiphase fluid becomes one of the basic theory for high efficient petroleum extraction, because the mechanic response of multi-porosity medium can be precisely depicted by the effective stress law. Based on the effective stress law of multi-porosity medium saturated single phase fluid, the effective stress problem of dual-porosity medium saturated multiphase fluid is closely studied, then the effective stress law of multi-porosity medium saturated multiphase fluid is established in this paper.

1 Introduction

The concept of effective stress for porous media is most often attribute to Terzaghi[1,2]. He realized that pore water pressure is the neutral pressure. For water saturated single porosity medium, The effective stress tensor is often expressed as the difference between the total stress and some fraction of the pore pressure:

$$\hat{\sigma}_{ij} = \sigma_{ij} + \kappa p_p \delta_{ij} \tag{1}$$

in which $\hat{\sigma}_{ij}$ is the effective stress tensor, σ_{ij} is the total stress, p_p is the pore pressure, δ_{ij} is Kronecker symbol, and κ expresses the fraction of the pressure that should be employed to make equation (1) express the effective stress. Geertsma[3], Skempton[4], Nur & Byerlee[5] proposed the expression for κ:

$$\kappa = 1 - \frac{k}{k_s} \tag{2}$$

in which k is the volumetric modulus of skeleton and k_s is the volumetric modulus of solid material. Carroll[6],Thompson & Willis[7] discussed the effective stress of single porosity medium.

As to double-porosity medium, the fracture permeability is much great than the pore permeability, as a result, two kinds of different pressure exist in one element: the fluid pressure in fracture and the pressure in pore. Barenblatt[8] deduced the double-porosity model applied to natural fracture-pore medium. Mian Chen[9] presented the formulation of the effective stress law for double-porosity medium and complex multi-porosity medium. The effective stress law expression of the complex multi-porosity medium is:

$$\hat{\sigma}_{ij} = \sigma_{ij} + (\gamma_n p_n + \gamma_{n-1} p_{n-1} + \cdots + \gamma_1 p_1)\delta_{ij} \tag{3}$$

in which $p_1, p_2, p_3, \ldots\ldots p_{n-1}$ and p_n are the pressure in one of the n kinds of pore, $\gamma_1, \gamma_2, \gamma_3, \ldots\ldots \gamma_{n-1}$ and γ_n are the coefficient of the effective stress, which is expressed:

$$\gamma_1 = \frac{k_{s,1,\cdots,n}}{k_{s,1}} - \frac{k_{s,1,\cdots,n}}{k_s} \qquad \gamma_2 = \frac{k_{s,1,\cdots,n}}{k_{s,1,2}} - \frac{k_{s,1,\cdots,n}}{k_{s,1}} \qquad \cdots$$
$$\gamma_{n-1} = \frac{k_{s,1,\cdots,n}}{k_{s,1,\cdots,(n-1)}} - \frac{k_{s,1,\cdots,n}}{k_{s,1,\cdots,(n-2)}} \qquad \gamma_n = 1 - \frac{k_{s,1,\cdots,n}}{k_{s,1,\cdots,(n-1)}} \tag{4}$$

In equation (4), $k_{s,1,2,\cdots,n}$ is the volumetric modulus of solid material and n parts of n kinds of pore.

The concept and application of effective stress law are developed from the effective stress of single porosity medium to the effective stress of multi- porosity medium. These effective stress laws are based the porosity medium saturated single phase fluid (water), and can not be depicted the porosity medium saturated multi-phase fluid (water, oil and gas).

2 The effective Stress of Dual-porosity Medium Saturated Multi-Phase Fluid

As to isotropic dual-porosity medium, fracture and pore are saturated two kinds of fluid without inter-dissolution, and the total stress of skeleton is σ_{ij}. The total porosity is ϕ, the porosity of porous medium is ϕ_p, and the porosity of fracture medium is $(1-\phi_p)$. The saturation of the two kinds of fluid in pore are s_{p_1} and s_{p_2}, and the corresponding pore pressure are P_{p_1} and P_{p_2}. The saturation of the two kinds of fluid in fracture are s_{f_1} and s_{f_2}, and the corresponding pore pressure are P_{f_1} and P_{f_2}. The relationships of the porosity and saturation between pore and fracture can be expressed:

$$s_{p_1} + s_{p_2} + s_{f_1} + s_{f_2} = 1 \tag{5}$$

$$s_{f_1} + s_{f_2} = [1-(s_{p_1} + s_{p_2})]\phi \tag{6}$$

$$s_{p_2} = (1 - s_{p_1})\phi_p \tag{7}$$

$$s_{f_2} = (1 - s_{f_1})(1-\phi_p) \tag{8}$$

According to the derivation of the expression of the effective stress of multi-porosity medium[9], the effective stress of dual-porosity medium saturated m phase can be expressed:

$$\hat{\sigma}_{ij} = \sigma_{ij} - (\gamma_{f_m} p_{f_m} + \ldots + \gamma_{f_1} p_{f_1} + \gamma_{p_m} p_{p_m} + \ldots + \gamma_{p_1} p_{p_1})\delta_{ij} \tag{9}$$

where

$$
\begin{aligned}
\gamma_{p_1} &= \frac{k_{s,p_1,\dots,p_m,f_1,\dots,f_m}}{k_{s,p_1}} - \frac{k_{s,p_1,\dots,p_m,f_1,\dots,f_m}}{k_s} \\
\gamma_{p_2} &= \frac{k_{s,p_1,\dots,p_m,f_1,\dots,f_m}}{k_{s,p_1,p_2}} - \frac{k_{s,p_1,\dots,p_m,f_1,\dots,f_m}}{k_{s,p_1}} \quad \cdots\cdots \\
\gamma_{p_m} &= \frac{k_{s,p_1,\dots,p_m,f_1,\dots,f_m}}{k_{s,p_1,p_2,\dots,p_m}} - \frac{k_{s,p_1,\dots,p_m,f_1,\dots,f_m}}{k_{s,p_1,p_2,\dots,p_{m-1}}} \\
\gamma_{f_1} &= \frac{k_{s,p_1,\dots,p_m,f_1,\dots,f_m}}{k_{s,p_1,\dots,p_2,f_1}} - \frac{k_{s,p_1,\dots,p_m,f_1,\dots,f_m}}{k_{s,p_1,\dots,p_2}} \\
\gamma_{f_2} &= \frac{k_{s,p_1,\dots,p_m,f_1,\dots,f_m}}{k_{s,p_1,\dots,p_2,f_1,f_2}} - \frac{k_{s,p_1,\dots,p_m,f_1,\dots,f_m}}{k_{s,p_1,\dots,p_2,f_1}} \quad \cdots\cdots \\
\gamma_{f_m} &= 1 - \frac{k_{s,p_1,\dots,p_m,f_1,\dots,f_m}}{k_{s,p_1,p_2,f_1,\dots,f_{m-1}}}
\end{aligned} \qquad (10)
$$

where

$$
\begin{aligned}
&k_s = \frac{1-2v_s}{E_s}, \qquad k_{s,p_1} = \frac{1-2v_{s,p_1}}{E_{s,p_1}}, \qquad \cdots \\
&k_{s,p_1,\cdots,p_m} = \frac{1-2v_{s,p_1,\cdots,p_m}}{E_{s,p_1,\cdots,p_m}}, \quad k_{s,p_1,\cdots,p_m,f_1} = \frac{1-2v_{s,p_1,\cdots,p_m,f_1}}{E_{s,p_1,\cdots,p_m,f_1}} \quad \cdots \\
&k_{s,p_1,\cdots,p_m,f_1,\cdots,f_m} = \frac{1-2v_{s,p_1,\cdots,p_m,f_1,\cdots,f_m}}{E_{s,p_1,\cdots,p_m,f_1,\cdots,f_m}}
\end{aligned} \qquad (11)
$$

3 The effective stress of multi-porosity medium saturated multi-Phase fluid

The multi-porosity medium can be supposed to consist of n kinds of pore $\Omega_1, \Omega_2, \Omega_3 \cdots\cdots$, and Ω_n, on which the pressure p_{sl} of m phase fluid acts. The effective stress of m-porosity medium saturated l-phase fluid is expressed:

$$
\hat{\sigma}_{ij} = \sigma_{ij} + \gamma_{sl} p_{sl} \delta_{ij} \qquad s = 1,\cdots,n \qquad l = 1,\cdots m \qquad (12)
$$

where

$$
\begin{aligned}
\gamma_{11} &= \frac{k_{s,\Omega_1^1,\Omega_1^2,\ldots\Omega_1^m,\cdots,\Omega_n^1,\cdots\Omega_n^m}}{k_{s,\Omega_1^1}} - \frac{k_{s,\Omega_1^1,\Omega_1^2,\ldots\Omega_1^m,\cdots,\Omega_n^1,\cdots\Omega_n^m}}{k_s} \\
\gamma_{sl} &= \frac{k_{s,\Omega_1^1,\Omega_1^2,\ldots\Omega_1^m,\cdots,\Omega_n^1,\cdots\Omega_n^m}}{k_{s,\Omega_1^1,\cdots,\Omega_1^l,\cdots,\Omega_s^1\cdots\Omega_s^l}} - \frac{k_{s,\Omega_1^1,\Omega_1^2,\ldots\Omega_1^m,\cdots,\Omega_n^1,\cdots\Omega_n^m}}{k_{s,\Omega_1^1,\cdots,\Omega_1^{l-1},\cdots,\Omega_s^1\cdots\Omega_s^{l-1}}} \\
&\ldots\ldots \\
\gamma_{nm} &= 1 - \frac{k_{s,\Omega_1^1,\Omega_1^2,\ldots\Omega_1^m,\cdots,\Omega_n^1,\cdots\Omega_n^m}}{k_{s,\Omega_1^1,\cdots,\Omega_1^{m-1},\cdots,\Omega_s^1\cdots\Omega_s^{m-1}}}
\end{aligned}
\tag{13}
$$

where

$$
\begin{aligned}
k_s &= \frac{1-2v_s}{E_s}, \qquad k_{s,\Omega_1^1,\cdots,\Omega_s^l} = \frac{1-2v_{s,\Omega_1^1,\cdots,\Omega_s^l}}{E_{s,\Omega_1^1,\cdots,\Omega_s^l}}, \qquad \cdots \\
k_{s,\Omega_1^1,\cdots,\Omega_n^m} &= \frac{1-2v_{s,\Omega_1^1,\cdots,\Omega_n^m}}{E_{s,\Omega_1^1,\cdots,\Omega_n^m}}
\end{aligned}
\tag{14}
$$

References

1. Terzaghi, K., Thearetical Soil Mechanics, New York, Wiley, 1943
2. Geertsma, J., The effect of fluid pressure decline on volumetric changes of porous rocks, Petroleum Transactions, AIME, Vol. 210, No. 2, 331-340, 1957
3. Skempton, A. W., Effective stress in soils, concrete and rock, in pore pressure and suction in soils, Butterworths, London, 1961
4. Nur, A. and Byerlee, J. D., An exact effective stress law for elastic deformation of rock with fluids, J. Geophys. Res., Vol. 76, No. 26, 6414-6419, 1971
5. Carroll, M. M., An effective stress law for anisotropic elastic deformation, Journal of Geophysical Research, Vol. 84, No. B13, 7510-7512, 1979
6. Thompson, M. and Willis, J. R., A reformation of the equations of anisotropic poroelasticity, Journal of Applied Mechanics, Vol. 58, No. 11, 612-616, 1991
7. Barenblatt, G. I., Zheltov, Iu. P. and Kochina I. N., Basic concepts in the theory of seepage of homogeneous liquids in fissured rock, J. Appl. Math. Mech., Vol. 24, No. 5, 1286-1303, 1960
8. Mian Chen, Zhida Chen, The effective stress law of multi-porosity medium, Journal of Applied Mathematics and Mechanics, Vol. 11,1999.

STUDY ON A NEW METHOD OF PREDICTION FOR NATURAL FRACTURE DISTRIBUTION OF THIN LAYER

KAIJUN GUO MIAN CHEN YAN JIN
University of Petroleum, China
E-mail: jiny@www.bjpeu.edu.cn

It is very important to understand the distribution of natural fracture of thin sand layer for improving the EOR of the low permeable reservoir. Based on subsurface contour, the terrain trend can be acquired by trend surface analysis theory, then the curvature of trend surface can be calculated. According to plate-shell mechanics theory, the tectonic stress field can be derived from the curvature of trend surface. The distribution of natural fracture can be predicted by Griffith principle. The field application testified the model to be feasible and practical.

1 Introduction

Hydrocarbon resources and their production of tight and low permeability fractured reservoir take approximately one third of the total amount in China. In the development of reservoir, the dimension and orientation of manmade fractures by fracturing, the determination and performance of water packing-off and promotion scheme, and the selection and adjustment measure of water-injecting and developing wells are all related to the distribution of natural fracture. Therefore, it is important to research the regulations of natural fracture distribution and that is also useful to ensure high production, stable production and recovery ratio enhancement for low-permeable reservoir. There are several methods for prediction natural fracture, including coring formation analysis, log information analysis, special instrument measurement and mechanics analysis.

Coring formation analysis method[1] is the one that mainly accords to the coring to observe the fracture length, width and amount, so the density and its distribution curve is summed up before the distribution regulation of the natural fracture is analyzed. This method is effected by many real terms and only can be used in determined description.

Log information analysis method[2~4] is the one that uses the different response of the natural fractures to the viscosity, amplitude and frequency of stratum sound wave. This method had a high precision but only can reflect the distribution regulation of near-wellbore natural fracture.

It is very convenient [5~7] to use special instrument to research natural fracture and it also has a high reliability. But sometimes by this method could not obtained good effect when it was used in special petroleum deposit because of the limitation of equipment performance.

The mechanics model of fractured rock [8] considers the question from mechanic view and uses rock breakup principle to put forward the numerical method predicting fracture distribution regulation by tectonic stress field. But this method didn't effectively combine the tectonic stress field with rock breakup principle.

In this paper, the topography trend surface, curvature distribution and tectonic stress field is numerically simulated from the view of geometry distortion using plate-shell mechanics theory and trend surface analysis theory according to the equal depth map of the district geology. After that, the trend of distribution of natural fractures is studied according the Griffith Principle.

2 Basic hypothesis and modeling

It is supposed that the stratum is continuous originally and hasn't broken up, and fractures form after the rock formation is broken due to tectonic movement. The fracture forms because the rock stresses are different in 3-D tectonic stress field, where the stress exceeds its strength fractures will form because of the rock breakup and where the stress does not exceed its strength fractures will not form.

The tectonic stress field is studied to find whether the rock breaks and fracture forms according Griffith Principle. The detail process is as below:

1. to compute the tectonic stress field combining with trend surface analysis;
2. to add the stratum 3-D stresses to the corrected 3-D formula of Griffith Principle to compute stress σ_t, then the computed σ_t distribution of this area is gained.
3. to compare the tensile strength distribution of studied areas, if the computed σ_t exceeds the rock tensile strength S_t, the rock will break.

3 Inverse model for tectonic stress field

Supposed the formation to be uniform, continuous, isotropic, perfectly elastic, and shaped in one tectonic movement. The formation is regarded as a crooked sheet and the middle of that sheet regarded as plane XY. According to plate-shell mechanics theory, the strain of one point in the sheet is:

$$\varepsilon_x = -z\frac{\partial^2 w}{\partial x^2}, \quad \varepsilon_y = -z\frac{\partial^2 w}{\partial y^2}, \quad \varepsilon_{xy} = -z\frac{\partial^2 w}{\partial x \partial y} \tag{1}$$

where $w = w(x, y)$ is the deflection of the formation. Because w is tiny, the curvature in x and y orientation of the sheet median plane may be expressed as below:

$$\frac{1}{r_x}=-\frac{\partial^2 w}{\partial x^2},\quad \frac{1}{r_y}=-\frac{\partial^2 w}{\partial y^2},\quad \frac{1}{r_{xy}}=-\frac{\partial^2 w}{\partial x \partial y} \tag{2}$$

Formula (1) may be rewritten as:

$$\varepsilon_x = z\frac{1}{r_x},\quad \varepsilon_y = z\frac{1}{r_y},\quad \varepsilon_{xy} = z\frac{1}{r_{xy}} \tag{3}$$

According to the constitutive equation of elasticity, the stress components can be written as:

$$\begin{cases} \sigma_x = \dfrac{E}{1-\mu^2}\left(\varepsilon_x + \mu\varepsilon_y\right) = \dfrac{E}{1-\mu^2}\left(\dfrac{1}{r_x}+\mu\dfrac{1}{r_y}\right) \\ \sigma_y = \dfrac{E}{1-\mu^2}\left(\varepsilon_y + \mu\varepsilon_x\right) = \dfrac{E}{1-\mu^2}\left(\dfrac{1}{r_y}+\mu\dfrac{1}{r_x}\right) \\ \tau_{xy} = \dfrac{E}{2(1+\mu)} r_{xy} = \dfrac{Ez}{1+\mu}\dfrac{1}{r_{xy}} \end{cases} \tag{4}$$

in which E and μ represent the elastic modulus and poisson ratio.

From formula (4), it is obvious that the principal stresses and direction of one point on the sheet may be determined absolutely by the its curvature.

4 Curvature

To solve the principal curvature and orientation of a certain point on the structural surface, the deflection equation $w = w(x, y)$ should be solved and gained first. The tectonic trend surface is a continuous space curve, which is the function of plane coordinates X and Y. Therefore, it is supposed that the trend surface equation (i.e. the displacement function or the deflection equation) is a binary Fourier series, which is combined by binary harmonic series. The detail is as:

$$w(x,y) = F(x,y) = \sum_{t=0}^{r}\sum_{k=0}^{s}\left[E_{tk}\cos\frac{2t\pi x}{L}\cos\frac{2k\pi y}{H} + G_{tk}\sin\frac{2t\pi x}{L}\sin\frac{2k\pi y}{H} + P_{tk}\cos\frac{2t\pi x}{L}\sin\frac{2k\pi y}{H} + W_{tk}\cos\frac{2k\pi y}{H}\sin\frac{2t\pi x}{L}\right] \tag{5}$$

The computation value of E_{tk}, G_{tk}, P_{tk} and W_{tk} in formula (9) is solved by given point (x_i, y_i, z_i) on the curve. Therefor, all the binary partial derivatives of $w = w(x, y)$ can be computed to calculate the curvature of each point.

5 Conclusions and further discussions

It's applicable to use rock mechanics to study the natural fracture distribution in low-permeable sheet sandstone formation. It provides a new research thought to study the tectonic stress field when the subsurface contour is known and the log information is fairly less. Therefore, this method has a great practicability in natural fracture distribution research.

In addition, the problem considered in this article is a static problem applying the above method studying natural fracture distribution from the configuration and mechanics of rock formation to set out. It must be pointed out that the natural fracture distribution of the rock formation has a good reflection from the field developing dynamic data, i.e. the natural fracture distribution can be analyzed from the field developing dynamic formation qualitatively.

References

1. D.J. Tillard, M.S. King, General report: Mechanical behavior of rocks and rock masses discontinuities at great depth, Rock at Great Depth, Maury & Fourmaintraux(eds), 1990: 1235-1241
2. Vinegar,H.J.,Wellington,S.l., Tomographic Imaging at Three-Phase Flow Experiments, Rev.Sci.Instrum, 1987: 58(1)
3. Wellington, S.L.,Vinegar, H.j. X-ray Computerized Tomography, JPT, 1987,Aug.
4. Sedgwick,G.E., Milesdixon,E.W.A. Application of X-ray Imaging Techniques to Oil Sands Experiments, 1987, CIM Paper No. 87-38-84
5. Gareet, C.H., Sharer, P.J. and Halsey, J.H., Nature Fractures and the Goniometer, The Informer, Core Laboratories, Irving, Texas, 1988
6. Noblett, Circumferential Acoustilog Finds Vertical Fracture, World Oil, November, 1987: 37~38
7. Fertl, Circumferential Acoustic Logs Detect Natural Fracture and Determine Their Orientation, 1988, Log Analysis Line, November.
8. Zeng Jinguang, LuoYuanhua, Study On Natural Fracture of Reservoir by Structure Principal Curvature, Journal of Mechanics, 1982,3(2):202~205.

TWIN T^2 STRENGTH THEORY OF TRI- PARAMETERS AND ITS APPLICATION IN THE ROCK MATERIALS

SILI CHEN[1,2] XIATING FENG[2,3] HUI ZHOU[3]

1 Department of Architectural Engineering, Shenyang University of Technology Shenyang 110023, China;

2 Institute of Resources and Civil Engineering, Northeastern University Shenyang 110006, China;

3 Key Laboratory of Rock and Soil Mechanics, Institute of Rock and Soil Mechanics the China Academy of Science, Wuhan 430071, China

The strength theory of rock is an important theory in the safety and stability designs of the civil, mining and hydraulic rock engineering et al, and it has important significance in the study of rock mechanics. At first, based on the twin T^2 strength theory in this paper, the twin T^2 strength theory of tri-parameters is derived under the complex stress state in this paper. The relations among the compressive strength, tensile strength and shear strength is proposed. Secondly, the brittle failure of rock is analyzed by using the twin T^2 strength theory of tri-parameters and compared with the available experimental data for some rock materials. The results of the twin T^2 strength theory of three parameters are shown to be satisfactory.

1 Introduction

Rock is a kind of widely used brittle material, the failure criteria of which have been taken seriously by researchers. Almost all rocks are in complex stress condition, therefore, it has important theoretical and practical significance to establish rock-strength theory under complex stress condition, for the analysis and design of safety and stability in civil, mining and hydraulic rock engineering et al. The author in paper[1-2] has put forward twin T^2 strength theory with twin-parameters and successfully applied it in the research of elastic and brittle metal material. Based on T^2 strength theory, the twin T^2 strength theory of tri-parameters was derived in this paper and the relationship among compressive strength, tensile strength and shear strength of rock material was proposed. With this theory the brittle failure of rock was analyzed and compared with experimental data of some rock specimen. The result had shown that twin T^2 strength theory of tri-parameters was satisfied in the strength analysis of rock specimen and, therefore, the strength criteria was perfected.

2 Expression of T^2 strength theory of tri-parameters

According to the T^2 strength theory of twin-parameters given in paper [1,2], the T^2 strength theory of tri-parameters can be written as

$$T_1^2 + T_3^2 + A\sigma_m^2 + B\sigma_m = C \tag{1}$$

In the formula

$T_1 = \tau_{12} + \tau_{13}$; $T_3 = \tau_{31} + \tau_{32}$; $\tau_{12} = \dfrac{\sigma_1 - \sigma_2}{2}$; $\tau_{13} = -\tau_{31} = \dfrac{\sigma_1 - \sigma_3}{2}$;

$\tau_{23} = -\tau_{32} = \dfrac{\sigma_2 - \sigma_3}{2}$; $\sigma_m = \dfrac{1}{3}(\sigma_1 + \sigma_2 + \sigma_3)$

where σ_1, σ_2 and σ_3 are main stresses, *A*, *B* and *C* are tri-parameters waiting for determined. In the force analysis of rock, the compressive stress is positive and tensile stress is negative. The stress σ_1, σ_2 and σ_3 are maximum main stress, middle main stress and minimum main stress respectively. Assuming that σ_c is the limiting compressive stress of single-axis, σ_t is the limiting tensile stress of single-axis and τ_k is the limiting shear stress, then we have following parameters :

$$A = \frac{9(18\tau_k^2 - 5\sigma_c\sigma_t)}{4\sigma_c\sigma_t}; B = \frac{27(\sigma_t - \sigma_c)\tau_k^2}{2\sigma_c\sigma_t}; C = \frac{9}{2}\tau_k^2 \tag{2}$$

substituting formula (2) and the main stress formulae in to (1), we get

$$\left[\sigma_1 - \frac{1}{2}(\sigma_2 + \sigma_3)\right]^2 + \left[\sigma_3 - \frac{1}{2}(\sigma_1 + \sigma_2)\right]^2 +$$

$$\frac{18\beta^2 - 5\alpha}{4\alpha}(\sigma_1 + \sigma_2 + \sigma_3)^2 - \frac{9(\alpha - 1)\beta^2}{2\alpha^2}(\sigma_1 + \sigma_2 + \sigma_3)\sigma_c = \frac{9\beta^2}{2\alpha^2}\sigma_c^2 \tag{3}$$

among them, $\alpha = \sigma_c / \sigma_t$ and $\beta = \tau_k / \sigma_t$ are called ratio of limiting stress. The limiting stress σ_c, σ_t and τ_k will be determined by experiment.

3 the application and analysis of result

Formula (3) is suitable for brittle material. It can also be applied to elastic material if $\sigma_c = \sigma_t$($\alpha = 1$)and $\tau_k = \sqrt{5/18}\sigma_t$ ($\beta = \sqrt{5/18}$). Now, the expression (3) is twin T^2 yielding criteria [1]. For rock, the compressive strength is far larger than the tensile strength (i.e., $\sigma_c > \sigma_t$), $\alpha > 1$. Now we discuss some general stress states.

3.1 Stress state of $\sigma_1 > \sigma_2 = \sigma_3$

This is an ordinary working condition for rock under load with tri-axis, from formula (3) we get

$$\left(\frac{\sigma_1}{\sigma_c} - \frac{\sigma_2}{\sigma_c}\right)^2 + \frac{18\beta^2 - 5\alpha}{4\alpha}\left(\frac{\sigma_1}{\sigma_c} + 2\frac{\sigma_2}{\sigma_c}\right)^2 - \frac{9(\alpha - 1)\beta^2}{2\alpha^2}\left(\frac{\sigma_1}{\sigma_c} + 2\frac{\sigma_2}{\sigma_c}\right)^2 = \frac{9\beta^2}{2\alpha^2} \tag{4}$$

Here, taking different limiting stress ratio α and value of β, the strength theory can be obtained for different material under the stress state of $\sigma_1 > \sigma_2 = \sigma_3$.

3.2 Twin-axis stress state of $\sigma_3=0$

Under the twin-axis stress state, because σ_3 is equal to 0, from formula(3) we get

$$\left(\frac{\sigma_1}{\sigma_c}\right)^2 - 0.4\left(\frac{\sigma_1}{\sigma_c}\right)\left(\frac{\sigma_2}{\sigma_c}\right) + 0.4\left(\frac{\sigma_2}{\sigma_c}\right)^2 + \frac{18\beta^2 - 5\alpha}{5\alpha}\left(\frac{\sigma_1}{\sigma_c} + \frac{\sigma_2}{\sigma_c}\right)^2 - \frac{18(\alpha-1)\beta^2}{5\alpha^2}\left(\frac{\sigma_1}{\sigma_c} + \frac{\sigma_2}{\sigma_c}\right)^2 = \frac{18\beta^2}{5\alpha^2} \tag{5}$$

Taking different limiting stress ratio α and value of β, the limiting stress curve can be obtained under the stress state of $\sigma_3=0$ for different material. Figure 1 has shown that, under the compressive stress state, if σ_3 is smaller, the limiting failure stress σ_1 will increase as σ_2 increases, if σ_2 is larger, σ_1 will decrease as σ_2 increases

3.3 Tri-axis stress state of $\sigma_1>0$, $\sigma_2>0$, $\sigma_3>0$

Under the tri-axis stress state , from formula (3) we get

$$\left[\frac{\sigma_1}{\sigma_c} - \frac{1}{2}\left(\frac{\sigma_2}{\sigma_c} + \frac{\sigma_3}{\sigma_c}\right)\right]^2 + \left[\frac{\sigma_3}{\sigma_c} - \frac{1}{2}\left(\frac{\sigma_1}{\sigma_c} + \frac{\sigma_2}{\sigma_c}\right)\right]^2 + \frac{18\beta^2 - 5\alpha}{5\alpha}\left(\frac{\sigma_1}{\sigma_c} + \frac{\sigma}{\sigma_c} + \frac{\sigma_3}{\sigma_c}\right)^2 - \frac{18(\alpha-1)\beta^2}{5\alpha^2}\left(\frac{\sigma_1}{\sigma_c} + \frac{\sigma_2}{\sigma_c} + \frac{\sigma_3}{\sigma_c}\right) = \frac{18\beta^2}{5\alpha^2} \tag{6}$$

Taking different limiting stress ratio α and value of β, the strength theory can be obtained for different material under tri-axis stress state of $\sigma_1>0$, $\sigma_2>0$, $\sigma_3>0$.

4 Comparison with experimental data

To verify the rationality and usability of twin T^2 strength theory of tri-parameters, we compare it with every kind of experimental results. Shown in Fig. 1, under the twin axis stress state, the experimental data is taken from paper [3], shown in Fig. 2 of the tri-axis stress state, which is the experimental result[4] of soft sandstone. Here, the experimental data is taken as $\alpha=5$, $\beta=1.2$ and $\sigma_c=30$MPa. In the figure of specimen test data, the white circle express the relationship between σ_1 and the surrounding while the black one express $\sigma_1 \sim \sigma_2$ relationship as $\sigma_3=5$MPa, and rhombus express $\sigma_1 \sim \sigma_2$ relationship as $\sigma_3=10$MPa.

5 Conclusion

The problem of rock failure was analyzed here with twin T^2 strength theory of tri-parameters. The result showed that the strength of rock under the state of $\sigma_1 = \sigma_2 > \sigma_3$ is higher than that under the state of $\sigma_1 > \sigma_2 = \sigma_3$. Furthermore, under the stress state of twin-axis ($\sigma_3 = 0$), when σ_2 is smaller, σ_1 will increase as σ_2 increases, when σ_2 is larger, σ_1 will decrease as σ_2 increases, which is identical with the experimental data. At the same time, the limiting stress ratio of rock α and the value of β are derived ,which has shown that the failure of rock is directly related with the material feature value of α and β . The twin T^2 strength theory of tri-parameters is not only suitable for certain rock material, but has its universality. The comparison of some experimental data with the analysis of this paper has identified that twin T^2 strength theory of tri-parameters is reasonable in the analysis of rock strength and the strength of rock material is thus improved.

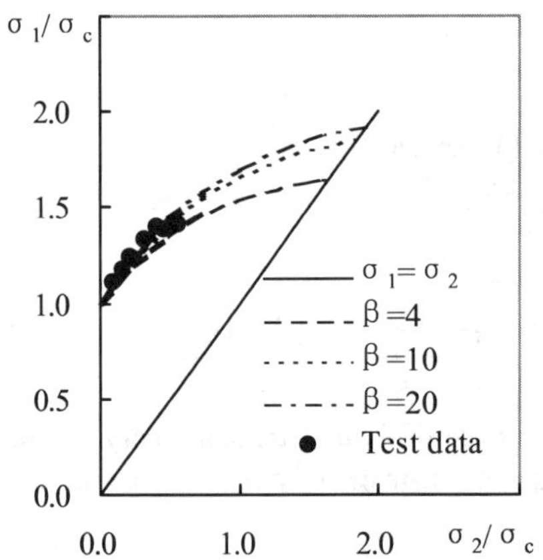

Figure 1. Twin-axis stress state of $\sigma_3 = 0$

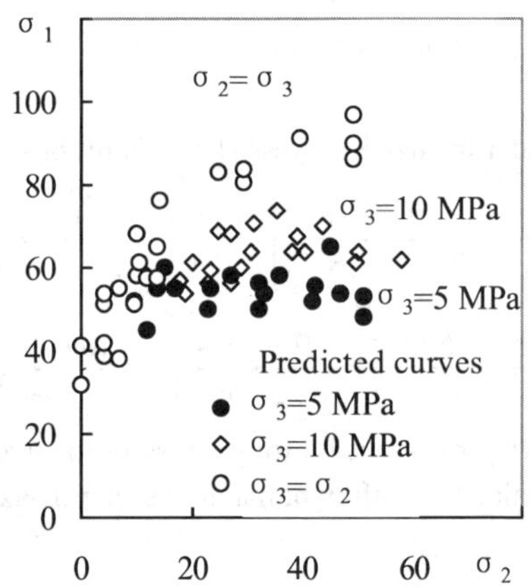

Figure 2. Comparison between the predicted curves and experimental data

References

1. Chen Sili and Yu bingyi, Twin T^2 and τ^2 yield criteria and their generalization, Mechanics and Practice, 1994; 16(5): 60-62.
2. Chen Sili, Twin T^2 strength theory and its generalization, Chinese Journal of Applied Mechanics, 1995; 12(5):118-121
3. Yu Maohong, Twin shear strength theory and application, Science Press, 1998.
4. Xu Dongjun, Geng Naiguang, The variation law of rock strength with increase of intermediate stress, 1985;(1): 72-80.

GENERALIZED TWIN SHEAR STRENGTH CRITERION OF ROCK

XUEQIANG YANG
Hubei Polytechnic University, Wuhan, 430068, China
ZUDE LIU
Wuhan University, 430071, China

Based on twin shear strength theory and some rock true triaxial strength testing data, a more reasonable generalized twin shear strength criterion is proposed and verified, corresponding parameters in the criterion are determined by traditional triaxial compressive and triaxial tensile tests, and some strength characteristics of rock are further revealed in this paper.

1 Introduction

Twin shear theory was proposed and developed by professor Yu Maohong [1], and can be expressed as follows:

$$\left.\begin{aligned} F&=\tau_{13}+\tau_{12}+A(\sigma_{13}+\sigma_{12})-D=0, & \tau_{12}+A\sigma_{12}&\ge\tau_{23}+A\sigma_{23} \\ F'&=\tau_{13}+\tau_{23}+A(\sigma_{13}+\sigma_{23})-D=0, & \tau_{12}+A\sigma_{12}&\le\tau_{23}+A\sigma_{23} \end{aligned}\right\} \tag{1}$$

where A is a influence coefficient of normal stress on rock failure, D is a strength parameter of rock, $\tau_{13}=(\sigma_1-\sigma_3)/2$, $\sigma_{13}=(\sigma_1+\sigma_3)/2$, $\tau_{12}=(\sigma_1-\sigma_2)/2$, $\sigma_{12}=(\sigma_1+\sigma_2)/2$, $\tau_{23}=(\sigma_2-\sigma_3)/2$, $\sigma_{23}=(\sigma_2+\sigma_3)/2$.

2 The Generalized Twin Shear Strength Criterion Of Rock And Its Forms

2.1 The generalized twin shear strength criterion of rock

Supposed compressive stress is positive, under $\sigma_1\geqslant\sigma_2\geqslant\sigma_3$ acting on rock, average principal stress, $\sigma_m=(\sigma_1+\sigma_2+\sigma_3)/3$, load angle θ_σ and octahedral shear stress τ_8 are expressed as follows:

$$\mathrm{tg}\theta_\sigma=\frac{1}{\sqrt{3}}\left(2\frac{\sigma_2-\sigma_3}{\sigma_1-\sigma_3}-1\right) \tag{2}$$

$$\tau_8=[(\sigma_1-\sigma_2)^2+(\sigma_2-\sigma_3)^2+(\sigma_1-\sigma_3)^2]^{\frac{1}{2}}/3 \tag{3}$$

As shown in Fig.1, rock triaxial experimental results show that the $\tau_8\sim\sigma_m$ relationships are non-linear. To formula (1), because $\sigma_{13}+\sigma_{12}=(3\sigma_m+\sigma_1)/2$ and

$\sigma_{13}+\sigma_{23}=(3\sigma_m+\sigma_3)/2$, including the linear part of σ_m, it is necessary to supplement $(\sigma_{13}+\sigma_{12})^2$ and $(\sigma_{13}+\sigma_{23})^2$ parts in formula (1). At the same time, rock is a material which triaxial compressive strength ($\theta_\sigma=-30°$) is not equal to triaxial tensile strength ($\theta_\sigma=30°$), its corresponding $\tau_8\sim\sigma_m$ strength curves under the

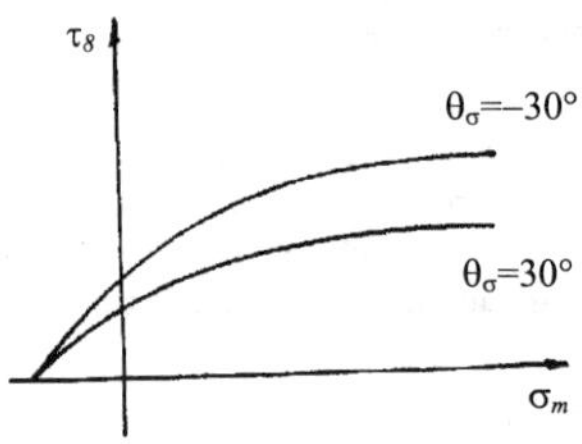

Figure1. Rock $\tau_8\sim\sigma_m$ strength curve

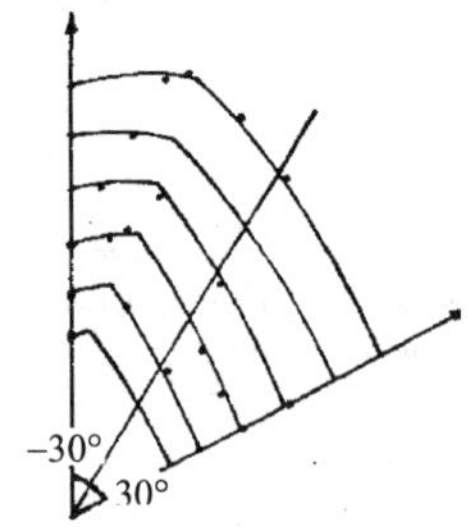

Figure2. Comparison between twin shear strength criterion and granite triaxial tests

two states are different. In order to stimulate the rock strength characteristics more correctly, formula (1) should be developed into formula (4):

$$\left.\begin{aligned} &F=\tau_{13}+\tau_{12}+A(\sigma_{13}+\sigma_{12})+B(\sigma_{13}+\sigma_{12})^2-D=0,\\ &\tau_{12}+A(\sigma_{13}+\sigma_{12})+B(\sigma_{13}+\sigma_{12})^2-D\geq\tau_{23}+A'(\sigma_{13}+\sigma_{23})+B'(\sigma_{13}+\sigma_{23})^2-D'\\ &F'=\tau_{13}+\tau_{23}+A'(\sigma_{13}+\sigma_{23})+B'(\sigma_{13}+\sigma_{23})^2-D'=0,\\ &\tau_{12}+A(\sigma_{13}+\sigma_{12})+B(\sigma_{13}+\sigma_{12})^2-D\leq\tau_{23}+A'(\sigma_{13}+\sigma_{23})+B'(\sigma_{13}+\sigma_{23})^2-D' \end{aligned}\right\} \quad (4)$$

2.2 Transforming forms of formula (4)

Formula (5) is a equivalent transforming forms of formula(4):

$$\left.\begin{aligned} &\tau_8=\frac{4}{B(\sqrt{3}\cos\theta_\sigma-\sin\theta_\sigma)}\left\{-\sqrt{2}[B\sigma_m+\frac{1}{4}(3+A)]+\sqrt{3B\sigma_m+\frac{1}{8}(3+A)^2+\frac{BD}{2}}\right\}, \quad -30°\leq\theta_\sigma\leq\theta_{\sigma cr}\\ &\tau_8=\frac{4}{B'(\sqrt{3}\cos\theta_\sigma+\sin\theta_\sigma)}\left\{-\sqrt{2}[-B'\sigma_m+\frac{1}{4}(3-A')]+\sqrt{-3B'\sigma_m+\frac{1}{8}(3-A')^2+\frac{B'D'}{2}}\right\}, \quad \theta_{\sigma cr}\leq\theta_\sigma\leq30° \end{aligned}\right\} \quad (5)$$

To formula (5), making the first equation equal to the second equation, then corresponding critical load angle $\theta_{\sigma cr}$ can be obtained. Based on formula (5), it can be seen that the generalized twin shear strength criterion has $\tau_8=g(\theta_\sigma)\cdot f(\sigma_m)$ function expression, and can be proved that the twin shear strength criterion is convex in three dimensional principal stress space.

2.3 Experiments and verifications

Literature [2] given granite true triaxial strength experimental results of Table 1:

Table 1. True triaxial strength experimental results of granite.

σ_3/σ_0	σ_2/σ_0, σ_1/σ_0, θ_σ						
0.00	0,1, −30°	0.3,1.2, −16.1°	0.63,1.23, 0.81°	0.77,1.18, 9.99°	0.9,1.1, 20.17°	1.08,1.08, 30°	
0.06	0.08,1.26, −29.17°	0.33,1.7, −21.17°	0.4,1.8, −19.38°	0.52,1.88, −15.94°	0.9,1.68, 1.23°	1.44,1.44, 30°	
0.19	0.2,1.7, −29.67°	0.38,2.08, −24.76°	0.58,2.47, −20.8°	0.8,2.8, −17.09°	1.03,3.0, −13.07°	1.13,2.87, −9.78°	1.33,2.4, 1.05°

Note: σ_0 is single axial compressive strength of granite.

Putting granite strength test results (0,0,1), (0.06,0.08,1.26) and (0.19,0.2,1.7)which have $\theta_\sigma \approx -30°$ into the first equation of formula (4) respectively, then A=-0.6232 , B=0.02035/σ_0 and D=0.3971 σ_0 can be obtained by solving corresponding three algebraic equations. Putting granite strength test results (0,0.63,1.23),(0.06,0.9,1.68)and (0.19,1.33,2.4)which have $\theta_\sigma \approx 1°$ into the second equation of formula (4), then $A' = -0.8828, B' = 0.0739/\sigma_0$ and $D' = 0.1729\sigma_0$ can be gotten. Then generalized twin shear strength criterion of granite can be expressed as follows:

$$\left.\begin{aligned} \tau_8 &= \frac{196.5216\sigma_0}{\sqrt{3}\cos\theta_\sigma - \sin\theta_\sigma}\left(-0.0288\frac{\sigma_m}{\sigma_0} - 0.8402 + \sqrt{0.7102 + 0.06106\frac{\sigma_m}{\sigma_0}}\right), \quad -30° \le \theta_\sigma \le \theta_{\sigma cr} \\ \tau_8 &= \frac{54.1272\,\sigma_0}{\sqrt{3}\cos\theta_\sigma + \sin\theta_\sigma}\left(0.1045\frac{\sigma_m}{\sigma_0} - 1.3726 + \sqrt{1.8909 - 0.2217\frac{\sigma_m}{\sigma_0}}\right), \quad \theta_{\sigma cr} \le \theta_\sigma \le 30° \end{aligned}\right\} \tag{6}$$

Based on Table1, compared test results of formula (3) with predicting results of formula (6), as shown in Fig.2, it can be observed that maximum relative error is not bigger than 5.9%, this shows that the formula (6) is reliable in predicting granite strength characteristics.

3 Rock Strength Characteristics

To formula (6), under triaxial compressive state ($\sigma_1 > \sigma_2 = \sigma_3$),granite strength curve corresponding to $\theta_\sigma = -30°$ can be obtained as follows:

$$\tau_{8c} = 98.2609\sigma_0\left(-0.0288\frac{\sigma_m}{\sigma_0} - 0.8402 + \sqrt{0.7102 + 0.06106\frac{\sigma_m}{\sigma_0}}\right) \tag{7}$$

Under triaxial tensile state($\sigma_1 = \sigma_2 > \sigma_3$)granite strength curve corresponding to $\theta_\sigma = 30°$ can be gotten as follows:

$$\tau_{8t} = 27.0636\sigma_0 \; 0.1045\frac{\sigma_m}{\sigma_0} - 1.3726 + \sqrt{1.8909 - 0.2217\frac{\sigma_m}{\sigma_0}} \qquad (8)$$

Putting the $\theta_{\sigma cr}$ into either equation of formula (6), corresponding maximum stress deviator τ_{8m} can be gotten on π plane under the σ_m stress level. According to Table 1, granite strength characteristics on π plane when $\sigma_m = (0.3 \sim 1.5)\sigma_0$ can be observed from Table2:

Table 2. Granite strength characteristics on π plane.

σ_m/σ_0	0.3	0.5	0.7	0.9	1.1	1.3	1.5
$\theta_{\sigma cr}$	-26.39°	-21.52°	-18.65°	-16.81°	-15.59°	-14.78°	-14.27°
τ_{8c}/σ_0	0.4611	0.5951	0.7234	0.8461	0.9633	1.0752	1.1818
τ_{8t}/σ_0	0.2558	0.3744	0.4875	0.5948	0.696	0.7909	0.8793
τ_{8m}/σ_0	0.462	0.6017	0.7379	0.869	0.9946	1.1143	1.2278
τ_{8c}/τ_{8t}	1.8026	1.5895	1.4839	1.4225	1.3841	1.3595	1.344

4 Conclusions

(1)Calculating results of Table 2 shows that τ_{8c}/τ_{8t} will be changed as σ_m is changed. τ_{8c}/τ_{8t} is bigger when σ_m is at lower stress level, and is smaller when σ_m is at higher stress level. This phenomenon demonstrates on one side that rock is developed toward to homogeneous material gradually as σ_m is increased, on the other side that the Drucker-Prager strength criterion is not suitable to rock material when σ_m is at lower stress level.
(2) Calculating results of Table 2 shows that $\theta_{\sigma cr}$ corresponding to τ_{8m} is not a constant, $\theta_{\sigma cr}$ will be increased as σ_m is increased, this agrees with the rock homogeneous tendency and experimental results as σ_m is increased.

References

1. M.H.Yu, *Chinese Journal of Geotechnical Engineering*. D16,98(1994)
2. Y.F.Gao, *Chinese Journal of Geotechnical Engineering*. D15, 312 (1990).

CONSTITUTIVE MODEL FOR SIMULATING EFFECTS OF STRAIN RATE ON STRENGTH AND ELASTIC MODULI OF ROCKS UNDER UNIAXIAL COMPRESSION

YONGLAI ZHENG[1] SHUGUANG LIU[1] XISHENG ZHOU[2]
1. *Dept. of Geotechnical Engineering, Tongji University, Shanghai 200092, P.R. China*
2. *Shanghai Tunnel Engineering Co.*

A viscoelastic damage constitutive model is presented that simulates effects of strain rate on the strength and elastic moduli of rocks under uniaxial compression. The model is based on viscoelasticity and damage in the process of deformation of rocks. In the rheological portions of the model, multiple relaxation times can be obtained by adding elements, so we can simulate the effects of strain rate on the strength and elastic moduli in a wider strain rate range (strain rates ranging from about 10^{-7} to 10^{6}/s.). From the comparison of the model predictions with experimental data for porphyritic tonalite (at temperatures of 25°C, -78°C and -191°C) and sandstone (at room temperature), it is shown that the model can not only simulate the rate effects, but also can predict extremely sharp increase in the strength of rocks at about strain rate 10^{3}/s. Besides, the model can exhibit the upper and lower limits of the rate effects on the strength and elastic moduli.

1 Introduction

Knowledge of rock deformation and strength around an opening or tunnel is of great importance in engineering practice, and it will be of some value in studying stress wave spread and design in nuclear explosion and chemical explosion, thus a full constitutive curve is necessary for analysis (Pawel 1992). The effect of strain rate due to viscosity is common in cases of great depth, and for particularly susceptible rocks such as shales, sandstones etc. The various rock compression tests showed that both elastic moduli and strength are rate dependent quantities (Bieniaski 1970,Zheng 1992, et al.)

In tests (Perkin & Green 1970, Brace 1971, and Kumer 1993, et al.) on relatively isotropic and fine-grained geologic materials it is indicated that their strength and elastic moduli are strain rate sensitive, and that the strength increases with increasing the rate with going through a very sensitive region at about 10^{3}/s. However, a general theory model has not been established to express these rate effects quantitatively.

In fact, in the process of the deformation for rocks, the rocks not only display rheological properties, but also damage properties, which may be regarded as enlarged microvoids, microcracks, or similar defects.

In this paper, a viscoelastic damage constitutive model is proposed by taking into account both viscoelasticity and damage. It can not only simulate the rate effects on the strength and elastic moduli for ordinary strain rate range ($\dot{\varepsilon} \leq 10^{2}$/s.), but also the sharp rate effects of the strength at about strain rate 10^{3}/s.; besides it can simulate both the upper and lower limits of the rate effects.

2 Theory Model

2.1 Viscoelastic Model

Differential equation of the Boydin model is

$$\dot{\sigma} + \frac{E_1}{\eta}\sigma = (E_0 + E_1)\dot{\varepsilon} + \frac{E_0 E_1}{\eta}\varepsilon \tag{1}$$

where $\dot{\sigma}$ and $\dot{\varepsilon}$ denote the derivative of stress and strain with respect to time t, respectively. Suppose that the initial condition is $\varepsilon(0)=0$ and $\sigma(0)=0$ (t=+0), thus, $\sigma(t)$ in equation(1) is

$$\sigma(t) = E_0\varepsilon + E_1\int_0^t \dot{\varepsilon}(\tau)\cdot\exp\left(-\frac{t-\tau}{\theta_1}\right)d\tau \tag{2}$$

where $\theta_1 = \eta / E_1$. In the equation(2), damage is not considered.

2.2 Viscoelastic damage constitutive model

For considering the time effects of the friction on the surface of the cracks and propagation of the cracks, which can be equivalent to defects development in the process of the deformation, author develops a viscoelastic damage model, which involves both viscoelasticity and damage under uniaxial compression. The model is

$$\sigma(\varepsilon) = \left[\left(\frac{\varepsilon}{\varepsilon_0}\right)^m + 1\right]\cdot e^{-\left(\frac{\varepsilon}{\varepsilon_0}\right)^m}\cdot\left[E_0\varepsilon + \dot{\varepsilon}E_0\cdot\frac{\alpha_1}{n}\sum_{i=1}^{n}\frac{1}{c_i}\left(1-e^{-c_i\cdot\frac{\varepsilon}{\dot{\varepsilon}}}\right) + \dot{\varepsilon}E_h\cdot\frac{1}{c_2}\left(1-e^{-c_2\frac{\varepsilon}{\dot{\varepsilon}}}\right)\right] \tag{3}$$

where n is determined by the strain rate range in which strength and elastic moduli increase with increasing strain rate, and the strength and elastic moduli increase about percentage α_1 per tenfold increase in strain rate. The increase of n can enlarge the rate-dependent range; $c_i = 10^{i-4}$, its unit is s^{-1}. α_2 determined by the percentage which the strength increases sharply per tenfold increase in strain rate at about strain rate 10^3/s. . $c_2 = 10^5$, its unit is s^{-1}. In the equation, all variables will be in international system of units (SI units). The equation only can be used when strain rate is constant in the condition of uniaxial compression.

3 Comparison of Experiment and Model Prediction

The porphyritic tonalite was tested in uniaxial stress at strain rate from 10^{-4} to 10^3/s. by Perkins(1970) et al. . The sandstone was tested in uniaxial at strain rates from 3.3×10^{-5} to 3.3×10^{-3}/s. by Zheng(1992) .

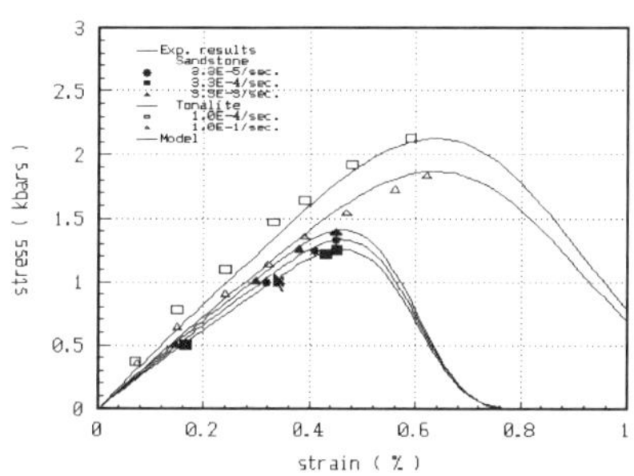

Figure1. Comparison of experimental results and the models about stress and strain under uniaxial compression. (The tonalite from Perkin, the sandstone from Zheng)

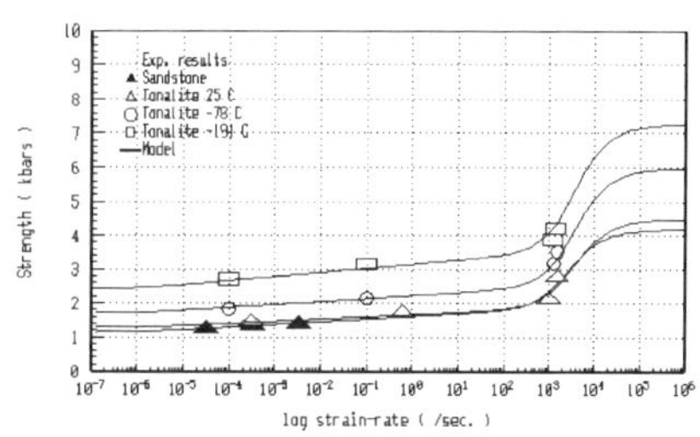

Figure 2. Comparison of experimental results and the model predictions about strength and strain rate under uniaxial compression

For the tonalite and sandstone at low and medium strain rate, the stress-strain theory curves predicted by the proposed model are in good agreement with the experimental results before the maximum stress, after the maximum stress the loading is uncontrolled, the strain rate increases and becomes so high that the stress and strain are not recorded, but in shape the model predictions (Fig.1)on stress and strain are similar to the typical experimental result recorded with servo-controlled system (Tang 1993,Pawel 1992).

The fig.2 shows that strength increases with strain rate at low and medium strain rate range, and behaves abrupt increase of rate-dependence near strain rate 10^3/s. ;.

It is reasonable that the strain rate effects on the strength and elastic moduli of rocks shouldn't be infinite. This means that the effects of strain rate on the strength and the elastic moduli have a upper and lower limits. The model predictions can be in good agreement with the property.

4 Conclusion

In the paper, the proposed viscoelastic damage constitutive model can simulated the effects of strain rate on the deformation under uniaxial compression, include that of the strength and the elastic moduli of rocks, especially on the sharp increase of the strength at strain rate about 10^3 /s . . The model is in good agreement with the experimental results. Besides, it can show the upper and lower limits of the rate effects on the strength and elastic moduli.

References

1. Bieniawski,Z.T.(1966): Mechanism of rock fracture in compression, Rep.Counc. Scient. Ind. Res. S. Afr., MEG pp459.
2. Brace,W.F., and A.H. Jones (1971): Comparison of Uniaxial Deformation in Shock and Static Loading of Three Rocks, J. Geophys, Res., 76(20), pp4913-4920.
3. Lemaitre, J.A. (1985): Continuous Damage Mechanics Model for ductile Fracture, J. of Eng. Material & Technology, 107, pp83
4. Nawrocki, P.A. and Mroz. (1992): Constitutive model for rocks accounting for viscoplastic deformation and damage, Rock Mechanics ,Balkema. Rotterdam., pp691-700.
5. Perkin.R.D. and S.J.Green (1970): Uniaxial stress behaviour of porphyritic tonalite at strain rate to 10^3/second, Int. J. Rock Mech. Min.Sci.,7,pp527-535.
6. Goodman,R.E. (1989): Introduction to Rock Mechanics,2nd.ed. New York pp 204-217
7. Tang,C.A.(1993): Catastrophe in Rock Unstable Failure,(in chinese),Coal Industry Press,20pp.
8. Zheng,Y.L. & Xia S.Y.(1997): Mechanism of Dynamic Elastic Modulus of Geotechnical materials,16(2) (in chinese),Journal of Vibration and Shock, pp56-61.
9. Zheng,Y.L.(1992): An Investigation of Rock strain Rate Sensitivity and Dynamical Mechanics Properties, M.S.Thesis, Department of earth and space science,Univ. of Science and technology of China,Hefei, P.R.China.

Author's address: Dr. Yonglai Zheng, Dept. Of Geotechnical Engineering, TongJi University, Siping road 1239 , Shanghai,200092, P.R. China.

THE RESEARCH OF MATHEMATICS MODEL FOR CONFIGURATION OF SALT CAVITY AND TRANSMUTATION DISCIPLINARIAN OF LIQUID-SOLID IN WASHED SALT CAVERNS

ZHICHENG ZHAO WEIYAO ZHU WENWEN SHAN
Institute of Porous Flow and Fluid Mechanics, China National Petroleum Corporation & Academia Sinica Langfang Hebei 065007

With the development of the natural gas industry and with the change of the energy resources structure, the research of underground natural gas storage take an important role on the rapid and persistent development of national economy and on the strategic reserve of energy source. This paper presents a general discussion of configuration of salt cavity and transmutation disciplinarian of liquid-solid in washed salt caverns. To design a stable configuration of cavity, a mathematics model is established for solute transmission and liquid flowage in the washed salt caverns based on the hydrodynamics and chemistry dynamics and the principle of material equilibrium in this paper. Also a dissolution velocity equation for halite is established in this paper based on the basic principle of hydrodynamics and theory of convection and diffusion and the basic disciplinarian of physics modeling. The numerical simulation results indicate that shape of the salt cavern is affected considerably by the brine producing rate and the brine concentration, as well as the injecting/producing site in the salt caverns. The physical properties and purity of halite considerably contribute to the parameter during the cavern shape is formed. The injecting/producing site need adjusting to acquire an optimum mechanics stability condition and a technical producing results. The theory foundation and programming means can be provided for the design of washed salt caverns.

1 Introduction

Underground natural gas storage can be constructed by drilling into a salt dome or salt bed and enlarged into a cavern with a designed shape and volume for storage of nature gas.

According to the mechanism of hydrodynamics and chemical dynamics, a model of solute transmission / fluid motion can be established to denote the coupling relationship of velocity field and concentration field. And a mathematical model of liquid-solid transmutation disciplinarian can be set up to compute the halite comprehensive dissolution velocity.

General assumptions:
The halite is homogeneous and isotropic, without intermediate;
The flow is incompressible and irrotational;
The insoluble residue will settle completely;
The influence of impurity of insoluble to diffusion is negligible;
The influence of temperature gradient is negligible;
The natural convection is negligible compared with forced convection.

2 Establishing of Mathematical Model

2.1 Concentration Field and Velocity Field

The course of washing a cavern in salt dome is a transmission process of solute in the flow system of solvent, which may be expressed by a multi-dimension non-stable convection-diffusion equation in tensor form[1]:

$$\frac{\partial C}{\partial t}+(\vec{v}\cdot\nabla)C=D\nabla^2 C \tag{1}$$

Where C standing for the concentration of material, D standing for the diffusion coefficient, which has the dimension of length square dividing time.

It has been proved that the concentration distribution depends on the velocity distribution when the boundary conditions are specified. Usually, velocity V should be a continuous function of time and space. So controlling equations of velocity field should be set up.

On the contrary, variation of physical characteristics of fluid resulting from the concentration distribution will affect the velocity distribution simultaneously. In fact, there is a function relationship between density and concentration of fluid as follows:

$$\rho=\rho_w+CM\left(1-\frac{\rho_w}{\rho_s}\right) \tag{2}$$

Where ρ is density of fluid, ρ_w is density of fresh water, C is mole concentration of fluid, ρ_s is density of the halite, M is mole mass of the halite.

For a flow with source, the mass conservation law (or continuity equation) of variable density fluid can be written as:

$$\frac{\partial \rho}{\partial t}+\nabla\cdot(\rho\vec{v})=q \tag{3}$$

The momentum conservation equation can be written as following according to Navier-Stokes fluid dynamics theory[2]:

$$\rho\frac{D\vec{v}}{Dt}=\rho\vec{f}-\nabla p+\frac{1}{3}\mu\nabla(\nabla\cdot\vec{v})+\mu\nabla^2\vec{v} \tag{4}$$

Where p is the pressure, f is the volume force, and μ is the viscosity of fluid.

2.2 Equation of dissolution velocity

The boundary of the cavern expands outward continuously when halite dissolving, which is called dynamic boundary problem in engineering.

We can assume that there is a extremely thin boundary layer attached to the surface of the cavern side in which the concentration hold constant of saturation concentration. According to material balance principle and Fick's diffusion law, the halite molecules dissolving from the surface of the cavern side will migrate into the boundary layer where the halite molecules migrate into the region of diffusion, while the concentration of fluid in the boundary layer hold dynamic balance.

When the halite of the surface of the cavern side resolve, the boundary of the cavern (together with the boundary layer attached) withdraw a little distance because of the material loss on the surface of the cavern side. The comprehensive dissolution velocity equation of the halite can be set up as following:

$$\frac{\partial R}{\partial t} = -D\frac{M}{\rho}\frac{\partial C}{\partial n}\Big|_{\Gamma_1} \qquad (5)$$

Where R is the boundary of the cavern.

The equations from (1) to (5) can be constructed to a complete model together with the definite conditions. Notice that there are two kind of Boundary conditions for the concentration field[3]: concentration on the surface of the cavern side holding constant; the top surface and the bottom surface of the cavern can be regarded as impermeable boundary (the blanketing and the settled insoluble residue)

3 Example

Based on the mathematical model, an example for numerical simulating is illustrated as follows:

A borehole of 100 meters can be selected in the salt dome with an initial diameter of 2 meters.

Typical injection/production tubing can be selected with a distance of 80 meters from inflow to outflow.

The numerical simulating result is shown in table 1.

Table 1. Numerical Simulating Result For The Mined Cavern

time (month)	Flow rate (l/s)	concentration (%)	Water injection (m^3)	Salt production (kg)	Cavern volume (m^3)
1	10	23.6	25920	8003100	4017
2	15	21.3	38880	10512180	8878
3	20	19.4	51840	12458880	14636

4 Conclusions

It has been proved by numerical simulation that:

1.The collection speed of the brine and the concentration of the brine take an important part on the dissolution speed.

2.The physical properties and the purity of the halite take a role of dissolving speed and the shape of the cavern.

3.The different production scheme will result in the different shape of the cavern. The position of injecting/producing should be adjusted according the requirement of the dynamic stability to acquire a optimum technology effect.

References:

1. University of Science & Technology of China, Computational Methods for Fluid Dynamics, Peking: science press, (1982).
2. Benluo Yang, Classical Analysis of Fluid Motion, Beijing: science press, (1986).
3. Thoms, R.L. and Gehle, R.M. Survey of Caverns in Salt Domes of the U.S., presented at the Solution Mining Research Institute Meeting, Denver, CO, October 16-19,(1983).

STUDY ON CONSTITUTIVE EQUATIONS OF A TRANSVERSELY ISOTROPIC FOUNDATION AND ITS EARTH STRESS

SHIFAN QIAO BAOCHEN LIU
College of Civil Engineering, Central South University, Changsha 410075, China
E-mail: qiaoshifan@163.com

ZHILIN DUN
Department of Civil, Jiaozuo Institution of Technology, Jiaozuo 454000, China

Based on the constitutive equations of a transversely isotropic foundation whose isotropic plane is horizontal, those of a transversely isotropic foundation whose isotropic plane is inclined are induced. And on the basis of certain postulates, computational methods about earth stress of a transversely isotropic foundation in the above two conditions are obtained. The results of examples in this paper accord with measured earth stress. According to the results of this article, those important actual measurements of earth stress can be explained that horizontal stress is larger than vertical stress and horizontal stress is anisotropic in theory, which can be used to predict earth stress.

1 Introduction

Geotechnical medium is very complex. In order to solve the engineering problems, many kinds of geotechnical medium models are presented under many assumptions. In some cases, because of tropism in the sediment process of the foundation's particles, the horizontal and vertical properties are different. For example, horizontal elastic modulus is larger than vertical one, and the horizontal plane is isotropic, which are the properties of the transversely isotropic foundation. For a rock foundation, which is composed of such mineral as shale, mica-schist, gneiss or phyllite, its mechanics behaviors have obvious transverse isotropic phenomena and its relation of stress and strain can be described with the transversely isotropic foundation model presented in this paper. Some researches in this field have be done before [1~5]. Authors have done some further study[6~7]

2 The constitutive equations of a transversely isotropic foundation

When the isotropic plane of a transversely isotropic foundation is horizontal, it is described in a ortho-axis coordinate system. In Figure 1, Axis Z is vertical and xoy coincides with its isotropic plane. If the three axes point to primary stresses , the coordinate system is called a ortho-axis coordinate system. The constitutive equation of a transversely isotropic foundation whose isotropic plane is horizontal and

$$\{\varepsilon\} = [A]\{\sigma\} \qquad (1)$$

where $\{\sigma\}$ is stress matrix that is $\{\sigma_x \ \sigma_y \ \sigma_z \ \tau_{xy} \ \tau_{yz} \ \tau_{zx}\}^T$ and $\{\varepsilon\}$ is strain matrix that is $\{\varepsilon_x \ \varepsilon_y \ \varepsilon_z \ \gamma_{xy} \ \gamma_{yz} \ \gamma_{zx}\}^T$, and $[A]$ is flexibility matrix which is

$$[A]=\begin{bmatrix} \frac{1}{E_1} & -\frac{\mu_1}{E_1} & -\frac{\mu_2}{E_2} & 0 & 0 & 0 \\ -\frac{\mu_1}{E_1} & \frac{1}{E_1} & -\frac{\mu_2}{E_2} & 0 & 0 & 0 \\ -\frac{\mu_2}{E_2} & -\frac{\mu_2}{E_2} & \frac{1}{E_2} & 0 & 0 & 0 \\ 0 & 0 & 0 & \frac{2(1+\mu_1)}{E_1} & 0 & 0 \\ 0 & 0 & 0 & 0 & \frac{1}{G_2} & 0 \\ 0 & 0 & 0 & 0 & 0 & \frac{1}{G_2} \end{bmatrix}$$

where E_1 and μ_1 are deformation modulus and Possion's ratio in the horizontal plane, E_2 and μ_2 are deformation modulus and Possion's ratio in the vertical plane. G_2 is shear modulus in the plane which is vertical with Plane xoy.

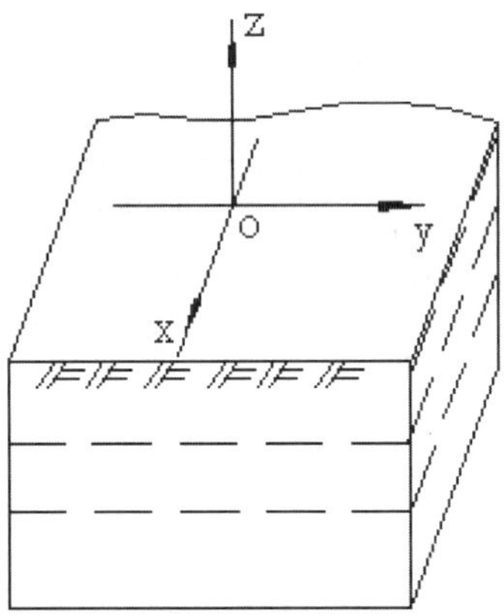

Figure 1. A transversely isotropic whose isotropic plane is horizontal

Table 1. Computing Table of q_{mi} and q_{nj} (It is 6×6 rank conversion matrix Q)

m(n)	i(j)					
	1	2	3	4	5	6
1	α_1^2	α_2^2	α_3^2	$2\alpha_2\alpha_3$	$2\alpha_3\alpha_1$	$2\alpha_1\alpha_2$
2	β_1^2	β_2^2	β_3^2	$2\beta_2\beta_3$	$2\beta_3\beta_1$	$2\beta_1\beta_2$
3	γ_1^2	γ_2^2	γ_3^2	$2\gamma_2\gamma_3$	$2\gamma_3\gamma_1$	$2\gamma_1\gamma_2$
4	$\beta_1\gamma_1$	$\beta_2\gamma_2$	$\beta_3\gamma_3$	$\beta_2\gamma_3+\beta_3\gamma_2$	$\beta_1\gamma_3+\beta_3\gamma_1$	$\beta_1\gamma_2+\beta_2\gamma_1$
5	$\gamma_1\alpha_1$	$\gamma_2\alpha_2$	$\gamma_3\alpha_3$	$\gamma_2\alpha_3+\gamma_3\alpha_2$	$\gamma_1\alpha_3+\gamma_3\alpha_1$	$\gamma_1\alpha_2+\gamma_2\alpha_1$
6	$\alpha_1\beta_1$	$\alpha_2\beta_2$	$\alpha_3\beta_3$	$\alpha_2\beta_3+\alpha_3\beta_2$	$\alpha_1\beta_3+\alpha_3\beta_1$	$\alpha_1\beta_2+\alpha_2\beta_1$

Table 2 .The Cosine of Angles between relative axes in the ortho and inclined-axis coordinate systems

Orth-axis coordinate system	Inclined-anxis coordinate system		
	Axis x'	Axis y'	Axis z'
Axis x	α_1	α_2	α_3
Axis y	β_1	β_2	β_3
Axis z	γ_1	γ_2	γ_3

When isotropic plane of a transversely isotropic foundation is not coincident with the horizontal plane, it is often described in the inclined-axis coordinate system whose character is that three axes do not all point to primary stress directions. Because inclined-axis coordinate system cannot be built only one, we often do not measure elastic constants in it and often get them by converting rules.

Suppose a_{mn} is independent parameter in the coordinate system o-xyz and a'_{ij} is in the new coordinate system $o-x'y'z'$ whose value depends on a_{mn}, the conversion equation is

$$a'_{ij}=\sum_{m=1}^{6}\sum_{n=1}^{6}a_{mn}q_{mi}q_{nj} \qquad (i, j=1, 2, 3, 4, 5, 6) \qquad (2)$$

where q_{mi} and q_{nj} are conversion symbols which can be got according to Table 1, α_i, β_i and γ_i are the cosine of angles between relative axes in the two coordinate systems which are shown in Table 2.

The above discussions are suitable for all kinds of anisotropic body

For computing conveniently, building principles of inclined-anxis coordinate system are shown as follows:

(1) The origins of the orth-axis and inclined-axis coordinate systems are the same point .

(2) Plane $x'oy'$ is horizontal and Axis y' coincides with the strike of isotropic plane; Axis x' meets the line acting by loads at right angle.

(3) Plane xoy and Axis x are coincident with isotropic plane and its strike respectively.

(4) Suppose that θ is the obliquity of isotropic plane and α is the angle made by the strike of isotropic plane with the line acting by loads.

According to the above principle, the inclined-axis coordinate system is built like Figure 2. From Equ. (2), the flexibility matrix of transversely isotropic foundation in the inclined-axis coordinate system is

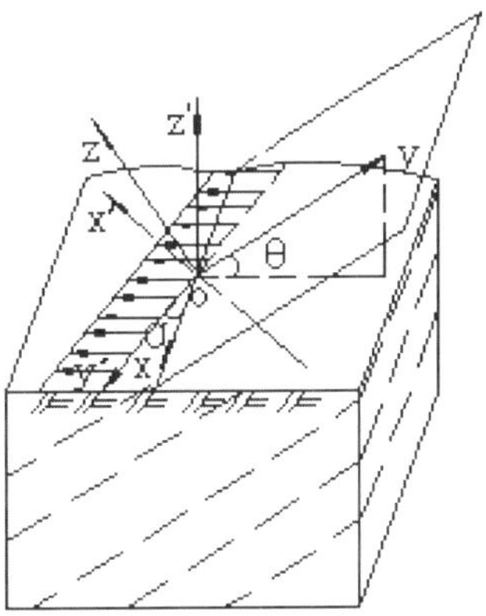

Figure 2. A transversely isotropic whose isotropic plane is inclined

$$A'=Q^T AQ \qquad (3)$$

where A is flexibility matrix in a orth-axis coordinate system and Q is conversion matrix of elastic parameters (Table 1). After computing, from Table 2, we can get

$$\left.\begin{array}{lll}\alpha_1=-\sin\alpha; & \alpha_2=\cos\alpha; & \alpha_3=0\\ \beta_1=-\cos\theta\cos\alpha; & \beta_2=-\cos\theta\sin\alpha; & \beta_3=\sin\theta\\ \gamma_1=\sin\theta\cos\alpha; & \gamma_2=\sin\theta\sin\alpha; & \gamma_3=\cos\theta\end{array}\right\} \qquad (4)$$

Stipulate that α which is got by turning Axis y' to Axis x counterclockwise is positive, and θ is positive if it is from horizontal plane to isotropic plane. Insert Equ. (4) into Table 1, then got Matrix Q as follows

$$Q=\begin{bmatrix} \sin^2\alpha & \cos^2\alpha & 0 & 0 & 0 & -\sin 2\alpha \\ \cos^2\theta\cos^2\alpha & \cos^2\theta\sin^2\alpha & \sin^2\theta & -\sin 2\theta\sin\alpha & -\sin 2\theta\cos\alpha & \sin 2\alpha\cos^2\theta \\ \sin^2\theta\cos^2\alpha & \sin^2\theta\sin^2\alpha & \cos^2\theta & \sin 2\theta\sin\alpha & \sin 2\theta\cos\alpha & \sin^2\theta\sin 2\alpha \\ -\frac{1}{2}\sin 2\theta\cos^2\alpha & -\frac{1}{2}\sin 2\theta\sin^2\alpha & \frac{1}{2}\sin 2\theta & -\cos 2\theta\sin\alpha & -\cos 2\theta\cos\alpha & -\frac{1}{2}\sin 2\theta\sin 2\alpha \\ -\frac{1}{2}\sin\theta\sin 2\alpha & \frac{1}{2}\sin\theta\sin 2\alpha & 0 & \cos\theta\cos\alpha & -\cos\theta\sin\alpha & \sin\theta\cos 2\alpha \\ \frac{1}{2}\cos\theta\sin 2\alpha & -\frac{1}{2}\cos\theta\sin 2\alpha & 0 & \sin\theta\cos\alpha & -\sin\theta\sin\alpha & -\cos\theta\cos 2\alpha \end{bmatrix} \quad (5)$$

Therefore, constitutive equation of a transversely isotropic foundation whose isotropic plane is inclined shown in figure 2 is

$$\begin{Bmatrix} \varepsilon_x' \\ \varepsilon_y' \\ \varepsilon_z' \\ \gamma_{xy}' \\ \gamma_{yz}' \\ \gamma_{zx}' \end{Bmatrix} = \begin{bmatrix} a_{11}' & a_{12}' & a_{13}' & a_{14}' & a_{15}' & a_{16}' \\ a_{21}' & a_{22}' & a_{23}' & a_{24}' & a_{25}' & a_{26}' \\ a_{31}' & a_{32}' & a_{33}' & a_{34}' & a_{35}' & a_{36}' \\ a_{41}' & a_{42}' & a_{43}' & a_{44}' & a_{45}' & a_{46}' \\ a_{51}' & a_{52}' & a_{53}' & a_{54}' & a_{55}' & a_{56}' \\ a_{61}' & a_{62}' & a_{63}' & a_{64}' & a_{65}' & a_{66}' \end{bmatrix} \begin{Bmatrix} \sigma_x' \\ \sigma_y' \\ \sigma_z' \\ \tau_{xy}' \\ \tau_{yz}' \\ \tau_{zx}' \end{Bmatrix} \quad (6)$$

Let $a_{11}=\frac{1}{E_1}$, $a_{12}=-\frac{\mu_1}{E_1}$, $a_{13}=-\frac{\mu_2}{E_2}$, $a_{33}=\frac{1}{E_2}$ and $a_{55}=\frac{1}{G_2}$, then from Equ. (3) and (5), a_{ij}' in Equ. (6) is

$$a_{11}'=a_{11}(\sin^4\alpha+\frac{1}{2}\sin^2 2\theta\cos^4\alpha+\cos^4\theta\cos^4\alpha)+\frac{1}{2}a_{12}(\cos^2\theta\sin^2 2\alpha-\sin^2 2\theta\cos^4\alpha)$$
$$+a_{33}\sin^4\theta\cos^4\alpha+a_{13}(\frac{1}{4}\sin^2\theta\sin^2 2\alpha+\sin^2\theta\cos^2\alpha$$
$$-\sin^4\theta\cos^4\alpha+\frac{1}{4}\sin^2 2\theta\cos^4\alpha)+\frac{1}{4}a_{55}\sin^2\alpha$$

$$a_{21}'=\frac{1}{4}a_{11}\sin^2 2\alpha(1+\frac{1}{2}\sin^2 2\theta+\cos^4\theta)+a_{12}(\cos^2\theta\cos^2\alpha-\frac{1}{8}\sin^2 2\theta\sin^2 2\alpha+\cos^2\theta\sin^4\alpha)$$
$$+\frac{1}{4}(a_{33}\sin^4\theta-a_{55})\sin^2 2\alpha+a_{13}\sin^2\theta(\cos^4\alpha+\frac{1}{2}\cos^2\theta\sin^2 2\alpha+\sin^4\alpha)$$

$$a_{31}'=\frac{1}{4}(a_{33}-2a_{11})\cos^2\alpha\sin^2 2\theta+a_{12}(\sin^2\theta\sin^2\alpha+\frac{1}{2}\sin^2 2\theta\cos^2\alpha)$$
$$+a_{13}(\cos^2\alpha-\frac{1}{2}\sin 2\theta\cos^2\alpha+\cos^2\theta\sin^2\alpha)$$

$$a_{41}'=(a_{33}-a_{11})\cos^2\alpha\sin^2\theta\sin\alpha\sin 2\theta+a_{13}(\cos 2\theta\cos^2\alpha+\sin^2\alpha)\sin 2\theta\sin\alpha$$

$$a_{51}'=(a_{33}-a_{11})\cos^3\alpha\sin^2\theta\sin 2\theta+(a_{13}-a_{12})(\sin^2\alpha+\cos^2\alpha\cos 2\theta)\cos\alpha\sin 2\theta$$

$$a'_{61} = a_{11}\sin 2\alpha(\cos^4\theta\cos^2\alpha + \frac{1}{2}\cos^2\alpha\sin^2 2\theta - \sin^2\alpha) - a_{55}\sin 4\alpha$$

$$+ a_{12}(\sin^2\alpha\cos^2\theta - \frac{1}{2}\cos^2\alpha\sin^2 2\theta - \cos^2\alpha\cos^2\theta)\sin 2\alpha$$

$$+ a_{13}(\cos^2\alpha\cos 2\theta + \sin^2\alpha)\sin 2\alpha\sin^2\theta + a_{33}\cos^2\alpha\sin^4\theta\sin 2\alpha$$

$$a'_{22} = a_{11}\left[(\cos^4\theta + \frac{1}{2}\sin^2 2\theta)\sin^4\alpha + \cos^4\alpha\right] + \frac{1}{2}a_{12}(\sin^2 2\alpha\cos^2\theta - \sin^4\alpha\sin^2 2\theta)$$

$$+ \frac{1}{2}a_{13}(\sin 2\theta\sin^4\alpha + \sin^2 2\alpha\sin^2\theta + a_{33}\sin^4\alpha\sin^4\theta)$$

$$a'_{32} = \frac{1}{4}(a_{33} - a_{11})\sin^2 2\theta\sin^2\alpha + a_{12}(\cos^2\alpha\sin^2\theta + \frac{1}{2}\sin^2\alpha\sin^2 2\theta)$$

$$+ a_{13}(\cos^4\theta\sin^2\alpha + \sin^4\theta\sin^2\alpha + \cos^2\theta\cos^2\alpha)$$

$$a'_{42} = (a_{33} - a_{11})\sin 2\theta\sin^2\theta\sin^3\alpha + (a_{12} - a_{13})(2\sin^2\theta\sin^3\alpha - \sin\alpha)\sin 2\theta$$

$$a'_{52} = (a_{33} - a_{11})(\sin 2\theta\sin^2\theta\sin^2\alpha\cos\alpha + (a_{13} - a_{12})(\cos^2\alpha + \sin^2\alpha\cos 2\theta)\cos\alpha\sin 2\theta$$

$$a'_{62} = -a_{11}(\sin^4\theta\sin^2\alpha + \cos 2\alpha)\sin 2\alpha + a_{12}(\cos 2\alpha - \sin^2\alpha\cos^2\theta)\sin 2\alpha\cos^2\theta$$

$$+ a_{33}\sin^4\theta\sin 2\alpha\sin^2\alpha + a_{13}\sin^2\theta\sin 2\alpha(\sin^2\alpha\cos 2\theta + \cos^2\alpha) + \frac{1}{4}a_{55}\sin 4\alpha$$

$$a'_{33} = a_{11}\sin^2\theta(1 + \cos^2\theta) + \frac{1}{2}(a_{13} - a_{12})\sin^2 2\theta + a_{33}\cos^4\theta$$

$$a'_{43} = (a_{33} - a_{11})\sin 2\theta\cos^2\theta\sin\alpha + \frac{1}{2}(a_{12} - a_{13})\sin 4\theta\sin\alpha$$

$$a'_{53} = (a_{33} - a_{11})\sin 2\theta\cos^2\theta\cos\alpha + \frac{1}{2}(a_{12} - a_{13})\sin 4\theta\cos\alpha$$

$$a'_{63} = \frac{1}{4}(a_{33} - a_{11})\sin^2 2\theta\sin 2\alpha + (a_{12} - a_{13})\cos 2\theta\sin^2\theta\sin 2\alpha$$

$$a'_{44} = a_{11}\sin^2\alpha(1 + \cos^2 2\theta) - 2a_{12}\cos^2 2\theta\sin^2\alpha + (a_{33} - 2a_{13})\sin^2 2\theta\sin^2\alpha + a_{55}\cos^2\alpha$$

$$a'_{54} = \frac{1}{2}a_{11}\sin 2\alpha(1 + \cos^2 2\theta) - a_{12}\cos^2 2\theta\sin 2\alpha + \frac{1}{2}(a_{33} - 2a_{13})\sin^2 2\theta\sin 2\alpha + \frac{1}{2}a_{55}\sin 2\alpha$$

$$a'_{64} = (2a_{12} - a_{11} + a_{33} - 2a_{13})\sin 2\alpha\sin\alpha\sin 2\theta\sin^2\theta$$

$$a'_{55} = a_{11}\cos^2\alpha(1 + \cos^2 2\theta) - 2a_{12}\cos^2 2\theta\cos^2\alpha + (a_{33} - 2a_{13})\sin^2 2\theta\cos^2\alpha + a_{55}\sin^2\alpha$$

$$a'_{65} = (2a_{12} - a_{11} - 2a_{13} + a_{33})\cos\alpha\sin 2\alpha\sin 2\theta\sin^2\theta$$

$$a'_{66} = a_{11}(2 - \sin^4\theta)\sin^2 2\alpha - 2a_{12}(1 + \sin^2\theta)\sin^2 2\alpha\cos^2\theta$$

$$+ a_{13}(\cos 2\theta - 1)\sin^2\theta\sin^2 + a_{33}\sin^2 2\alpha\sin^4\theta + a_{55}\cos^2 2\alpha$$

$$a'_{ij} = a'_{ji}$$

3 Analysis of the earth stress in a transversely isotropic foundation

For computing, the transversely isotropic foundation is postulated that the directions of primary earth stresses are horizontal and vertical, and vertical stress equals with gravity of the above rock and soil mass. Because the foundation extends very long in horizontal direction, its horizontal strain equals zero. As shown in figure 1 and figure 3, induced that

$$\tau_{xy} = \tau_{yz} = \tau_{zx} = 0, \quad \varepsilon_x = \varepsilon_y = 0 \tag{7}$$

$$\tau'_{xy} = \tau'_{yz} = \tau'_{zx} = 0, \quad \varepsilon'_x = \varepsilon'_y = 0 \tag{8}$$

If the isotropic plane is horizontal, from Equ. (1) and (7), the three primary earth stresses are

$$\left.\begin{aligned} &\sigma_z = \gamma z \\ &\sigma_x = \sigma_y = \frac{E_1 \mu_2}{E_2 (1-\mu_1)} \gamma z \\ &\lambda = \frac{\sigma_x}{\sigma_z} = \frac{\sigma_y}{\sigma_z} = \frac{E_1 \mu_2}{E_2 (1-\mu_1)} \end{aligned}\right\} \tag{9}$$

Let $E_1 = E_2 = E, \mu_1 = \mu_2 = \mu$, the Equ.(9) becomes into $\sigma_x = \sigma_y = \frac{\mu}{1-\mu}$; $\sigma_z = \gamma z$, which coincides with the analytic solution of this problem in isotropic theory.

If the isotropic plane is inclined like Figure 3, then induced $\alpha = 0$. According to Equ.(6) and (8), the three primary earth stresses are

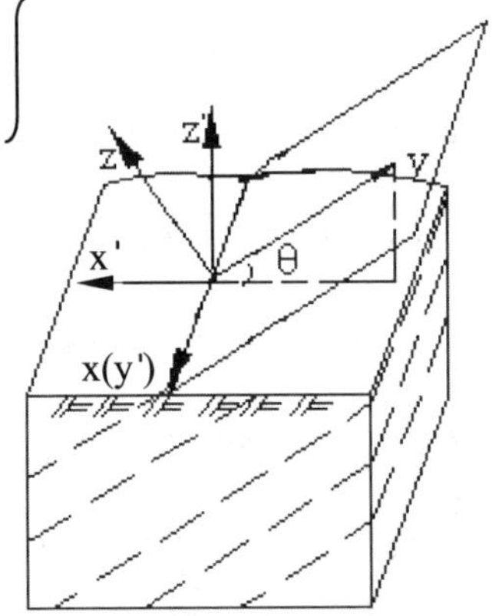

Figure 3. A Model of Computing Earth Stress in A Transversely Isotropic whose Isotropic Plane is Inclined

$$\left.\begin{aligned} &\sigma'_z = \frac{a'_{21} a'_{32} - a'_{22} a'_{31}}{a'_{11} a'_{22} - (a'_{21})^2} \gamma z' \\ &\sigma'_y = \frac{a'_{21} a'_{31} - a'_{11} a'_{32}}{a'_{11} a'_{22} - (a'_{21})^2} \gamma z' \\ &\sigma'_z = \gamma z' \end{aligned}\right\} \tag{10}$$

where a'_{ij} is the same as before, which can be got from Equ.(6). a'_{ij} is

$a'_{11} = a_{11}(1-\sin^4\theta) + \frac{1}{2}(a_{13} - a_{12})\sin^2 2\theta + a_{33}\sin^4\theta$

$a'_{21} = a_{12}\cos^2\theta + a_{13}\sin^2\theta$

$a'_{22} = a_{11}$

$$a'_{31}=\frac{1}{4}(a_{33}-a_{11}+2a_{12})\sin^2 2\theta+a_{13}(\sin^4\theta+\cos^4\theta)$$

$$a'_{32}=a_{12}\sin^2\theta+a_{13}\cos^2\theta$$

$$a'_{51}=2(a_{33}-a_{11})\sin^3\theta\cos\theta+\frac{1}{2}(a_{13}-a_{12})\sin 4\theta$$

$$a'_{52}=(a_{13}-a_{12})\sin 2\theta$$

If let $\theta=0$, it means that the isotropic plane is horizontal, then Equ. (10) becomes into Equ. (9).

Example: In a large scale of geochenical engineering, its foundation is transversely isotropic body and its parameters are measured as follows $E_1=26GPa$, $E_2=8.16GPa$, $\mu_1=0.202$, $\mu_2=0.32$, $G_2=0.466GPa$ and $\gamma=27kN/m^3$. The question is to compute earth stress of the place which is about 500m deep. (suppose θ epuals $\frac{\pi}{18}$, $\frac{\pi}{9}$, $\frac{\pi}{6}$ $\frac{\pi}{4}$, $\frac{\pi}{3}$ and $\frac{\pi}{2}$)

According to Equ. (10), the computing results are as follows :

$$\sigma_z'=13.5MPa$$

θ	0	$\frac{\pi}{18}$	$\frac{\pi}{9}$	$\frac{\pi}{6}$	$\frac{\pi}{4}$	$\frac{\pi}{3}$	$\frac{\pi}{2}$
σ_x'	17.2490	16.9478	15.7278	12.8808	7.7206	6.8044	7.7074
σ_y'	17.2490	17.2732	17.1548	16.2401	12.9616	11.0334	10.5855
λ	1.2777	1.2674	1.2179	1.0786	0.7660	0.6607	0.6775
k	1.0000	1.0190	1.0907	1.2608	1.6788	1.6215	1.3734

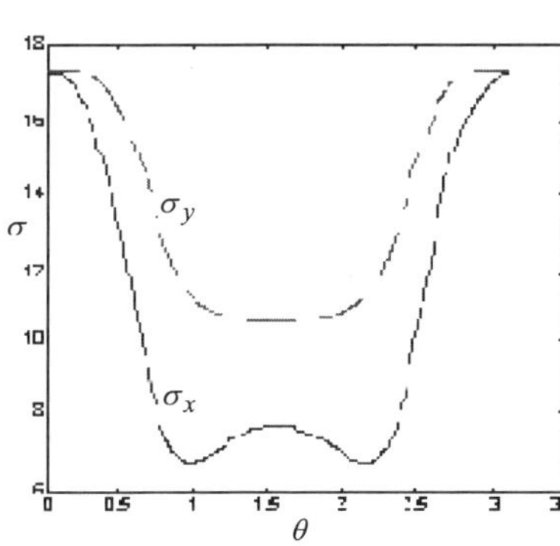

Figure 4. Comparison between σ_x' and σ_y'

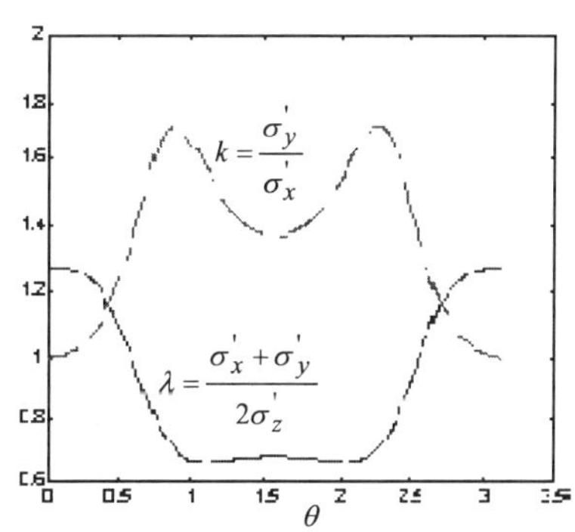

Figure 5. Earth stress anisotropy in the medium

From Figure 4 and Figure5, the four curves are symmetrical and their symmetrical axis is $\theta = \frac{\pi}{2}$, which correspond with engineering. When θ changes from 0 to $\frac{\pi}{2}$, the average horizontal earth stress is larger than vertical stress.

4 conclusion

(1) The incline of isotropic plane is mostly caused by tectonic movement, so it is proved that the structure of rock mass influence the transfer of earth stress in geotechnical medium.

(2) According to the computing results of earth stress, those important actual measurements of earth stress can be explained that horizontal stress is larger than vertical stress and horizontal stress is anisotropic in theory, which can be used to predict earth stress.

(3) Because the model of transversely isotropic foundation considers anisotropism, the model is more suitable to the anisotropic geomechnical mass .

References

1. Chen W T. On Some Problems in Transversely Isotropic Elastic Matetials. Journal of Applied Mechanics, 1966,33, 347~355
2. Chen W T. Axisymmetric Stress Field Around Spherical Inclusions and Cavities in a Transversely Isotropic Material. Journal of Applied Mechanics, 1968,35(4),
3. Hooper, J.A. Elastic Analysis of a Circular Raft in Adhesive Contact with a Transversely Isotropic Medium. J.Geotechnique, Vol.25 No.4, 1975
4. Birman V. On Three Dimensional State of Thermal Stress in a Transversely Isotropic Plate with a Circular Hole. International Journal of Science in Engineering, 1995,33(1),
5. Borodachev A N. Parametric Class of Solutions of Thermoelasticity Equations for Transversely Isotropic Bodies. Soviet Applied Mechanics, 1998,24(7),
6. Shifan Qiao. The analysis on stress and displacement of transversely isotropic foundation acted by horizontal loads. Jiaozuo Institute of Technology, 2000,9
7. Zhilin Dun, Ganbin Liu, Shifan Qiao. The analysis on stress and subside of transversely isotropic foundation for axis symmetry problem. Joural of Jiaozuo Institute of Technology, 2002, 21(1).

PHENOMENOLOGICAL STUDY ON VISCOSITY IN THERMAL-ACTIVATED RELAXATION FOR SATURATED ROCK

XIAOYAN LIU, DAOYING XI, CHENGYUAN ZHANG
Dept. of Earth & Space Science, University of Sci.&Tech. of China
HeFei, 230026, China
E-mail: liuxy@mail.ustc.edu.cn

A phenomenological model of macroscopic viscosity is gained for nonlinear relaxation law of fluid saturated rock, which presumes that solid frame and pore fluid are regarded as an entirety. Complex viscosity is computed from inner-friction experiment data. The contribution of pore fluid to response of rock properties to frequencies and temperatures is analyzed.

1 Introduction

Rock is porous visco-elastic media. Most rocks are saturated or partly saturated with different fluids under different depth, temperature and pressure conditions. Laboratory data show that pore fluid has important effect on rock attenuation. Theoretical and experimental researches have been done for finding the interaction between rock frames and pore fluid. Most commonly used concepts are the inertial coupling of Biot's and the local fluid or squirt flow mechanism. Best et al.[1] suggests that both mechanisms operated in a porous rock at the same time, with global flow dominantly in low viscosity fluid saturated sandstone, and local flow in low permeability sandstone. It seems that effect of pore fluid on rock attenuation is non-monotonic.

Attenuation peak in saturated rock has the property of typical thermal activated relaxation. Moreover, fluid viscosity generally depends on temperature; hence, seismic attenuation and dispersion are strongly affected. In seismic prospecting, different frequency bands are used to investigate pore fluid information. Sometimes, there exists a relation between frequency- and temperature- dependence on rock's seismic properties. Therefore, it is important to study the characteristics of viscosity during thermal-activated relaxation process and to study viscosity variation of the rock entirety caused by the conditions of pore fluid (such as frequency or temperature) changing.

The purpose of this paper is to develop a phenomenological model of entirety rock for macroscopic viscosity .

2 Phenomenological model for viscosity

Rheological models, most common of which is three-parameter model (or called Zener model [2]), are often adapted to describe phenomenological dissipation in rock. In order to study the elastic and visco-elastic behaviors of saturated rock, Burgers model consisting of a Viogt body and a Maxwell body is selected.

In terms of the total strain ε and total stress σ, we obtain

$$\ddot{\sigma} + \left(\frac{1}{\tau_1} + \frac{M_1}{M_2}\frac{1}{\tau_2} + \frac{1}{\tau_2}\right)\dot{\sigma} + \frac{1}{\tau_1\tau_2}\sigma = M_1\ddot{\varepsilon} + \frac{M_1}{\tau_2}\dot{\varepsilon} \quad . \tag{1}$$

If $\sigma = \sigma_0 \exp(i\omega t)$, $\varepsilon = \varepsilon_0 \exp[i(\omega t - \varphi)]$, then the complex viscosity of entire rock system is

$$\tilde{\eta} = \eta' - i\eta'' = \frac{\sigma}{\dot{\varepsilon}} = \frac{\tilde{M}}{i\omega} = \frac{M''}{\omega} - i\frac{M'}{\omega}$$

$$= \omega^{-1}\left[\frac{1}{M_1\omega\tau_1} + \frac{1}{M_2}\frac{\omega\tau_2}{1+(\omega\tau_2)^2}\right] - i\omega^{-1}\left[\frac{1}{M_1} + \frac{1}{M_2}\frac{1}{1+(\omega\tau_2)^2}\right] . \tag{2}$$

Attenuation has two relaxation times corresponding to anelasitc (τ_a) and visco-elastic (τ_v) behavior in this model:

$$Q^{-1} = \frac{M''}{M'} \approx \frac{M_1}{M_2}\frac{\omega\tau_a}{1+(\omega\tau_a)^2} + \frac{1}{\omega\tau_v} \quad .$$

Supposing that relaxation behaviors of the materials satisfy Arrhenius equation:

$$\tau = \tau_0 \exp(H/kT) \quad .$$

The formula (2) can be modified and the phenomenological model for viscosity containing frequency and temperature effect is obtained:

$$\eta' = \omega^{-1}\left\{\frac{1}{M_1\omega\tau_{10}\exp(H_1/kT)} + \frac{1}{M_2}\frac{\omega\tau_{20}\exp(H_2/kT)}{1+[\omega\tau_{20}\exp(H_2/kT)]^2}\right\}$$

$$\eta'' = \omega^{-1}\left\{\frac{1}{M_1} + \frac{1}{M_2}\frac{1}{1+[\omega\tau_{20}\exp(H_2/kT)]^2}\right\} \tag{3}$$

It is shown in equation (3) that complex viscosity nonlinearly dependents on frequency and temperature. The real and the imaginary parts are mostly decreasing with frequency and temperature. Experiment on rocks was conducted to testify this model.

3 Experimental method and result

The experimental method consists of exerting small strain oscillatory torsion (small sinusoidal torque) deformation on bars samples (5 mm * 2 mm * 50 mm). Dry and saturated samples include Zigong sandstone, Dali marble and Bole granite. Four saturants: water, crude oil, pure pump-oil and pump-oil (with 0.35% asphalt) are used. Liquid viscosity at about 20℃of the first three are 0.01, 0.02 and $0.7*10^{-3}$ Pa S.

The results for dry and pump-oil saturated sandstone with porosity 12% are shown to state the viscosity response to frequency and temperature. Figure 1 shows the complex viscosity versus temperature (a) and frequency (b).

The values of η' and η'' of saturated rock decrease with frequency. At lower frequency, small change in frequency will cause obvious variation in viscosity.

However, as frequency rising, viscosity is less dependent on frequency and trends to be stable at higher frequency. In most common temperature domain, same conclusion can be drawn with a peak in η' at lower temperature. Obvious peak appears at lower frequency. It can be found from Figure 1 that viscosity in dry rock is a little more than that in pump-oil saturated rock at the same conditions. When fluid entering pores, differential movement between pore fluid and frame of rock leads to inertial coupling. The entire body may get changes in some physical and dynamic properties.

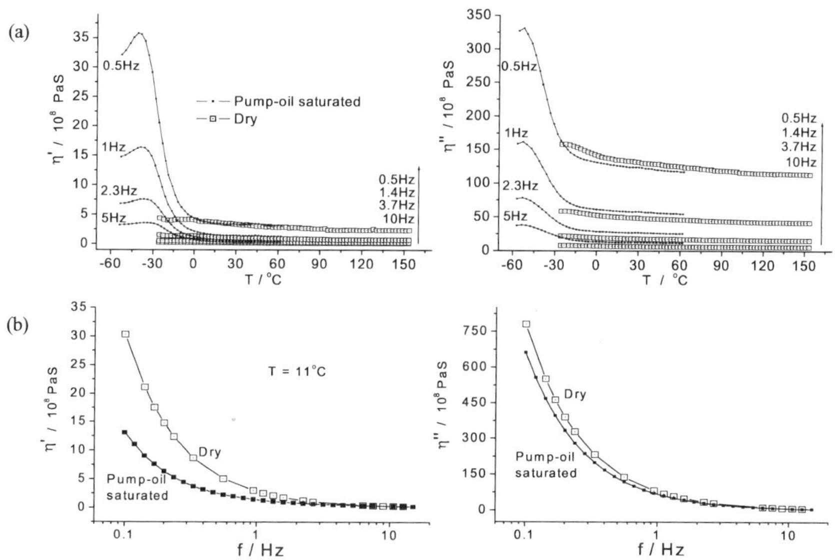

Figure 1. Complex viscosity versus temperature (a) and frequency (b)

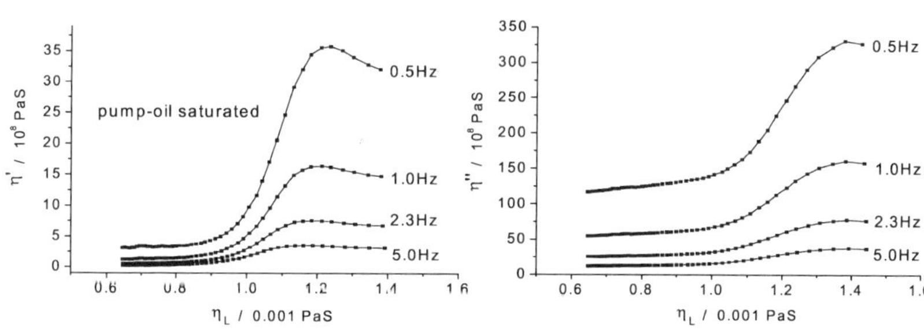

Figure 2. Complex viscosity in sandstone entirety versus pore fluid's viscosity

Figure 2 shows the complex viscosity in rock entirety (pump-oil saturated) as a function of pore fluid's viscosity. Liquid viscosity is affected by temperature. Change on small or large liquid viscosity have a slight effect on rock entirety viscosity, while it is mostly sensitive to the medial liquid viscosity especially at low frequency.

The calculation with phenomenological model on complex viscosity conforms to dynamic viscosity property of rocks. We can further analyze individual contribution of fluid to the visco-elasticity of rock through parameter inversion.

4 Discussion and conclusions

In this paper a phenomenological model of entirety rock for macroscopic viscosity is developed and Complex viscosity is computed from our inner-friction experiment data on dry and saturated sandstone with temperature changing from –60℃to 20℃ at low frequency (0.01-15Hz). The contribution of pore fluid to response of rock properties to frequencies and temperatures is analyzed.

The phenomenological model for viscosity has single value rather than a wide distribution of relaxation time. The distribution of relaxation time can be involved in future to include two different mechanisms in viscosity-dependent attenuation.

The physical mechanisms of visco-elasticity and elasticity for rock are related to many geophysical problems. Another potentially useful application of wave attenuation is the study of rock long-term creep, because both attenuation and rheology involve viscous effect. There exist some relationship between these two phenomena. Daoying Xi et al.[3] studied the time-temperature equivalence for rocks on the basis of the mechanical relationship between attenuation and creep. However, if such a frequency dependence is extrapolated for much lower frequencies (i.e. longer time scales), one would predict a significantly lager viscosity for long-term deformation than for shorter-term deformation. Therefore, understanding the variation law of macroscopic dynamic viscosity of saturated rock help us to consider the limit of extrapolation of non-elastic properties obtained from Q^{-1} to long-term phenomena, and is necessary to develop a physically sound model to link attenuation with creep in rock.

References

1. Best A.I, Geophysics, 60, 1386, (1995).
2. Nowick A.S. and Berry B.S. (Academic Press, New York, 1972).
3. Daoying Xi et al Computing techniques for geophysical and geochemical exploration, 22, 127, (2000).

The authors thank the Chinese National Natural Science Fund for the support under contract No. 40174050.

DILATANCY THEORY OF IMPACT GROUND PRESSURE

LIYOU PAN, YONGKUI SHI, XINGWEI LI
Shandong University of Science & Technology, Tai'an 271019, China

JIANXIN LIU
Northeaster University, Shenyang, 110004, China

Dilatancy character of impact coal mass is studied. The descriptive manner of destructive process of bumping coal mass is raised; the prognosis pattern of impact ground pressure is introduced; mechanical mechanism of coal-body stress and dilatancy regularity is established.

1 Introduction

Coal dilatancy means nonlinear growth of volume on bias stress action. Because coal is inhomogeneous medium, under action of bias stress, its deformation is disharmony. When bias stress comes at some value, volume of coal turns from compression into inflation and coal mass generates crack. This process of mechanics is termed as dilatancy. In geology fields, S'cholz etc. made impact-diffusion model as the physical base of earthquake prediction. Author mainly deals with impact dilatancy theory of coal.

2 Dilatancy Property under Monopodium Stress State

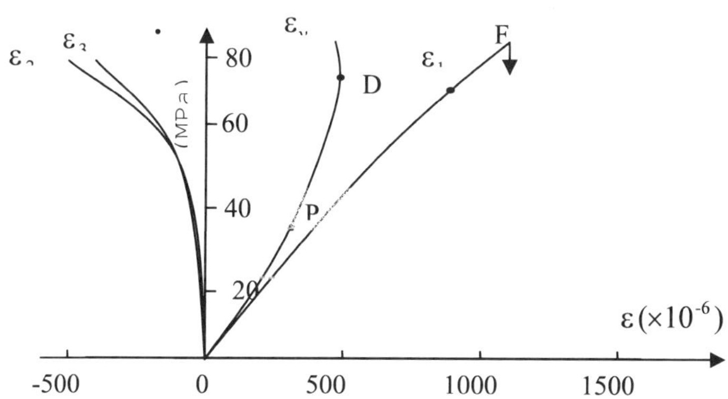

Figure 1. Marble's $\sigma_1 - \varepsilon_v$ dependence under monopodium stress

As shown in Figure1, curve is marble's $\sigma_1 - \varepsilon_v$ dependence under monopodium stress. The curve of strain of volume (ε_v) shows transition from compaction to inflation, this case notes that coal turns out dilatancy. Both lateral strains nearly equal shows that the medium is homogeneous and the two allotment of dilatancy of lateral are nearly equal. The curve of $\sigma_1 - \varepsilon_v$ shows two specific points: one is terminal point of linear portion (P); another is start point (D) of dilatancy, namely it is the start point that strain of volume transforms from uniaxial compression into inflation.

3 Mechanics Mechanism of Dilatancy of Impact Coal Mass

According to stress state of coal mass, the coal mass is divided into three realms: elastic area, plastic area and seam gap area

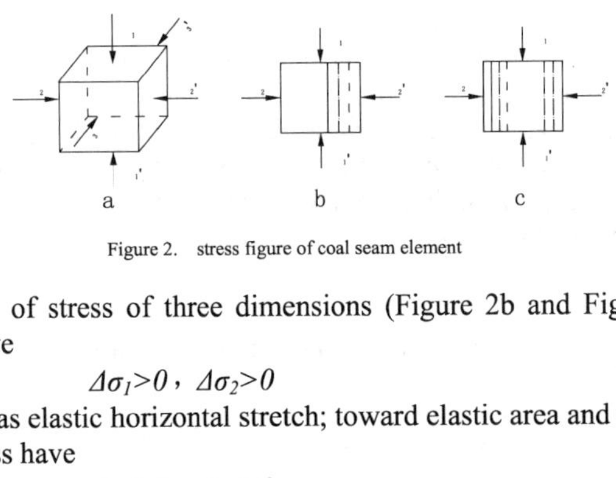

Figure 2. stress figure of coal seam element

Toward state of stress of three dimensions (Figure 2b and Figure 2c), these states of stress have

$$\Delta\sigma_1>0, \ \Delta\sigma_2>0$$

It represents as elastic horizontal stretch; toward elastic area and seam gap area, their states of stress have

$$\Delta\sigma_1>0, \ \Delta\sigma_2>0,$$

It represents as the unit has horizontal displacement and vertical displacement. In three states of stress, the change regularity of the value of stress difference has some dissimilarity, generally we have

$$\Delta\sigma_1(c)>\Delta\sigma_1(b)>\Delta\sigma_1(a), \ \Delta\sigma_2(b)>\Delta\sigma_2(a)>\Delta\sigma_2(c).$$

We can conclude that different value of stress is decided by the character of distribution of bearing pressure.

4 The description of process of coal fracture

According to mechanics performance curve that was acquired by routine loading experiment using the sample of coal (rock) of 2337 face from Kailuan Zhaogezhuang Mine, to process of fracture before coal (rock) reaches compression breaking strength (R_c) there are three parameter of feature stress to qualify the inching fracture process. Coal (rock)'s monopodium loading's stress, axial strain, transverse strain, volume strain are showed as figure3.

(1) Dilatancy stress

Volume deformation (ε_v) increases with increment of σ_1 linear transformation range (σ_v). Impact is led by fine crack's fissure and extension in rock. Before reaching the point of F, face of fine crack closed then rock represents elastic deformation, and volume of rock behaves as linear constringe.

(2) Critical pressure F_0

When critical pressure runs to F_0, induce equation $\partial(\varepsilon_\theta/\sigma_1)=0$, F_0 is called as critical stress. F_0 has explicit physical sense: after $\sigma < F_0$, expansion of volume begins to increase sharply (At this time lateral deformation of unloading arises apparent residue value), as the curve of $\mu-\sigma$ shows: when $\sigma <$F, the curve of $\sigma-\mu$ represents linearity;

When $\sigma > F_0$, Poisson ratio acutely raises with σ, rapid development of ε_v generates at the same time nearly as stress led to F_0. The strain of volume has

$$\Delta\varepsilon_v = \Delta\varepsilon_v^T + \Delta\varepsilon_v^f$$

$\Delta\varepsilon_v^T$ denotes volume compression of rock medium; $\Delta\varepsilon_v^f$ denotes volume inflation by the expansion of fine crack; ε_v^f is principal occasion of test piece's transverse inflation.

When $\sigma > F_0$, $\Delta\varepsilon_v^f > \Delta\varepsilon_v^T$.

As described above, noting that the inflation generating by fine crack's expansion takes the predominant status.

(3) True strength σ_F

When stress leads to σ_F, $\varepsilon_v = 0$. Here σ_F is the strength that test piece's volume begins to excess its volume before loading then destroy and inflate; also the critical strength that no creep of rock begins to boom. Showing that under low wall-compression there is the brittle coal (rock)'s fracture viz. splay and spread of fine crack.

From that we can conclude the forecast pattern of impact ground pressure as: When $\sigma > F_0$, coal takes into nonlinear state and appears fine crack and acoustic emission; when $\sigma > F_0$, acoustic emission gradates from weak to strong, AE of time F_σ can be made as critical point of forecasting.

5 Conclusion

1. impact coal's fractural process can be described with three specific stress quantity.
2. When $\sigma > F_0$, coal takes into nonlinear state and generates fine crack.
3. Use dilatancy regularity of volume to build up impact ground pressure forecasting pattern.

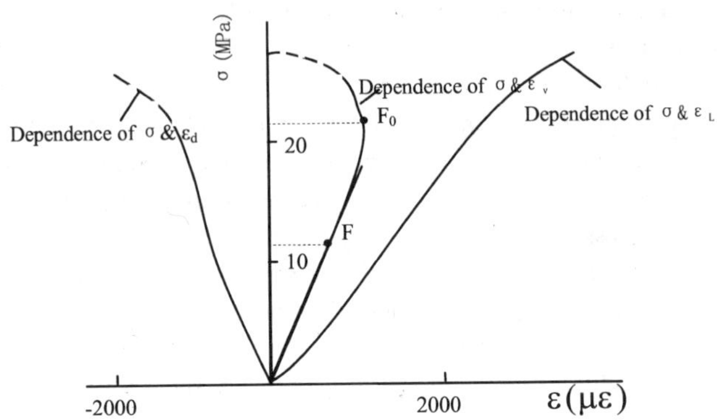

Figure 3. Dependence of stress & strain

References

1. Vardoulakis I. Rock bursting as a surface instability phenomenon. Int. J. Rock Mech. Sci. & Geomech. Abstr.,1984,21(3): 137-144
2. Germanovich L N, Fyskin A V, Tsyrulnikov N M, et al. A model of the deformation and fracture of brittle materials with cracks under uniaxial compression. Mechanics of solids, 1993, 28 (1):116-128
3. Myer L R, Kemeny J M. Extensile cracking in porous rock under different compressive stress. Appl. Mech. Rve. , 1992m, 45(8):263-280
4. Baud, Reuschle T, Charlez P. Technical note: an improved wing crack model for the deformation and failure of rock in compression. Int. J. Rock Mech. Sci. & Geomech. Abstr., 1996, 33: 539-542
5. Papamichos E, Vardoulakis I. The coupling effect of surface instabilities and surface parallel Griffith cracks in rock. In: Maury Fourmaintraux ed. Rock at Great Depth. Rotterdam: A. A Balkema, 1989, 481-487
6. Bazant Z P, Lin Feng-bao, Lippmann H. Fracture energy release and size effect in borehole breakout. International Journal for Numerical and Analytical Methods in Geomechanics, 1993, 17:1-11
7. Dyskin A V, Germanovich L N. Model of rockburst caused by cracks growing near free surface. In: Yoang ed. Rockburst and Scismicity in Mines. Rotterdam: A. A. Balkema, 1993, 169-174

INFLUENCE OF HYDRAULIC PRESSURE ON INITIAL DAMAGE AND DAMAGE EVOLUTION OF FRACTURED ROCK MASS

SHAOHE ZHENG HONG SUN XIURUN GE

School of Civil Engineering and Mechanics, Shanghai JiaoTong University, Shanghai, 200030, P.R.China

An elastic damage constitutive equation based on initial flexibility tensor and damage evolution of rock mass considering hydraulic pressure is developed. A concept of additional hydraulic pressure-induced flexibility tensor is introduced to indicate the influence of hydraulic pressure on the mechanical characteristics of fractured rock mass. The results obtained from theoretical analysis show that the influence of hydraulic pressure on initial damage and damage evolution could not be ignored when there is fluid flow in rock mass.

1 Introduction

Damage mechanics is used to analyze mechanical characteristics of fractured rock mass in terms of initial damage and damage evolution, which can quantitatively study effects of cracks. It should be pointed out that subjects of most previous studies (Germanovich, 1988; Zhang, 1998) are dry rock mass. It indicates that effects of hydraulic pressure on damage are not considered. But there often exits fluid flow in many geotechnical engineering. Hydraulic pressure due to fluid flow increases stress intensity factor of cracks and then modify damage characteristics of fractured rock mass, which is called hydraulic pressure-induced damage here. The hydraulic pressure-induced damage will further worsen mechanical characteristics. So the influence of hydraulic pressure on damage behavior is deserved to study.

According to Betti energy reciprocal theorem and damage mechanics, an elastic damage constitutive model based on initial flexibility tensor and damage evolution equation of rock mass considering hydraulic pressure are reported in the following.

2 Initial equivalent flexibility tensor considering hydraulic pressure

Macroscopic mechanical characteristic is determined by flexibility of fractured rock mass, which can be obtained from derivation of strain energy density.

Based on Betti theorem, initial equivalent strain energy density u_e can be expressed in terms of strain energy density of undamaged rock u_0 and additional strain energy density u_d due to cracks and hydraulic pressure as (Zheng, 2000)

$$u_e = u_o + u_d = \frac{1}{2}\sigma_{ij}(C^o_{ijkl} + C^d_{ijkl} + C^{\mathrm{w}}_{\mathrm{ijkl}})\sigma_{kl} \tag{1}$$

where σ_{ij} is stress tensor, C^{0}_{ijkl} is elastic flexibility tensor of undamaged rock, C^{d}_{ijkl} is additional crack-induced flexibility tensor without considering hydraulic pressure, C^{w}_{ijkl} is used to indicate effects of hydraulic pressure on flexibility tensor, which is called additional hydraulic pressure-induced flexibility tensor here.

To undamaged rock, the elastic flexibility tensor is defined as

$$C^{o}_{ijkl} = \frac{1+v_o}{2E_o}(\delta_{ik}\delta_{jl} + \delta_{il}\delta_{jk}) - \frac{v_o}{E_o}\delta_{ij}\delta_{kl} \tag{2}$$

where E_0 、 v_0 are Youngs' modulus and Poisson ratio of undamaged rock respectively, δ_{ij} is Kronecker's delta.

To fractured rock mass, if there are N groups principal cracks , whose average volume density is $\rho_v^{(m)}$, statistic specific length is a and hydraulic pressure is p, additional strain energy density u_d can be expressed as (Zheng, 2000)

$$u_d = \sum_{m=1}^{N} u_d^{'(m)} = \frac{1}{E_0}\sum_{m=1}^{N}\left\{a^{(m)^3}\rho_v^{(m)}\left(G_1^{(m)}\sigma_{ne}^2 + G_2^{(m)}\tau_{ne}^2\right)\right\} \tag{3}$$

where σ_{ne} and τ_{ne} are effective normal stress and shear stress, G_1 and G_2 are shape factors.

However, hydraulic pressure p has no effect on parts of contact area. A coefficient β is introduced to express the ratio of connected area to whole fractures face area. So the actual effect of hydraulic pressure p is defined as βp .

Subsequently taking account of contact area of fractures, σ_{ne} and τ_{ne} are written in a tensor form as

$$\begin{cases}(\sigma_{ne})_i = (1-C_v)\sigma_{jk}n_jn_kn_i + \beta p n_i \\ (\tau_e)_i = (1-C_s)\sigma_{jk}n_j(\delta_{ki} - n_in_k)\end{cases} \tag{4}$$

where C_v and C_s are coefficients of transferring normal stress and shear stress.

Hydraulic pressure p can be expressed as

$$p = p\frac{\sigma_{ss}}{\sigma_{ss}} = \sigma_{st}\delta_{ts}\frac{p}{3\overline{\sigma}} = \frac{1}{3}\sigma_{st}\delta_{ts}R \tag{5}$$

where $\overline{\sigma}$ is average stress, $\overline{\sigma} = \sigma_{ii}/3, \sigma_{ii}$ is the first stress invariant, $R = p/\overline{\sigma}$.
Substituting Eqs. (4) and (5) into (3) yields additional strain energy density

$$\begin{aligned} u_d &= \frac{1}{2}\sigma_{kj}C^{d-w}_{kjst}\sigma_{st} \\ &= \frac{1}{E_o}\sum_{m=1}^{N}\Big\{a^{(m)^3}\rho_v^{(m)}\Big\{\sigma_{kj}^{(m)}G_1^{(m)}\Big[(1-C_v^{(m)})^2 n_k^{(m)}n_j^{(m)}n_s^{m}n_t^{(m)} + \frac{1}{3}\beta \mathrm{R}^{(\mathrm{m})}(1-C_v^{(m)})(n_j^{(m)}n_k^{(m)}\delta_{st} + n_s^{(m)}n_s^{(m)}\delta_{kj}) \\ &+ \frac{1}{9}(\beta^{(m)}R^{(m)})^2\delta_{kj}\delta_{st}\Big]\sigma_{st}^{(m)} + \sigma_{kj}^{(m)}G_2^{(m)}(1-C_s^{(m)})^2(\delta_{kt}n_j^{(m)}n_s^{(m)} - n_j^{(m)}n_k^{(m)}n_s^{(m)}n_t^{(m)})\sigma_{st}^{(m)}\Big\}\Big\} \end{aligned} \tag{6}$$

Obtained from derivation of strain energy density, additional crack-induced and hydraulic pressure-induced flexibility tensor are defined as respectively

$$C_{ijkl}^{d}=\frac{1}{E_o}\sum_{m=1}^{N}\Big\{a^{(m)^3}\rho_v^{(m)}\Big[2G_1^{(m)}(1-C_v^{(m)})^2 n_i^{(m)}n_j^{(m)}n_k^{(m)}n_l^{(m)}+\frac{1}{2}G_2(1-C_s^{(m)})^2(\delta_{il}n_j^{(m)}n_k^{(m)}+\delta_{ik}n_j^{(m)}n_l^{(m)}+ \tag{7}$$

$$+\delta_{il}n_i^{(m)}n_k^{(m)}+\delta_{jk}n_i^{(m)}n_l^{(m)}-4n_i^{(m)}n_j^{(m)}n_k^{(m)}n_l^{(m)})\Big]\Big\}$$

$$C_{ijkl}^{w}=\frac{2}{3E}\sum_{m=1}^{N}\Big\{a^{(m)^3}\rho_v^{(m)}\Big[G_1^{(m)}(1-C_v^{(m)})\beta^{(m)}R^{(m)}(n_i^{(m)}n_j^{(m)}\delta_{kl}+n_k^{(m)}n_l^{(m)}\delta_{ij})+{}^1\!/_3\,G_1^{(m)}\delta_{ij}\delta_{kl}(\beta^{(m)}R^{(m)})^2\Big]\Big\} \tag{8}$$

Based on Eq. (2), (7) and (8), initial flexibility tensor considering hydraulic pressure can be defined as

$$C_{ijkl}^{o-d-w}=C_{ijkl}^{o}+C_{ijkl}^{d}+C_{ijkl}^{w} \tag{9}$$

From Eq. (9), it shows that the influence of hydraulic pressure on stress field not only is acted as effective nodal force in coupled analysis for stress and seepage, but also can reduce its corresponding stiffness matrix. Because additional hydraulic pressure-induced flexibility tensor C_{ijkl}^{w} is changed with the value of hydraulic pressure, mechanical characteristics are affected by flow behavior.

3 Damage evolution of fracture rock mass considering hydraulic pressure

Crack evolution makes Young's modulus decrease and Poisson ratio increase. The dynamic characteristics of rock mass can be described by damage evolution.

After crack evolution, total strain energy density is

$$u=u_e+u_{ad} \tag{10}$$

where u_e is strain energy density before crack evolution, u_{ad} is additional strain energy density due to crack evolution.

Additional elastic strain energy density as a result of wing crack evolution can be written as (Zheng, 2000)

$$u_{ad}=\sum_{m=1}^{N}\left[\frac{4}{E_o}\rho_v^{(m)}a^{(m)^2}\left(\frac{2}{\sqrt{3}}\tau'^{(m)}+\frac{5}{2}\sigma_1^{(m)}L^{(m)}\right)^2\right] \tag{11}$$

where L is the ratio of evolution length and initial length, $L=l/a, l$ is the evolution length under stress, other symbols are mentioned above.

So the additional flexibility tensor C_{ijkl}^{ad} due to crack evolution considering hydraulic pressure is obtained, which is defined as

$$C_{ijkl}^{ad}=\frac{1}{E_o}\sum_{m=1}^{N}\Big\{\rho_v^{(m)}a^{(m)^2}\Big[B_1^{(m)}n_i^{(m)}n_j^{(m)}n_k^{(m)}n_l^{(m)}+B_2^{(m)}(n_i^{(m)}n_l^{(m)}\delta_{kj}+n_j^{(m)}n_l^{(m)}\delta_{ki}$$

$$+n_i^{(m)}n_k^{(m)}\delta_{jl}+n_j^{(m)}n_k^{(m)}\delta_{il})+\frac{B_3^{(m)}}{3}\beta^{(m)}R^{(m)}\delta_{ij}\delta_{kl}\Big]\Big\} \tag{12}$$

where $B_1^{(m)}, B_2^{(m)}, B_3^{(m)}, A_1^{(m)}, A_2^{(m)}, A_3^{(m)}$ and $A^{(m)}$ are shown in references (Ping, Zheng & Bai).

Based on initial damage and damage evolution due to stress and hydraulic pressure, a damage evolution equation of fractured rock mass considering hydraulic pressure is expressed as

$$C_{ijkl}^{o-ad} = C_{ijkl}^{o} + C_{ijkl}^{d} + C_{ijkl}^{w} + C_{ijkl}^{ad} \tag{13}$$

where C_{ijkl}^{o} is elastic flexibility tensor of undamaged rock mass, C_{ijkl}^{d} is the initial crack-induced flexibility tensor, C_{ijk}^{w} is additional hydraulic pressure-induced flexibility tensor, C_{ijkl}^{ad} is additional crack evolution-induced flexibility tensor.

4 Constitutive model of fractured rock mass considering hydraulic pressure

Assume that the deformation of fractured rock mass is brittle elastic, the damage constitutive model of brittle rock mass considering hydraulic pressure based on initial damage and damage evolution is

$$\varepsilon_{ij} = C_{ijkl}^{o-ad} \sigma_{kl} \tag{14}$$

where C_{ijkl}^{o-ad} is determined by Eq. (13)

According to above theoretical analysis, underground fluid flow is exited in many engineering, so effects of hydraulic pressure on mechanical characteristic can not be ignored, especially when value of hydraulic pressure is very high.

References

1. L.N. Germanovich and A.V. Dyskin. A Model of Brittle Failure for Material with Cracks in Uniaxial Loading. *Mechanics of Solid*, 23(2): 111～124 (1988).
2. Y. Ping, S.H. Zheng and S.W. Bai. Research on Fracture-Damage Model of Fractured Rock Mass Considering Hydraulic Pressure. *6th National Conf. of Rock Mechanics and Engineering,* Wuhan (2000).
3. Q.Y. Zhang, W.S. Zhu. Analysis on elastoplastic damage for high jointed slope of the Three Gorges project shiplock during unloading due to excavation. Journal of Hydraulics, 20 (8), 19-22(1998).
4. Zheng Shaohe. Research on Coupling Theory Between Seepage and Damage of Fractured Rock Mass and its Application in Engineering. Ph.D Dissertation. Wuhan: Institute of Soil and Rock Mechanics, The Chinese Academy of Science (2000).

CALCULATION OF THE DRILLABILITY VALUE FOR BOTTOMHOLE ROCK

SHIBIN LI TIE YAN
Department of Petroleum Engineering Daqing Petroleum Institute
Anda, Heilogjiang Province, 151400, P.R.China

It is difficult to measure the rock drillability under the enclosing pressure. The difference between the rock drillability in atmosphere and in enclosing pressure is extremely large, So the rock drillability in atmosphere doesn't meet the demand of the practical engineering. Based on indoor experiment results, the model about the relationship between rock drillability and mechanical properties is established. By the use of mechanical properties in enclosing pressure, the rock drillability in enclosing pressure can be determined. The practical application indicates that this problem cab be solved by this method.

1 Introduction

The rate of penetration (ROP), while a well is being drilled, generally shows a steady decline as the depth increases. The causes of the reduction in ROP are: Processes that affect on the unbroken rock. Processes that act on rock which is broken into chips. The second cause is critical, and other people have researched chip removal in detail. But the strength of the unbroken bottom-hole rock is not negligible. Our research work is focused on the drillability of rock as it is affected by bottom-hole pressure.

Practice has demonstrated the severe reduction in ROP as the bottom-hole pressure is increased. At present it is a problem to select the correct drillbit, based on the drillability value from laboratory atmospheric tests with microbits. Based on indoor experimental results, a model of the relationship between rock drillability and mechanical properties in atmospheric tests was established. By the use of mechanical properties tested with enclosing pressure that is similar to bottom-hole pressure, the drillability value of rock under enclosing pressure can be determined accurately.

2 Establishment of a relationship Model Between Drillability Value and Mechanical Properties

2.1 Laboratory Test of The Rock Drillability and Mechanical Properties in Atmosphere

More than 100 atmospheric rock drillability tests were done on different types of rock, and the mechanical properties of these rocks was also atmospherically tested. Table 1 is part of the test results

Table 1. Test results of rock drillability value and mechanical properties

Serial number	Drillability value	Compression Strength (Mpa)	Poisson Ratio	Young's Modulus (MPa)	Plastic Coefficient	Hardness (Gpa)
1	4. 70044	49. 610	0. 2400	32. 978	1. 70	73. 42
2	4. 95420	37. 517	0. 2300	31. 392	1. 00	73. 04
3	6. 47735	61. 167	0. 2562	34. 768	2. 34	119. 50

2.2 A Relationship Models Between Drillability Value and The Rock Compression Strength and Hardness in Atmosphere

Regression analysis of the rock drillability values and compression strength and hardness tested in laboratory in Table 1, resulted in the regression equations:

$$k_d = 3.445 + 0.038\sigma_c \tag{1}$$

$$k_d = 3.136 + 0.0266 p_y \tag{2}$$

Where k_d is rock drillability value; σ_c is compression strength, MPa; p_y is rock hardness, GPa, the correlation factor of equation (1) is 0.81. the correlation factor of equation (2) is 0.78.

From the equation described above, it is clear that the drillability value and the hardness and strength of rock are closely correlated.

2.3 Relationship of Drillability Value and Mechanics Parameters of Rock

The following is the correlative model derived from multiple regression analysis using applicable rock mechanical property parameters (rock compression strength, hardness, Poisson ratio, Young's modulus and plastic coefficient).

$$k_d = 4.0023 + 0.0257 p_y + 0.0105\sigma_c - 0.0186E - 0.5782\mu - 0.2256k \tag{3}$$

Where, k_d is roller bit drillability; σ_c is rock compression strength; p_y is rock hardness; E is Young's modulus;μ is Poisson ratio; k is the plastic coefficient.

In the regression analysis above, the multiple correlation factor is 0.575, significance inspection value: F=22.8>>7.47. The T inspection result shows that the correlativity of drillability and hardness is significant, but it is relatively weak with the plastic coefficient.

As the significance of the plastic coefficient is weak, we delete this factor and create a new multiple regression analysis, the model is:

$$k_d = 3.912 + 0.0266 p_y + 0.0113\sigma_c - 0.01846E - 0.5790\mu \tag{4}$$

3 Calculation of Rock Drillability value With Enclosing Pressure

It is difficult to directly measure the rock drillability value with enclosing pressure in the laboratory. But due to the high correlation between drillability value, the mechanics parameters and the elastic parameters of the rock, the enclosing pressure drillabilty can be calculated. When simulating downhole pressure conditions to measure these parameters, the drillability value may be calculated by using the correlation model.

3.1 Indoor Triaxial Test of Rock

Rock compression strengths in enclosing pressure are tested, also we measure the elastic modulus and Poisson ratio of the rock. Table 2 lists the tri-axial test result of sectional rock sample.

Table 2. tri-axial test result of the rock

Depth(m)	Enclosing Pressure (MPa)	Young's modulus (MPa)	Poisson Ratio	Compression Strength (Mpa)
1288	5	5780	0. 17	58
	10	5460	0. 13	81
	15	7880	0. 20	108
	20	7660	0. 20	115

3.2 Rock Mechanical Parameters vs. Enclosing Pressure

1. Static Poisson Ratio and Young's modulus vs. enclosing pressure correlation models

According to rock mechanical parameters measurements at different enclosing pressures (shown in Table 2), we can apply mathematical statistics to analyze the rock with both the Poisson Ratio and Young's modulus vs. enclosing pressure correlation with regression analysis. After making a regression analysis of Poisson Ratio and Young's modulus vs. enclosing pressure correlation for a single rock (the correlation coefficient is 0.95), total test data resulted in a similar analysis. The regressed mathematical models for enclosing pressure vs. Poisson Ratio or Young's modulus correlation can be expressed as equations 5and 6:

Poisson Ratio vs. enclosing pressure:

$$\mu = 0.0029P_c + 0.1209 \quad (5)$$

Where μ is the static Poisson Ratio at a given enclosing pressure, *Pc* is the enclosing pressure, MPa. In equation (5), the correlation coefficient is 0.76. Young's modulus vs. enclosing pressure:

$$E = 293.29P_c + 14490 \tag{6}$$

Where E is the Young's modulus at a given enclosing pressure, MPa, Pc is the enclosing pressure, MPa. The correlation coefficient in equation (6) is 0.68.

The regression results suggest how the enclosing pressure affects Poisson Ratio and Young's modulus, depends on rocks own properties. However, when a regression analysis is made of the total test data, the correlation coefficient is relatively small.

2. The rock strength vs. enclosing pressure correlation model

When judging whether the rock could collapse, we should draw different Mohr circles using triaxial test data at various enclosing pressures, so the rock collapse curve can be derived, we may calculate the angle or internal friction (φ) and the cohesive force (C_0) of the rock at different depths. The follow equation is the Mohr strength theory:

$$\tau = C_0 + P_c \tan\varphi \tag{7}$$

Where τ is the shear strength under enclosing pressure P_c.

The collapse curve can be applied to directly estimate rock strength. If the stress circle is below the collapse cure, rock will be stable.

If the circle is over the curve, rock will collapse. If the curve is a tangent of the circle, the rock will be in a precarious balanced condition. The Mohr strength diagram demonstrates that the straight line is the tangent of the maximum main stress circle. The compression strength in enclosing pressure is expressed as:

$$\sigma_c = \frac{1+\sin\varphi}{1-\sin\varphi} + \frac{2C_0\cos\varphi}{1-\sin\varphi} \tag{8}$$

Where σ_c is the rock strength under enclosing pressure. Partial test results are shown as Table 3.

3.3 The Relationship Between Enclosing Pressure vs. Hardness

The enclosing pressure increases the rocks hardness for drilling purposes. The test results suggest that the change is the same as that of the strength parameters. Soviet experts have made many tests in order to find the relationship between rock hardness vs. enclosing pressure. According to their test results, hardness is related to the enclosing pressure through this expression:

$$P_y = P_{y0} + k\sigma \tag{9}$$

Where P_{y0} is rock hardness without enclosing pressure, k is the coefficient, the correlation factor is 0.788.

Table 3. the angle φ of internal friction and cohesive strength Co

<table>
<tr><th>Well name</th><th>Rock Depth (m)</th><th>P_c (Mpa)</th><th>σ_c (Mpa)</th><th>C_0 (MPa)</th><th>φ (°)</th></tr>
<tr><td rowspan="6">Bu 5</td><td rowspan="6">1225.9</td><td>0.0</td><td>47.3</td><td rowspan="2">16.88</td><td rowspan="2">46.8</td></tr>
<tr><td>14.0</td><td>94.0</td></tr>
<tr><td>0.0</td><td>47.3</td><td rowspan="2">18.93</td><td rowspan="2">42.12</td></tr>
<tr><td>30.0</td><td>186.0</td></tr>
<tr><td>14.0</td><td>94.0</td><td rowspan="2">40.50</td><td rowspan="2">29.06</td></tr>
<tr><td>30.0</td><td>186.0</td></tr>
</table>

3.4 Computation Model of Rock Drillability Value Under Enclosing Pressure

Rock drillability value are related to rock compression strength, hardness, Poisson Ratio and Young's modulus, therefore, rock properties at a given enclosing pressure can be applied to the drillability correlation model, and then we can obtain different expressions of drillability.

1. Drillability value denoted by cohesive force and the rocks angle of internal friction:

$$k_d = 3.445 + 0.038\left(\frac{1+\sin\varphi}{1-\sin\varphi}\sigma_3 + \frac{2C_0\cos\varphi}{1-\sin\varphi}\right) \tag{10}$$

Where φ is the angle of internal friction for the rock; C_0 is rock cohesive strength (MPa); σ_3 is the enclosing pressure for the rock (MPa).

2. Drillability expressed by hardness at a given enclosing pressure:

$$k_d = 3.316 + 0.026(p_{y0} + k\sigma) \tag{11}$$

3. Comprehensive relation:

$$k_d = 3.912 + 0.0266 p_y + 0.0113\sigma_c - 0.01846 E - 0.5790\mu \tag{12}$$

Where: $\sigma_c = \frac{1+\sin\varphi}{1-\sin\varphi}\sigma_3 + \frac{2C_0\cos\varphi}{1-\sin\varphi}$, $\mu = 0.0029 P_c + 0.1209$

3.5 Calculation of The Bottomhole Rock Drillability Value

In the actual drilling process, bottom-hole enclosing pressure exists, it depends on the weight of the drilling fluid in borehole and the formation pressure. So, if we want to calculate rock drillability value under the bottom-hole enclosing pressure, first, we must calculate the rock's actual enclosing pressure under the synthetic interaction of the borehole pressure of the drilling fluid and formation pressure.

At the bottom of the hole, there isn't shear stress between the fluid and the rock, so the main stress must be vertical to the bottom of the hole. For drilling purposes, the minimum main stress depends on the differential between the borehole pressure and the formation pressure.

The two other main stresses are parallel to the surface of borehole, as previously discussed, rock strength mainly depends on the minimum main stress. Then, for permeable rock, the increased strength due to the increases in depth, is mainly caused by the difference between borehole pressure and theoriginal formation pressure. So the enclosing pressure of the borehole can be calculated:

$$\Delta P = P_w - P_p = H \cdot \rho_m / 100 - P_p \qquad (14)$$

If equation (14) is substituted into equation (10), (11) and (12), the drillability value of the bottom-hole rock will be achieved.

4 Conclusion and Recommendations

(1). The research results suggest that rock drillability values are large relative to the rocks mechanical parameters and the multiple correlation factors are larger than a single correlation factor.

(2). The calculations suggest that the differential between rock drillability values at a given enclosing pressure of 3 MPa and those at normal pressure is 0.6.

(3). We suggest that rock drillability values at a given enclosing pressure should be the base for the optimization of bits and drilling design. We have applied these research results to optimize bits and have achieved improved results.

References

1. Robert T. Fabian, Confined compressive strength analysis can improve PDC bit selection, May 16,1994 Oil and Gas Journal.
2. J.R. Spaar and L.W. Ledgerwood, Formation Compressive Strength Estimates for Predicting Drillability and PDC Bit Selection, SPE/IADC 29397.

THE STUDY OF THERMO-RHEOLOGY OF THREE GORGE GRANITE

CHONGGE WANG ZHENQI SONG
Shangdong University of Science and Technology, Taian 271019,China

QUANSHENG LIU
Institute of Rock and Soil Mechanics, The Chinese Academy of Sciences, Wuhan 430071,China

Under the condition of uniaxial compression, the behaviors of the rheology in rock are discussed and the thermo-visco-elastic equations of the constitutive, creep, unload and relax are obtained. Meanwhile, the significances of this kind of model in practical use are explaine

1 Introduction

In recent years, the research, considering the effect which temperature has on rock mechanics behavior, has drawn the attention of a number of experts and scholars home and abroad. It has got a vast range of applied prospect in the fields of geological structure's movement, nuclear waste's disposal, the exploitation and utilization of terrestrial heat and deep mining etc. If the rock's complex property, such as mineral crystal, pore, water and joints is considered, it will be extremely sophisticated and difficult to apply in engineer projects. Hence the effect of micro crack or discontinuity will be average. The research of rock's mechanical and mathematical characters can be given a macroscopic and approximate account, to major rock mass project under a general condition. In this essay we imitate the behavior of rock's complicated rheology, by using spring, dashpot and other different combinability with the rock mediums under the condition of uni-axial compression.

The essay introduces temperature effect to Hooke model & Kelvin model. It sets up a thermo-visco-elastic rock model, which considers the effect of temperature as well, so as to make the mathematic descriptions about the behavior of rheology rock access to simplicity and generalization.

2 thermo-rheology rock model

2.1 Constitutive Equations

To study the characters of thermo -visco-elastic rock, we can adopt the simplified model, Hooke and Kelvin element in series, shows in figure 1.

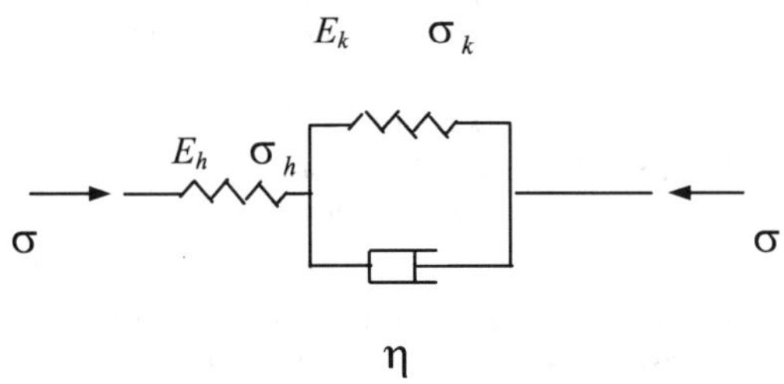

Figure1. Thermo-visco-elastic rock model

In this figure, E_h represents the spring Young's modulus in Hooke, which E_k in Kelvin; σ_h, σ_k stand for the spring's thermal expansion coefficients in Hooke and Kelvin respectively, when temperature's effect is considered. η is the viscosity coefficient of dashpot model, in which h stands for Hooke, k for Kelvin.

Assume σ stands for stress, and ε represents stain. Since both elements are connected in series, σ, ε, obey the law

$$\sigma = \sigma_h + \sigma_k \tag{1}$$

$$\varepsilon = \varepsilon_h + \varepsilon_k \tag{2}$$

$$\frac{d\varepsilon}{dt} = \frac{d\varepsilon_h}{dt} + \frac{d\varepsilon_k}{dt} \tag{3}$$

Considering the thermal expansion of spring due to the temperature change

$$\sigma_h = E_h \varepsilon_h - E_h \alpha_h T_c \tag{4}$$

where T_c is reference temperature variety comparatively room temperature T_0, $T_c = T - T_0$, and

$$\sigma_k = E_k \varepsilon_k - E_k \alpha_k T_c + \eta \frac{d\varepsilon_k}{dt} \tag{5}$$

From upward we can derive the constitutive equation:

$$\frac{1}{E_h}\frac{d\sigma}{dt} + \frac{E_h + E_k}{\eta E_h}\sigma + \alpha_h \frac{dT_c}{dt} + \frac{E_k}{\eta}(\alpha_h + \alpha_k)T_c = \frac{d\varepsilon}{dt} + \frac{E_k}{\eta}\varepsilon \tag{6}$$

2.2 creep equation

Under the condition of constant temperature, $dT_c / dt = 0$, when $\sigma = \sigma_0$ is constant, $d\sigma / dt = 0$, .Substituting them into constitutive equation (6), we can derive the creep equation (t)

$$\frac{d\varepsilon}{dt}+\frac{E_k}{\eta}=\frac{E_h+E_k}{\eta E_h}\sigma_0+\frac{E_k}{\eta}\ddot{\alpha}_h+\alpha_k\ T_c \tag{7}$$

assume $$\begin{cases} E^*=\dfrac{E_h+E_k}{E_hE_k} \\ \alpha^*=\alpha_h+\alpha_k \end{cases} \tag{8}$$

we now separate the variable of equation(7)to both side and integral, and substitute the initial condition, t=0 $\varepsilon_h=\varepsilon_{h0}=\sigma_0/E_h+\alpha_hT_c$ into Hooke. Because of the effect of dashpot in Kelvin model $\varepsilon_k=\varepsilon_{k0}=0$, we can derive the creep equation $\varepsilon(t)$

$$\varepsilon(t)=J(t)\sigma_0+\gamma(t)T_c \tag{9}$$

where $$\begin{cases} J(t)=E^*-\dfrac{1}{E_k}e^{-\frac{t}{t_{ret}}}; \\ \gamma(t)=\alpha^*-\alpha_k e^{-\frac{t}{t_{ret}}}; \\ t_{ret}=\dfrac{\eta}{E_k} \end{cases}$$

$J(t)$ is creep compliance, $\gamma(t)$ is creep coefficient affected by temperature. and t_{ret} represents delay time, from equation (9), we can see that the creep phase is extended to a horizontal line, when time tends to infinite. So the creep of rock mechanic model is steady creep.

2.3 unloading equation

After unloading at $t=t_u$, $\sigma=0$, Hooke model's deformation doesn't completely disappear due to the influence of temperature, so Kelvin's deformation plays a leading role. Substituting it into equation (7), we can derive

$$\frac{d\varepsilon}{dt}=-\frac{E_k}{\eta}\ddot{\varepsilon}-\alpha^*T_c \tag{10}$$

when, $t=t_u, \varepsilon=\varepsilon_u$, we can derive the unloading equation

$$\varepsilon=\varepsilon_u e^{\frac{E_k}{\eta}(t_u-t)}+\alpha^*\ddot{1-}\ e^{\frac{E_k}{\eta}(t_u-t)})T_c \tag{11}$$

2.4 relaxation equation

Under the condition of constant deformation, $\varepsilon=\varepsilon_0; d\varepsilon/dt=0$, substitute it into constitutive equation（ 6 ）

$$\frac{\eta}{E_h+E_k}\frac{d\sigma}{dt}+\sigma=\frac{1}{E^*}\varepsilon_0-\frac{\alpha^*}{E^*}T_c \qquad (12)$$

Separate the variable of equation（ 12 ）to both side and integral, and consider the initial condition, when t=0, only Hooke model plays an important role due to the influence of dashpot. So wc gct thc rclaxation equation

$$\sigma=\beta(t)\varepsilon_0+\zeta(t)T_c \qquad (13)$$

$$\begin{cases} \beta(t)=\dfrac{1}{E^*}-(\dfrac{1}{E^*}-E_h)e^{-\frac{t}{t_{rel}}}; \\ \zeta(t)=-\dfrac{\alpha^*}{E^*}+(\dfrac{\alpha^*}{E^*}-E_h\alpha_h)e^{-\frac{t}{t_{rel}}}; \\ t_{rel}=\dfrac{\eta}{E_h+E_K} \end{cases}$$

where t_{rel} is relaxation time, which represents the speed grade of rock stress relaxation. From equation（ 13 ）,we can also see that $\beta(t)$ is relaxation modulus， $\zeta(t)$ is relax coefficient affected by temperature. This relax curve is descendant exponential curve， but the relaxation is imperfect.

3 Study of experiments

The sample is fresh grey-black granite which was exploited in the area of Three Gorges. Its composition includes quartz, feldspar, hornblende and biotite. The sample was processed by the way of drilling to get the heart of rocks, cutting and polishing with the water drilling method being adopted. The size of the sample is as follows: diameter is 49.64 mm to 49.82 mm, height, 99.78 mm to 100.50 mm; its moisture content in savageness containing water, and it has a capacity of 3.36 g/cm^3. We can get the creep curves in figure 2 from the creep under uniaxial loading experiment, in which the sample is kept the constant stress of 120 MPa under the temperature of 20℃, 60℃, 80℃, 100℃, 150℃

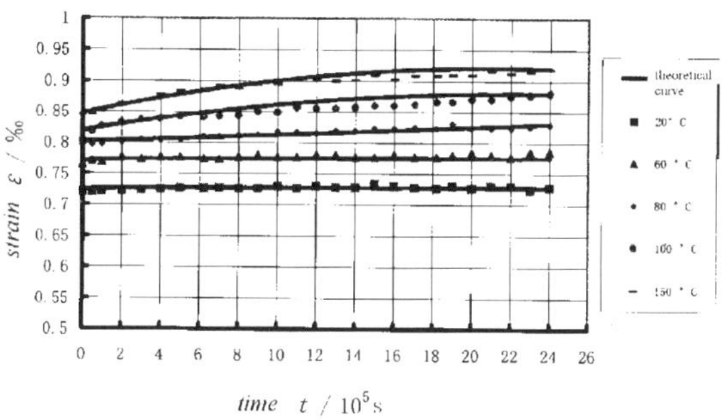

Figure2. Creep curves of granite under uniaxial loading

Table.1 Model parameter under different temperatures

tempe-rature (℃)	E_h (Gpa)	E_k (Gpa)	η 10^5 (Mpa. d)	α_h 10^{-6} (/℃)	α_k 10^{-6} (/℃)
20	82.67	1.15	6.935		
60	81.93	6.23	5.827	0.22	10.27
80	81.82	10.34	4.327	0.31	14.26
100	78.77	25.32	4.151	0.42	14.79
150	74.01	29.18	3.718	0.55	20.11

The specific thermo-visco-elastic mechanical model has been proposed for this kind of granite and its parameters have been determined by fitting the curves.

As the temperature get higher, E_h becomes lower, while E_k, stiffness coefficient of Kelvin model, is rising, but E_k is smaller than E_h, dashpot's viscosity coefficient is reducing. The rock's elasticity is lowering while the viscosity growing. The spring's thermal expansion coefficients in Hook and in Kelvin are growing as the temperature rising.

4 Conclusions

Based upon the theoretical analyses and the experimental studies, the thermo-visco-elastic equations of the constitutive, creep, unload and relaxation are obtained, and the parameters of this kind of mechanic model have been determined at different temperatures. Meanwhile, we find that the creep equation and relaxation equation, considering the effect of temperatures, have a similar structure form. furthermore, thermal expansion coefficients can be positive (in high temperature) or negative (in low temperature), therefore this model can reflect the rock or rock mass average thermo-visco-elastic character. It has been found that there exist broad prospect in application to such fields as nuclear waste's disposal, exploitation and utilization of geothermal resources and deep mining, etc. However this model has certain limitations, for example, there is no plasticity element, imitating merely rock's thermo- visco-elastic character and it can't reflect further rock's thermal-visco-elastic -plasticity. In the future, more inhomogeneous rock's experiments need doing to determine the model's coefficients at different temperatures.

References

1. Christensen,R.M. TheoryofViscoelasticity[M]. NewYork, 1982,pp.273-294
2. N.Cristescu. Rock Rheology[M]. Oxford, 1985,309~321
3. Carter, N. L. and Hansen, F. D., Creep of Rock Salt[M], Tectonophysics, 1983, pp.275-333
4. Wawersik,W.R.and Hannum,D.W. Mechanic Behavior of New Mexico Rock Salt in Triaxial Compression up to 200 ℃ [J]. J.Geophys Res.,1980,B2,pp.891-900
5. Ziegler, H., An Introduction to Thermodynamics, North Holland Pub. Co., 1997
6. Paterson, M. S., Experimental Rock Deformation-Brittle Field, Springer-Verlag, 1978
7. Roger, I., Tanner, Engineering Rheology, Oxford, 1985
8. Biot, M. A., Theory of Stress-strain Relations in Anisotropic Viscoelasticity and Relaxation Phenomena, J. Appl. Phys., 1954, 25(11), pp.1385-1391

EFFECT OF CONTINUITY ON CRACK PROPAGATION AND COALESCENCE IN HETEROGENEOUS BRITTLE MATERIALS UNDER UNIAXIAL COMPRESSION

TAO XU CHUN'AN TANG XINGDONG ZHAO ZENGTIAN MA
CRISR, Northeastern University, Shenyang 110004, China

By using Material Failure and Process Analysis code (MFPA^{2D}), the effect of crack continuity in heterogeneous brittle materials containing two pre-existing inclined parallel flaws under uniaxial compression on crack propagation and coalescence is numerically studied. Numerical simulations visually duplicate stress fields around crack tips and the dynamic process of crack propagation and coalescence. The numerically simulated results also show that stress fields between the tips of internal cracks and the propagation and coalescence mode of cracks gradually change with different crack continuity. In addition, the initiation stress on the tips of cracks is about 1/2 ~1/3 of ultimate failure strength of the specimens.

1 Introduction

An important problem in the constitutive description of heterogeneous materials is their mechanical behavior when subjected to compressive loads. Heterogeneous brittle materials contain a large number of different defects. These defects are responsible for the inelastic behaviors of materials and they finally lead to macroscopic failure under compressive or tensile loads. Extensive research has been carried out on the mechanism of crack propagation from single preexisting flaw in compression. More recently crack propagation and coalescence in rocklike materials containing two inclined flaws has also been investigated. Especially, Wong and Chau et al. [1] studied crack propagation and coalescence in rock-like materials containing three flaws and observed cracks coalescence occur between only two flaws. It can be concluded that the mechanism of propagation and coalescence between two cracks is essential and fundamental to understand the deformation and failure process and behavior at mesostructure level in compression.

Correspondingly, numerical experiments were conducted to investigate the mechanism of crack propagation and coalescence in heterogeneous brittle materials containing two inclined parallel cracks in uniaxial compression.

2 Numerical Model

Numerical simulation is currently the most popular method used for modeling deformation behavior of rock-like materials before failure. The demand for new tools, which may contribute to a better understanding of the failure mechanisms of heterogeneous brittle materials, initiated the development of MFPA^{2D} (Material

Failure Process Analysis code). The code has been developed based on similar basic principle of its companion, Rock Failure Process Analysis code (RFPA^{2D}). Details of RFPA^{2D} code have been described in many published literatures [1-2].

In this paper, numerical experiments on heterogeneous brittle materials containing two inclined parallel cracks with different continuity (c) are conducted to investigate crack propagation and coalescence in uniaxial compression. Fig. 1 shows the geometries of two cracks in heterogeneous brittle specimen and table 1 gives the configurations of preexisting cracks in model specimens with different continuity.

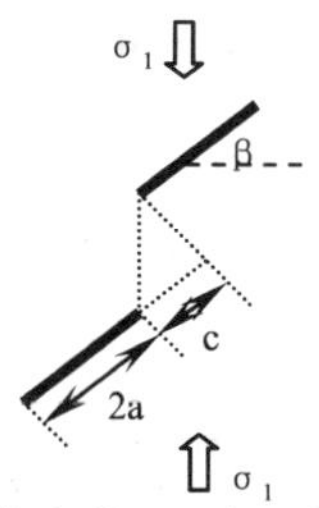

Fig.1 Geometries of two cracks in brittle specimen

Table 1. Configurations of cracks in brittle specimens

No.	D-2	D-3	D-4	D-6	D-7	D-9	D-8
Specimens for different continuity	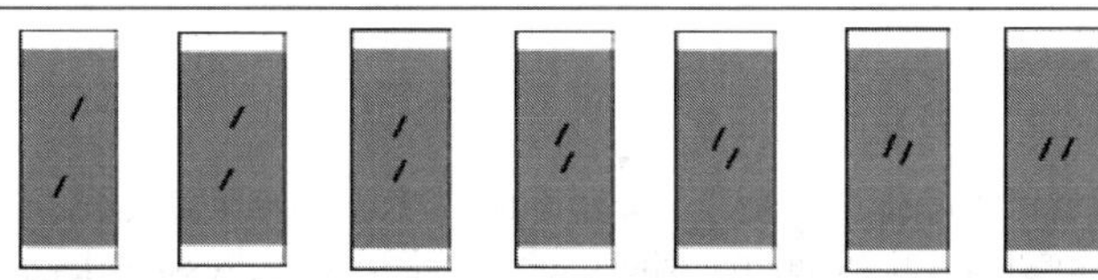						
c/mm	28.9	20.1	9.4	0	-8	-11	-16.6

In order to eliminate the effect of different specimens on the numerical experiments, a numerical specimen was duplicated or cloned as other specimens after setting up specific parameters, that is, the specimens are identically the same except for the location of cracks. It is noticing that it is impossibly achieved in laboratory; however, it is rightly one of the advantages of numerical simulations. The material parameters for specimens are Φ_E(5,6000MPa)and Φ_o(5，50MPa), receptively. In addition, considering the mismatch between load platens and specimen, the Elastic modulus of load platens is the same as that of specimens while the scale strength of load platens is much higher than that of specimens [2].

3 Modeling and discussion

Table 2 shows variations of stress fields around cracks tips and crack coalescence modes during uniaxial compression with the decrease of crack continuity. It can be seen from Table 2 that stress fields between cracks gradually transit from tensile stress to compressive stress, and to tensile stress with the decrease of crack continuity. The tensile stress fields between internal crack tips reach to the maximum when crack continuity is equal to zero, which means there is strong interaction around the internal tips of cracks. The crack propagation and coalescence modes are greatly dependent on the stress fields between cracks.

Table 2. Propagation and coalescence of cracks in model specimens

No.	D-1	D-2	D-3	D-4	D-5	D-6	D-7
Initiation and propagation of cracks	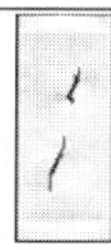		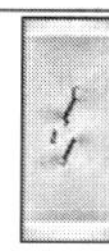		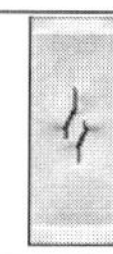	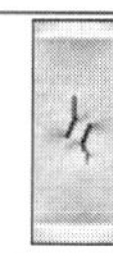	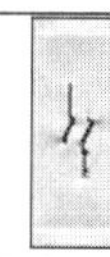
Initiation of secondary cracks				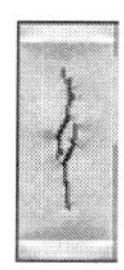	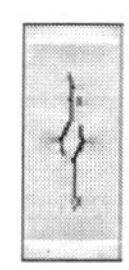	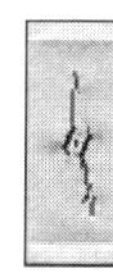	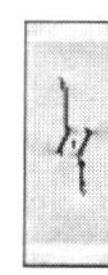

In Table 2, wing cracks propagate and two branch cracks along the preexisting crack forms when there is large crack continuity. Wing cracks initiate at 0~10° with the direction of uniaxial stress and grow in a curvilinear path that aligns with the compression direction. Moreover, the branch cracks are smooth under tensile stress, the crack continuity decreases and the stress fields around internal crack tips gradually transform into compressive stress fields. In this case, the faulting occurs around cracks under higher compressive stress and form serrated cracks mingled with crushed grains. During the complete process of crack propagation and coalescence, tensile and shear failure (T-S) successively occurs, which is good agreement with the WI type of crack propagation and coalescence[1]. However, when the continuity is small, the initiation of internal crack is prohibited under compressive stress fields while the external cracks propagate quickly and form branch cracks with the initiation angle of 10~30. With the increase of loads, the stress fields around internal crack tips change and higher tensile stress concentration occurs in the middle of rock bridge in the course of crack propagation. Meanwhile, clusters of secondary cracks initiate and coalesce, as shown specimens D6 and D7 in Table 2. This type of crack propagation can be defined as compressive mode (C), which is very similar with Fig. 2 in literature [3]. In addition, the initiation angle gradually increases with the decrease of continuity.

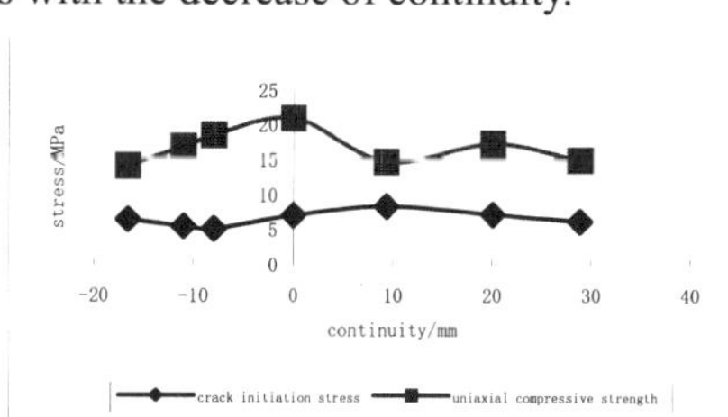

Figure2. Crack initiation stress and ultimate failure strength of specimens with different continuity

Crack initiation stress and ultimate failure strength for specimens containing two preexisting cracks with different continuity are variable from the numerically simulated results. Initiation stress and ultimate failure strength are both higher than that under other cases. On the whole, crack initiation stress is about 1/3~1/2 of ultimate failure strength of specimens with preexisting cracks. The initiation stress and ultimate failure strength change with continuity are shown in Fig. 2.

4 Conclusions

The following conclusions can be drawn based on numerical results about crack propagation and coalescence in pre-cracked specimens under uniaxial compression.

Stress fields around crack tips change from tensile stress fields to compressive stress fields, and back to tensile stress fields with the decrease of crack continuity. Crack coalescence transits from tensile and shear mode (T-S) to compressive mode (C) with the decrease of crack continuity. The branch cracks are smooth in the propagation and are rough with crushed grains in the coalescence under the tensile and shear mode; however, secondary cracks initiate and coalesce to form a faulting zone in the end due to high compressive stress under the compressive mode. The initiation angle gradually increases from approximately parallel to the compressive stress direction to about perpendicular to the preexisting crack with the decrease of crack continuity. Crack initiation stress and ultimate failure strength also vary with continuity. The initiation stress is about the 1/3~1/2 of ultimate failure strength of specimens with preexisting cracks.

Acknowledgement

The work in this paper was supported by NSFC (No.49974009 and 50174013).

References

1. Wong R.H.C., Chau K.T., Tang C.A., et al., Analysis of crack coalescence in rock-like materials containing three flaws-Part I : experimental approach, Int. J. Rock Mech. Min. Sci., 2001, 38(7): 909-924
2. Tang CA., Tham LG et al. Numerical studies of the influence of microstructure on rock failure in uniaxial compression—Part II: Constraint, slenderness and size effect. Int J Rock Mech Min Sci , 2000,37:571-583
3. Bobet A. Num. simulation of initiation of tensile and shear cracks. Rock Mech. in the National Interest, Elsworth, Tinucci & Heasley (eds), 2001:731-738

2. Experimental technique and field Measurement

THE STATE OF THE ART OF ROCK STRESS MEASUREMENT AT GREAT DEPTH BY BOREHOLE PRESSURIZATION AND THE POINTS AT QUESTION

YOSHIAKI MIZUTA
Professor, Dr., Faculty of Engineering, Yamaguchi University, Tokiwadai, Ube 755-8611, Japan

The state of the art of rock stress measurement by hydraulic fracturing is reviewed and a comparative study is made of single dry fracturing technique, which has been developed as the potential method for stress determination. The general review is followed by individual summaries of each method and the main characteristics of its validations and applications.

1 Introduction

The stress-relief method of overcoring is widely applied in field stress measurement, which is based upon the displacement relief of surrounding rock through overcoring process, and is applicable only for low stress estimation within hard rocks.

The hydraulic fracturing method was developed about 50 years ago along with the petroleum production in oil field, and has been researched as a potent method for determination of in-situ stress at great depth by laboratory and field tests afterwards. The hydraulic fracturing technique as a method for determination of in-situ stresses has the advantage over other stress measuring methods in that it can be used at considerably greater depths from a point of access. This is of great significance in earthquake regions where stresses at hundreds or thousands of meters are of interest. Despite of the increasing use of hydraulic fracturing, however, there are uncertainties associated with the interpretation of the resulting data. In particular, confidence in the calculated maximum principal stress is less than in the minimum principal stress, although the former is often of great moment. Furthermore, the equation established to calculate the maximum principal stress is unclear. More recently, two kinds of dry fracturing techniques, double fracture method and single fracture method, which mean that the loading fluid is prevented from penetrating into the induced fractures, have been developed and examined as the new potential method for determination of in-situ stress at great depth, through studies both in the laboratory and in the field. In single dry fracturing, the borehole wall is subjected the shear load as well as normal load, which are induced by internal pressurization of the probe.

2 Hydraulic fracturing

The basic provision of stress field determination at plane perpendicular to borehole axis in hydraulic fracturing is introduced as follows:

(1) The artificial fracture (longitudinal fracture) shall be initiated perpendicular to the minimum principal stress σ_h.

(2) Shut-in pressure p_s is presented by stress component perpendicular to the fracture plane as follows: $\sigma_h = p_s$ (1)

Since the injected fracturing fluid might permeate into the surrounding rocks, in most cases it is hard to determine p_s definitely (In Fig. 1(a), the shut-in pressure stays steady while in practice it has exponentially decrease). Practically, instantaneous shut-in pressure p_{inst} is used mostly as p_s.

(3) The fracture shall reopen when tangential stress $\sigma_{\theta r}$ on borehole wall at fracture plane induced, by fluid pressure, reaches $\sigma_{\theta r}$, induced by effective rock stress related with pore pressure p_p in borehole wall fracture, i.e.

$$\sigma_{\theta f} = -\sigma_H + 3\sigma_h - p_p = \sigma_{\theta r},\quad \sigma_{\theta r} = p_r$$

(2)

where p_p is not explicit in practice, with the following assumptions (Bredehoeft et al. 1976, Ito et al. 1999):

1) $p_p = p_o = 0$ (unsaturated field surrounded by voids)

where Eq. (2) turns to $p_r = 3\sigma_h - \sigma_H$ (3)

2) $p_p = p_o = \rho_w gh$ i.e. $p_r = 3\sigma_h - \sigma_H - p_p$ (4)

3) $p_p = p_{ro} = p_r$ where $p_r = (3\sigma_h - \sigma_H)/2$ (5)

The reopening pressure is generally taken as the pressure at the point parting from the straight-line in pressure-time curve during the primary stage of injection cycle by constant rate of flow after artificial fracture initiated (see Fig. 1 (a)). Bredehoeft (1976) uses this point to express the opening pressure of borehole fracture which is related to the magnitudes of σ_H and σ_h. However, it has been known before that the reopening pressure can not be defined under large deviation stress because the borehole fracture is opened before fluid injection. Furthermore, the reopening pressure p_r approached by the above process is close to the shut-in pressure p_s.Hence, $\sigma_H/\sigma_h=2$ is obtained by putting $p_r = p_s = \sigma_h$ in Eq.(3) and it means that the stress field measured anywhere shall be the same.

Ito and Hayashi (1993) demonstrated through numerical simulation that p_r can be equal to p_{inst} by considering that fluid permeate into the fracture before it is opened. High permeability of the fracture can be explained in that there are residual fracture opening permeated by fluid even when the fracture is closed, which had been proved by laboratory test. Ito et al. (1999) pointed out that the apparent fracture reopening pressure $p_r^{(a)}$ and the true opening pressure p_{ro} are different since $p_r^{(a)}$ is affected by water volume V and flow rate q in the pressure system. For numerical simulation, $p_r^{(a)}$ decreases when q decreases, and becomes closed to p_{ro} when V decreases. The situation that, $p_r^{(a)}$ becomes close to p_s, can be explained if V is large. The reopening pressure $p_r^{(a)}$ approaches true reopening pressure p_{ro} if the pressure is measured by low injection rate with pressure system of quite small effective volume. In this case, σ_H can be evaluated according to Eq. (5). It is valid for σ_H / σ_h<3 but not valid for stress field $\sigma_H / \sigma_h \geqq 3$, which has great disadvantages against application of hydrofracturing method in earthquake stress measurement.

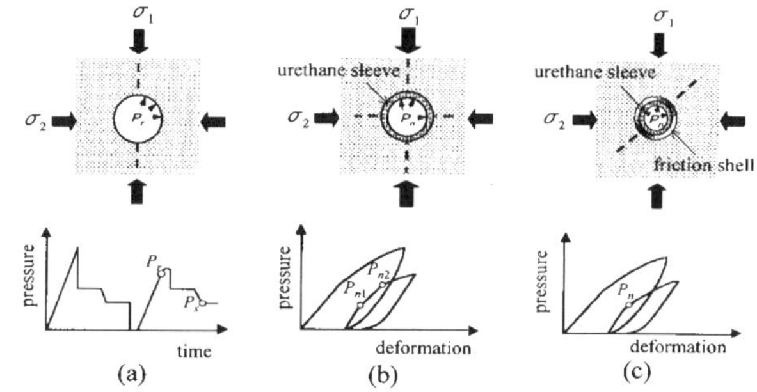

Figure1. Detection of re-opening pressure from hydrofracturing record (a) double dry fracturing record (b) and single dry fracturing record (c).

3 Single fracture method

The single fracture method is applicable to greater depth as same as hydrofracturing method, moreover it has the same confidence in the calculated maximum principal stress and in the minimum principal stress. But the interpretation of data detected is not explicit; especially the measurement process for fracture reopening pressure lacks objectivity. Pressure-diametral deformation curve of single fracture method is illustrated in Fig. 1(c) and the orientation of diametral deformation measured is perpendicular to the artificial fracture. The following elements are needed in the determination of rock stress by single fracture method:

1) The artificial fracture is developed on the appointed orientation.
2) The borehole wall fracture is reopened when the tangential stress $\sigma_{\theta p}$ on borehole wall at fracture plane, induced by fluid pressure, reaches the tangential stress $\sigma_{\theta r}$ induced by rock stress, i.e.,

$$\sigma_{\theta r} = \sigma_1 + \sigma_2 - 2(\sigma_1 - \sigma_2)\cos 2\theta_f - p_0 = \sigma_{\theta p} = (1+k)p_i \qquad (6)$$

Where θ_f is the angle between σ_1 direction and the orientation of the fracture, and k is stress concentration factor induced by friction shell, which is calculated by numerical analysis described later. Different from hydrofracturing, loading pressure shall not propagate into the fracture and so, p_p in Eq. (2) is always p_o, which makes the equation definitely constructed.

3.1 Theory of single fracture stress measurement

In Fig. 2, σ_1、σ_2($\sigma_1 > \sigma_2$) are principal stresses in rock; p_i is the fluid pressure in the probe; θ_f is the orientation of the fracture plane counted anticlockwise from σ_1 direction.

If p_o=0, the tangential stress on borehole wall at fracture plane is calculated as follows:

$$\sigma_{\theta\,pr} = \sigma_1 + \sigma_2 - 2(\sigma_1 - \sigma_2)\cos 2\theta_f - (1+k)p_i \qquad (7)$$

In general, direction of σ_1 is unknown. Put ts direction be ψ and $\theta_f + \psi$ be θ_x as Fig. 2. The following equation can be achieved with $p_i = p_n$, $\theta_x = \theta_p$ at reopening moment:

$$0 = \sigma_1 + \sigma_2 - 2(\sigma_1 - \sigma_2)\cos(\theta_p - \phi) - (1+k)p_n \qquad (8)$$

If three fractures at different orientation are produced and those are reopened, the three unknown quantities σ_1, σ_2 and ψ can be calculated with the reopening pressures measured.

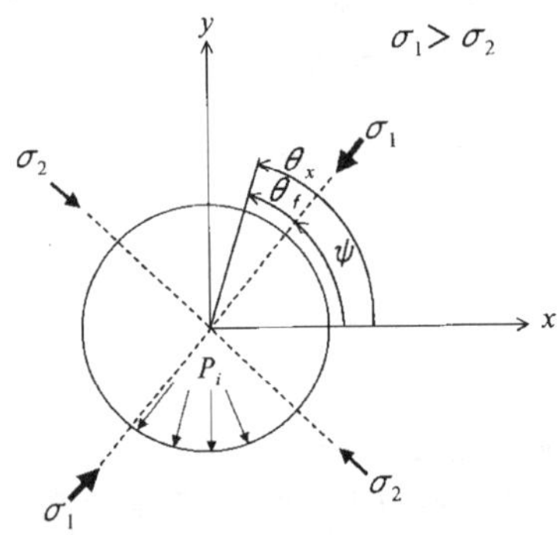

Figure 2. Designation of the parameters and the coordinates.

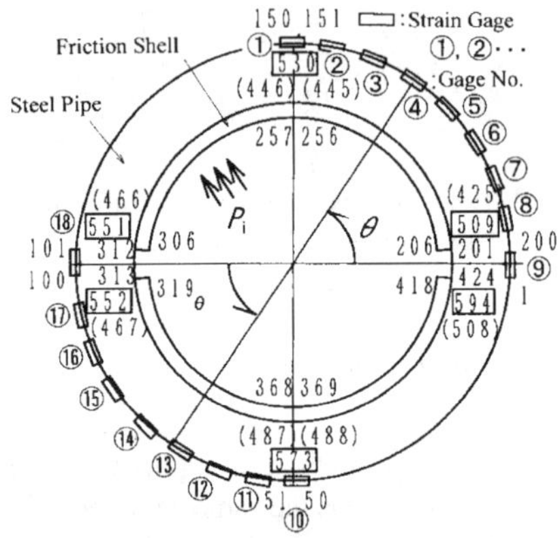

Figure 3. The cross section of two friction shells being pressed against the inside wall of the steel pipe. BEM modeling and locations of strain gages are shown in the figure.

3.2 Loading test in steel pipe

As illustrated in Fig. 1 (c), the friction shell pasted around the probe periphery has a jagged shape to ensure the occurring of shear stress on borehole wall. The transection of the friction shell pressed on the wall of the steel pipe is illustrated in Fig. 3. Nine strain gauges with No.1~9 are pasted around the first quadrant along the steel pipe and strain gauges with No.10~18 are pasted around the third quadrant along the steel pipe. The diamond and quadrangle in Fig. 4 show the tangential stresses σ_θ around the steel pipe periphery obtained from the strains measured. It is obvious in the figure that σ_θ is small at $\theta = 0$, which means that the tangential stress on the inner wall of the steel pipe shall be great tensile stress at $\theta = 0$ and the artificial fracture shall occur along this direction.

3.3 Numerical simulation of the loading test in steel pipe

In single fracture method, normal stress as same as fluid pressure p_i acts on the borehole wall and the shear stress induced by friction shell and normal stress p_i is distributed around borehole wall. The simulation of the effect of friction shell is represented by numerical simulation using two kinds of BEM element in two-dimensional boundary element method .The transection of the numerical model of the steel pipe test is shown in Fig.1.

The boundaries between friction shells and pipe wall are divided into 86 DDM joint elements (convex) and 84 DDM crack elements (concave). The other boundaries are divided into 412 FSM (Fictitious Stress Method) elements. The black balls in Fig. 4 represent the tangential stress distribution along the steel pipe periphery, which are calculated with the same elastic modulus for the steel pipe and friction shell. The figure shows that practical situation can be represented precisely by numerical calculation. The numerical simulation values of $\tau_{r\theta}$ and σ_r along the inner wall of the steel pipe is distributed as illustrated in Fig. 5. It is found from the figure that the distribution of $\tau_{r\theta}$ is of exponential function and so, its values become extremely big at the portion close to $\theta = 0$, which shows the effect of the friction shell.

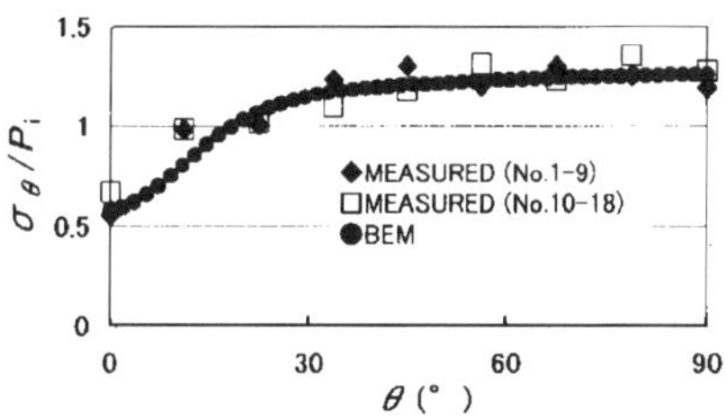

Figure 4. The measured distribution of tangential stresses along the outer surface of the steel pipe, compared with the results calculated by BEM.

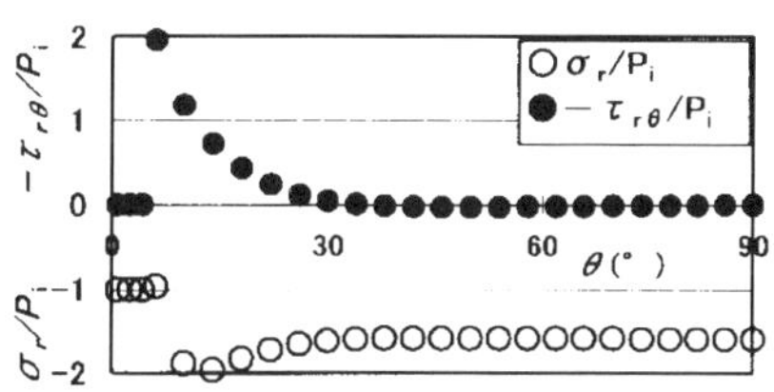

Figure 5. The calculated distributions of $\tau_{r\theta}$ and σ_r along the boundary between the friction shell and inner wall of the steel pipe.

4 Wire-line System

4.1 Hydraulic fracturing

Wire-line system is developed for low cost hydrofracturing stress measurement. It is shown in Fig. 6. The wire-line hydrofracturing stress system is first developed by Rummel et al.(1981). The new system recently developed in Japan has a precise flowmeter which is installed in the probe. It can measure the reopening pressure correctly.

4.2 Single dry fracturing

Loading pressure control device of the hydrofracturing stress measurement system shown in Fig. 6 is installed on the ground surface, while the single fracture stress measurement system introduced here by Serata Geomechanics Inc. are not only a wire-line system, but also has the loading pressure control device which is installed in the probe (Fig. 7). The control device for angle regulation of the friction shell is also installed in the probe.

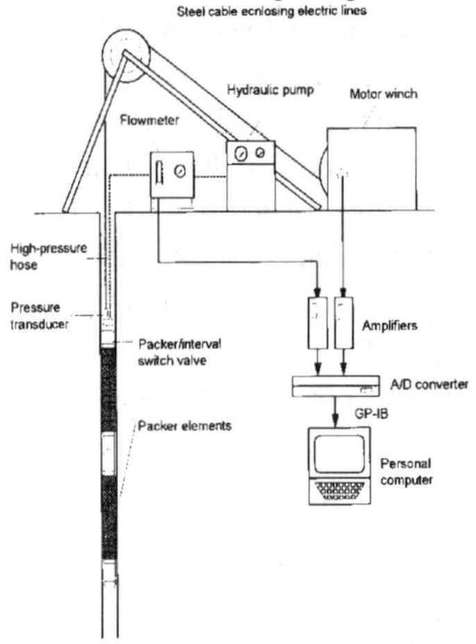

Figure 6 Wire-line hydraulic fracturing system.

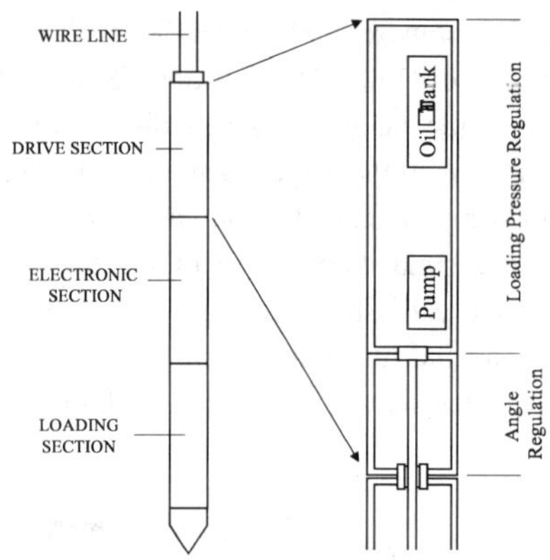

Figure 7 Outline of the dry fracturing probe including the devices for loading pressure regulation and angle regulation.

5 Conclusion

The rock stress measurements at great depth by wire-line system including hydrofracturing and single dry fracturing methods are discussed above. For hydraulic fracturing method, since the loading pressure is related to the pore pressure in artificial fracture and the apparent reopening pressure affected by the pressure system compliance C and flow rate q, true reopening pressure cannot be measured precisely. Furthermore, it can not be measured under the stress field of $\sigma_H / \sigma_h > 3$, which concerns the earthquake stress. The single dry fracturing method does not have such problem since the pore pressure in the artificial fracture has no relation with the loading pressure. But still the true reopening pressure p_r cannot be read precisely from the pressure-displacement curve.

To solve this problem, development of a new system is expected.

References

1. Y. Mizuta, Shigen to Sozai, 118, 5/6 (2002).
2. J.D. Bredehoeft, R.G. Wolff, W.S. Keys and E. Shutter, Geol. Soc. Amer. Bull. Vol.87, (1976).
3. T. Ito and K. Hayashi, Int. J Rock Mech. Min Sci. & Geomech. Abstr Vol.30, (1993).
4. T. Ito, K. Evans, K. Kawai et al., Int. J. Rock Mech. Min. Sci. & Geomech. Abstr. 36, (1999)
5. F. Rummel J. Baumgartner etc., Technical Report to the U.S. Geological Survey, Open-file Report 82-1075(Proc. of workshop on hydraulic fracturing stress measurement, Vol.1) (1981).

EXPLORATION AND APPLICATION OF SEISMIC ANALYSIS IN THE WAVE VELOCITY TEST OF ROCK MASS IN SITU

JINZHONG SUN CHENGHU WANG
College of Engineering and Technology, China University of Geosciences, Beijing, 100083

JINGRU LIU
Unit 92181, People's Liberation Army of China, Gaopeidian, Hebei Province, 074000

Wave velocity test in situ is an important approach to explore the engineering properties of the rock mass. In practice, many factors may affect the waveforms, so precise traveling time of seismic phases is seldom measured from the wave data. In order to overcome this obstacle, a series of rules and criteria for seismic phase analysis are put forward in this paper. The rational travel time of P and S wave can be obtained in the light of these criteria and rules. The method has been applied to the test of the rock mass in situ for a project, and a very good effect of the rock mass classification has been obtained.

1 A main problem in the wave velocity test of rock mass in situ

Wave velocity test of rock mass in situ is an important approach to explore the engineering properties of rock masses by elastic waves. Drawing the travel time of P wave and S wave correctly from wave data is the kernel of the wave velocity test. In practice, however, many factors such as stimulating source, coupling of the detector to the well wall, sampling interval of wave data etc. may affect the waveforms so that precise traveling time of seismic phases is seldom measured from the wave data. Figure 1 gives us an example waveform that reveals the influence of different factors on the travel time of seismic phases. The waveform is recorded on the same depth by a detector once coupled in a borehole of a rock mass in situ. Waveforms A and B are stimulated on the ground surface in the same direction, and waveform C is stimulated in the opposite horizontal direction. We can clearly see that there is about 0.5ms difference in the arrival time of P and S waves between waveforms C and B. The error in the travel time will lead to unreliable wave velocities so that the properties of the rock mass reflected by wave velocities will also be distorted. To resolve this problem, an approach of

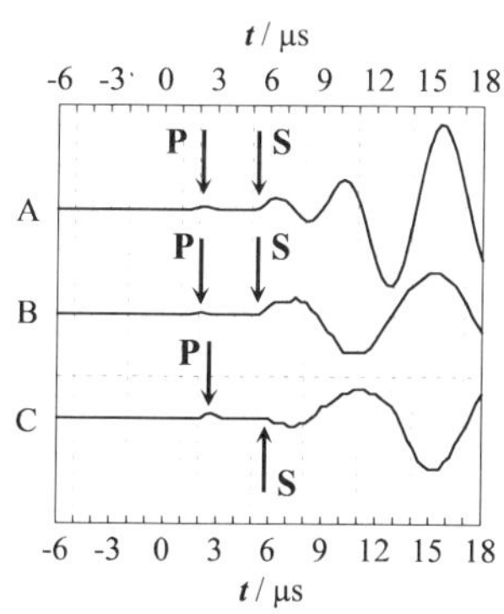

Figure 1 Example waveforms of bore wave velocity measurement

$\Delta t = 0.3$ ms, $t_P = 1.6$ms, $t_S = 5.4$ms

A: vertical component of measured wave; B, C: horizontal component of measured waves stimulated in two opposite directions seismic phase fitting is put forward below.

2 Fitting analysis of seismic phase in wave velocity test

2.1 *Basic features of P and S waves and relative rock mass parameters*

A seismic phase is a wave that arrived at the measuring point by a certain propagating paths with a certain particle vibration manner.[1] In the wave velocity test of rock mass in situ, both P and S waves all come directly from the stimulating point, but their vibration manners are different, and P wave is always ahead of S wave. The particle vibration of P wave is in the same direction with its propagating path, but the particle vibration of S wave is in the direction vertical to its propagating direction. If we stimulate the waves on the ground surface with a hammer hitting in two opposite horizontal directions, we shall record two kinds of S waves with opposite initial motions of particle vibrations as shown in Figure 1, whereas we cannot find any other initial motion of P wave but going down.

The velocities of P wave and S wave V_P and V_S can be written as follows:

$$V_P = \sqrt{\frac{E(1-\nu)}{\rho(1+\nu)(1-2\nu)}}, \quad V_S = \sqrt{\frac{\mu}{\rho}} = \sqrt{\frac{E}{2\rho(1+\nu)}} \tag{1}$$

in which E is the elastic modulus, μ is the shear modulus, ν is the Poisson's ratio, and ρ is the density. Therefore, the parameters E, μ and ν, which can describe the mechanical properties of rock masses, can be calculated from V_P and V_S as follows:

$$E = \frac{\rho V_S^2(3V_P^2 - 4V_S^2)}{V_P^2 - V_S^2}, \quad \mu = \rho V_S^2, \quad \nu = \frac{V_P^2 - 2V_S^2}{2(V_P^2 - V_S^2)} \tag{2}$$

2.2 *rules for seismic phase fitting*

So called "seismic phase fitting" is to adjust the travel time of a seismic phase slightly around the initial values of the travel time read from recorded waveforms so that proper wave velocities can be obtained. "Rational in physics" is the basic principle for the seismic phase fitting.

Several physical parameters such as the ratio V_P/V_S, the integrality index K_V, and Poisson's ratio ν etc. should be considered in the fitting analysis.

According to formula (1), the ratio V_P/V_S can be written as follows:

$$\frac{V_P}{V_S} = \sqrt{\frac{2(1-\nu)}{1-2\nu}} \tag{3}$$

When ν varies from 0 through 0.5, the ratio will take values from 1.414 to ∞. Generally, the value of ν for rocks is around 0.25, so the referential value of V_P/V_S for rock masses is about 1.732. On the other hand, the Poisson's ratio calculated by formula (2) with measured wave velocities V_P and V_S should also be $0 < \nu < 0.5$.

K_V is a parameter to assess the quality of rock masses, and it is defined as:

$$K_V = (V_{Pm})^2/(V_{Pr})^2 \tag{4}$$

where V_{Pm} is the V_P value from the wave velocity test of rock mass in situ, and V_{Pr} is the V_P value of rock specimens. In general, $V_{Pm} < V_{Pr}$, so that $0 < K_V < 1$.[2]

Another important rule in the fitting analysis is for the travel time. We should recognize that for the same seismic phase stimulated on the ground surface, no travel time at a relative small depth be greater than that at a larger depth.

In the light of rules discussed above, we can revise the error of the travel time and keep the physical parameters of rock masses in their rational ranges.

3 Application of seismic phase fitting to a project

3.1 General situation of the project

This method is applied to a project in Beijing. Boreholes Z1 and Z2 with depths more than 100m are drilled on a hillcrest to assess the quality of rock mass. A detector is put into the borehole and tightly coupled with the hole wall by a rubber bag full of air to measure the waves from the stimulating point on the ground surface about 3.8m away from the borehole. The interval between measuring points in the hole is 1m. The waves are generated with a hammer of 18lb hitting on two opposite vertical surfaces of rock mass so that the S waves can be clearly identified according to the opposite initial motion in the horizontal component as shown in Figure 1.

3.2 Result of wave velocity test

From the waveforms measured in site, the initial travel time data of P and S waves in the two test holes are obtained. By the approach of seismic phase fitting, the proper travel time of P and S waves in the borehole Z2 is determined as shown in Figure 2, and the wave velocity distribution in the hole is shown in Figure 3.

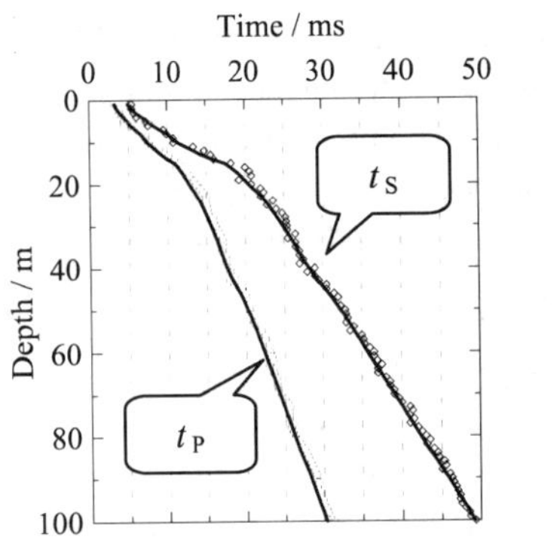

Figure 2 Curves of travel time vs. depth

o measured t_P, ◇ measured t_S, — fitting value

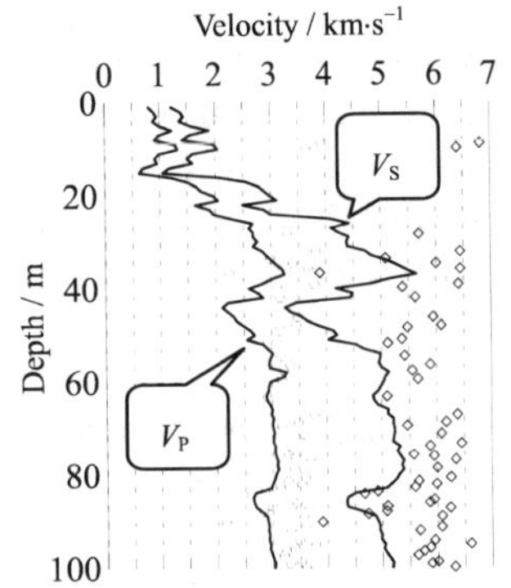

Figure 3 Curves of velocity vs. depth

o V_P of rock specimens

◇ V_S of rock specimens

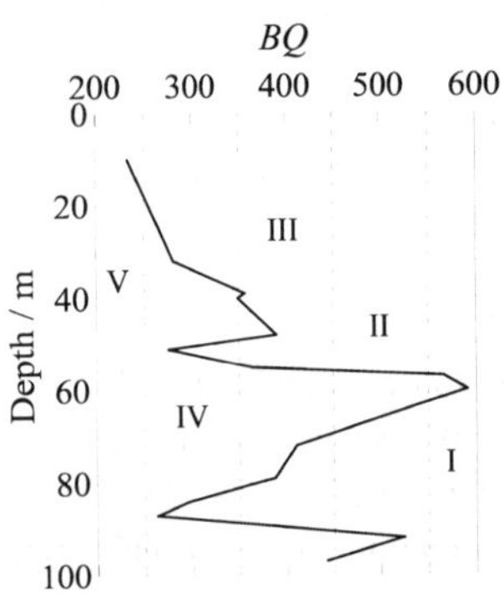

Figure 4 Curves of *BQ* vs. depth and classification of rock mass

In the light of the integrality index K_V of the rock mass and the uniaxial compressive strength *Rc* of rock specimens, the basic quality index *BQ* of the rock mass[2] is calculated as follows:

$$BQ = 90+3Rc+250\,K_V \tag{5}$$

The classification of the rock mass near the borehole Z2 by *BQ* is as shown in Figure 4.

The application result shows us that the method of seismic phase fitting is a feasible way to enhance the precise of the travel time in the wave velocity test of rock mass in situ.

References

1. Zhao Hongru, Sun Jinzhong, Tang Wenbang et al. *FULL SEISMIC PHASE ANALYSIS*. Beijing: Earthquake Press, 1991
2. State Standard of P. R. China. *STANDARD FOR ENGINEERING CLASSIFICATION OF ROCK MASSES* (GB 50218—94). Beijing: China Press of Planning, 1995

USING LASER HOLOGRAPHIC INTERFEROMETRY TO OBSERVE BEHAVIOUR OF ROCK DEFORMATION AND FRACTURE UNDER UNIAXIAL COMPRESSION

DONGMEI LIU
Department Of Construction Engineering, The Institute Of Southern Metallurgy Ganzhou city,Jangxi province ,China,Liudm21@163.net

JINPING XIE
Department Of Physics,Gannan Teachers College,Ganzhou city,Jiangxi province,China

YONGSHENG GONG YUBIN ZHOU
Department Of Construction Engineering, The Institute Of Southern Metallurgy Ganzhou city,Jiangxi province,China

This paper uses double exposure holographic interferometry to observe the wholly process of deformation and fracture of rock samples under uniaxial compression. The interference fringes patterns on holograms show that the behavior of deformation and fracture in samples is remarkably different. In being loaded process, the macro-deformation on serpentine sample has been contributed by linear elastic deformation of parent serpentine, the slate's deformation is caused by the deformation of linear elasticity and original closed defects, and the macro-deformation on metasandstone sample is dominantly due to the extension of cracks, obviously the granite's macro-deformation is due to the close of original defects and the extension of cracks. The experimental observation provide a proof that the various forms of rock deformation and failure is derived from rock heterogeneity and complexity of crack evolving path from the germination to growth and coalescence

1 Introduction

It is well known that rock micro-cracks from germination to growth and coalescence determine its properties of deformation and failure, and the rock fracture is highly interrelated to microstructures and fundamental flaws in it. Based on the stress-strain curves for rocks, the macroscopic nonlinear deformation is divided into three parts: linear elastic deformation in native rock, nonlinear deformation resulted by primitive defects closure or propagation and cracks growth. Owing to the complexity of rock's component and its structure, each part contributing to total deformation is not ascertained effectively at present. Therefore it is interesting and necessary that the newly test method for rock mechanics should be probed. As we know, laser holographic interferometry with the unique advantages has been widely used in the nondestructive test of materials, especially is suitable for the micro-deformation measurements in materials.[1] Using double exposure holographic interferometry, authors have observed the wholly process of deformation and failure for rock samples under uniaxial compression. Holographic interferograms corresponding to the stress-strain curves for rock samples are obtained.

2 Experimental Measurement

Based on the principle of holographic interferometry,[2] the optical layout is designed by use of He-Ne laser as light source and holographic emulsion plate as a record medium. Each sample is compressed in the loading frame which has enough rigidity and is placed in optical system with oil jack whose maximum pressure value measured by a sensor is 15 ton under the condition of maintaining a constant loading rate and controlling a rational loading difference between double exposure. Meanwhile the strain gage rosette stuck on samples is connected with strain wire gauge in order to measure the strain value simultaneously. Granite, meta-sandstone, slate and serpentine are used as rock samples. The dimension of these samples is 20 (mm)×40(mm)×50(mm),20(mm)×40.2(mm)×50.1(mm),21.7 (mm)×41.6(mm)×47.8(mm),20.6(mm)×40.8(mm)×49.5(mm) respectively. The homogeneity degree is serpentine>slate> meta-sandstone> granite.

3 Experimental result and analysis

The holographic interferograms and the stress-strain curves on the samples are obtained in experiment. Based on the explanation of holographic interference fringes for materials[2] the experimental result can be interpreted as follows.

The interference fringes on the serpentine sample are continuous and well distributed, not being distortion. The strike changes gradually from inclination to perpendicular and horizontal as shown in figure 1(1-1.1-2.1-3.1-4.1-5). It shows that there is no internal cracks growth on the sample during the process of the loading and the local deformation is fundamentally equal. On the analysis of the holograms and the stress strain curve in figure2, the macro-deformation for serpentine sample is essentially composed of linear elastic deformation in native rock.

As seen in figure 1(1-6.1-7.1-8.1-9) and the corresponding curve points in figure3, the distortion zone of the interference fringes initially emerges from the right below region located on the slate sample and developed gradually towards the right above and central region with the increase of loading. It represents that the crack growth path in sample preferentially takes along the primitive crack on the right below region, then the rock micro-cracks on the right above region are propagated and evolved, finally both coupling cause splitting fracture of rock. So that the macroscopic nonlinear deformation on the sample is contributed by linear elastic deformation in native rock and the deformation from the original flows closed. It caused by the macro-cracks extension accounts for minimum.

On the basis of references3,4 and figure1(1-10.1-11.1-12.1-13) & figure4, the size of domain of micro-crack growth on meta-sandstone sample grow larger continuously

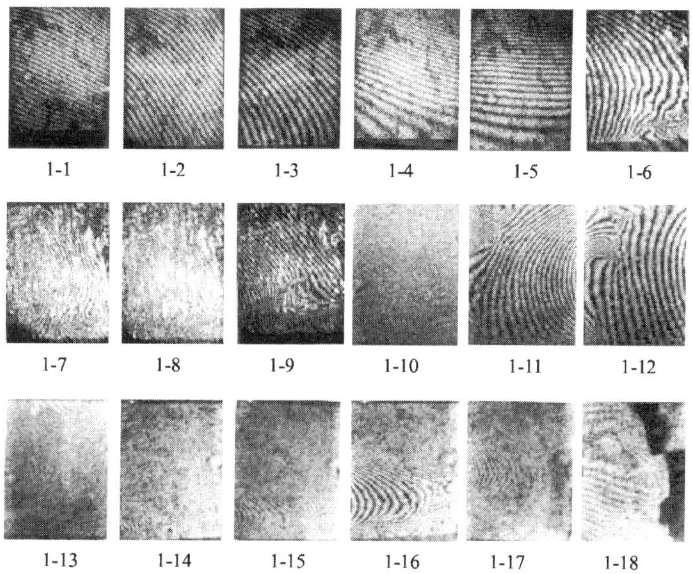

Figure 1. Holographic interferograms of rock samples

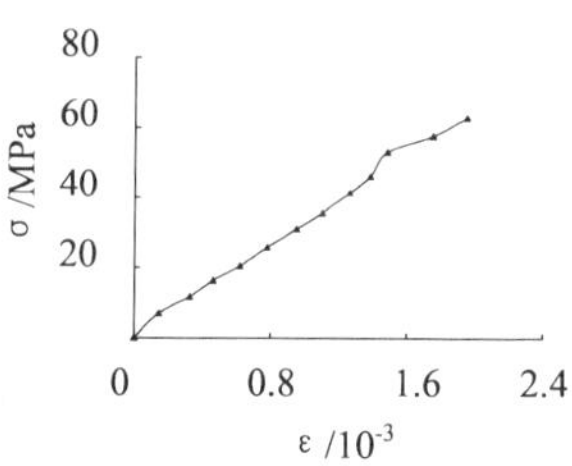

Figure 2. Stress-strain curve for serpentine

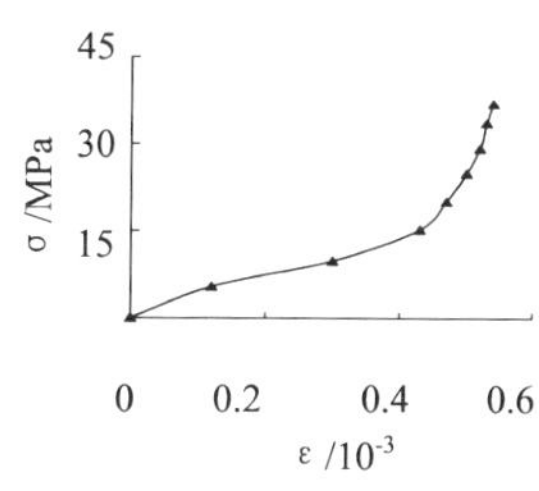

Figure 3. Stress-strain curve for slate

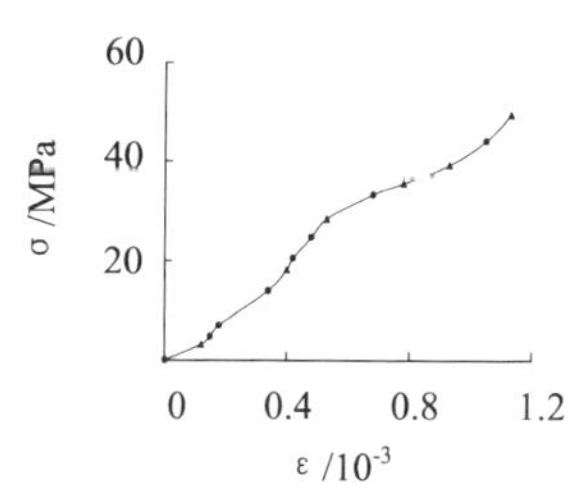

Figure 4. Stress-strain curve for meta-sandstone

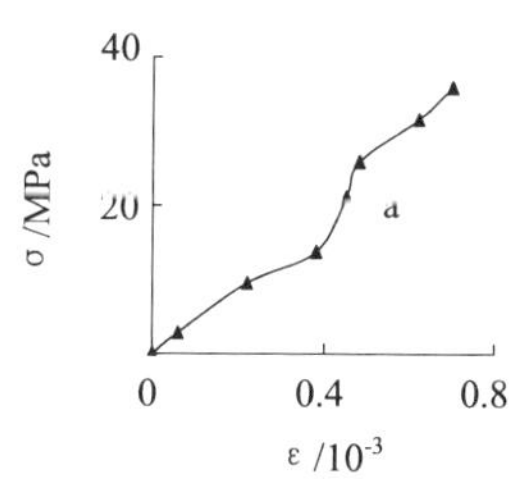

Figure 5. Stress-strain curve for granite

and orderly , and the steady development of the main crack takes along weak join, simultaneously the extension micro-cracks germinate and the bifurcation cracks evolve. When the ratio of compressive stress to rupture stress equals 93.7percent., the main crack penetrates quickly. Then the macro-deformation is dominantly due to the extension of cracks. However, there is an abrupt mutation of cracks within granite sample as shown in figure1 (1-14.1-15.1-16.1-17.1-18). Where the main crack formed is corresponding to the curve point a in fig.5. The macro-deformation preceding point a is resulted by primitive flaws closure, that behind point a is derived from the main crack developing and gliding along itself.

4 Conclusion

The holographic interference fringes reappear visually rock cracks evolving path and deforming behavior, the fringes distortion patterns and the corresponding variable tendency in the stress-strain curves for rock samples indicates the proportion of three components to rock macro-deformation. The experimental results provide a proof that rock cracks evolving path which is influenced remarkably by the heterogeneity of component and structure as well as the difference of primitive damage scale result in the local deformation, it restrain the quantity of rock deformation and the failure patterns. Therefore it is feasible to use laser holographic interferometry to test behavior of rock deformation and fracture under loading.

Acknowledgements

This work is supported by a grant for national natural science foundation of china. The item number is 50164004 .The authors also thank our colleagues for their help.

References

1. GuoguangYang, Modern optical testing technology (Zhejiang University Press, Hangzhou, 1997).
2. Junzhe Zhang, Nondestructive monitor tech.(Academic Press , Beijing, 1993)
3. Dongmei Liu et al, Application of holographic interferometry to measurement of rock damage growth, J. Univ. Sci.& Tech. China, D 31, 199 (2001).
4. Dongmei Liu et al, Double-exposure holographic interferometry testing study on process of deformation and failure for rock, Chinese J. Rock Mech. & Eng. D 20, 944 (2001).

EXPERIMENTAL INVESTIGATIONS INTO FLOOR BEARING STRENGTH OF JOINTED AND LAYERED ROCK MASS

DHEERAJ KUMAR, DR. SAMIR KUMAR DAS
Department of Mining Engineering, Indian Institute of Technology, Kharagpur, W.B, India
E-mail:samirdas@mining.iitkgp.ernet.in

This paper presents an evaluation of the floor bearing strength of soft floor strata and the influence of various parameters on it. A series of model plate loading test on massive and jointed floor strata were carried out under confinement, in which the floor rock were simulated with the help of sand- cement mixture using equivalent material modeling technique (Chug et al., 1997). Physical modeling technique has been used for the laboratory investigation purpose. All the testing were done under confinement in the self developed testing blocks resembling the in-situ condition which are contrary to the work done earlier (Ramamurthy, 1995). Plate loading tests were carried out for varying size and shape of footing plates. The nature, orientation and the location of joints present in the floor rocks were also varied. Based on the findings it was seen that for a given condition floor bearing strength is strongly related to footing plate size and shape. The investigation has indicated that for a given condition ultimate bearing strength decreases with the increase in the plate size and normalized settlement also decreases considerably. Based on the result obtained from the scaled model tests on jointed floor rocks it was found that bearing strength of floor rock is also related to location and orientation of the joints present there in. Bearing strength of jointed rock mass is almost 0.6 times the bearing strength of intact rock mass for joint located vertically. The ratio reduced to almost 0.3 for joint located at an angle of 80° with the direction of loading. In this paper the practical consequences of the investigation are also discussed.

1 Model Studies

The strength of firm and soft layers are directly related to the bearing strength of the mine floor. The investigations of the bearing strength of mine floor become an integral part of the study of mine stability. Scale model test offers another approach by which the variability of site conditions and rock properties can be studied to achieve an economical design. To arrive at a realistic value of floor bearing strength, laboratory tests on both intact as well as jointed floor rock strata are conducted with physical scaled model in the case of a surface footing. The prime objectives of this paper are the influence of the following parameters on floor bearing strength

- Footing plate geometry [size (d) and shape(circular, square, rectangular)]
- Presence and location of joint plane
    a) on the edge of the footing
    b) in the middle of the footing

To accomplish these objectives, scaled model specimens are prepared in the laboratory with the help of sand – cement mixture (fulfilling the similarity conditions for intact rocks [Mostyn et al., 1995]) in different proportion so as to achieve the uniaxial compressive strength of the range 2MPa – 20MPa to represent soft floor strata. Prior to the construction of physical model specimen, the composition and properties of material to be used for the purpose were studied in details.

Figure 1. Experimental setup for the determination of floor bearing strength of model strata

Test frames are specially fabricated for containing the model floor strata (Fig.1). These test frames are developed to offer confinement to the scaled model strata as to simulate the actual field condition. The size of the samples and the footing plates are decided on the basis of the zone of influence of load exerted on the footing plate and the space available in the compression testing machine. For each test of bearing strength, parameters like, longitudinal and lateral strains as a function of applied load, strain at failure and the vertical settlement of footing at the time of floor strata failure, were also recorded.

2 Result and Analysis

Physical modeling of underground soft floor have been carried out with a Geo-technical modeling scale (C) of 1:25 which assisted to model a floor strata up to 2m in thickness and 4.25 mX4.25m in size (Kumar et al. , 2002).

3.1 Effect of Footing Geometry on the Bearing Strength of Floor Strata.

The tests for the effect of footing geometry (both size and shape) were conducted on the various model strata having same physico-mechanical properties by varying the size of the footing. For each of the footing sizes three sets of tests were conducted and the average bearing strength was calculated. The Table 1 shows the Physico-mechanical properties of model strata. Table 2 shows floor bearing strength for different size and shapes of footing plates.

Table 1. Physico-mechanical properties of model strata

Physico-mechanical properties	Model strata.[experimental]	Prototype Strata.[calculated]
Shape	Square.	Square
Dimension (LxW)	17cmX17cm.	4.25mX4.25m
Thickness (t)	4cm,8cm,12cm,16cm,20cm.	1m,2m,3m,4m,5m
Strata condition	Massive.	Massive.
UCS (σ_c)	0.6 MPa	16 MPa
Bulk density	1785 kg/m^3	2000 kg/m^3
Strata composition Ratio (Cement: sand) (Cement: water)	Sand, cement, water 1:1.70 1:1.80	-------
Footing size/strata thickness (d/t)	0.625	0.625
Footing plate shape	Circular, square, rectangular	--------
Footing plate size	2.5cm, 5.0cm, 7.5 cm, 10.0cm, 12.5 cm	0.625m, 1.25m, 1.875m, 2.5m, 3.125m

Table 2. Floor bearing strength for different size and shape of footing plates (central footing)

Footing geometry(d), cm	Bearing Strength(Bs), MPa		
	Circular footing	Square footing	Rectangular footing (aspect ratio 1:1.33)
2.5	10.5	11.05	7.24
5.0	7.9	8.20	4.63
7.5	4.56	4.96	4.19
10.	3	3.75	2.82
12.5	2.32	2.98	2.01

Figure 2 shows the effect of footing size and shape on the bearing strength of floor strata for different types of footings.

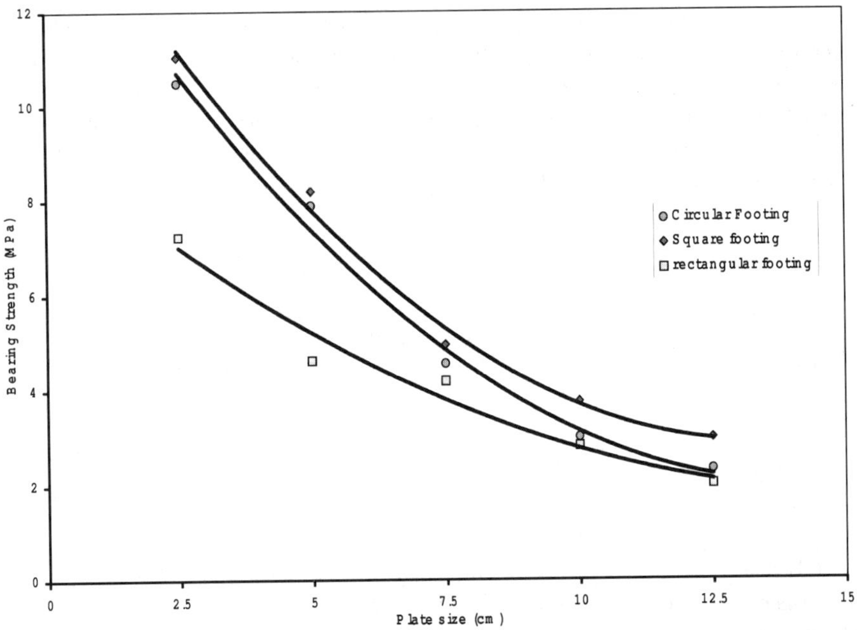

Figure 2. Effect of footing geometry on bearing strength for circular, square and rectangular footing

From the graph it is clear that the bearing strength reduces polynomially with the increase in the plate size for all the shapes of footing plate. Square plate footings offer maximum floor bearing strength as compared to other shapes of footing plates.

3.2 Influence of joint (orientation and location) on floor bearing strength

The tests for knowing impact of presence and location of weakness plane (joint) on the bearing strength of the floor strata were carried out on various sets of simulated floor strata having same dimensions and physico-mechanical properties as mentioned in table 1. The thickness of model floor strata was kept constant to 8cm. The single set of joint was introduced into the floor strata with the help of thin mica sheets at the time of casting. The strength and aperture of the joint was kept constant for all the tests. The same circular footing of size 5cm was used for all the tests. The orientations of joint were varied in step of 20° with respect to the direction of loading i.e. vertical. The tests were carried out for both the following cases of joint location:

(a) when the joint located at the middle of surface strip footing.
(b) when the joint located at the edge of surface strip footing.

Table 3 shows the joint properties of model floor strata and corresponding prototype representation.

Table 3. Properties of the joint introduced in the model floor strata

Joint properties	Model strata	Prototype floor rock
Joint surface	Smooth	Smooth
Joint filling	Soft	Soft
Joint aperture	0.2mm	5mm.
Joint spacing	Remote (7cm)	Remote (>2m)

Figure 3 shows the graph of Bs_j/Bs_i versus Joint inclination with respect to direction of loading for central and edge footing (strip footing).

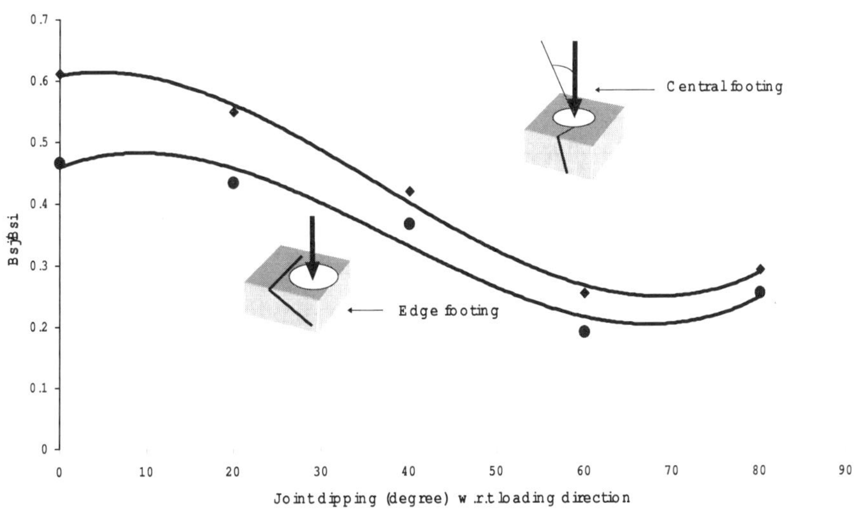

Figure 3. Influence of joint and their location on floor bearing strength

It is evident from the graph that the Bearing Strength (Bs_j) for jointed rock is about 0.6 times the Bearing Strength of the intact rock (Bs_i) when the joint is located almost vertical ($\beta = 0^o$).

The said ratio (Bs_j/Bs_i) decreased to about 0.3 at $\beta = 80^o$ in case of footing located at the center. For the joint located at edge of the footing the ratio (Bs_j/Bs_i) is reduced to almost 0.45 at $\beta = 0^o$ and it is 0.25 for $\beta = 80^o$.

3 Conclusion

From the experimental outcomes following conclusions may be drawn.

- The floor bearing strength (Bs) decreases considerably with increase of footing plate size for all the shape of footing plates .For the same plate size of diameter (d) for circular footing, length (l) or width (b) for square footing and length (l) for rectangular footing (aspect ratio of 1:1.33), square plate footings offers maximum floor bearing strength .
- The mode of failure of solid dry rock under low pressure and room temperature under a flat-ended punch (footing) is dependent upon the presence of weakness planes and intrusion.
- Tensile cracks initiates at the rim of the footing forcing a depression beneath the footing with stable crack growth with increasing load. With increased loads, the depression may eventually crush with growing unstable cracks and with fracture reaching the surface.
- In most of the cases the bearing strength (Bs) is not limited with characteristics of homogeneous and isotropic rock mass, but with characteristics of joints and layers which form rocks. The bearing strength (Bs_j) for jointed rock is about 0.6 times the bearing strength of the intact rock (Bs_i) for the joint located almost vertical ($\beta = 0^o$) and it decreases to about 0.3 at $\beta = 80^o$ in case of footing located at the center.
- For the joint located at edge of the footing the ratio (Bs_j/Bs_i) is reduced to almost 0.45 at $\beta = 0^o$ and it is 0.25 for $\beta = 80^o$.

References

1. Chugh, Yoginder P. Singh, T.N Singh, V.K. Ober, S. Development of an Equivalent Material Facility, *Department of Mining Engineering, Southern Illinois University*, Carbondale, IL 62901,9-83, (1997).
2. Ramamurthy, T. Bearing Capacity of Rock Foundation, *Rock Foundation*. Balkema, 1995, 311- 318, (1995).
3. Mostyn, G. R. Bagheripour, M. H. New model material to simulate rock. *Mech. of Jointed and Faulted Rock*, Rossmanith (ed.), Balkema, Rotterdam, 225- 230, (1995).
4. Kumar, D. and Das, S. K. Evaluation of Floor Bearing Strength of Indian Coal Measure Strata by Physical Modeling, *Proceedings of RockEx 2001, National seminar on Rock Excavation Techniques*, 28 -38, 21 -22 September (2001).

TUNNEL EXCAVATION USING THE GEOLOGICALCONDITIONS BY REFLECTION SEISMIC SURVEY

KAZUHIRO ONUMA, HIROYOSHI KASA, HIROYUKI YAMAMOTO
Hazama Corporation, 2-5-8, Kita-Aoyama, Minato-Ku, Tokyo, 107-8658, Japan
E-mail: onumakaz@hazama.co.jp

When the tunnel is long such as some few kilometers in length, the tunnel tends to situate deeply under the ground. We can't investigate the geological conditions correctly on the ground surface, so try to investigate them in the tunnel. We studied the tunnel about 1,100m long before excavation, and investigated surrounding rock by using the reflection seismic survey. TSP (Tunnel Seismic Prediction) is one of some reflection seismic surveys, we set several shot points (24 points usually), one receiver and a recorder (PC), and pick up the reflection seismic wave occurred at reflector . Then we analyze the reflection seismic wave, and we can predict the geological conditions by reflector in the front of tunnel face about 100 m. in 1.5 day. At last, the geological conditions have been understood and the tunnel could be constructed successfully by using the reflection seismic survey in the aspect of tunnel supports and the safe control, also in the prediction of the physical properties.

1 Outline of Reflection Seismic Survey in The Tunnel

In the plan and design of a tunnel, the geological conditions are one of the important factors for construction of the tunnel in quality, delivery, cost etc. We must investigate geological conditions of tunnel by indoor test, geophysical survey and geological survey etc. When the tunnel is long such as some few km in length, the tunnel tends to situate deeply under the ground. In such case the geological conditions could not be correctly investigated on the ground surface, so investigation always fulfilled in the tunnel. Some methods such as level bowling survey and reflection seismic survey, underground radar etc. can be used in the tunnel. Among these, the reflection seismic survey carries out TSP in the tunnel . TSP[1)] (Tunnel Seismic Prediction) is one of some reflection seismic surveys as shown in Fig. 1, we set several shot points (24

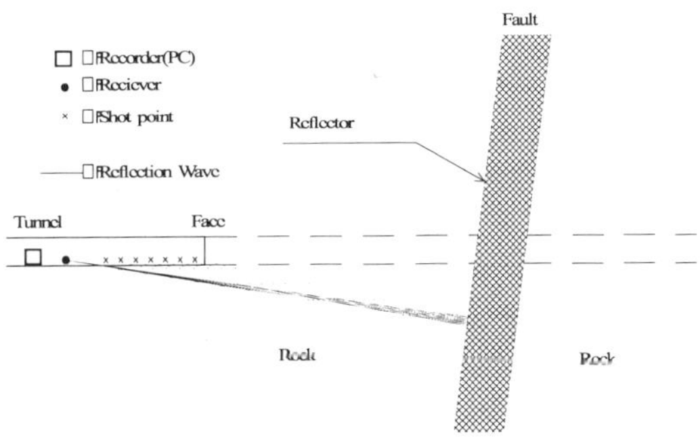

Figure 1. Reflection seismic survey (TSP)

points usually), one receiver and a recorder (PC), then record the reflection seismic wave occurred at reflector like fault etc. After analyzing the reflection seismic wave, it is possible to discover the change of the sound impedance at the reflector from the phase of the reflection wave. We can predict the geological conditions by reflector in the front of tunnel face about 100 m. in 1.5 day. Therefore the geological conditions could be estimated by using the combination of the reflection pattern shown in Table 1[2].

However, TSP only gave the relative evaluation. The physical quantity could not be decided directly from the P wave velocity at the front of tunnel face. We use the TSP mainly for the reasonable design and construction of the tunnel, but not for the special investigation of the geological parameters.

In this study, we carried out reflection seismic survey (TSP)[3], in order to construct the tunnel rationally.

Table 1. Geological condition estimated by TSP[2]

Reflection Type	Rock Strength Changes (Face Side)			Estimated Geological Structure		
Fault	Increasing	Decreasing			Fault	
Dyke	Decreasing	Increasing			Dyke	
Boundary A	Decreasing			Hard Rock		Soft Rock
Boundary B	Increasing			Soft Rock		Hard Rock

2 Tunnel construction to use reflection seismic survey (TSP)

2.1 Survey plan

Studied tunnel has a length of about 1,100 m, and constructed by the NATM. The rock of tunnel is composed mainly by rhyolite formed at Paleogene , and partly by granite. The physical properties of the rhyolite are as follows: density 2.65g/cm^3 (mean), unconfined compression strength 1.3 MPa (mean) by indoor tests, and P wave speed
(Vp) is 4.0~4.5 km/sec by seismic survey on the ground surface.

Before tunnel construction, we knew that there were 11 sections of poor geological conditions like faults and low velocity zones (Vp = 2.1~3.6km/sec). So we decided to predict the geological conditions by TSP, altogether we arranged 5 times of surveys. The objects in each survey were shown in table 2, and the survey positions were shown in Figure 2.

Table2. Investigation Plan

Survey No.	Available Survey Length	Survey Objects
1	50m	• Faults: f-2 • f-3 • f-4 • Low velocity Zone (Vp=2.3km/s)
2	100m	• Low velocity Zone(Vp=3.6km/s)
3	100m	• Faults: f-5 • f-6 • Low velocity Zone(Vp=2.1km/s)
4	70m	• Low velocity Zone(Vp=3.6km/s)
5	80m	• Low velocity Zone(Vp=3.6km/s)

According the results from TSP, the tunnel supports[4] were designed for the section B~CI mainly, as well as for the section CII~D, where the geological conditions were
very poor.

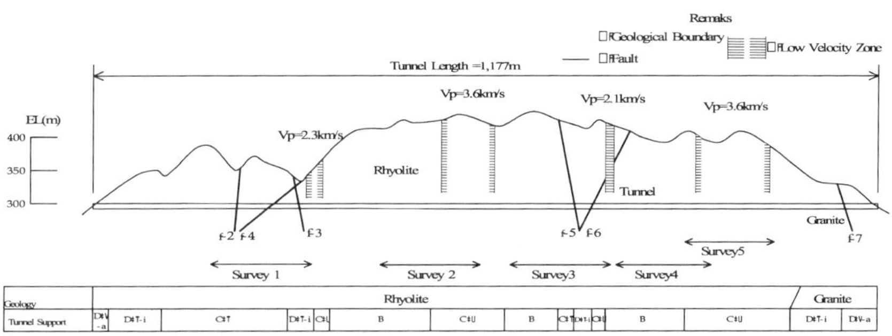

Figure 2. Geological section along tunnel axis (in design)

2.2. Survey results

According the results from surveys by TSP in the tunnel, 29 sections (No. 1-1 • 5 4) with poor geological conditions have been discovered later. The geology of the tunnel faces has been recorded in the process of excavation, and compared with the prediction results obtained from the TSP survey sections. The comparison of poor geological conditions in practice to TSP predictive results is shown in table 3.

From table 3, we can obtain following rules:

1) 29 poor geological conditions were discovered in the process of excavation, and 20 sections were predicted by TSP. So, the probability of hitting in prediction of poor geological conditions was about 70% by TSP.

2) the errors of location predicted by TSP were within±5 meters almost.

Table 3. TSP investigation result

*: Available Survey Length

Sur. No. (*m)	Design Support	Excavation Result: Bad Geological Conditions	Excavation Result: Width (m)	Excavation Result: Remarks	Excavation Result: Support	TSP Survey: Reflector	TSP Survey: Width (m)	TSP Survey: Difference of Location(m)
1 (50)	C•	1-1	1	Joint	D•-i	×	•	•
		1-2	1	Joint		•	<5	-2.5
		1-3	2	Cracky zone		•	<5	+9.0
		1-4	1	Joint		•	<5	-3.5
		1-5	1	Fault		•	<5	-4.0
2 (100)	B	2-1	1	Joint	C•	•	<5	+0.5
		2-2	1	Joint		•	<5	+0.5
	C•	2-3	2	Fault		•	<5	-4.5
		2-4	2	Cracky zone		•	<5	-5.0
		2-5	2	Cracky zone		×	•	•
		2-6	2	Cracky zone		×	•	•
		2-7	2	Cracky zone		×	•	•
		2-8	1	Joint		•	<5	-0.5
		2-9	**20<**	**Fault**		**•**	**10<**	**-5.0**
3 (100)	B	3-1	1	Fault	C•	•	<5	+2.0
	C•	3-2	2	Cracky zone		•	5•10	+2.0
	D•-i	3-3	1	Fault		×	•	•
	C•	**3-4**	**20<**	**Fault**	**D•-i**	**•**	**20<**	**-3.5**
	B	3-5	4	Cracky zone		•	<5	-2.5
4 (70)	B	4-1	1	Cracky zone	C•	×	•	•
		4-2	1	Cracky zone		×	•	•
		4-3	5	Cracky zone		×	•	•
		4-4	**20**	**Fault**	D•-i	**•**	**20<**	**-6.0**
	C•	4-5	1	Fault		•	<5	-5.0
		4-6	6	Cracky zone		•	10<	+2.5
5 (80)	C•	5-1	8	Cracky zone	D•-i	•	<5	+3.5
		5-2	8	Cracky zone		•	<5	+3.5
		5-3	13	Fault		×	•	•
		5-4	**30<**	**Fault**		**•**	**10<**	**0.0**
Total		29				20		

2.3 Change of the tunnel supports

TSP survey gave the relative evaluation for the geological conditions, and the tunnel supports have been designed referred to a flowchart shown at figure3. Especially, the design of tunnel support should be changed while the width of poor geological conditions was 10m or over 10m (table 3, bold style). When we excavated the tunnel, we predicted the wide band fault and attendants due to the help of reflector. In No.3-4 and No.4-4, the small-scale rock falls were occurred; therefore, we changed the design of the tunnel supports from CII to DI-i.

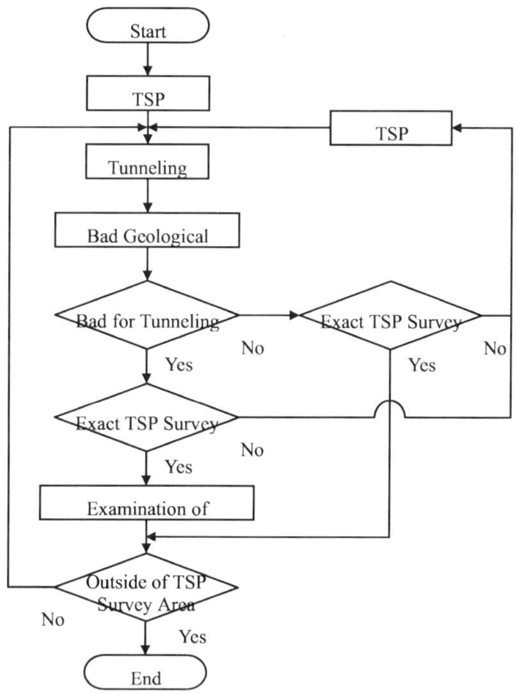

Figure3. Flow chart of tunneling

2.4 Effects of reflection seismic survey (TSP)

The effects of the reflection seismic survey (TSP) in the tunnel are as follow.

(1) Speedy and economical research method

We carried out the TSP 5 times all together, and every time the report could be finished within the time merely about 1.5 day, Therefore we could grasp the geological conditions of a long distance section within very short period. This is showing that we can investigate and obtain results quickly in comparison with the other geophysics surveys and bowling surveys etc. So we can construct the tunnel safely and economically

(2) The change of tunnel supports

We could change the design of tunnel supports by predicting the change of geological conditions.

(3) The security of safety

While poor geological condition would be reached in excavation according the prediction of reflectors, we checked the practical rock condition by drilling at the face. So we could get the security of safety in the tunnel construction.

(4) The confirmation of the geophysical properties

The geophysical properties of rock could be checked by comparison with the values calculated according the measured P wave velocity in the direct wave of TSP, and could be used more accurately in the design.

3 Conclusion

A lot of long tunnels situated deeply at the underground, the geological conditions of these tunnels were difficult to predict in practice, therefore an effective predicting tool and method should be investigated for the excavation of tunnel.

We studied the tunnel about 1,100m long before excavation, and investigated the geological condition by using the reflection seismic survey (TSP) for a long time. The results of application show that the geological conditions could be easily grasped, and the construction of the tunnel was success due to the use of reflection seismic survey not only in the change of design of tunnel supports and the safe production, but also in the prediction of physical properties.

References

1. G. Sattel, P. Frey and R. Amberg, Prediction ahead of tunnel face bu seismic methods – pilot project in Centovalli Tunnel, Locarno, Switzerland. First break vol.10, No.1, 1992
2. H. Kasa, T. Igari, H. Yamamoto and N. Maeda, A study on exploration accuracy of tunnel seismic prediction method, Proceeding of Tunnel Technology Conference of JSCE, Vol.6, pp.95-100 (in Japanese), 1996
3. H. Yamamoto, H. Kasa, K. Onuma and A. Kirihara, Utilization of TPS's data for the determination of the best support pattern, Proceeding of 56th Annual Conference of JSCE, Part 6,pp.458-459 (in Japanese), 2002
4. Japan Road Assoc., Stand. for road tunnel, Structures, p.273 (in Japanese), 1989

A STRENGTH MODEL FOR JOINTED ROCKS UNDER UNIAXIAL COMPRESSION

KAI DUAN[1], YUEMEI LIN[2], XIAOZHI HU[1]

[1] *Department of Mechanical & Materials Engineering, University of Western Australia 35 Stirling Highway, Crawley, WA 6009, Australia*

[2] *College of Resources and Civil Engineering, Northeastern University, Shenyang, P.R. China*

A strength model is proposed to characterise the strength behaviour of jointed rocks subjected to uniaxial compression. The model consists of several equations to describe different failure mechanisms and associated strengths of jointed rocks, which vary with the relative angle between the applied load and weak planes. Furthermore, the model is found to provide a reasonable prediction and explanation when applied to the experiments on a jointed rock.

1 Introduction

Natural rock masses often contain various weak geological structures such as faults, joints and bedding planes[1]. These weak structures or planes substantially alter the mechanical responses of rock masses and therefore, play a very important role in determining the performance and stability of rock structures built in, on or of rock masses. As jointed rocks are basic elements of rock masses and also, are structurally similar to rock masses[1], their strength and deformation properties have attracted considerable attention in both fundamental study and applications of rock mechanics.

Joint planes often lead to the direction dependence of the strength and deformation of jointed rocks, i.e. the strength and deformation exhibit anisotropy[2]. The joint-induced anisotropy is particularly strong for uniaxial compression, and decreases with increasing confining pressure for triaxial compression[2-3]. Therefore, uniaxial compression tests are particularly suitable for studying the joint-induced anisotropic properties.

Extensive experimental studies on the compression failure of jointed rocks have showed four basic failure mechanisms, i.e. tension-cleavage and shear failure of both the weak planes and intact rock material[4-7]. The presence of a particular failure mode or a combination of these failure modes is dependent on the angle between the compression load and a joint plane. Generally, compressive load parallel to weak planes often leads to tension-induced failure along the weak planes while perpendicular compressive load induces either shear or cleavage failure of the rock material[4-7]. For the angles of 20° to 70°, the shear-induced sliding is often observed[4-7].

The first comprehensive strength model for jointed rocks was proposed by Jaeger[8] in 1960, which consists of two Coulomb-Mohr formulae to describe the shear failure in a joint plane and intact rock material respectively. Subsequently, various strength models have been developed to describe the failure mechanisms and associated strengths of jointed rocks[5, 9-10]. Although supported by experimental observations, these models apply basically only to the transversely isotropic strength behaviour associated with isotropic shear strength in a joint plane.

However, observations have shown the in-plane anisotropy of the shear strength of joint planes, i.e. the shear strength of joints changes with the shear direction even if the normal pressure is kept a constant[6-7, 11-12]. This paper aims at establishing a strength model for the in-plane anisotropy-associated shear strength of jointed rocks. The strength model consists of several failure criteria, and is used to analyse strength data measured from a jointed rock.

2 Strength Model

For a jointed rock consisting of homogeneous isotropic rock material and one set of weak planes, the anisotropic mechanical properties result from the weak planes, and

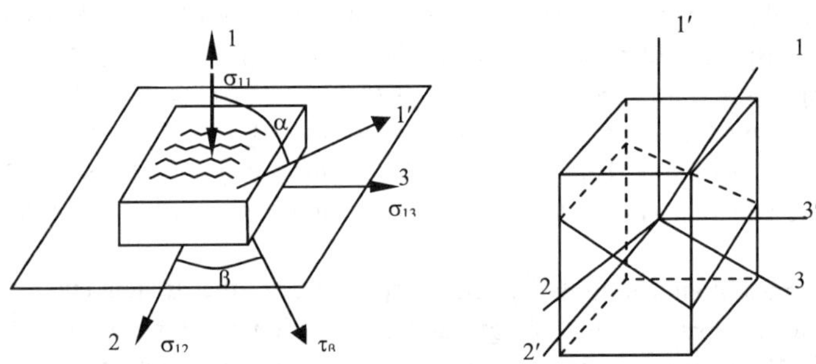

Figure 1. The in-plane shear anisotropy of joint plane and designated Cartesian coordinates

the orientation-dependence of joint surface textures often leads to the direction dependence of the shear strength of the weak planes. For the convenience of the following analysis, two Cartesian coordinate systems of O-123 and O-1′2′3′, corresponding to the joints and applied stress respectively, are attached to the jointed rock where a uniaxial compression load is applied in the 1′ direction (Fig.1).

For the rock given in Fig.1, the uniaxial strength σ_f due to the tension-cleavage failure of the weak planes is related to the stiffness of joint plane-formed plate, Q_w

$$\sigma_f = \sigma_{1'1'} = \kappa_w \cdot Q_w \tag{1}$$

where κ_w is a material constant depending on the joint frequency. Similarly, the strength associated with the tension-cleavage failure in rock material is related to the stiffness of the cleavage-produced plate, Q_r by

$$\sigma_f = \sigma_{1'1'} = \kappa_r \cdot Q_r \tag{2}$$

where κ_r is a material constant related to the crack frequency. As many previous studies, the classical Coulomb-Mohr criterion is used here to describe the relationship between the shear strength τ_f of rock material and the normal stress σ on the friction plane,

$$\tau_f = c_r + \mu_r \cdot \sigma \tag{3}$$

where c_r and μ_r are the cohesive strength and friction coefficient of the rock material, respectively. The uniaxial compression strength σ_f can be derived from Eq (3)

$$\sigma_f = 2c_r \Big/ \left(\sqrt{1+\mu_r^2} - \mu_r\right) \tag{4}$$

Jaeger[8] extended Eq (3) to the shear failure of a joint plane by replacing c_r and μ_r with the cohesive strength c_w and friction coefficient μ_w for the weak plane. Because of the in-plane shear anisotropy, c_w and μ_w must change with the shear orientation, and therefore, the shear strength of a weak plane in the β direction (Fig.1a), τ_β can be written as

$$\tau_\beta = c_w(\beta) + \mu_w(\beta)\sigma_{11} \tag{5}$$

To derive the uniaxial compression strength from Eq (5), the stresses $\sigma_{i'j'}$ given in O-1′2′3′ have to be transformed to the stresses σ_{ij} in O-123 (Fig.1b)

$$\sigma_{i'j'} = l^i_{i'} l^j_{j'} \sigma_{ij} \quad \& \quad \sigma_{ij} = l^{i'}_i l^{j'}_j \sigma_{i'j'} \qquad (i, j, i', j' = 1,2,3) \tag{6}$$

where $l^i_{i'}$, $l^j_{j'}$, $l^{i'}_i$ and $l^{j'}_j$ are the cosines for coordinate transforming, which are related each other by

$$l^i_{i'} l^{j'}_i = \delta^{j'}_{i'} \quad \& \quad l^{i'}_i l^j_{i'} = \delta^j_i \tag{7}$$

where $\delta^{j'}_{i'}$ and δ^j_i are Kronecker delta. Substituting $\sigma_{i'j'} = 0$, $\sigma_{11} \neq 0$ and Eq (6) into (5), the uniaxial compression strength σ_f is expressed as

$$\left.\begin{aligned} &\sigma_f = \sigma_{11} = c_w(\beta) \big/ \left[l^1_1 l^1_2 \cos\beta + l^1_1 l^1_3 \sin\beta - l^1_1 l^1_1 \mu_w(\beta)\right] \\ &\tan\beta = l^1_3 / l^1_2 \end{aligned}\right\} \tag{8}$$

For a jointed rock containing n sets of joints, the coordinate O-123 is attached to one set of joints, O_k-123 (k = 1, 2, … n-1) are attached to the other n-1 sets of joints. The stresses σ_{kmn} in O_k-123 are related to the stresses σ_{ij} in O-123 and applied stress $\sigma_{i'j'}$ by

$$\sigma_{kmn} = m^i_{km} m^j_{kn} \sigma_{ij} = m^i_{km} m^j_{kn} l^{i'}_i l^{j'}_j \sigma_{i'j'} \tag{9}$$

where m^i_{km}'s are the transforming cosines between these Cartesian coordinates.

In a multi-set joint rock, the plate formed due to the tension-cleavage failure of kth set of joints includes other joints. Therefore, the compression strength due to this mechanism is anisotropic, and Eq (1) for the kth set of joints is re-written as

$$\sigma_f = \kappa_{kw} \cdot Q_{kw}(\beta_k) \tag{10}$$

The compression strength due to the shear failure of the *k*th set of joints is obtained by substituting $\sigma_{i'j'} = 0$, $\sigma_{11} \neq 0$ and Eq (9) into (5),

$$\left.\begin{aligned} &\sigma_f = \sigma_{11} = c_k(\beta_k) \Big/ \left[m_{k1}^i m_{k2}^j l_i^1 l_j^1 \cos\beta_k + m_{k1}^i m_{k3}^j l_i^1 l_j^1 \sin\beta_k - m_{k1}^i m_{k1}^j l_i^1 l_j^1 \mu_k(\beta_k) \right] \\ &\tan\beta_k = m_{k1}^i m_{k3}^j l_i^1 l_j^1 \Big/ m_{k1}^i m_{k2}^j l_i^1 l_j^1 \end{aligned}\right\} \quad (11)$$

The compression strength of a jointed rock is determined by the minimum value from Eqs (2), (4), (10) and (11).

3 Application of the Strength Model to a Jointed Rock

Uniaxial compression tests were completed using a graphitic phyllite from Dagusan Iron Mine in Ansan, China[6-7], which contains two sets of joints, cleavage and flake joints, intersecting with an angle of 70° (Fig.2a). For the further analysis, two Cartesian coordinates, O-123 and O′-1′2′3′ are attached to the flake and cleavage joints (Fig.2). Totally 61 prisms with size of 5×5×10 cm^3 in 16 directions were tested. The measured compression strength data are plotted in Fig. 3.

The experiments showed that the Dagusan graphitic phyllite under uniaxial

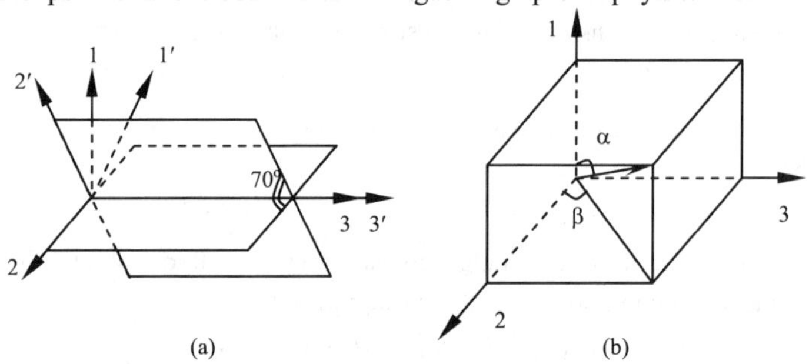

Figure 2. The sketch of the flake and cleavage joints in the Dagusan Graphitic Phyllite and relevant Cartesian coordinates.

compression failed mainly due to the tension-cleavage and shear failure of the flake and cleavage joints, and the combinations of these mechanisms[6-7]. The limited damage of the rock material was found due to the stress concentration induced by the rotation of the blocks resulted from joint failure. To determine the parameters in the strength equations, the measured strengths were firstly divided into four groups based on the failure modes, and then used to fit these formulae. Some data may appear repeatedly in different groups because the specimens exhibit duplicate failure modes. For the tension-cleavage failure of the cleavage joints, only an average strength of 32.5 MPa is obtained because of limited strength data, and the compression strength due to the tension-cleavage of the flake joints is

$$\sigma_f = 29.29 - 4.82\cos 2\beta + 2.39\cos 4\beta \tag{12}$$

The compression strengths due to the shear failure of the flake joints is expressed as

$$\sigma_f = 2\left(4.93 + 0.45\sin^2 2\beta\right) / \left\{\left[1 - \left(0.69 + 0.28\sin^2 2\beta\right)/\tan\alpha\right]\sin 2\alpha\right\} \tag{13}$$

and for the shear failure of the cleavage joints

$$\left.\begin{aligned} \sigma_{11} &= \frac{(5.55 - 0.49\sin 4\beta')/(\sin 20^o\cos\alpha - \cos 20^o\sin\alpha\cos\beta)}{\sin\alpha\sin\beta/\sin\beta' - (\sin 20^o\cos\alpha - \cos 20^o\sin\alpha\cos\beta)(1.15 + 0.52\sin 4\beta')} \\ \tan\beta' &= \sin\alpha\sin\beta/(\cos 20^o\cos\alpha + \sin 20^o\sin\alpha\cos\beta) \end{aligned}\right\} \tag{14}$$

The predictions from the above equations are plotted in Fig.3 to compare with those measured strength values, and show a reasonable agreement. For some specimens, the predicted strengths from two criteria are close and therefore, both failure modes have been observed. For example, the rock specimens with $\alpha = 45^o$ and $\beta = 0^o$, shear failure in both the flake and cleavage joints were observed.

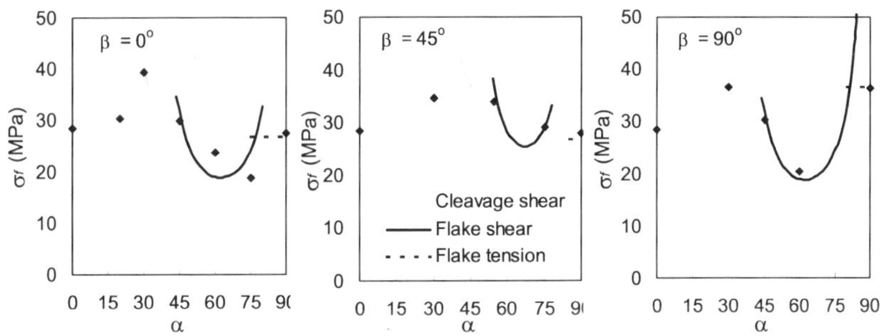

Figure 3. The comparison of the compression strengths calculated using Eqs (12) to (14) with those measured values.

4 Conclusion Remarks

The strength model developed in this paper has focused on two important characteristics of jointed rocks. Firstly, the failure mode and strength of jointed rocks vary with the loading direction relative to joint planes. This characteristic has been widely studied, and various models have been developed. However, Jaeger's two formula model[8] is still one of most well received for shear failure of jointed rocks because of its clear physical explanation. The present paper extended further Jaeger's model to include the tension-cleavage failure into the failure criteria for jointed rocks.

The second focus of this paper is on the in-plane anisotropy of the shear failure of joint planes. Although it was reported as early as in 1960s[11], few attempts have been made to combine the in-plane anisotropy into the failure criteria for jointed rocks. In this paper, this characteristic has been considered by transforming the applied stresses to the normal and shear stresses in the joint coordinates.

The present strength model having considered the above two characteristics has been applied to the uniaxial compression tests of the Dagusan graphitic phyllite, and has provided reasonable predictions for the measured strength values. More importantly, the model can provide a reasonable explanation for the observed failure mechanisms and associated strength values.

References

1. E.T. Brown in *Comprehensive Rock Engineering: Principles, Practice & Projects, Vol.1 Fundamentals*, ed. J.A. Hudson & E.T. Brown (Pergamon Press, Oxford, 1993).
2. J.C. Jaeger & N.G.W. Cook, *Fundamentals of Rock Mechanics* (Chapman & Hall, London, 1969).
3. F.G. Bell, *Engineering Properties of Soils and Rocks* (Blackwell Science, Oxford, 2000).
4. R.D. Lama & V.S. Vutukuri, *Handbook on Mechanical Properties of Rocks*, Vol.4 (Trans Tech Publications, Clausthal, Germany, 1978).
5. G. Duveau *et al*, *Mech. Cohs-Fric Mater.* **3**, 1 (1998).
6. K. Duan, *Failure Modes and Strength Analysis of Jointed Rocks Under Uniaxial Compression* (ME Thesis (in Chinese), Northeastern University, Shenyang, China, 1984)
7. K. Duan & Y.M. Lin in Proc. 3rd Rock Mech. Conf. (in Chinese), (Chinese Metal Soc. Shenyang, China, 1985).
8. J.C. Jaeger, *Geol. Mag.* **97**, 65 (1960).
9. E. Hoek & E.T. Brown, *J. Geomech. Eng. Div. ASCE* **106**, 1013 (1980).
10. E. Hoek, *Geotech.* **33**, 187 (1983).
11. J.F. Uff & J.K.T.L. Nash in *Proc. Geotech. Conf. Shear Strength of Properties of Natural Soils and Rocks* (Oslo, 1967)
12. L. Jing et al, Int. J. Rock Mech. Min. Sci. & Geomech. Abstr. **29**, 535 (1992)

REGRESSION ANALYSIS ON SIZE EFFECT IN IRREGULAR LUMP ROCK TESTING

JAAK J. K. DAEMEN, LUMIN MA, YONGHONG WANG
Department of Mining Engineering,
Mackay School of Mines,
University of Nevada, Reno, Nevada, USA

Point load testing is used to estimate the uniaxial compressive strength of rock. Rock specimens have been collected from Yucca Mountain, a candidate site for a High Level Nuclear Waste Repository in the United States. Most specimens tested have smaller sizes than recommended in ASTM D 5731 and ISRM (1985). Due to the size effect on rock properties, the formula for calculating the size-corrected point load strength index of a rock specimen affects the results. This paper gives evidence of this distortion and demonstrates size effect. A linear regression approach is used. A regression model is derived and modified from the formula for the size-corrected point load strength index in ASTM D 5731 and ISRM (1985). Rock specimens were collected from various locations in Yucca Mountain. Location variability is considered when modeling. After comparison and analysis of regression, a final regression equation is obtained. A sensitivity analysis for size effect is performed based on this final regression equation.

1 Introduction

Point load testing is an economical way to test rock. It is used to estimate the uniaxial compressive strength of rock. Among point load tests, irregular lump testing is the one with minimal need for specimen preparation. Conversely, because the test is performed on specimens that may take on a variety of sizes and shapes, this test tends to generate large variations in results. Figure 1 shows load configuration and specimen shape requirement for the irregular lump test from ASTM D 5731 and ISRM Suggested Method for Determining Point Load Strength (ISRM, 1985).

The point load strength is highly dependent on the size of the specimen (Thuro *et al*, 2001). Size correction has been taken into account in the point load strength calculations in ASTM D 5731 and ISRM (1985). A range of specimen sizes is recommended in ASTM D 5731 and ISRM (1985). The range of D is between 30 mm and 85 mm.The recommendation in ASTM D 5731

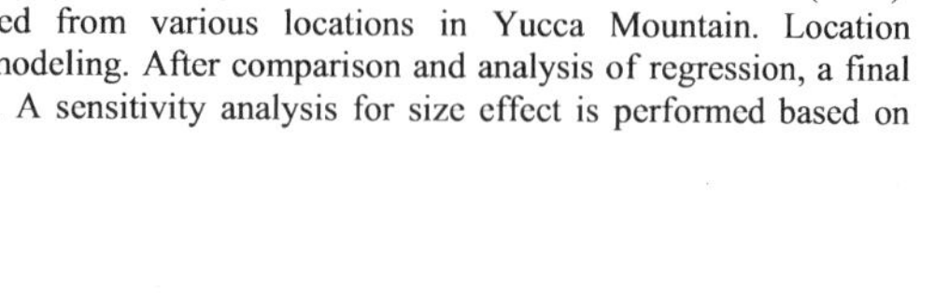

Figure 1. Load configuration and specimen shape requirement for the irregular lump test (after ASTM D 5731 and ISRM, 1985)

and ISRM (1985) for specimen size implies that specimens with sizes outside of the recommended range may distort the results. In our study most of the specimens collected and tested are smaller than recommended in ASTM D 5731 and ISRM (1985). Table 1 gives the distribution of D of the specimens. The total number of the specimens is 79.

Table 1. Distribution of D in irregular lump tests

Interval (mm)	$D \leq 10$	$10 < D \leq 20$	$20 < D \leq 30$	$30 < D \leq 40$	$40 < D \leq 50$	$50 < D \leq 60$	$60 < D \leq 70$
Frequency	1	21	34	12	7	3	1

Locations from which the specimens were collected are spread along the Enhanced Characterization Repository Block (ECRB) drift as well as from several exploratory boreholes located at various points across the potential repository block. In each borehole, specimens were collected from different formations. The formations on which our study mainly focuses include Upper Lithophysal, Middle Nonlithophysal, Lower Lithophysal and Lower Nonlithophysal units of the Topopah Spring. Mechanical properties of rocks from different formations are found to be quite variable. Statistical studies on rock strength in each formation show wide distributions of properties. Specimens that have been tested for analysis in this paper come from five boreholes and two formations. Table 2 gives the details.

Table 2. Summary of specimens tested

	Borehole	Formation (All Topopah Spring)	Short form	# of tests
1	USW NRG-7/7A	Lower Lithophysal	NRG7ALL	10
2	USW NRG-7/7A	Lower Nonlithophysal	NRG7ALN	8
3	USW SD-6	Lower Lithophysal	SD6LL	8
4	USW SD-7	Lower Lithophysal	SD7LL	10
5	USW SD-9	Lower Lithophysal	SD9LL	18
6	USW SD-9	Lower Nonlithophysal	SD9LN	17
7	UE-25 UZ14	Lower Lithophysal	UZ14LL	8

Dimensions (Figure 1) of each specimen were measured before testing. Load at failure was recorded. These parameters basically control the mechanical properties of specimens in irregular lump testing. The tests were performed as close to required by ASTM D 5731 and ISRM (1985) as possible. In ASTM D 5731 and ISRM (1985), uncorrected point load strength is calculated as:

$$I_s = P / D_e^2 \quad (1)$$

where P = failure load, D_e = equivalent core diameter. $D_e^2 = 4A/\pi = 4WD/\pi$. .

The size-corrected point load strength index $I_{s(50)}$ is calculated using the formula:

$$I_{s(50)} = F \times I_s \tag{2}$$

where F is the size correction factor, obtained from:

$$F = (D_e / 50)^{0.45} \tag{3}$$

Results of test specimens from each location deviate quite widely. To perform statistical analysis efficiently, the data is refined by the following rule: For each location, if the number of tests is greater than 5 and less than 11, delete the results with highest and lowest values of $I_{s(50)}$. If the number of tests is greater than 10 and less than 16, delete the results with two highest and two lowest values of $I_{s(50)}$. If the number of tests is greater than 15 and less than 21, delete the results with three highest and three lowest values of $I_{s(50)}$. In this study, the total number of test results is thus reduced to 57.

Figure 2 shows the box plot of $I_{s(50)}$ by location. The $I_{s(50)}$ is calculated using the formula (2). This plot reveals the distribution of size-corrected point load strength index for each location.

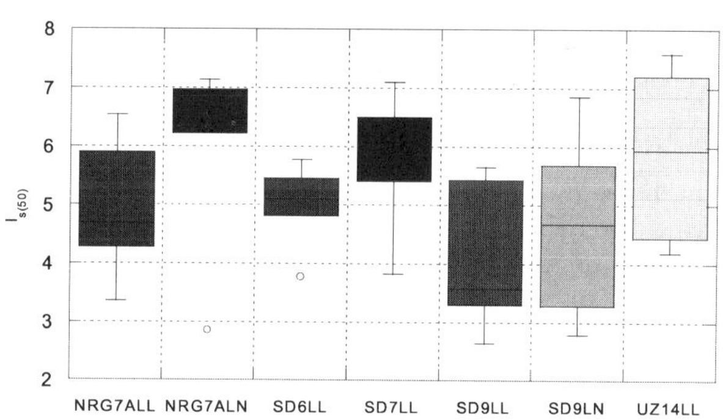

Figure 2. Box plot of size-corrected strength index by location

2 Modeling

The relation between $I_{s(50)}$ and D_e^2 for lump test results can be established from equations (1), (2) and (3) as:

$$I_{s(50)} = \left(\frac{D_e}{50}\right)^{0.45} \left(\frac{P}{D_e^{\,2}}\right) \tag{4}$$

Recall that $D_e^2 = 4WD/\pi$ for lump test. Replacing $4WD/\pi$ for D_e^2, equation (4) can be written as:

$$I_{s(50)} = \frac{P}{50^{0.45} \times \left(\frac{4WD}{\pi}\right)^{0.775}} \quad (5)$$

According to equation (5) we expect the failure load P to be directly proportional to $(WD)^{0.775}$: $P \propto (WD)^{0.775}$, which can be written as:

$$P = k(WD)^{0.775} \quad (6)$$

where k is a constant. Take the natural logarithm on both sides of equation (6):

$$\ln P = \ln k + 0.775 \ln WD \quad (7)$$

A basic regression model in this study can be derived from equation (7) as:

$$\ln P = \beta_0 + \beta_1 \ln WD + \varepsilon \quad (8)$$

Test specimens in this study are from seven locations. It is reasonable to assume that the mechanical properties of rock specimens from different locations may not be the same. So a categorical variable should be added to model (8). The categorical variable has seven categories, the same number as the number of locations in this study. By including the categorical variable the full model can be written as:

$$\ln P = \beta_0 + \beta_1 \ln WD + \beta_2 Location + \beta_3 (\ln WD \times Location) + \varepsilon \quad (9)$$

where $\ln P$ = response variable, $\ln WD$ = continuous predictor variable, *Location* = categorical predictor variable.

SAS, Statistical Analysis System (Der *et al*), is used in analysis. By applying General Linear Modeling (GLM) procedure, the output is obtained as in Table 3.

Table 3. SAS output for regression of model (9)

R-Square	Coeff Var	Root MSE	lnP Mean
0.818866	2.556440	0.227695	8.906713

Source	DF	Type III SS	Mean Square	F Value	Pr > F
lnWD	1	3.26900238	3.26900238	63.05	<.0001
Location	6	0.65229467	0.10871578	2.10	0.0738
lnWD×Location	6	0.68658314	0.11443052	2.21	0.0611

The R-Square, coefficient of determination, is 0.818866. This means that about

81.9% of the variability in the response variable is explained by the predictor variables. The P-values of F-tests for predictor variables indicates that ln*WD* is significant. Neither *Location* nor ln*WD*×*Location* is significant.

To facilitate for the modeling work without distorting the study, the categorical variable can be removed from the model. Model (9) is thus reduced back to model (8). When regressing model (9), a nonlinear relation between ln*P* and ln*WD* is significant. This problem remains unchanged in regressing model (8). Table 4 shows the evidence from the SAS result of model (8). The quadratic relation is significant (P-value = 0.0028).

Table 4. SAS output for regression of model (8)

R-Square	Adj R-Sq	Root MSE	lnP Mean	Coeff Var
0.6810	0.6752	0.27458	8.92423	3.07684

Regression	DF	Type I Sum of Squares	R-Square	F Value	Pr > F
Linear	1	8.853922	0.6810	136.26	<.0001
Quadratic	1	0.638058	0.0491	9.82	0.0028
Total Model	2	9.491980	0.7301	73.04	<.0001

To fit the data better, two models have been tried. One is a quadratic model; the other one is an exponential model. The quadratic model contains terms ln*WD* and $(\ln WD)^2$. These terms are highly correlated, which result in a perfect collinearity problem in the regression. The choice of an exponential model is better.

Taking exponential of ln*WD* in model (8):

$$\ln P = \beta_0 + \beta_1 e^{\ln WD} + \varepsilon \text{ or } \ln P = \beta_0 + \beta_1 WD + \varepsilon \tag{10}$$

Regress ln*P* on *WD* in (10), the SAS output is given in Table 5.

Table 5. SAS output for regression of model (10)

R-Square	Adj R-Sq	Root MSE	lnP Mean	Coeff Var
0.7286	0.7237	0.25327	8.92423	2.83804

Variable	Label	DF	Estimate	Standard Error	t Value	Pr > |t|
Intercept	Intercept	1	8.34431	0.05833	143.05	<.0001
WD	WD	1	0.00044102	0.00003629	12.15	<.0001

R-square of model (10) is 0.7286. It is higher than the one of model (8), 0.681. The estimated equation is:

$$\ln P = 8.34431 + 0.00044102 \times WD \quad (11)$$

The relation between P and WD can be established by taking anti-natural logarithm on both sides of equation (11):

$$P = e^{8.34431+0.00044102\times WD} \text{ or } P = 4206.18 \times e^{0.00044102\times WD} \quad (12)$$

3 Analysis

Equation (12) is the final regression equation in this study. The next step is to validate our regression and perform a size effect analysis. A relation between uncorrected point load strength I_s and WD can be obtained from formula (1), equation (12) and the relation $D_e^2 = 4WD/\pi$ as:

$$I_s = \frac{3303.53 \times e^{0.00044102\times WD}}{WD} \quad (13)$$

Figure 3 shows the scatter plot of I_s-WD from our laboratory tests and a fitting curve from the regression analysis. When WD becomes smaller, the size effect gets more obvious. After the point where $WD \approx 1600$ mm^2, the curve tends to be stable. It proves the reasonability that ASTM D 5731 and ISRM (1985) set the range of D as 30 mm $\leq D <$ 85 mm, although $D >$ 40 mm probably would be preferable in order to minimize size effects.

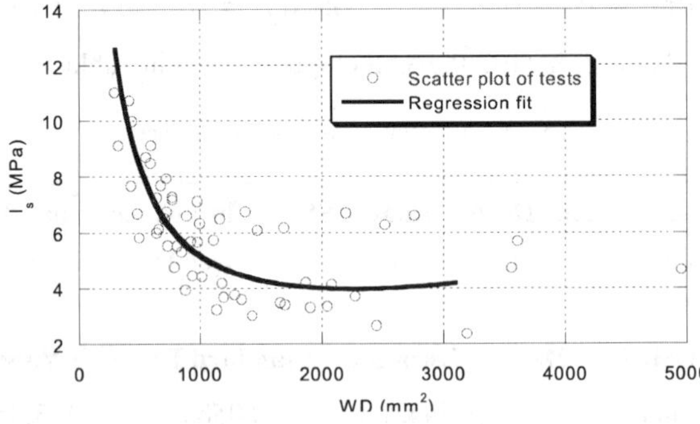

Figure 3. Variation of uncorrected point load strength as a function of WD

Let's take a sensitivity analysis on size effect in our lump test. Sensitivity of I_s to WD can be expressed as $S(I_s, WD)$, and calculated by:

$$S(I_s, WD) = \frac{dI_s}{dWD} \times \frac{WD}{I_s} \tag{14}$$

An example for calculating $S(I_s, WD)$ is given below to demonstrate the application of (14). Figure 4 gives a graphical show of the sensitivity calculation.

Results of sensitivity analysis in Figure 4 are explained as: when WD = 500 mm^2, an increase of 1% of WD will result in a 0.78% decrease of I_s.

At the point where $WD \approx 2270$ mm^2, $S(I_s, WD) = 0$. It means that at that point the sensitivity of I_s to WD becomes zero. At that point, the equivalent core diameter D_e is 53.8 mm. ASTM D 5731 and ISRM (1985) consider that when D_e = 50 mm, size correction is not needed.

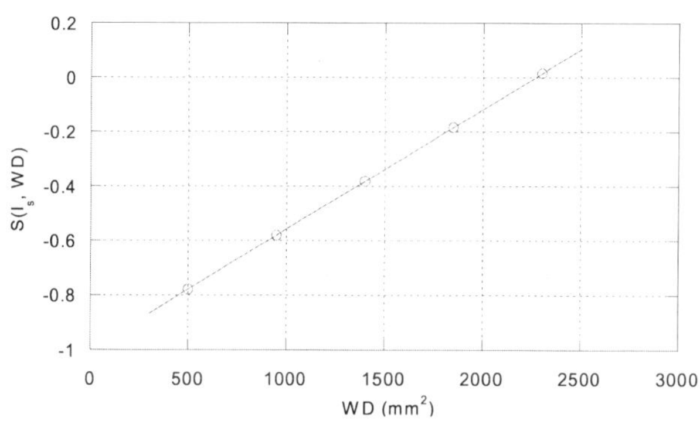

Figure 4. Plots of sensitivity of I_s to WD

From the modeling work in section 2, we can see that the linear log-log relation between P and WD is not the best for our data. Instead, a semi-log relation between P and WD has a better fit. This relation is equivalent to an exponential relation. It is different from the formula in ASTM D 5731 and ISRM (1985). We assume that size effect is the main cause of the difference. This is the foundation for our sensitivity analysis.

Since a linear log-log relation between P and WD is not the best for our data, the calculation of size correction factor shown as (3) is not the best to be applied here. To derive a suitable size correction factor, we need more tests and further exploration.

4 Conclusions

A 79 irregular lump tests data set is used in this study. The test specimens are from various locations in Yucca Mountain. Most of the specimens have a smaller size than recommended in ASTM D 5731 and ISRM (1985). A series of statistical methods are applied to this data set performing data refinement, location variability analysis and analysis of relation between failure load *P* and the product of *W* and *D*. A relation between I_s and *WD* is then established. This relation is the foundation on which our size effect analysis is built. The regression analysis indicates that the formula of calculation for size-corrected strength of irregular lump tests holds only approximately for our data. Sensitivity analysis indicates the size effect appears more obvious when *WD* becomes smaller. When $WD > 1600$ mm^2, the I_s-*WD* curve tends to be flat. When D_e, the equivalent core diameter, is 53.8 mm, the size effect disappears.

References

1. ASTM D 5731 – 95, "Standard Test Method for Determination of the Point Load Strength Index of Rock." *Annual Book of ASTM Standards*, Section 4, Construction, Volume 04.08 Soil and Rock; Building Stones. American Society for Testings and Materials, Philadelphia.
2. ISRM 1985, "Suggested method for determining point load strength." ISRM Commission on Testing Methods, Working Group on Revision of the Point Load Test Method. *Int. J. Rock Mech. Min. Sci. & Geomech. Abstr.* 22: 51-60.
3. Broch, E. and J. A. Franklin, "The Point-Load Strength Test." *Int. J. Rock Mech. Sci.* Vol. 9, pp 669-697, 1972.
4. Thuro, K. and R.J. Plinninger, "Scale effects in rock strength properties. Part 2: Point load test and point load strength index," ISRM Regional Symposium: EUROCK 2001, Rock Mechanics-a challenge for society.
5. Neter, J., Kutner, M. H., Nachtsheim, C. J. and W. Wasserman, *Applied Linear Regression Models*, 3rd Ed. The McGraw-Hill Companies, Inc., 1996.
6. Meerschaert, M. M., "*Mathematical Modeling*." 2nd Ed., Academic Press, 1999.
7. Der, G and B. S. Everitt, *A Handbook of Statistical Analysis using SAS*, 2nd Ed., Chapman & Hall/CRC, 2001.

CT REAL TIME TESTING OF FAILURE PROCESS OF ROCK

REN JIANXI [1,2], GE XIURUN[1]
1 Shanghai Jiaotong University, Shanghai 200030,China
2 Xi'an University of Science & Technology , Xi'an 710054,China

The computerized tomography(CT) real time testing of rock meso-damage evolution law under uniaxial compression loading has been completed using the specified loading equipment corresponding to the CT machine developed by authors .The preliminary results of rock meso-damage evolution law under uniaxial compression are obtained. Based on the CT testing, the complete rock stress-strain process curve is divided into several segments.

1 INTRODUCTION

Recently, CT becomes an advancing tool for meso-damage examination of rock. Over the scanned electronic microscope (SEM) and the optical microscope, CT technique has advantages in non-destructive scanning designation of multiple planes and so on. Therefore it interests many researchers accordingly[1-2]. This paper develops a real-time CT scanning equipment that is used specially for studying the damage evolution law of rock subject to triaxial compression .

2 LOADING EQUIPMENT FOR SPECIAL USE OF CT MACHINE

The loading equipment can bear a maximum designed axial compression and a confining pressure of 400 kN and 20 MPa separately. The chamber will be positioned at the scanning area of the tomograph to reduce the effect of chamber material on scanning results, so it is made of top-quality light metal of LY12—a kind of carbide for aeroplane manufacture. The triaxial load applying unit is specially designed for the complete use of the CT machine of the State Key Laboratory of China Frozen Soil Engineering.

3 CT TESTING OF SANDSTONE UNDER UNIAXIAL COMPRESSION

The sandstone blocks were sampled from Tongchuan Coal Mining of Shaanxi Mining Bureau. In testing, five tomographic planes were arranged for each specimen. During loading, the real-time CT scanning was performed on the tomographic planes in various loading levels and the variation in image showing in the complete unloading process of the initiation, branching, development and fracture of each tomographic plane have been carefully observed.

An example of 8# specimen is given. The rate of strain was 10^{-5}/s as a quasi-static loading. When σ_1 from 0MPa to 16.66MPa, both CT number of 5 scan sections and value of rock specimen are risen (table 1,2). In nature, the rock microcrack begin to be closed under compression,it is the weakness stage of damage.

Table 1. Testing results of each section

Scan Order	1st scan section CT umber/variance	3rd scan section CT number/variance	5th scan section CT number/variance
1	1703.7/35.36	1702.9/35.26	1700.8/42.44
2	1703.8/36.97	1704.5/35.57	1704.4/34.65
3	1705.5/34.19	Non-scan	1704.0/34.82
4	1705.1/35.38	1705.9/33.96	1703.9/48.79
5	1704.4/38.64	Non-scan	1703.0/33.14
6	1704.2/36.95	1704.8/34.12	1703.0/37.25
7	1703.8/37.01	Non-scan	1702.8/39.52
8	1702.5/38.76	1704.4/34.22	1700.3/32.79
9	1690.0/67.55	1675.2/98.20	1693.3/44.92
10	1688.7/72.43	1671.5/105.35	1692.3/46.79
11	1687.3/76.62	1669.5/109.22	1691.7/48.83
12	1680.4/94.01	1662.2/127.45	1690.2/53.43
13	1676.0/97.81	1638.4/171.11	1642.8/292.79

Table2. Testing results of rock samples

Scan Order	Rock specimen CT number/variance	σ_1 /MPa
1	1702.5/37.69	0
2	1704.2/35.73	16.66
3	1704.8/34.51	20.42
4	1705.0/39.38	27.50
5	1703.7/35.89	32.29
6	1704.0/36.11	36.17
7	1703.3/38.27	36.62
8	1702.4/35.26	38.45
9	1686.2/70.22	39.36
10	1684.2/74.86	40.96
11	1682.8/78.22	43.92
12	1677.6/91.63	49.17
13	1652.4/187.24	58.53

σ_1 is the representative of axial stress

In the stage of σ_1 from 16.66MPa to 36.17Mpa (not include σ_1=36.17MPa), both CT number of each scan sections and CT number of specimen are small descending from the above rising state . A little white (low density) occurs in the CT images, but the distinct meso-scale crack is not occurring. It is stage of linear.

When σ_1=36.17MPa, a 5.8mm length meso-crack is occurring in the left-up of 1st scan section(Fig.1a).The damage become to evolution ,this stress value is the threshold value of specimen, its value is equal to 59.59% of peak stress strength (σ_{c1}=60.7MPa, σ_{c1} is the representative of peak stress strength of specimen).

When σ_1 from 36.17MPa to 36.62MPa, the above meso-crack (this crack is called 1st crack) begins to lengthen and widen (length:10.4mm)(Fig.1b).When σ_1= 38.45MPa,the 1st crack is continuous to lengthen and widen ,and nearly to penetrated (length:42.6mm)(Fig.1c).When σ_1=39.36MPa,the 1st crack is thoroughly penetrated and continuous to broaden (length:43.3mm,Fig.1d).In the meantime ,a bifurcation crack(length: 29.8mm,the angle of horizontal direction is 40.5°) is occurring(Fig.1 e).When σ_1=40.96MPa and σ_1=43.92MPa,the 1st crack and the bifurcation crack in 1st scan section is continuous broaden.(Fig.1f,g).When σ_1 from 36.17MPa to 43.92MPa, both of CT number of each scan section and specimen are fall, the variance of CT number is continuous to increasing . It is beginning of damage evolution and stability propagation.

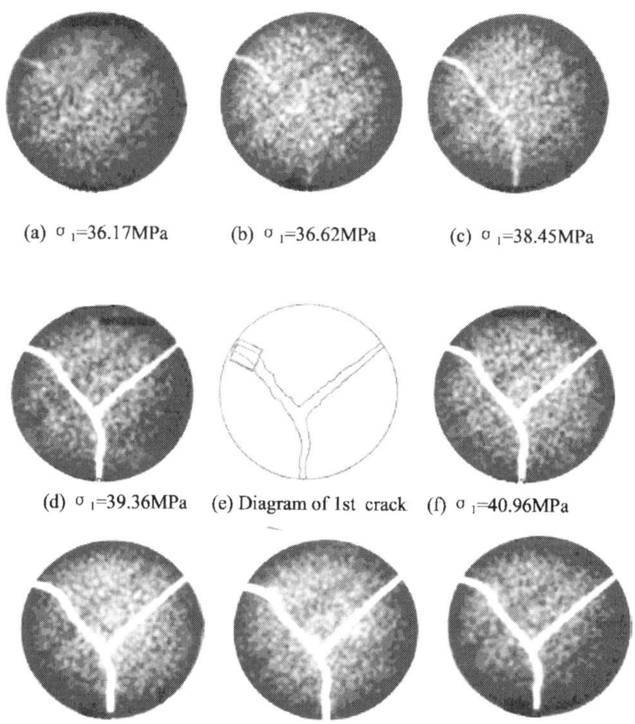

(a) σ_1=36.17MPa (b) σ_1=36.62MPa (c) σ_1=38.45MPa

(d) σ_1=39.36MPa (e) Diagram of 1st crack (f) σ_1=40.96MPa

(g) σ_1=43.92MPa (h) σ_1=49.17MPa (i) σ_1=58.53MPa

Figure1. CT images of 1st scan section of 8# specimen

When σ_1 from 49.17MPa to 58.53MPa (Fig.1 h-i), the damage is fast descend, the speed of crack propagation is fast too. The width of 1st crack and the bifurcation crack is fast increase. When loading axial stress $\sigma_1=\sigma_{c1}$=60.70Mpa (corresponding to the peak axial strain is 16.40‰),the strength come to the peak strength, the specimen is failure .Because the speed of specimen failure is very fast, the CT images of sections are not obtained, only the topography is obtained .

From the results of above CT testing of rock damage of 8# (Table 2) and that of other specimens, it shows that there is a relationship between the CT number change law and the stress-strain curve. Based on the relationship between the CT number change law and the complete process curve of rock, the complete process curve of rock can be divided into 5 segments. For example, the complete process curve of 8# specimen can be divided into 5 segments:(1)compress density segment: 0MPa$<\sigma_1 \leqslant$ 16.66MPa,i.e. σ_1=0~27.45% σ_{c1};(2)linear segment: 16.66MPa $<\sigma_1<$ 36.17MPa,i.e. σ_1= 27.45%~59.59% σ_{c1};(3)beginning of damage evolution and stability propagation segment :36.17MPa $\leqslant \sigma_1 <$ 49.17MPa,i.e. σ_1 = 59.59%~81.00% σ_{c1};(4)fast evolution segment of damage: 49.17MPa$\leqslant \sigma_1 \leqslant$ 60.70MPa,i.e. σ_1=81.00%~100% σ_{c1};(5) post-peak segment.

4 CONCLUSIONS

CT real-time experiments on the rock specimens subject to uniaxial compressive stress state for the meso-damage evolution law in the complete failure process have been conducted using a new load-applying apparatus, which is developed for CT machine .The preliminary law of meso evolution of rocks has been obtained.

The above uniaxial experiment have given clear images of the micro voids development of rock material in such stages as closing-up of voids, initiation and branching and development of micro cracks, rupture, failure .The results have shown that the damage evolution of rocks is a non-linear dynamic process, whereas rocks are a typical material that has been damaged originally and then the damage evolution manifests itself in locality. The testing results are satisfactory, showing a successful design of the load-applying apparatus.

Based on CT testing of meso-damage evolution of rock, the stress threshold value of rock damage is given, and the stress-strain curve is divided into 5 segments.

REFERENCES

1. G. S.,Yang et al, " CT Anal. on Mech. Characteristics of Damage Propagation of Rock." Chinese J. of Rock Mech. and Eng.18(3): 250-254,1999.
2. H.,Kawakata, et al, "Three-dimensional Observation of Faulting Process in Westerly Granite Under Triaxial Compression by X-ray CT Scan, " Tectonophysics ,313:293-305,1999.

The support from National Natural Science Foundation of China under grant No.10172057 and China Postdoctoral Science Foundation (2001 Year) is gratefully acknowledged.

STUDY ON 3-D SEISMIC RESPONSE OF UNDERGROUND STRUCTURES

JIANYUN CHEN ZHIQIANG HU JING LI
State Key Lab. of Coastal and Offshore Eng., Dalian Univ. of Technol., Dalian 116024, China
Tel: 0411-4707805-210, Fax:0411-4708501, E-mail:eerd001@dlut.edu.cn

The effects of some factors in the 3-D underground structure-wall rock dynamic interaction analysis on the seismic response of underground structures are studied using proposed practical dynamic interaction analysis method. And the deficiency of several approximate methods often used for underground structures seismic response analysis are also studied.

1 Introduction

Because of the interaction with wall rock, the dynamic response of underground structure is very complex. Various approximate methods have been used for the dynamic analysis of underground structures. And the methods can be placed in three categories: wave methods with artificial transmitting boundary, dynamic interaction methods with the interaction force applied on the soil-structure interface and the inertial force methods with the viscous dampers enforced on the outer boundary.

In this paper, the interaction analysis method is adopted and the effects of main factors in interaction motion equation have been studied incorporated with the analysis of the seismic response of Xiluodu large scale underground caverns.

2 Interaction analytical method for seismic response of underground structures

Based on the method of DSEM(Wolf,1994), the dynamic stiffness matrix of unbounded rock medium can be obtained by following procedure.

Intercepting the limited domain in the infinite rock medium and introducing artificial high damping ζ into bounded domain, the waves reflected by the outer boundary will be eliminated by the artificial high damping and can't reach the structures. So the motion on the structure-rock interface is determined only by the outgoing waves. And the dynamic stiffness on the interface of bounded domain with damping can be approximation to the dynamic stiffness of unbounded damped domain, Then extracting the artificial high damping from the dynamic stiffness on the interface of bounded domain leads to the dynamic stiffness matrix $[S^{\infty}(\omega)]$ of unbounded un-damped domain.

The analysis of dynamic stiffness characteristics shows that the dynamic stiffness of unbounded rock dependent on frequency can be approximated as

$$[S^{\infty}(\omega)] = [K_0] + i\omega[C_0] - \omega^2[M_0] \qquad (1)$$

In which, $[M_0]$、$[K_0]$ and $[C_0]$ may be equivalent to the mass, stiffness and damping attached to the structure-rock interface.

The equation of dynamic interaction for underground structure can written as

$$\begin{bmatrix} S_{ii} & S_{ib} \\ S_{bi} & S_{bb}^{s} + S^{\infty} \end{bmatrix} \begin{Bmatrix} u_i^t \\ u_b^t \end{Bmatrix} = \begin{Bmatrix} 0 \\ S^{\infty} u_b^g \end{Bmatrix} \qquad (2)$$

Where $\{u_b^t\}$ and $\{u_i^t\}$ represent the displacements of structure-rock interface nodes and those except on the structure-rock interface respectively. $\{u_b^g\}$ represents the earthquake input on the interface.

Substituting Eq.(1) into (2) and performing inverse Fourier transformation leads to the equation of dynamic interaction of the underground structure

$$\begin{bmatrix} M_{ii} & \\ & M_{bb} + M_0 \end{bmatrix} \begin{Bmatrix} \ddot{u}_i^t \\ \ddot{u}_b^t \end{Bmatrix} + \begin{bmatrix} C_{ii} & C_{ib} \\ C_{bi} & C_{bb} + C_0 \end{bmatrix} \begin{Bmatrix} \dot{u}_i^t \\ \dot{u}_b^t \end{Bmatrix} + \begin{bmatrix} K_{ii} & K_{ib} \\ K_{bi} & K_{bb} + K_0 \end{bmatrix} \begin{Bmatrix} u_i^t \\ u_b^t \end{Bmatrix} = \begin{Bmatrix} 0 \\ K_0 u_b^g + C_0 \dot{u}_b^g \\ + M_0 \ddot{u}_b^g \end{Bmatrix} \qquad (3)$$

Where $[M]$、$[C]$、$[K]$ represents the mass, stiffness and damping matrix of underground structure respectively.

In order to study the extent how the seismic response of underground structure is affected by the various factors, decomposing Eq. (3) into the following form

$$\begin{aligned} &[M]\{\ddot{u}^t\} + [C]\{\dot{u}^t\} + [K]\{u^t\} + [M_0]\{\ddot{u}_b^t\} + [C_0]\{\dot{u}_b^t\} + [K_0]\{u_b^g\} \\ &= [M_0]\{\ddot{u}_b^g\} + [C_0]\{\dot{u}_b^g\} + [K_0]\{u_b^g\} \end{aligned} \qquad (4)$$

à á

Where the first three and last three items on the left hand of Eq.(4) denote the inertial, damping and elastic effect acted on nodes in the structure and acted on the interface nodes by the unbounded wall rock respectively. And the items on the right hand denote the external seismic input corresponding to seismic acceleration, velocity and displacement on the interface respectively.

For comparison with the interaction analysis method, the inertial force method is adopted with two kinds of boundary conditions which are fixed boundary corresponding to Eq.(5) and viscous boundary $[C_1]$ to Eq.(6) are considered.

$$[M]\{\ddot{u}^t\} + [C]\{\dot{u}^t\} + [K]\{u^t\} = -[M]\{\ddot{u}_b^g\} \qquad (5)$$

$$[M]\{\ddot{u}^t\} + [C]\{\dot{u}^t\} + [K]\{u^t\} + [C_1]\{\dot{u}_b^t\} = -[M]\{\ddot{u}_b^g\} \qquad (6)$$

In order to determine which factor have greater effect on the seismic response of underground structure, the results obtained from ten cases are studied and compared with that obtained by Eq.(3). The cases are listed in detail in table1.

Table 1. Table of different computation condition

case	S-M ①	S-D ②	S-E ③	B-M ④	B-RD ⑤	B-VD ⑤'	BI-E ⑥	BI-A ⑦	BI-V ⑧	BI-D ⑨	SIL ⑩
1	●	●	●	●	●	○	●	●	○	○	○
2	●	●	●	●	●	○	●	○	●	○	○
3	●	●	●	●	●	○	●	○	○	●	○
4	○	●	●	●	●	○	●	●	●	●	○
5	●	●	●	○	●	○	●	●	●	●	○
6	●	○	●	●	●	○	●	●	●	●	○
7	●	●	●	●	○	○	●	●	●	●	○
8	●	●	●	●	●	○	○	●	●	●	○
9	●	●	●	○	○	○	○	○	○	○	●
10	●	●	●	○	○	●	○	○	○	○	●

In which, ● included； ○ not included; S- structure; B- unbounded medium ;
M-mass; E-stiffness; D-material damping; RD-radiation damping; VD-viscous damping;
BI- interface seismic input ; SIL-structural inertia load; A-accelerate ; V-velocity ; D-displacement

3 Analysis Results

The seismic response analysis of a large underground plant power caverns located in south-west china were performed. The scope of discretization is 400 meters long in the flow direction, 400 meters in height and 78 meters long in direction cross to the river. The seismic wave is taken from the site. The amplitude of the horizontal acceleration is 0.18g, 3 times that of the vertical acceleration.

The stresses at the vault of main plant (point A) and tail water tunnel (point B) corresponding to each case are listed in table 2.

Table 2. Maximum main stresses in total time history (Unit: Mpa)

case	1	2	3	4	5	6	7	8	9	10	Eq (3)
A	0.24	1.24	0.173	0.28	1.36	1.62	2.13	1.10	1.64	4.08	1.34
B	0.16	0.84	0.051	0.14	0.92	1.10	2.49	1.63	1.10	5.02	0.90

In case 1 to 3, the effects of seismic input, i.e. interaction forces induced by displacement, velocity and acceleration from the boundary are studied. In case 4 and 5, the effects of the mass of structure and the equivalent mass on boundary are studied. In case 6 and 7 the effects of the material damping of structure and the radiation damping of unbounded rock are studied. In case 8, the effect of stiffness of unbounded rock is studied. In case 9 and 10, the extent of exactness of the inertial force method is studied.

4 Conclusion

The results show that, for large scale underground caves, the dominating seismic input on interface is equivalent damping force corresponding to velocity. The effect of the elastic force input and the inertial force input are relatively small; Elastic constraint of unbounded rock have great effect on the seismic response, If this effect is omitted, the maximum stress decrease by 81% at the vault of main plant and increase by 21% at the vault of tail water tunnel.; The equivalent inertial of unbounded rock is of no effect and can be omitted, while the inertial effect of structure is great and can't be ignored; The effect of radiation damping of unbounded rock is much greater than that of structural material damping, if this effect is omitted the maximum stress at the vault of main plant and tail water tunnel increase by 59% and 167% respectively. The structural material damping has no effect on stress distribution and has relatively small effect on the amplitude of stress comparing with the former. If the structural material damping is not considered, the amplitude of stress increase by 21 percent.

The fixed boundary condition may results in unacceptable result in the inertial force method while the results with the viscous dampers attached to the boundary may be approximation to the seismic response of structure to some extent.

Acknowledgements

The support of the State Key Laboratory of Coastal and Offshore Engineering to this work under grant LP9804 is gratefully acknowledged.

References

1. L C Mathews and T L Geers. A doubly asymptotic nonreflecting boundary for ground-shock analysis[J]. J.Appl.Mech.ASCE,54(1987):489-497.
2. Wolf J P & Song C M. Finite Element Modeling of Unbounded Media. John Wiley & Sons,1996..

RELATION BETWEEN YOUNG'S MODULUS AND STRENGTH OF WEAK ROCK

MINGQING YOU SHENGQI YANG CHENGDONG SU
Jiaozuo Institute of Technology, Jiaozuo City, Henan Province, China
E-mail: youmq@jzit.edu.cn

This paper reports the results of the experiments for weak rocks. Specimens obtained from weathered or weak rock usually have many fissures and therefore the measuring values of their modulus and strength have great dispersion. The mechanical property of weak rock was studied by the way of ultrasonic waves and axial compression in servo-controlled testing machine. The effect of opening fissure on the ultrasonic wave may be minimized while the specimens have been saturated with water, no matter the value of ultrasonic waves varies in experiments. So the static average modulus and the dynamic modulus still have a good linear relation. Research indicated that there existed a good linear relation between the strength and average modulus in the specimens obtained from the same strata, and the linear regressive coefficient proved that the elastic deformation of material in all specimens is approximately the same at the strength.

1. Introduction

Rock is heterogeneous. For unweathered rock, such as quartz sandstone, marble, norite and granite with Φ50mm×100mm, it may be considered to be homogeneous in macrostructure. The yield stress of the specimens increases with the increasing of confining pressure, but deformation property is basically the same before their yield. And in linear deformation phase, the tests of cyclic loading in axial direction show that the average modulus, the slope of linear parts in axial stress-stain curve, has no obvious variance [1,2]. The stress-stain curves of fifteen marble specimens with different length and diameter in uniaxial compression are the same before yield [3].

There are evident fissures in sedimentary rock. However, the specimens from intact rock blocks are compact and homogenous except some local flaws. The yield stress and the deformation after specimens yield is diverse, but the stress-strain relation is still the same before yield. The specimen has a practical Young's modulus, and its strength and flaw degree can be confirmed from the stress-strain curve [4].

Specimens from weathered or weak rock usually have plenty of bedding planes and tiny fissures. When drilling in strata, rock will unload step by step in the direction of drilling, resulting in stress concentration in horizon. So the fissures in rock cores will produce from up to down, distributed along the length direction. The mechanical property of such specimens has great dispersion, and the relation between Young's modulus and strength is studied in this paper.

2. Experimental Results

Many sandstone cores, 105mm in diameter, were obtained from five drilling holes in one bridge base, and their mineral ingredient and grain have no evident difference except fissures distribution. The normal direction of the fissures is 22° to the axial direction of specimen. Sixteen specimens with the diameter 50mm and different length from 60 to 105mm were obtained from the above rock cores. Two specimens were detached completely by the fissures, and one specimen has so many fissures

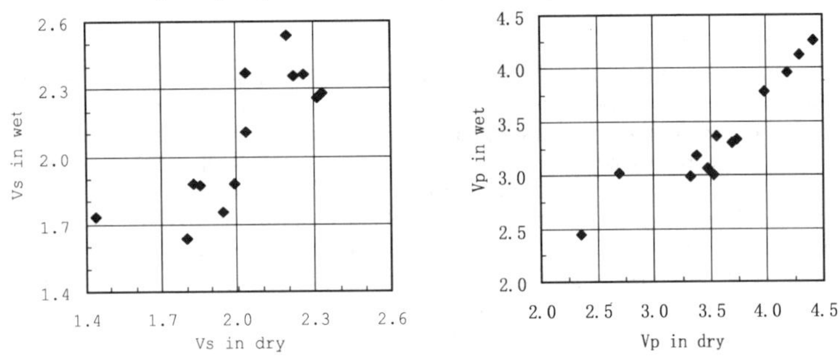

Figure 1. Ultrasonic wave speed of sandstone specimens in wet and dry

that ultrasonic wave cannot pass completely although it has no rupture.

The site needs the uniaxial compression strength of specimens in the water-saturated state. The soakage is lower than 0.5g, and it is found that the sopping is mainly the fissures after specimens failed. The longitudinal and transverse wave speeds through the specimens in wet and dry state have some variance, as shown in Figure 1. For 13 specimens saturated, the longitudinal wave speed of two specimens increased a little, but that of other specimens decreased, and the variance may reach to ±12%. The specimens' numbers of transverse wave speed increased or decreased are approximately equivalent, and the variance is in the range of –10%~20%. This may be understood from the following two aspects. Water can soften rock and make the deformation easier. For example, there is one specimen fractured completely after water saturated. Therefore, the supersonic wave speed becomes low. On the other hand, after the fissures were filled by water, specimen's stiffness becomes higher, and then the time and the resistance of the traveling wave decrease, so the wave speed increases.

Figure 2 is the stress-stain curves of four specimens in uniaxial compression. The strength and average modulus for all the specimens have a great dispersion up to 10 times. The average modulus is defined as the secant modulus between two points, stresses of which are 30% and 70% of the strength. Moreover, the fissures closing process influenced the initial deformation of specimens. In other words, the deformation modulus cannot show directly the fissures' distribution in the specimens, which is not related to the strength, such as specimen C3 and C4 in Figure 2.

The stress may make opening fissures close, whether the fissures open or not can only influence the deformation modulus, but do not influence the average modulus.

In the water-saturated state, the static average modulus and the dynamic modulus have good linear relation because they all reflect the numbers of fissures in the specimens, as shown in Figure 3. However, in the dry state, the dynamic modulus has no clear pertinence with the static modulus of specimens.

Figure 4 is the stress-strain curves of some coal specimens in uniaxial

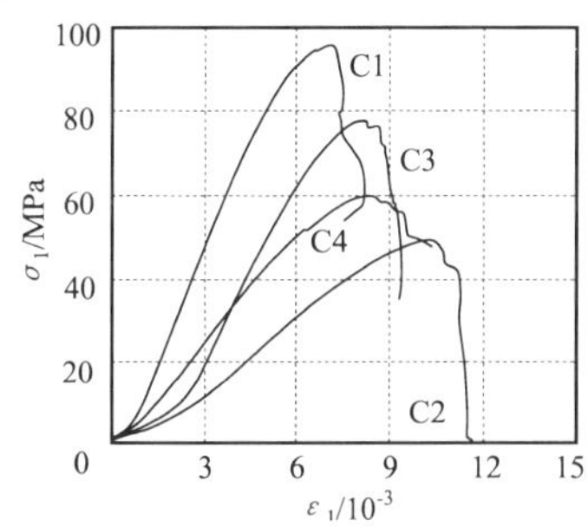

Figure 2. Stress-strain curves of sandstone specimen in uniaxial compression

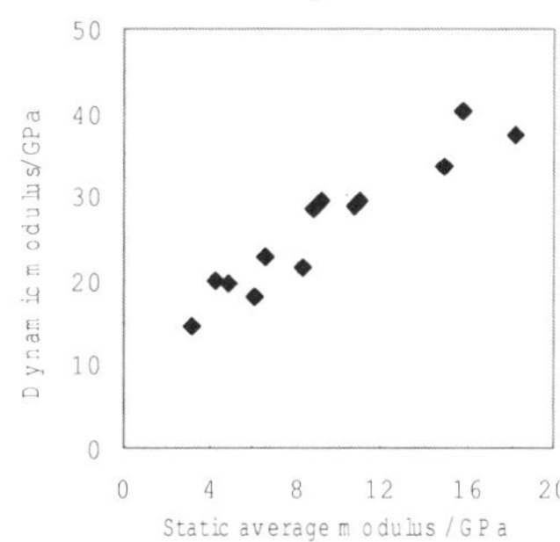

Figure 3. Dynamic modulus and average modulus of specimens saturated by water

compression, and specimens' strength and Young's modulus have great dispersion. The weakening course after peak stress of coal block A and B is not monotonous, which shows that failing and yielding of material in the specimens is a stepwise process. When a certain weak face in the specimen reaches its supporting limit and fails in local part, the specimen can keep on the supporting capacity.

The ultrasonic wave measurement is also carried out for coal specimens. But there

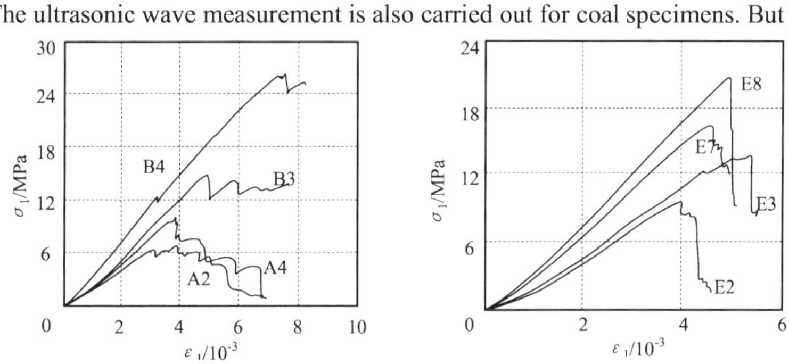

Figure 4. Stress-strain curves of coal specimen in uniaxial compression

is no evident correlation between the dynamic modulus and the static parameters, such as the Young's modulus and strength. For example, if a weak face with a little cohesion has a large inclination angle through the whole specimen, its strength and Young's modulus will be very low, but the supersonic wave speed may be high. The coupling between the probe and the end face of specimens, which is grease or tin foil for measuring the longitudinal and transverse wave respectively, is actually a weak face. On the other hand, an opening fissure vertical to the axis may prevent the propagation of the supersonic wave, but it can support load after closed in the uniaxial compression, which does not influence the average modulus.

3. Relation Between Young's Modulus and Strength

There are abundant fissures in the specimen, which induces internal friction force to influence deformation of the specimen. The normal stress and shear stress on one fissure face are relative to its direction and the confining pressure, so is the strength of specimen. Therefore, the average modulus and strength vary with direct ratio for they reflect the fissures in the specimen. The relation of sandstone in Figure 2 has given in Figure 5. Regressing with the equation

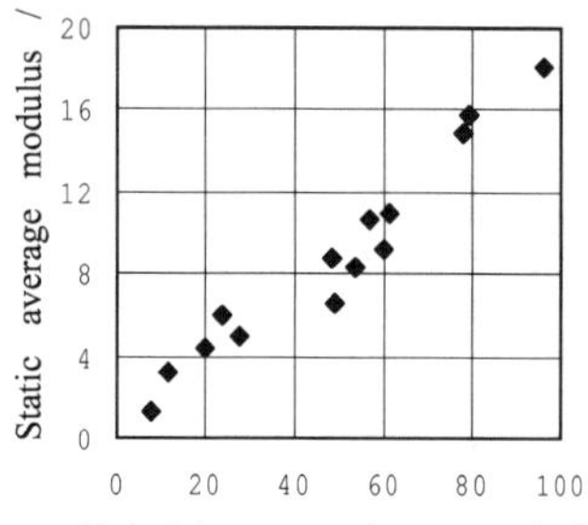

Figure5. Strength and Young's modulus of sandstone

$$\sigma_S = \varepsilon_0 E \tag{1}$$

the coefficient $\varepsilon_0 = 5.25\times10^{-3}$, and the correlation coefficient R=0.971.

Drilled from one coal seam with a depth of 150m, 56 specimens of bauxite and six kinds of sandstone were obtained and tested in uniaxial and triaxial compression. The confining pressure was from 0 to 15MPa[5]. Test results show that the strength of specimens and average modulus has a good linear relation, as shown in Figure 6.

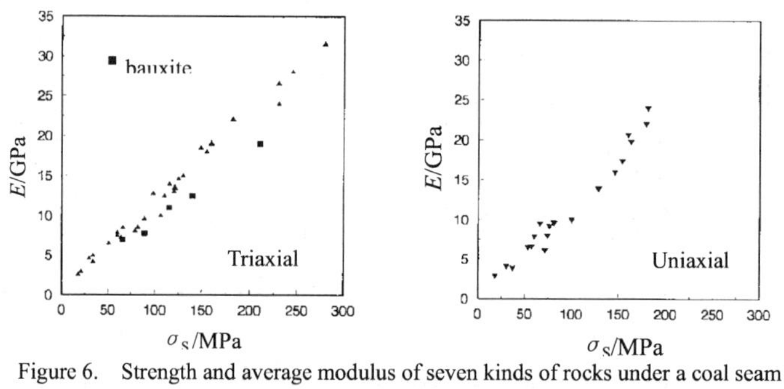

Figure 6. Strength and average modulus of seven kinds of rocks under a coal seam

The uniaxial compression results show more clearly that the increment of average modulus results from the supporting capacity of material. Confining pressure increases the supporting capacity of abundant fissures in the specimen, and their slipping is restricted by friction force, therefore the average modulus gets increment.

In Figure 6, the specimens' strength and average modulus have a good linear relation. Regressing with equation $\sigma_S = \varepsilon_0 E$, for 21 specimens in uniaxial compression, $\varepsilon_0 = 8.34\times10^{-3}$; for all the 56 specimens compressed in the uniaxial and triaxial, $\varepsilon_0 = 8.57\times10^{-3}$, and the correlation coefficient R =0.997.

The mechanical parameters of specimens from different coal blocks in Figure 4 are distributed in different ranges. However, the strength of specimens has same linear relation with average modulus when it is lower than 25MPa. After that, the average modulus keeps 4.5GPa or so, as shown in Figure 7. The strength of coal specimen B4 is 26.2MPa, and the average modulus as 3.72GPa is evidently lower, which results from the local failure and slipping. If the average modulus is calculated by subsection, it still can reach 4.5GPa or so. In triaxial compression, the average modulus of all specimens is (4.9 ± 0.4) GPa, larger than that in uniaxial compression, which varies slightly with different coal blocks. For sixteen coal specimens of strength lower than 25MPa, the result of regression is

$$\sigma_S + 8.05\text{MPa} = 6.23\times 10^{-3} \quad (2)$$

The correlation coefficient R=0.905. There are sedimentary weak faces with steep inclination in coal specimen, which weakly influences on the deformation, but greatly on the strength. From the stress-strain curve in Figure 4, it can be seen that coal specimens from block A or B locally fail gradually. Eq. (2) shows that local flaws in specimens weaken the strength.

4. Deformation Property of Fissure Specimens

From Figure 5, 6 and 7, the strength and average modulus of the specimens have great dispersion, but they conform to the linear relation $\sigma_S = \varepsilon_0 E$. The ratio

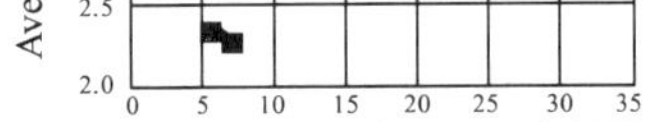

Figure 7. Strength and Young's modulus of specimens from various blocks

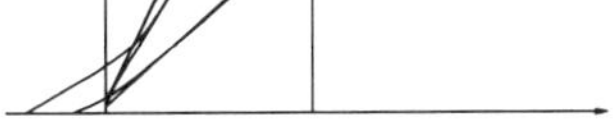

Figure 8. The deformation property of specimens with distributing fissures

coefficient ε_0 is the property parameter of strata, as shown in Figure 8. The strength and corresponding strain of specimen are different with grains and fissures, so is confining pressure. But when the specimen reaches its strength, the elastic strain is the same, ε_0. That is to say, no matter how slipping and extending of fissures in the specimen happen, the specimen will not reach its strength until deformation of rock material reaches ε_0. In other words, although there are lots of fissures in the specimen, there is a contact supporting structure, which will begin to weaken when the strain reaches ε_0.

The fissure faces may slip frictionally or not during axial compression. The friction force between the fissures increases gradually during its slipping. In this process, the deformation and stress of specimen has also linear relation. However, linear does not mean elastic. The new fissures result in the nonlinear relation between deformation and stress. Obviously, the slipping fissures are less at higher confining pressure and the slipping resistance larger as well, so the axial deformation of the specimen is less. Therefore, the specimen possesses higher Young's modulus. When the confining pressure increases the supporting capacity of fissures so highly that the axial stress at the strain ε_0 is still not enough to make the fissures slip, the analysis above would not come into existence any longer. Young's modulus will be a constant as the property of rock material compression, and will not increase with confining pressure or strength.

5 Conclusion

(1) The dynamic modulus and static average modulus of specimens saturated by water give a good linear relation for the eliminating of opening fissures. However, the opening fissures in dry specimens have a clear influence on their ultrasonic wave speeds, and the pertinence between the dynamic and static parameters is worse.

(2) Strength and average modulus of specimens from the same strata have great dispersion, but they give a linear relation. The ratio coefficient ε_0 means that the material has the same compression deformation when specimen reaches peak stress.

References

1. You Mingqing. The effect of confining pressure on the Young's modulus of rock specimen. Chinese Journal of Rock Mechanics and Engineering, 2002, 21(11)
2. You Mingqing, Su Chengdong, Xu Tao. Loading or unloading process in axial direction and Young's modulus of rock specimen. Chinese Journal of Geo-technical Engineering, 2001, 23(5): 588-592
3. Hudson J A, Crouch S, Fairhurst C. Soft, Stiff and Servo-controlled Testing Machines. Eng. Geol., 1972, 6(3): 155-189
4. You Mingqing et al. A study of triaxial strength and deformation of flawed specimen. Chinese Journal of Geotechnical Engineering, 1998, 20(2): 98-102
5. Guo Zhi. Rock mechanical testing for the prediction of water burst in coal mine.. In: Xu Xuehan, Wang Jie and et al. Study on the prediction of water burst in coal mine. Geology Press, Beijing China, 1991. 61-70

LABORATORY INTERPRETATION OF ROCK FRACTURE TOUGHNESS MEASUREMENT AT GREAT DEPTH

GUANGQING ZHANG MIAN CHEN
University of Petroleum, China
E-mail: crml@bjpeu.edu.cn

In this paper, finite element analysis (FEA) and empirical methods are employed to determine the fracture toughness. It is proven experimentally that the analytic procedures are usable to calculate the fracture toughness only when fracture tip radius is a small. And it is required that the failure pressure as an input of calculation correspond to the real magnitude of fracture tip radius. The conclusions drawn provide theoretical grounds as well as feasible experimental procedures for correctly determining the fracture toughness.

1 Introduction

The fracture toughness is widely accepted as the criterion for fracture propagation in the hydraulic fracturing simulation, that is, the fracture is supposed to extend further when $K_I \geq K_{IC}$ [1]. If the fracture toughness could not be determined properly, the simulation would be affected greatly in terms of magnitude and accuracy, thus receiving increasing attention throughout this field in the world.

As a common method, the stress intensifying factor (SIF or fracture toughness) is obtained by the elastic fracture formulas into which the experiment fracturing pressure is substituted. To decide the fracture toughness of rocks, three testing methods[2] have been suggested by the International Society for Rock Mechanics. Due to extreme difficulty of getting the cores out of great underground depth, thick-wall cylinder specimen with symmetric fractures is chosen in this work. In fact, experiments are required to get the failing pressure of specimens. However, the analytic formulas can not take real factors into account when calculating K_{IC}, such as complex geometry and boundary conditions, thus probably leading to noticeable errors in results. In this paper, with respect to the specimen dimensions and loads, the finite element models are established to calculate the fracture toughness on various specimens. Several conclusions were drawn based on numerical computation as well as experiment verification.

2 Basic hypothesis and modeling

In two-dimensional strain problem, suppose zero fracture-tip radius only in fracture formula, elastic material behavior with small deformation.

2.1 Analytical solution [3]

$$K_I = \left[1 + (R_i/R_0)^2\right] / \left[1 - (R_i/R_0)^2\right] \cdot p\sqrt{\pi a} \cdot F \tag{1}$$

where K_I is SIF, R_i and R_0 outer diameter and inner diameter of the thick wall cylinder respectively, a pre-fracture length, p the internal pressure, and F is a coefficient decided by R_0/R_i and $a/(R_0 - R_i)$ from some specified graphs. The critical SIF, i.e. fracture toughness, is calculated when the inner pressure corresponds to the failure pressure.

2.2 Finite element model

The finite element models are established with nonzero fracture tip, which permits continuous change of fracture tip radius to facilitate quantitative investigation. A-quarter models are built as shown in Figure 3, according to the symmetrical principle. The singular elements around the fracture tip are configured so that the stress singularity around the fracture tip could be described with acceptable accuracy. The stresses within these singular elements are singular with $1/\sqrt{r}$, and the enlarged locally is shown as in Figure 4.

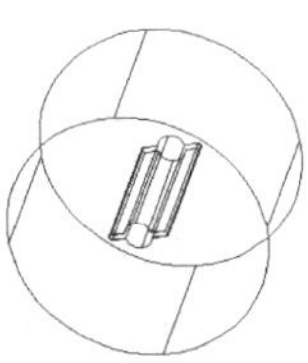
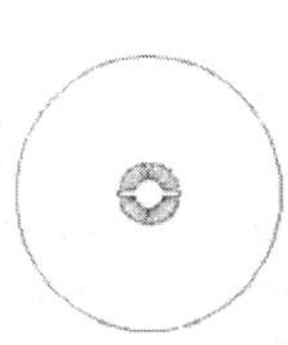
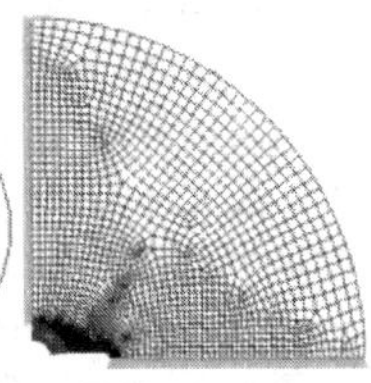
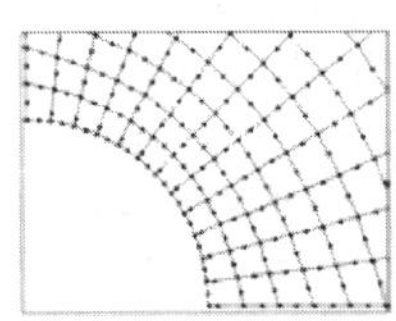

Figure1. Three-dimension schematics of the specimen
Figure2. Top view of the specimen
Figure3. FEA model with nonzero fracture tip radius
Figure4. Singular elements configured around the tip with nonzero radius(enlarged partly)

2.3 Fracture toughness calculation procedures for J-integrate scheme

After the calculation of displacements and stresses from the finite element models, the J-integrate scheme is implemented around the fracture tip shown as below[4].

$$J = \int_{\Gamma} \left(w dy - \vec{T} \bullet \partial \vec{U} / \partial x \, ds \right) \tag{2}$$

Γ is the arbitrary integrating path around the fracture tip, w strain energy density(i.e. stain energy per unit volume), $\vec{T}$ surface vector of the outer normal direction along the integrating path, $\vec{U}$ displacement vector, and ds is the distance along the integrating path.

Ultimately, the fracture toughness value is calculated following the relationship between J-integrate and the SIF in elastic mechanics.

$$K_I = \sqrt{E' J} \tag{3}$$

3 Experiment verifications

As shown in the Figure 1, thick-wall cylindrical specimens with two symmetric fractures are made in advance out of rock or cement materials. The purpose for taking the radius of fracture tip as zero is to utilize the present analytical formulas. However, as shown in the Fig2, the practical fracture tip radius is inevitably nonzero.

The experimental specimens are cut out from the underground rock formation and the artificial rocks in the form of cemented specimens, with the dimensions of the specimens and the necessary material parameters listed in Table1 and Table2. The artificial specimens are made of cement and fine soil with different ratio, intending to expand the effective range of the study. The values of fracture toughness in Table1 by the two methods are close for each artificial rock specimen, of which the FEA model with 0.8mm of fracture width is considered accurate. It should be noted that the failure pressure of specimen with fracture width 0.8mm is used in the analytical model. As a result, the analytical method gives a satisfactory accuracy under conditions presented in this paper.

Table1. Fracture toughness for cement specimens

No.	Outer dia/mm	Inner dia/mm	Fracture length/mm	failure pressure/MPa	Analytic model	FEM model with 0.8 mm width
1	97.85	10.36	5.40	4.76	0.2873	0.2797
2	96.53	10.55	4.035	4.02	0.2397	0.2433
3	98.23	10.17	4.005	3.50	0.2007	0.2057
4	97.1	10.40	4.305	3.27	0.1935	0.1951
5	98.02	10.05	4.72	3.01	0.1743	0.1739
6	95.71	10.68	5.03	2.40	0.1497	0.1456
7	96.75	10.00	4.40	1.38	0.0793	0.0799

Table2. Fracture toughness for rock specimens

Spec.	Outer dia/mm	Inner dia/mm	Fracture length/mm	Failure pressure/MPa	Analytic model	FEM model with 0.2mm width
sandstone	97.1	10.0	3.5	20.5	1.10	1.16
sandstone	101.2	9.7	3.5	14.8	0.75	0.814
shale	100.0	10.2	3.5	11.9	0.65	0.68

As shown in Table2, the analytic results are less than those by the FEA model with 0.2mm model, which however satisfies general engineering needs. The results indicate an acceptable error existing with the analytic model when the fracture width is small compared with 0.8mm or 0.2mm, and the error for the fracture toughness increases with the analytical model and FEA model with 0.2mm when fracture width becomes larger.

4 Conclusions and discussions

1) The analytical formulas can be used to calculate fracture toughness when the width is in the effective fracture radius range, based on experiment verification.
2) Finite element model is established for determining fracture toughness, and feasible experiment procedures are provided.

The conclusions drawn above are applicable to the models presented in this paper, and further attempts are expected for other type specimen and load conditions.

Reference

1. ZhixiChen, MianChen, Internal Pressure Measurement of Fracture toughness, Rock Mechanics & Engineering Annual Academy of China, Publishing press of University of Chengdu Science & Technology, 1995
2. R.D. Dwivedi, A.K. Soni, R.K. Goel, A.K. Dube, Fracture toughness of rocks under sub-zero temperature conditions, International Journal of Rock Mechanics and Mining Sciences 37(2000): 1267~1275
3. .ChengkangShen, Fracture mechanism, the publishing press of Tongji University, 1996.3: 110~128
4. Aerospace Institute of China, Handbook of stress intensifying factors (Revised Version), Beijing, Science Publishing House, 1993: 426~428; 430~431

STUDY ON THE FRACTURING PRESSURE MODEL OF FORMATION BY HYDRAULIC FRACTURING TREATMENT

XUDONG ZHANG YAN JIN MIAN CHEN
University of Petroleum, China
E-mail: zxdfracture@263.net

The accurate prediction of fracturing pressure for pay zone is the very important guidance to hydraulic fracturing design and operation. The pore pressure around the wellbore happens to change variously as the fracturing fluid enters the pay zone. The change in pore pressure affects the stress-state and the fracturing pressure around the wellbore. In this paper, a new concept of the effective membrane pressure coefficient is presented according to the wall building capacity of the fracturing fluid, then the change of pore pressure around the wellbore is studied, and it is proven that the prediction model for fracturing pressure is improved.

1 Introduction

The accurate prediction of fracturing pressure for pay zone is the very important guidance to hydraulic fracturing design and operation. Before the fracturing fluid enters the pay zone, the fracturing pressure mainly depends on the stress-state around the wellbore, rock mechanical property and pore pressure. As the fracturing fluid enters the pay zone, the influence of the change in pore pressure on the fracturing pressure must be considered. Hubbert-Willis formula [1-2] can not predict the fracturing pressure of the permeable zone. Haimson-Fairhurst formula[3] can not predict fracturing pressure of the impermeable zone. Bezalel Haimson[4] considered closely the influence of permeability on the fracturing pressure, and the formula for fracturing pressure is applied widely, but the formula can not interpret the pressure abrupt change from impermeability to permeability. As a result, the formula can not estimate the fracturing pressure of low permeable zone. In this paper, a new concept of the effective membrane pressure efficient is presented according to the wall building capacity of the fracturing fluid, then the change in pore pressure around the wellbore is studied, finally the prediction model for fracturing pressure is proven improved.

2 The pore pressure around the wellbore

The pore pressure around the wellbore happens to change variously as the fracturing fluid enters the pay zone. The change of pore pressure affects the stress-state around the wellbore, so it is necessary to determine the pore pressure distribution caused by the pressure difference between the pressure fracturing fluid p_m and original formation pore pressure p_0. The equation of the changed formation pore pressure p can be expressed:

$$\frac{1}{r}\frac{\partial}{\partial r}\left(r\frac{\partial p}{\partial r}\right)=\frac{k_{rock}}{\phi\mu C}\frac{\partial p}{\partial t} \tag{1}$$

where, ϕ is the porosity, μ is the viscosity of the fracturing fluid, and C is the formation compressibility. The boundary condition and initial condition are expressed as following:

$$r=r_w\,\text{(at wellbore)},\quad p=\zeta p_m \tag{2}$$

$$r\to\infty,\quad p=p_0 \tag{3}$$

$$t=0,\quad p(r)=p_0 \tag{4}$$

the solution of the equation (1) can be expressed:

$$p(r,t)=\zeta p_m+(p_0-\zeta p_m)erf\left(\frac{r}{\sqrt{4C_p t}}\right) \tag{5}$$

where $erf()$ is error function. There is a new coefficient ζ in the equation (2) and (5), which is the effective membrane pressure coefficient and defined:

$$\zeta=\frac{p_c}{p_m}=\frac{k_{cake}+k_{rock}\dfrac{p_0}{p_m}}{k_{cake}+k_{rock}} \tag{6}$$

where, k_{rock} and k_{cake} are the permeability of the formation and the membrane caused by the fracturing fluid, p_c is pore pressure around the wellbore.

Following the membrane becomes tight ($k_{cake}\to 0$), $\zeta=\dfrac{p_0}{p_m}$, $p_c=p_0$. The fracturing fluid does not enter the formation, and the pore pressure around the wellbore does not change.

If there is no membrane around the wellbore, ($k_{cake}\to\infty$), $\zeta=1$, $p_c=p_m$; The fracturing fluid enters directly the formation without any obstruction, and the pore pressure around the wellbore changes to the maximum degree.

3 The stress-state around the wellbore

As the fracturing fluid enters the formation, the changed pore pressure induces the excessive stress around the wellbore.

$$\sigma_r = \left[\frac{\alpha(1-2\nu)}{2(1-\nu)}\frac{(r^2 - r_w^{\ 2})}{r^2} - \phi\right]\left[p_m - p(r,t)\right]$$

$$\sigma_\theta = \left[\frac{\alpha(1-2\nu)}{2(1-\nu)}\frac{(r^2 + r_w^{\ 2})}{r^2} - \phi\right]\left[p_m - p(r,t)\right] \tag{7}$$

$$\sigma_{z\theta} = \left[\frac{\alpha(1-2\nu)}{2(1-\nu)} - \phi\right]\left[p_m - p(r,t)\right]$$

where, ν is Poisson's ratio, α is effective stress coefficient, r is the distance from the formation to the well axis, r_w is the well radius. When $r = R$, the tangent stress σ_θ can be expressed:

$$\sigma_\theta = -p_m + (1 - 2\cos 2\theta)\sigma_H + (1 + 2\cos 2\theta)\sigma_h + [\frac{\alpha(1-2\nu)}{1-\nu} - \phi](1-\zeta)p_m \tag{8}$$

where, σ_H is the maximum horizontal in-situ stress, σ_h is the minimum horizontal in-situ stress.

4 The model of fracturing pressure

The fracturing pressure principle can be expressed,

$$\sigma_\theta - \alpha\varsigma p_m = -S_t \tag{9}$$

where S_t is the tensile strength of the formation. According to equations (8) and (9), as $\theta = 0$, the new model of fracturing pressure can be obtained,

$$p_f = \frac{3\sigma_h - \sigma_H + S_t}{1 + \alpha\zeta - \left[\frac{\alpha(1-2\nu)}{1-\nu} - \phi\right](1-\zeta)} \tag{10}$$

5 Field application

For the dolomite formation of the well F443 at 3808 meters depth, the maximum horizontal in-situ stress is 102.56MPa, the minimum horizontal in-situ stress is 66.83 MPa, with the original formation pore pressure of 41.19Mpa. The tensile strength is 11.24MPa, and Poisson's ratio is 0.21. According to Haimson formula, the estimated fracturing pressure is 67.98 MPa, while the fracturing pressure predicted by formula (10) is 83.05MPa, which is in accordance with field fracturing pressure of 84.5MPa.

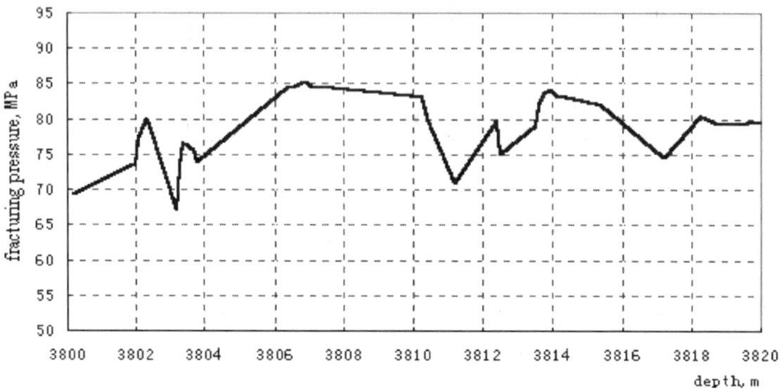

Figure 1. the predicted fracturing pressure of well F433

6 Conclusions

The accurate prediction of fracturing pressure for pay zone is very important guidance to hydraulic fracturing design and operation. The pore pressure around the wellbore happens to change variously as the fracturing fluid enters the pay zone. The change of pore pressure affects the stress-state and the fracturing pressure around the wellbore. The effective membrane pressure coefficient can be effectively used to describe the contribution of the change in pore pressure around the wellbore caused by fracturing fluid. The prediction model for fracturing pressure is successful in field application.

References

1. Hubbert M. K., Willis D. G., Mechanics of hydraulic fracturing, Trsns., AIME(1957)210:153~166
2. Detournay E, Carbonell R, Fracture-mechanics analysis is of the breakdown process in minifracture or leakoff test, SPE Production & Facilities, August 1997:195~199
3. Haimson B, Faihurst C, Initiation and Extension of Hydraulic Fractures in Rocks. SPEJ, Sept.1967: 310~318.
4. Haimson B, Faihurst C., Hydraulic Fracturing in porous-Permeable Materials. SPEJ, July.1969: 811~817

REVIEW OF MINI-FRAC DEVELOPMENTS

LINA YANG MIAN CHEN XUDONG ZHANG
Petroleum University, Beijing , 102249, China

Mini-frac plays an important role in the fracturing design and oil field exploration. The developments of mini-frac are reviewed in this paper. The growth of fracture height, the effect of loss coefficient dependent on pressure and the mathematical model of closure pressure are concluded in this paper. A method estimating near wellbore frictions is also proposed. The review provides a systematic methodology and guidelines for application.

1 Introduction

Mini-frac analysis is a method provides an estimate of fracture extension pressure, closure pressure, fracture dimensions, fracturing efficiency, leak-off coefficients and near-wellbore frictions prior to designing a large-scale fracture treatment. As a result, the mini-frac has become an industry standard for evaluating and improving the design of treatments.

Mini-frac pressure decline analysis as presented by Nolte[1] first became popular in the 1980's and is used routinely today. However the assumptions of the basic analysis are seldom met in practice. Some of these assumptions are that the fracture height is constant, fracture propagation stops as soon as shut in, fluid loss independent of pressure and so on. In addition to that the near well-bore frictions and closure pressure can't be obtained directly in mini-frac analysis. Although detailed description of theoretical framework for pressure decline has been given by Notle[1], the developments of research work haven't been concluded yet. This paper presents a systematic review of developments of mini-frac so that the field engineers can make better decisions before treatments.

2 Fracture height variation model

The traditional pressure decline assumes that the fracture height is constant. However it is not the case in the fields. Zhang[2] provided a good evaluation of pay zone parameters based on pseudo-three dimensional pressure decline analysis. Fracture extension after shut-in and the in-situ difference effect are both considered in his numerical model. Meyer[3] couples a multi-frac simulator with a three-dimensional hydraulic fracture simulator to predict fracture propagation and pressure response. The multi-stage mini-frac work well where the formations with pay zones less than 10 meters and low permeability. The main advantage of the simulator is that the governing mass and momentum equation are satisfied for both phases.

3 Closure pressure

Closure pressure is a significant factor affecting fracture conductivity. Usually it is obtained from conventional field tests or from the pressure decline vs. time plot. The error can't be neglected in field method.

Meng[4] provided a mathematical model for closure pressure. In the model the match pressure is calculated by coupling volume conservation equation with width equation as well as the fracture extension rule. Equations (1)-(2) are concerned in his method:

$$(P-P_c)*\beta_s >= \sigma_T \qquad (1)$$

For which P is pressure at any time after shut-in; P_c is closure pressure, β_s is the ratio of average pressure in the fracture and the bottom pressure, σ_T is tensile strength.

The equation means fracture extension doesn't stop until the fracture extension rule is satisfied.

Coupling the pressure decline equation and the width equation result in the following equation:

$$P-P_c = (ISIP-P_c)/T_D^m - P^* \cdot F_1(T_D) \qquad (2)$$

For which $ISIP$ is the instantaneous pump shut down pressure; $T_D = t/t_p$; t_p is pump time, $F_1(T_D)$ is a function of time.

Closure pressure can be obtained by solving the equations. The results have a good accordance with field test.

4 Pressure dependent leak off

Traditional pressure decline analysis developed by Nolte[1] assumed that leak off coefficient is independent of pressure. The assumption is valid when fluid loss is controlled by compressible filter cake. When fluid loss is mostly controlled by incompressible filter cake or by filtrate viscosity or by the reservoir, leak off coefficient is dependent on pressure. This is well reviewed by Castillo [1,7].

The general form of the leak off coefficient as a function of fracture pressure can be expressed as:

$$C(z,t) = M \cdot (\Delta P_{f,r}(z,t))^{\alpha} \tag{3}$$

For which c is leak off coefficient. $\Delta P_{f,r}(z,t) = P_f - P_r$ is the difference between fracture pressure and well-bore pressure.

Figure 1 is the G-Function derivative plot with leak off dependent on pressure. It indicates that the G-Function plot isn't horizontal during the closure period. This implies that the linear relation between decline pressure and the G-Function doesn't hold when leak off is pressure dependent. Castillo[5] showed that ignoring pressure dependence can result in an optimistic estimate of the leak off coefficient.

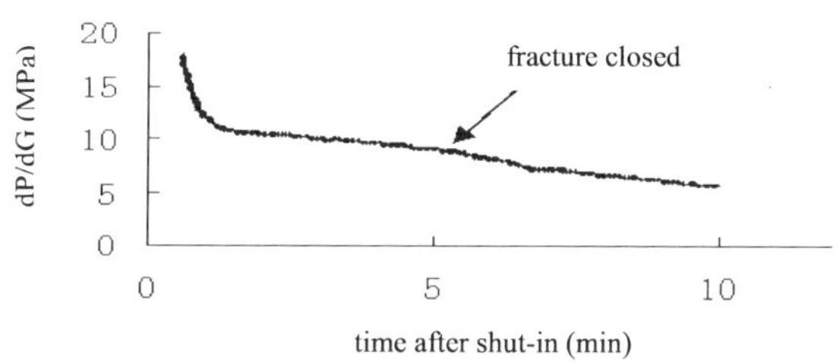

Figure1. Leak of coefficient dependent on pressure

5 Near-wellbore effects

The high near-wellbore pressure decline has been reported in fracturing treatment is an indicative of ineffective communication between the well bore and the fracture.

Guidelines for distinguishing near-wellbore friction have been given by J.Romero[6]. The Meyer & associates[8] provided a step down analysis to get the pressure drop near well bore. The analysis is performed after fracture initiation and propagation has been established.

During shut down if the injection rate decreased in a stair-step fashion for a short period of time, the pressure also decreases because of perforation and near well bore pressure losses. The relationship between the decreasing rate and pressure results in determination the total near well-bore pressure drop.

$$f(Q) = A * Q^{\alpha} + B * Q^{\beta} \tag{4}$$

For which $A * Q^{\alpha}$ is perforation frictions only， $B * Q^{\beta}$ is near-wellbore friction only. Fracture data from TUO22-1A well in JiangHan field are regressed to get α and β. The result is shown as figure 2, the equation regressed is:

$$f(Q) = 0.38 * Q^{2} + 1.26 * Q^{1.06} \tag{5}$$

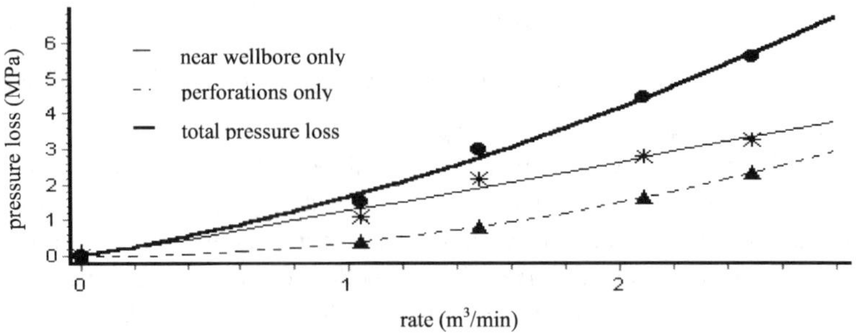

Figure 2. Pressure loss vs. pump rate

6 Conclusions

1. The result of three-dimensional numerical model of pressure decline analysis is more reasonable than that of two-dimensional model.
2. Loss coefficient is dependent on pressure. Ignoring pressure can result in an optimistic estimate of leak-off coefficient
3. A numerical model of closure pressure has been introduced which is based on consideration of the fluid volume conservation and extension rules. The method provides an important way for fracturing optimization design.
4. Guidelines have been proposed for calculating the near-wellbore pressure loss. A field case has been demonstrated for the application of the numerical model.

References

1. K.G. Nolte, Determination of Fracture Parameters from Fracturing Pressure Decline, SPE 8341 (1979)
2. Zhang Shicheng et al, The P-3D Anal. Method of Fracturing Pressure Decline Curve and Its Appl., ACTA PETROLEI SINICA, 11(3): 60-67 (1990)
3. B.R. Meyer, Three Dimensional Hydraulic Fracturing Simulation on Personal Computers: Theory and Comparison Studies, SPE 19329 (1989)
4. Meng Weihong et al, A Method for Determining of Fracture Parameters by means of The First Section of Pressure Decline Curve, J. of the University of Petroleum, 4~5:70-80 (China, 1988)
5. J.L. Castillo, Modified Fracture Pressure Decline Analysis Including Pressure-Dependent Leak off, SPE 16417 (1987)
6. J.Romero et al, Theoretical Model and Numerical Investigation of Near-Wellbore Effects in Hydraulic Fracturing, SPE 30506 (1995)
7. Meyer Associate, Mini Fracturing User's Manual (2000)

FULL-SCALE MODELING TEST
ON THE PRESTRESSED BRIDGE TOWER ANCHOR ZONE

HOUXIANG LI YUESHUN CHEN SINIAN LI
Hubei PolytechnicUniversity, Wuhan 430068, China

Be through with a full-scale test about prestressed bridge tower anchor zone, confirming the stress distribution of the model in the process of the action by the load, synthetically the law of the crack come into existence and expand, by the way of the load and the deformation of the model, getting the ultimate bearing capacity of the anchor zone.
Based on the testing result, the large-scale nonlinear finite element software Msc.Marc has been utilized to analyze the reaction of the model under the pressure, on one hand, it proves the reliability of the test, otherwise, it provides the tool for the research about the spatial structure. And the plasticity theory has been used to analyze the orifice plate, and come to its plasticity ultimate bearing capacity.

1 Introduction

The anchor zone of the cable stayed bridge tower is the primary component supporting the load, almost all of the load on the bridge will be transfer to the anchor zone by the cable, then to the base. The mono-cell box pre-stressed anchor zone, as a new-style of the cable bent tower, the cables were anchored directly on the wall of the tower, and the internal force of the cable was balanced by the pre-stressing force of the pre-stress steel in the tower. For the section of the cable was smaller than the cable bent tower, the internal force of the cable was transfer to the wall as the concentrated force acted. The carrying capacity of anchor zone is the key factor for design. So the actual distributing state of stress is studied not only by the finite element analysis, but also by modeling test.

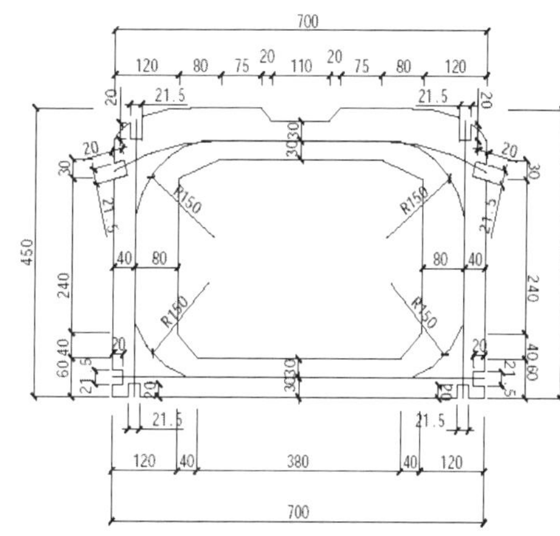

Figure1. The Anchor Zone Ichnography

2 Test Project and Testing Content

1). Test Project

The aim of the test is to make certain about the stress distributing in the anchor zone acting by the prestressing force and cable force together, the security degree, the utmost carrying and resisting crack capacity of the anchor zone.

According to the aim above and the blueprint, one segment of the cable bent tower was modeled with full-scale size. The interaction between the border upon cables were ignored based on the experiences ago. The prestressed steel was tensioned by YCW250 jack on its two ends, when the concrete strength reached 85 percent of its yield strength, and the magnitude of the tensile force was controlled by the distortion and the tensile force of the cable. The tensile force of each bind was 2343.6 kilo-Newton. The load act on the anchor zone was the horizontal component of forces of the internal force of the cable, it was provided by jack YGL-1000.

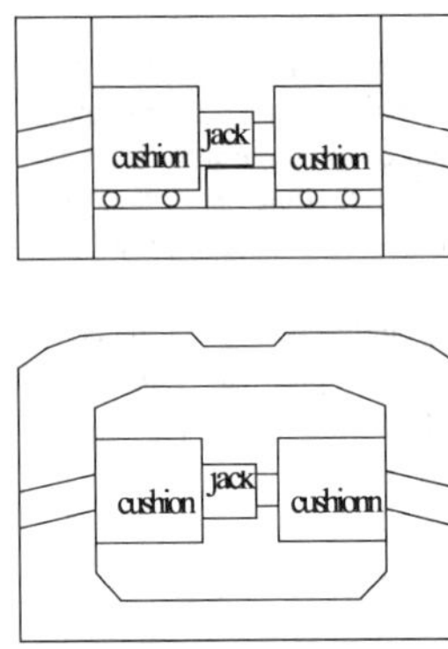

Figure2. Loading Sketch Map

2). Testing Content

Testing contents include the exterior and internal stress, the distortion and the utmost bearing capacity of the model.

3). Measuring Point Settings

The external strain of concrete was measured by strain gauge with resistance 152 Ω ,10*50 millimeter. The internal strain was measured by homemade reinforcing steel bar (be demarcated by MTS).

According to the character of the anchor zone and the result of likeness, the measuring point was settled in the anchorage wall, the side wall and the top of the model in the form of strain-gauge rosette. The deformations that parallel to the bridge and vertical to the bridge were measured by the instruments (precision 0.01micrometer) which setting on the anchor wall and the border of the model.

In order to get the load when cracks' appearance in the model, macrography and graduated magnifier was used to examine the crack's birth, breadth and distributing. On the other hand, the changes on the strain gauge can also reflect it.

4). Test Equipments and loading Manner

Two jacks YGL-1000 (with max. measurement 10000 kilo-Newton) bring pressure to bear inside the model to simulate the inner force of the cable projection in horizontal. For the measurement range of the jack is limited, two cushions were designed specially. The jack, the cushions and the model were composed to a self-balance system, and the model was in the operating condition as the cable bent tower being. In order to simulate the local stress state, a circular steel board was putted between the cushion and the anchor. It was shown in Fig.2. The free-roll steel tubes under the cushions were to reduce the friction.

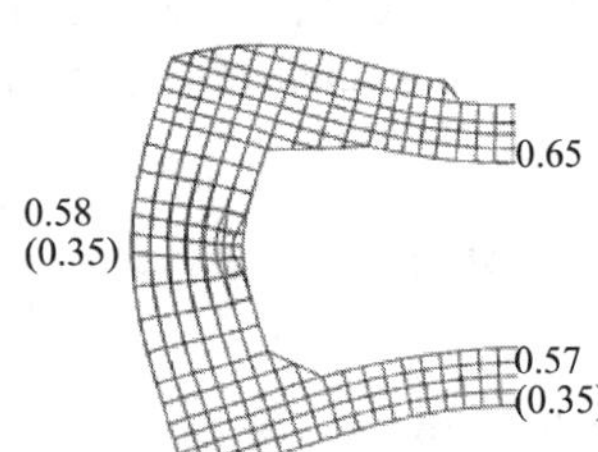

Figure 3. Model Deformation

3 Testing Result and Analysis

1). Checking Computations of Outside Force's Balance

During the loading, one section was chosen to converse its stress to inner force to check out the veracity of loading. When the load reaches to 5400 kilo-Newton, the conversed inner force was 5279 kilo-Newton, they were close highly, and also say the testing data was credible.

2). Load-Deformation Relationship of The Model

Fig. 3 shown the deformation of the model when load reaches to 5400 kilo-Newton, data between the brackets were measured in testing.

Fig. 4 shown the load-strain relationship been measured in the testing, obvious nonlinear characteristics were shown in the curve, and were in accord with the enlargement of the crack. In the testing, as the load reached to 5400 kilo-Newton, small cracks appeared, but quintuple magnifier was used to examine. As the load reached 6000 kilo-Newton, the enlargement of cracks was distinct, as the load reached 7000 kilo-Newton, the cracks in the wall with hole enlarged to the top of the model. Lastly, the load reached 16500 kilo-Newton, the breadth of the crack was 2.01 millimeter, when the load was removed, cracks close again. It can be seen that 16500 kilo-Newton is the utmost bearing capacity.

Bear the weight of the cable force, distinct bent character appears in the anchor wall, namely the vertical stress σ_x is small, transverse stress σ_Y is great; on the other hand, distinct tension-bent character appears in the side wall. And the stress outside the anchor wall was $\sigma_{Y\max}$=6.2Mpa. For the shear stress is small, it is not necessary to be considered as the key factor for the design.

3). Checking Out by MARC

In order to verify the testing result, using the finite element program Msc.Marc to calculate the model.

The element Rebar was used to simulate the steel in reinforced concrete, for the prestressed steel, the effective prestressing force was equivalent to element concrete, on the other hand, contact algorithm was used to check out its influence. The parameters of every material adopted were those measured in the testing. The results indicated that high coherence existed between the result of testing and calculation.

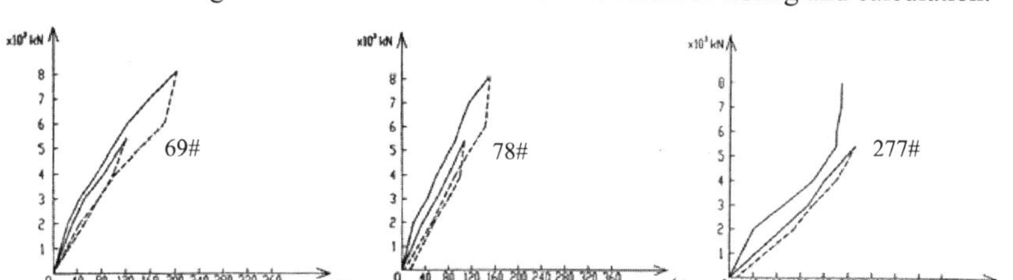

Figure 4. Load-Strain Curve

4 Conclusion

The following conclusions can be drawn on the basis of the testing and the numerically simulated results about the mono-cell box prestressed anchor zone.

Under the circumstance without regard to the creep strain and the long-term shrink of concrete, as load reached 5400 kilo-Newton, the first crack appear, it can be seen as the cracking load. As load reached 16500 kilo-Newton, the maximal breadth of the crack was 2.01 millimeter, when the load was removed, cracks close again, so it can be concluded that the utmost bearing capacity is more than 16500 kilo-Newton.

The results of finite element analysis were almost identical with the testing, the cracking load be measured was consistent with the theoretic solution. Acting with the local stress around the cable hole, the primary tension stress would be regarded as the key factor like the vertical stress and the horizontal stress.

References

1. Fenwick, R. C.; Lee, S. C. ANCHORAGE ZONES IN PRESTRESSED CONCRETE MEMBERS. Author Affiliation: Univ of Auckland, Auckland, NZ Source: Magazine of Concrete Research , v 38, n 135, Jun, 1986, p 77-89 ISSN: 0024-9831 CODEN: MCORAV (In English)
2. Ibell, T.J.; Burgoyne, C.J. Experimental investigation of behaviour of anchorage zones. Author Affiliation: SASOL Technology Source: Magazine of Concrete Research , v 45, n 165, Dec, 1993, Publ by Thomas Telford Services Ltd, London, Engl, p 281-291 ISSN: 0024-9831 CODEN: MCORAV(In English)
3. Fujii, Manabu; Miyamoto, Ayaho; Mizuno, Hideo .FUNDAMENTAL STUDY ON REINFORCEMENT OF RIB-ANCHORAGE ZONE IN PRESTRESSED CONCRETE MEMBER. Author Affiliation: Kobe Univ, Jpn Source: Journal of the Society of Materials Science, Japan/Zairyo , v 29, n 318, 1980, v 29, nMar, 1980, p 299-305 Sponsored by: , p 299-305 ISSN: 0514-5163 CODEN: ZARYAQ (In Japanese)

STUDY TO THE CREEP LIFE OF GRANITE UNDER UNIAXIAL TENSION STRESS

JINXING WANG
Jiaozuo Institute of Technology, Jiaozuo NJ454000, China

XIAOLIN YANG
Jiaozuo Institute of Technology, Jiaozuo NJ 454000, China

JUNICHI KODAMA
Muroran Institute of Technology, Muroran, NJ 050-8585 Japan

Series of creep tests under uniaxial tension stress using granite have been carried out and the following results are obtained: Under uniaxial tension creep stress, the minimum value of strain rate is almost linear to creep life on the log-log plot and the relation- ship between it and creep life after axial stain rate reached the minimum magnitude is also linear on log-log plot; In the tertiary creep stage, there are inversely-proportional relationship between axial strain and residual life. According to these relationships, the creep life of granite can be estimated.

1 Introduction

In the past others test methods, such as crushing test and bend test, have been used to conjecture the properties of rock under tension stress. But in some conditions, there are great errors between them. To deeply understand and grasp the over-all mechanical properties of rocks, so as to give a correct direction to the design, construction, maintenance and administration of the underground engineering, the mechanical properties and the law of deformation of rocks under directly tension stress should be studied. This paper concerned on behaviors of deformation and methods to predict the creep life of rocks under tension stress.

2 Test method[1]

The test specimens are made of granite excavated from Ibarakiken in Japan. They were made cylindrical with 30mm in diameter and 60mm in length and the parallel degree between two end faces reached ± 1/20mm. In the experiment, servo-controlled testing equipment made by MTS Corporation was used.

In experiment, the specimens were directly cemented to the platens by series of epoxy cohesive material. The cementing bond of it has been tested before and confirmed to be adequate. The load control function of test machine is utilized in the uniaxial tension creep experiments. The loaded creep stresses were 5.0Mpa, 5.3Mpa, 5.6Mpa and 5.9Mpa.They were 80%, 85%, 90%, and 95% of the tensile strength at the axial strain velocity of $\dot{\varepsilon}_a = 10^{-5} / s$. That is to say the creep stresses were selected every five percent of the tensile strength. As the time limitation of indoor experiment, if a specimen was not ruptured when they were loaded three days, they were thought to be safety under the creep stresses.

In the experiments, loads, axial displacement and acoustic emission waves were measured and the data captured were processed by a computer. In the experiments, four strain gauges were paralleled pasted around a specimen. It was indicated from the experiments that the four strain values are almost the same and the tension stress were equally loaded on specimens.

3 Results of experiment

3. 1 Creep curves

Axial strain and acoustic emission event rate varied with time. The results are shown in Fig.1. From Figure 1, it can be obviously seen the three stages of creep, which is called primary stage, secondary stage and tertiary stage. The tendency of acoustic emission rate is as follows: Just after the creep stress is loaded on, the rate came to much higher; At the secondary stage the variation of it is little; At the tertiary stage it come to a rapid increase until the rupture of the specimen. The change of it is well correspondence to the three stages of creep.

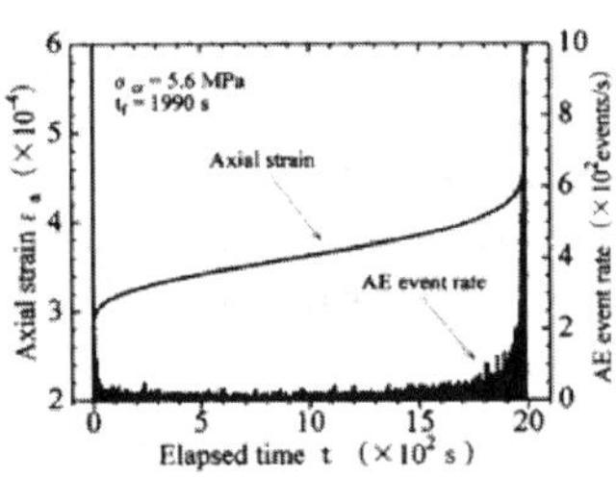

Figure1. Strain and AE event rate in a creep

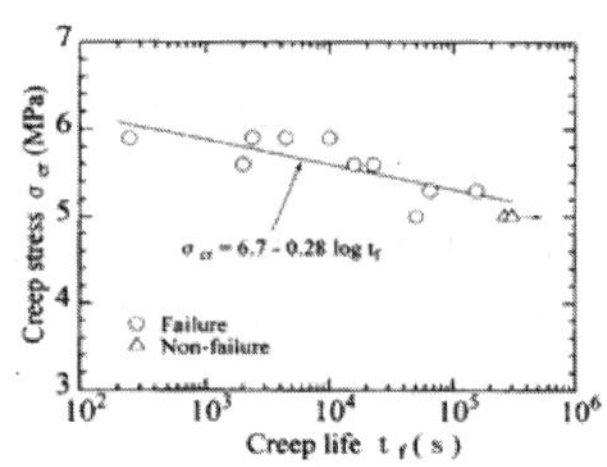

Figure 2. Distribution of creep life

3. 2 Relationship between creep stress and creep life

Relationship between creep stress (σ_{cr}) and creep life (t_f) is plotted in Fig. 2. Entirely, the lower is the creep stress the longer is the creep life, as shown in Eq. (1).

$$\sigma_{cr} = -6.7 - 0.28 \log t_f \qquad (1)$$

Using formula (1) creep lives can be coarsely estimated. While observing the four creep lives at creep stress σ_{cr}=5.9Mpa,we can find that they are 248s, 2379s, 4403s and 10011s.There are great difference among them. That is to say, only considering the creep stress, we cannot give an accurate estimation to the tension creep life of rocks. To find a more accurate way to estimate creep lives, it is necessary to study the behaviors of creep deformation.

3.3 Relation between creep deformation and creep life

The relationship between axial strain rate $\dot{\varepsilon}_a$ and creep time is shown in Figure 3. On the log-log plot, at first the strain rates straightly decrease with time. After it reached a minimum rate it turns to increasing until the specimen was ruptured. Totally, the longer is the creep life the value of strain rate is smaller and the minimum value of it is smaller.

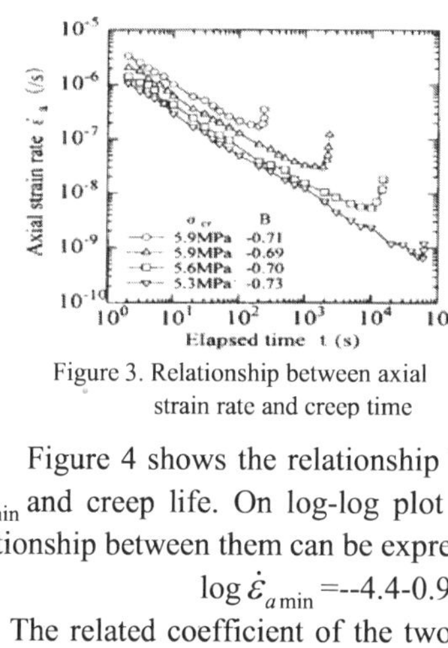

Figure 3. Relationship between axial strain rate and creep time

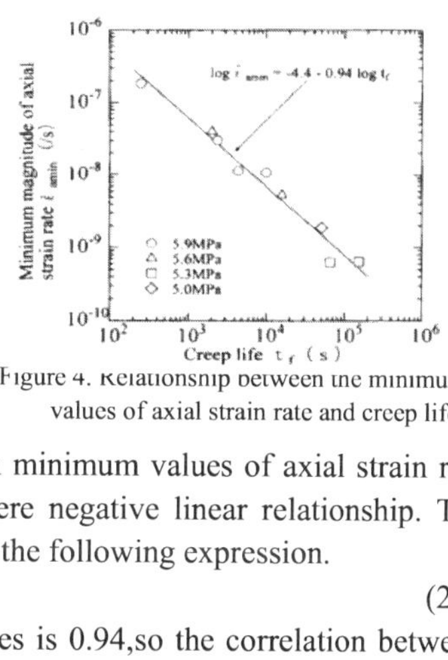

Figure 4. Relationship between the minimum values of axial strain rate and creep life

Figure 4 shows the relationship between minimum values of axial strain rate $\dot{\varepsilon}_{a\min}$ and creep life. On log-log plot they were negative linear relationship. The relationship between them can be expressed by the following expression.

$$\log \dot{\varepsilon}_{a\min} = --4.4-0.94\log t_f \qquad (2)$$

The related coefficient of the two variables is 0.94,so the correlation between them is high. What should pay attention to is that the upper expression include phenomena of great differences at the same stress which shown in Figure 2. That is to say, even if there are great difference in creep life under the same load, there are linear relationship between minimum strain rate and creep life. When using the expression (2) to estimate tensile creep lives, the minimum strain rate must be known before and the time creep began shall also be known. In some events, to know the time creep began is difficult. So this is a shortcoming of the expression.

Here, we define the time from the strain rate reached the minimum value to the rupture of specimen as remaining life t_r. The relation can be expressed by the following expression.

$$\log \dot{\varepsilon}_{a\min} = -8.98-0.67\log t_r \qquad (3)$$

If the minimum strain rate has been known, the time from that moment to rupture can be predicted by expression (3) and we have no need to know the begin time of creep. When comparing the data from experiment with the results calculated by the expression, we can see that the former is within 0.5 to 2 times the later. There is no error more than order of magnitude.

According to the upper results, the minimum axial strain rate has close relationship with the creep life. But if the creep deformation has reached the tertiary stage, the method cannot be used to predict the creep life. To realize the prediction in this stage, deformations in the stage should be studied.

Here, we define residual creep life (T) as the difference between creep life and the pasted creep time ($T = t_f - t$). At tertiary stage, there is about linear relation between them and their relationship can be expressed by the following equation.

$$\varepsilon_a = B_a \log T + C_a \tag{4}$$

In this equation, B_a and C_a are invariables. When make differential calculus on this equation, it can be turned into the following equation.

$$T = D_a / \dot{\varepsilon}_a \tag{5}$$

The D_a in the equation is an invariable. From this equation, it can be seen that at tertiary stage there are inversely-proportional relationship between residual life and axial strain rate. Using this equation the residual life can be estimated simply.

4 Conclusions

Four methods are put forward to estimate uniaxial tension creep lives, residual life and remaining life for granite.

(1) Estimate creep life by creep stress. This method can only give a coarse prediction to the creep life because the dispersion is much larger;

(2) Estimate creep life by the linear relationship between it and minimum strain rate. This method requires knowing the minimum strain rate, so a long period of measure time is needed. But the result of this method is more accurate.

(3) Estimate remaining life with the relationship between it and minimum strain rate. Using this method we can predict the rupture time from that moment only by the minimum strain rate.

(4) At the tertiary stage, there is inversely-proportional relationship between residual time and axial strain rate. We can predict the residual life easily by this simple relationship.

Reference

1. Jinxing Wang, Study to the deformation of Granite under Uniaxial Creep test[D].Jiaozuo:Jiaozuo Institute of Technology,2000

EXPERIMENTAL STUDY ON THE INTENSITY OF COAL AND GAS OUTBURST

KAI WANG, QIXIANG YU, CHENGLIN JIANG, JIANZHI FANG
China University of Mining & Technology, Xuzhou, Jiangsu 221008, China

Aimed at the four main factors which control the occurrence and intensity of coal and gas outburst: gas pressure (gas content), ground stress, coal mass strength and thickness of soft slice in coal seam, laboratory simulating tests have been conducted to investigate their influence on outburst intensity. Then by applying poly-elements regression analysis, the sequence of these four factors is obtained according to the degree of their influence on outburst intensity. Finally, the inherent physical mechanism that determines such sequence is analyzed.

1 Introduction

Based on the synthetical hypothesis of outburst mechanism and statistical data from the spot, it can be drawn that the factors which control the occurrence and intensity of outburst mainly include: (1) gas pressure (gas content), (2) ground stress, (3) coal mass strength, (4) thickness of soft slice in coal seam. Moreover, the factors such as geological structure and pitch angle of coal seam, passage tunnel for outburst and the action of mining and drivage activity also influence the occurrence and intensity of outburst. In order to effectively prevent and control outburst, the study must be deeply strengthened on possibility and intensity of outburst occurrence.

2 Simulating Tests for Outburst Intensity

2.1 Simulating Tests

The experimental setting-up installed horizontally for outburst simulating tests is shown in Fig.1. In the tests, the forenamed first four parameters are controlled as follows.

(1) The coal mass strength is controlled by sampling coal from mines

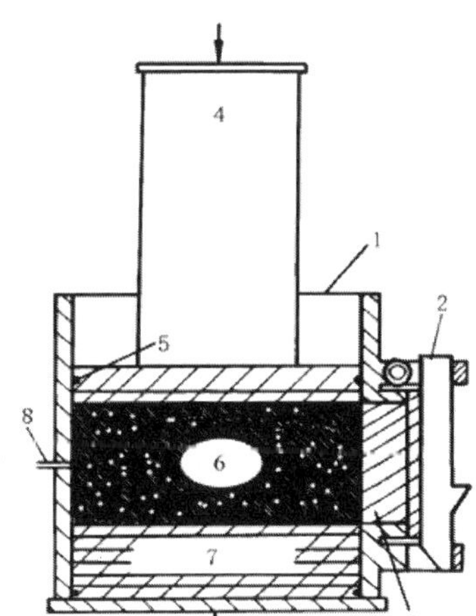

Figure 1. Diagram of experimental setting-up
1-vessel; 2-shield; 3-outburst passageway; 4-press; 5-sealing circle; 6-coal sample; 7-steel plates; 8-hole for injecting gas

with outburst danger in China, and the strength of coal mass is expressed as sieving modulus. The larger sieving modulus is, the smaller coal mass strength is.

(2) Through adjusting the thickness of steel plates, reserve a certain fixed space in the vessel to accommodate coal samples. Take certain suitable weight coal and grind it into powder of approximate 1mm diameter. Mix even the powdered coal with a small amount of water in the vessel, then press it by TYS-500 Press with 30MPa pressure. Repeating above steps several times, we can get an artificial coal sample with fixed thickness that can be regarded as the thickness of soft coal body.

(3) Put a certain fixed pressure on the coal sample by TYS-500 Press, and let it keep 60 hours for simulating the ground stress acted on coal seam, so that the coal sample can be even pressed and completely deformed.

(4) While the coal sample is pressed in the vessel, gases in it are also drawn by vacuum pump for 12 hours. Then the experimental gas CO_2 is injected into and this process should last 48 hours to ensure that the experimental gas can be fully absorbed by coal sample and the gas pressure is well distributed in coal sample.

2.2 Results of the Tests

After the process of pressing coal sample and injecting experimental gas into it finished, turn off the gas resource and then take off the shield quickly. Two typical classifications of dynamic phenomena have occurred in the tests: (1) no dynamic phenomenon; (2) coal and gas outburst. For the tests of No.2 classification, the intensity of outburst, i.e. the weight of coal thrown out from the vessel via the outburst passageway has been measured.

Use P_0, P_g, T_h, S_c, Q to represent ground stress, gas pressure, thickness of soft coal body, sieving modulus of coal mass and outburst intensity respectively. The concrete experimental conditions and results of the tests are listed in Table 1.

3 Poly-elements Regression Analysis of the Relation Between Outburst Intensity and the Four Factors

The relationship between Q and P_0, P_g, T_h, S_c is achieved according to Eq.1, 2:

$$Q(i) = b_0 + b_1 P_0(i) + b_2 P_g(i) + b_3 T_h(i) + b_4 S_c(i) + \varepsilon(i) \quad (1)$$

$$Q(i) = e^{b_0 + \varepsilon(i)} P_0(i)^{b_1} P_g(i)^{b_2} T_h(i)^{b_3} S_c(i)^{b_4} \quad (2)$$

where b_k (k=1, 2, 3, 4) are fitting coefficients; $\varepsilon(i)$ is normal random variable of which the mean value is 0.

Table 1. Experiment Data and Results of Outburst Simulating Tests

No.	Site for Sampling	P_0(MPa)	P_g(MPa)	T_h(mm)	S_c	Classification	Q(kg)
1	Luling 82	14.37	0.18	109.0	8.1	1	0.0
2	Luling 83	14.37	0.21	38.0	6.2		
3	Kongzhuang 8#	16.61	0.22	37.0	11.1		
4	Jiahe 2#	13.09	0.13	163.5	7.8		
5	Jiahe 9#	17.24	0.21	85.5	14.7		
6	Majiagou 9#	14.69	0.23	30.5	15.1		
7	Jiulishan Y	13.73	0.14	76.0	7.7		
8	Jinggezhuang	18.52	0.27	33.0	13.2		
9	Panji No.1 Mine	11.50	0.20	96.5	7.6		
10	Panji No.3 Mine	16.61	0.10	142.5	7.1		
11	Jiaoxi Hard Coal	14.05	0.32	105.0	4.9	2	4.0
12	Luling 83	14.37	0.72	85.5	6.2		1.9
13	Luling 9#	15.01	0.44	79.0	11.2		5.8
14	Luling 9#	15.01	0.36	62.0	11.2		1.8
15	Luling 10#	13.09	0.35	60.0	7.3		2.2
16	Kongzhuang 8#	16.61	0.35	127.0	11.1		4.3
17	Jiahe 7#	12.77	0.43	125.0	8.8		4.0
18	Majiagou 9#	14.69	0.48	68.0	15.1		4.8
19	Majiagou 12	11.50	0.52	33.5	12.1		0.7
20	Majiagou 12	11.50	0.55	67.0	12.1		3.6
21	Baishanping 6#	13.73	0.22	34.5	6.8		0.5
22	Baishanping 6#	13.73	0.41	34.5	6.8		1.6
23	Jiulishan Y	13.73	0.36	75.5	7.7		4.0
24	Xinzhuangzi	14.37	0.48	75.5	8.7		5.7
25	Jinggezhuang	18.52	0.25	118.5	13.2		4.0
26	Panji No.1 Mine	11.50	0.44	97.5	7.6		2.6
27	Panji No.2 Mine	13.41	0.49	73.0	12.5		4.2
28	Panji No.2 Mine	13.41	0.39	56.5	12.5		2.9
29	Panji No.3 Mine	16.61	0.48	142.5	7.1		6.1
30	Jiaoxi Hard Coal	14.05	0.74	105.0	4.9		14.8
31	Jiaoxi Soft Coal	12.77	0.30	191.5	11.8		16.2
32	Jiaoxi Soft Coal	12.77	0.75	74.5	11.8		11.7
33	Luling 82	14.37	0.65	110.0	8.1		14.7
34	Luling 10#	13.10	0.85	172.0	7.3		22.1
35	Kongzhuang 7#	17.24	0.82	118.0	5.9		15.1
36	Jiahe 2#	13.36	0.82	162.5	7.8		19.3
37	Jiahe 7#	12.77	1.0	37.0	8.8		4.3
38	Majiagou 9#	14.69	1.16	30.5	15.1		5.3
39	Mixed Coal 1#	11.82	0.91	161.0	7.6		23.0
40	Mixed Coal 2#	19.48	0.61	107.5	6.6		13.0
41	Xinzhuangzi	14.37	0.93	154.5	8.7		22.2

Before calculating the fitting coefficients b_k, the experiment data should be firstly unitized to eliminate the influence of their difference in dimension and quantity upon the calculating process. The process of unitization can be written as:

$$s_{ij} = \frac{x_{ij} - \overline{x_j}}{S_j}; \overline{x_j} = \frac{1}{n}\sum_{i=1}^{n} x_{ij}; S_j^2 = \frac{1}{n-1}\sum_{i=1}^{n}(x_{ij} - \overline{x_j})^2 \qquad (3)$$

where x_{ij} is the measurement of No.j variable of No.i sample; s_{ij} is the value of x_{ij} after unitization.

Use the experiment data of No.2 classification as sample data. The fitting coefficients b_k and the complex correlation coefficients R of Regression Eq.1 and 2 are obtained as shown in Table 2.

Table 2. Fitting Coefficients and Complex Correlation Coefficients

Variables	P_0	P_g	T_h	S_c	R
Fitting Coefficients	b_1	b_2	b_3	b_4	
Linear Eq.(1)	0.027	0.536	0.108	-0.244	0.84
Exponential Eq.(2)	0.015	0.577	0.479	-0.047	0.92

4 Discussion and Conclusion

(1) The result of regression in Table 2 indicates that the degree of the four factors' influence upon outburst intensity is as follows: the influence of gas pressure on outburst intensity is greatest, that of thickness of soft coal body and coal mass strength (sieving modulus) comes second, and that of ground stress is smallest.

(2) The negative coefficient corresponding to sieving modulus in Table 2 indicates that the smaller (larger) sieving modulus (coal mass strength) is, the greater outburst intensity is.

(3) It is well known that the smaller coal mass strength is, the greater outburst danger is. Considering this understanding, we can conclude that within certain range the larger the strength of coal mass is, the more difficult outburst occurs, but once outburst occurs, the greater outburst intensity is. This is because that in larger strength coal it is liable to form laminae (shells) spalling parallel to the free surface. Under the action of gas expansion energy, the laminae (shells) will be accelerated to move out, then a high gas pressure gradient zone will occur behind the laminae (shells), which can sustain the development of outburst.

References

1. Yu Qixiang. *Prevention of Methane in Coal Mine*. Xuzhou: CUMT Press, 1992
2. Jiang Chenglin, et al. *The Principle of Spherical Shell Losing Stability and Preventive Technology for Coal and Gas Outburst*. Xuzhou: CUMT Press, 1998
3. Wang Kai. *Research on Prediction of Coal and Gas Outburst by Boring Method* [*Doctorate Dissertation*]. Xuzhou: CUMT, 1997
4. Fang Jianzhi, et al. *A Lamination Separation and Fragmentation Model of Coal and Gas Outburst*. Journal of China Coal Society, 1995, 20(2): 149~153

EXPERIMENTAL STUDIES ON CRACKS' FRACTAL PROPERTY OF REINFORCED CONCRETE BEAM

SHUGUANG ZHANG
Liaoning Technical University, Fuxin, 123000, China

On this paper, in virtue of model experiment, cracks' distributing rule is simulated when the beam loaded. By using fractal geometry, research show that cracks' distributing rule has self-comparability. At the same time, the relation of fractal dimension and damage variable, the relation of fractal dimension and maximum deflection are developed.

1 Introduction

Concrete has very complex non-linear character. This happened due to inanition, cracks, crystal, cementation etc included in concrete. Reinforced concrete beams have to produce deflection under load, and the magnitude of deflection not only lie on design in initial stages, but also lie on beams' damage and cracks' state[2,3]. Though we have founded beams' mathematics model under different damage state[4], there is no precise solution to decide beams' damage degree or damage variable. This paper bring forward that the foundation of damage variable via to fractal dimension by using fractal property of external cracks' distributing.

2 Experiment Study of Crack

Cross section and properties of specimen are shown in Fig.1 and table 1. It only provided longitudinal reinforcement, and no hoop was provided.

Table1. properties of specimen

A/d	Length /mm	Compressive strength /Mpa	Maximum moment /kN m
4.5	1980	43.0	15.85

Load put on beam by setting hydraulic experiment machine at the top of beam, and the maximum capability of experiment machine is 100kN. Load is put gradually on beam by 50kN/min. Voltage displacement sensor is set at mid-span used to determine deflection.

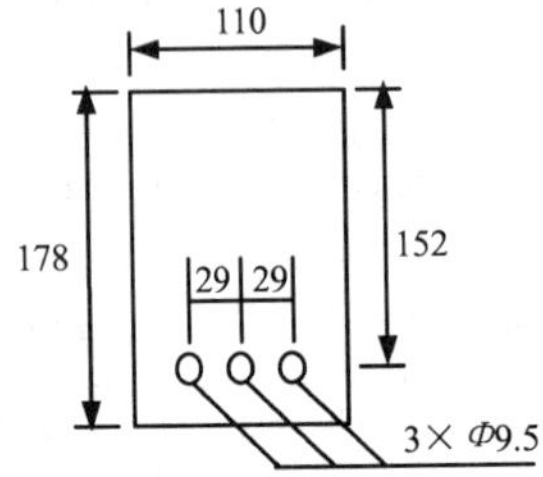

Figure1. Cross section

3 Distributing Rule of Cracks

From above experiment of different load stage, the figure of cracks' distributing state are plotted in Fig.2, while the load equal to 20kN, 30kN, 40kN, 45kN.

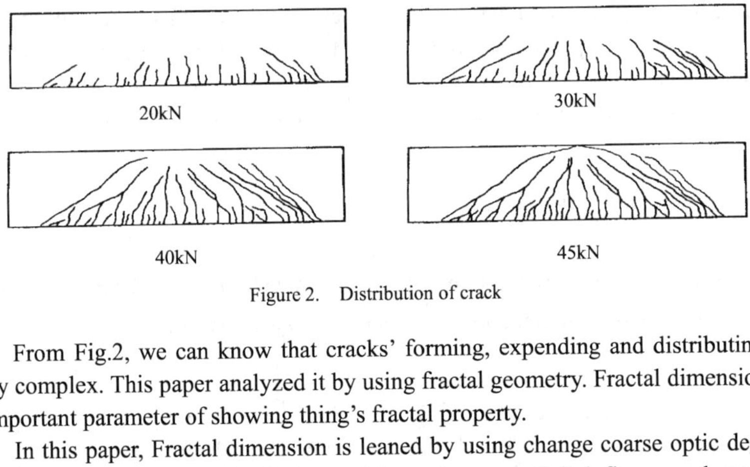

Figure 2. Distribution of crack

From Fig.2, we can know that cracks' forming, expending and distributing is very complex. This paper analyzed it by using fractal geometry. Fractal dimension is a important parameter of showing thing's fractal property.

In this paper, Fractal dimension is leaned by using change coarse optic degree method. Firstly, cracks' distributing of beam became digital figure and created automatically digital file. Secondly, digital file put into computational program (Fractal. For)[5, 6]. Lastly, the relation curve of $lgN(r)$~$lg(r)$ are plotted in Fig.3. i.e. double logarithm figure.

From Fig.3, we can obtain that double logarithm figure take on good coincidental degree. this shows beam's crack distributing has good self comparability and fractal character.

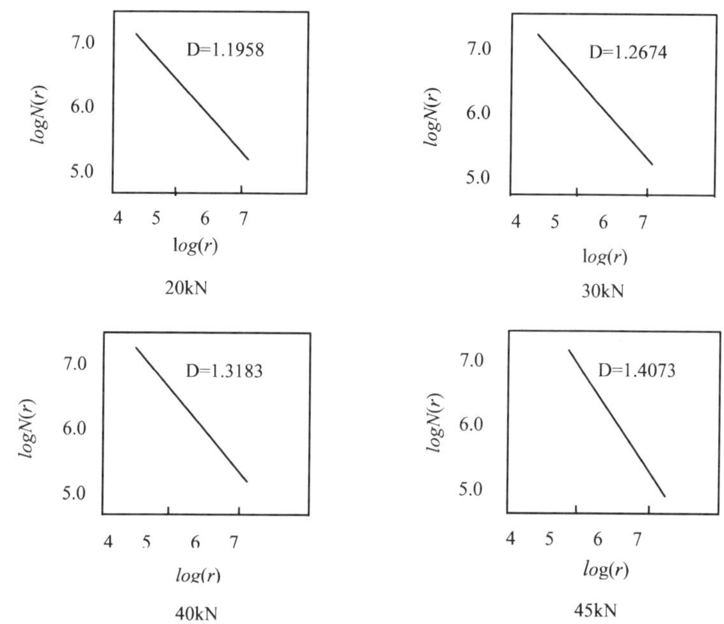

Figure3. Curve of double logarithm

4 Relation of Fractal Dimension and Damage Variable

Displacement gauge was set under mid-span of beam in order to measure maximum deflection at mid-span. From Damage Mechanics, we can deduce the following formula[5].

$$\frac{d_{\max}}{d_{0\max}} = \frac{1}{1-\Omega} \tag{1}$$

where d_{0max} is maximum deflection at mid-span of beam when don't bear damage and Ω is damage variable. Note that d_{max} is maximum deflection at mid-span when it bear damage. Thereby, Ω can be deduced by using d_{max}. The relation of Ω and D is shown in Table 2.

Table 2. relation of Ω and D

Load/kN	20	30	40	45
d_{max}/mm	3.6270	4.9820	6.8790	19.6370
Ω	0.1120	0.2080	0.4800	0.7980
D	1.1958	1.2674	1.3183	1.4073

The experience relation of Ω and D is well-regress as

$$\Omega = -1.5715D^3+3.6129D^2+3.4337D-7.0863 \quad (2)$$

Then

$$d_{max} = \frac{d_{0\,max}}{1.572D^3 - 3.613D^2 - 3.434D + 8.086} \quad (3)$$

5 Conclusions

1. Cracks' distributing of reinforcement concrete beam take on fractal character, and fractal dimension can describe its' fractal degree .
2. Cracks' distributing character can deduce beam's damage state via the founded mathematics model of fractal dimension and damage variable.
3. After knowing maximum deflection when beam don't bear damage, cracks' distributing character can deduce beam's maximum deflection under damage state.

References

1. Xie He-ping. Damage Mechanics of Rock and Concrete. Xuzhou: China Mining University Press. 1990,111~115
2. Shrikrishna M. Kulkarni and Surendra P. Shah Response of Reinforced Concrete Beams at High Strain Rates. ACI Structural Journal.1998,Vol.95(6):705~715
3. Stephen J. Foster and R.lanGilbert. Experiment Studies on High-Strength Concrete Deep Beams. ACI Structural Journal.1998,Vol.95(4):382~390
4. Xu Zhi-lun. Perspicuity Tutorial of Elasticity Mechanics (second edition). Beijing：Higher Education Press，1980，42~46
5. Chen Jian-ping. Computer Simulation of Fractal Dimension in Rock Crack Network. Engineering geology Journal. 1995,3(3):79~85
6. Yu Guang-ming. Test Study of Excavation Subsidence in Jointed Rockmass and Analysis on Damage Dynamics. Rock Mechanics and Engineering Journal,1998,13（2）：145~153

EXPERIMENTAL STUDY ON UNLOADING FAILURE OF RED SANDSTONE

GANG WU, XINGCHUN HUANG, JIANHUA PAN
Department of Civil Engineering, Shanghai Jiaotong University, Shanghai, 200240, China

A series of true triaxial tests were carried out to investigate on the unloading failure effect of red sandstone. Experiment adopted four types of tests that correspond to four excavation ways. By analysis of test results, attaining the characteristics of deformation, failure and strength of red sandstone in unloading states. Accordingly, revealing deformation and failure laws of engineering rock mass under engineering unloading conditions.

1 Introduction

It has been proved that failure or unstability of engineering rock mass is mainly due to the engineering excavation or the influence of environment (such as water, temperature etc.). It's very important realistic meaning that studying on deformation and failure effect of rock mass under excavation unloading states.

With the improvement of test machine and the development of test technique, the unloading experiment of rock has been carried through [1–4]. To reveal the deformation and strength character of the engineering rock mass, four kinds of unloading failure tests of red sandstone under the true triaxial condition in this paper.

2 Preparation of Experiment

2. 1 Rock Specimens

The red sandstones were taken from the Emei mountains in Sichuan, China. They were made of cuboid specimens (100mm×100mm×200mm). The red sandstone material was dense and had no macroscopic flaw. Before test, all specimens were dried in an oven at 110℃ for over 24 hours. The physics and mechanics properties of the red sandstone specimen are listed in table 1.

Table 1. The physics and mechanics properties of red sandstone specimen

Dry Unit Weight γ(kN/m^3)	Void Ratio N(%)	Poisson Ratio μ	Elastic Modulus E(MPa)	Uniaxial Strength σ_c(MPa)	Tensile Strength σ_t(MPa)
22.9	14.9	0.26	1.31×10^4	42.4	1.86

2.2 Method of Experiment and Procedure of Experiment

According to principle that corresponds with engineering practice, we made four types of unloading failure tests of red sandstone. They are as the follows:

(a)Type U1 Test $\sigma_1>\sigma_2>\sigma_3>0$, keeping σ_1 and σ_2 unchanged, unloading σ_3.

(b)Type U2 Test $\sigma_1>\sigma_2>\sigma_3>0$, keeping σ_1 unchanged, first unloading σ_3, then unloading σ_2.

(c)Type U3 Test $\sigma_1>\sigma_2>\sigma_3>0$, keeping σ_2 unchanged, first unloading σ_3, then loading σ_1.

(d)Type U4 Test $\sigma_1>\sigma_2>\sigma_3>0$, first unloading σ_3, then loading σ_1 to a certain value (not failure), finally unloading σ_2.

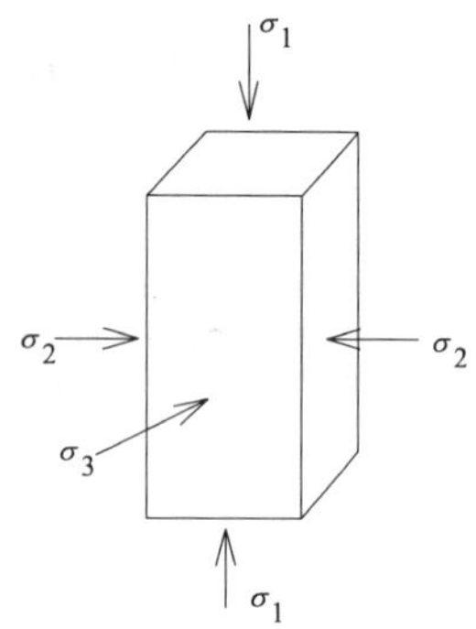

Figure 1. Loading diagram of rock specimen

Testing procedure: Firstly, we put the specimen into triaxial pressure pot and exerted lateral pressures up to a set point respectively. Lateral pressures were exerted by four hydraulic jack and controlled by three-passage pressure regulator. Afterwards, adding the axial pressure and starting all kinds of measurement. Unloading adopted successive unloading way (controlled by pressure regulator) until the unstable rupture of the specimen. The axial loading speed was at 0.01~0.283MPa/sec and unloading speed was at 0.01~0.015Mpa/sec during test.

In all unloading tests, the initial lateral pressures are respectively σ_2=3MPa, σ_3=1.5MPa. The load-bearing state of specimen is as shown in figure 1.

3 The Result of Experiment

We tested fourteen rock specimens using aforementioned unloading failure ways. The results of test are as follows.

3.1 *Typical Stress-Strain Curves*

According to the test data, we obtained the typical stress-strain curves in each type test shown in figure 2. Where, σ_e is stress intensity and ε_e is strain intensity.

$$\sigma_e = \frac{1}{\sqrt{2}}\sqrt{(\sigma_1-\sigma_2)^2+(\sigma_2-\sigma_3)^2+(\sigma_3-\sigma_1)^2}, \varepsilon_e = \frac{\sqrt{2}}{3}\sqrt{(\varepsilon_1-\varepsilon_2)^2+(\varepsilon_2-\varepsilon_3)^2+(\varepsilon_3-\varepsilon_1)^2}$$

3.2 *Contrast of Deformation and Failure in Different Unloading Way*

Relationship of stress difference($\sigma_1-\sigma_3$) versus volume-strain ε_V are shown in fig. 3.

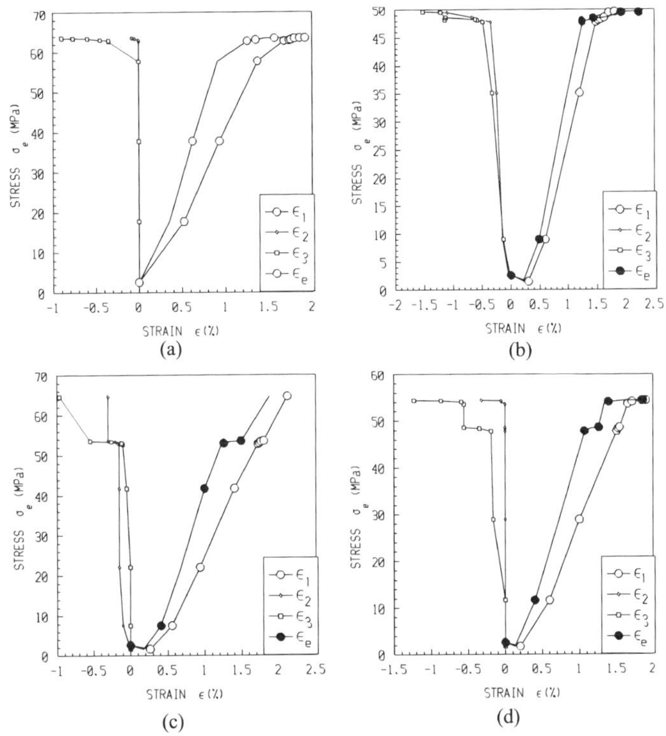

Figure 2 The typical stress-strain curves of red sandstone specimens in various unloading test
(a)Type U1 (b)Type U2 (c)Type U3 (d)Type U4

In fig. 3, all four unloading failure tests experienced so a process that volume of rock specimen was first reduce and then expand (dilatancy). The Dilatancy quantity is closely related to unloading quantity and way. Moreover, Dilatancy quantity of loading is far less than that of unloading. Dilatancy quantity of type U2 test is biggest, that of type U4 is second, that of type U3 is third and that of type U4 is least.

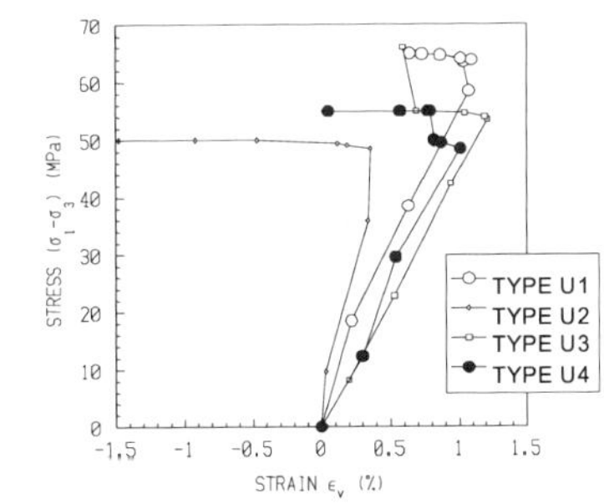

Figure 3. The relations of principal stress difference and voluminal strain in the four kinds of tests

3.3 The Strength Character of Each Type Test

The failed stress states of rock specimens in each type test are listed in table 2.

Table 2. Failing stress states of rock specimens in various tests

Type of Test	Number of Specimen	σ_1 (MPa)	σ_2 (MPa)	σ_3 (MPa)	σ_e (MPa)	Loading Rate Before Unloading (MPa/s)	Time of measure (s)
	R – 4	55	3	0.23	53.44	0.122	720
	R – 5	54.13	3	0	52.69	0.143	660
Type U1	R – 8	61	3	0.35	59.37	0.169	600
	R – 11	54.37	3	0.76	52.53	0.135	600
	R – 10	65	3	0	63.55	0.283	670
	R – 19	65	3	0.70	63.18	0.206	690
	R – 13	55	0.93	0	54.54	0.138	730
Type U2	R – 14	50	0.82	0	49.60	0.135	665
	R – 23	55	1.92	0	54.07	0.185	705
Type U3	R – 2	66.1	3	0	64.65	0.131	660
	R – 17	59.22	3	0	57.78	0.100	670
	R – 15	55	0.64	0	54.68	0.166	930
Type U4	R – 18	55	0.17	0	54.92	0.114	270
	R – 21	55	1.31	0	54.36	0.172	390

Table 2 shows when loading speed is approximate, the relationships of stress intensity of rock specimens in different test are: $(\sigma_e)_{U3} > (\sigma_e)_{U4} > (\sigma_e)_{U1} > (\sigma_e)_{U2} > \sigma_c$. Besides, the axial loading speed before unloading would produce more influence on the final strength of rock specimen.

4 Conclusion

(a) The most important effect of the unloading is that caused the rock mass produce tension failure. It is easier that the unloading cause the failure or unstability of the rock mass than the loading.
(b) Every unloading way possesses its own deformation and failure character.
(c) The strength and deformation of the rock mass relate with the unload way.
(d) The stress history affects the strength of the rock mass.

References

1. Swanson S.R. et al., *Int. J. Rock Mech. Min. Sci.*, 8(3):277-281(1971).
2. Chen Yong et al., *Science in China*, 11: 1093-1100(1979).
3. Xu Dongjun et al, *Chinese Jour.of Rock and Soil Mechanics*, 7(2):17-25(1986).
4. Yin Guangzhi, et al., *Chinese Journal of Geotechnical Eng.,* 9(2): 20-27(1987).

A TESTING METHOD FOR DETERMINING THE PERMEABILITY OF OVERBROKEN ROCK

WEIQUN LIU XIEXING MIAO ZHANQING CHEN
School of Sciences, CUMT, Xuzhou 221008, PRC
E-mail: wqliu@cumt.edu.cn

Traditional testing instruments and methods provide only the permeability of pre-broken rock. A self-designed seepage instrument for overbroken rock (STIOR) is suggested here to analyze the permeability of post-broken rock in compression after installed into the MTS815.02 Rock Mechanics Testing System with two independent sets of close loops exerting axial force and porosity pressure. A feasibility proof was given in the experiment of testing permeability of crushed sandstone in heavy compression, and some valuable results from such an experiment are shown here.

1 Introduction

The Seepage of water and methane in fractured or broken rock mass is a significant and thorny problem in mining engineering. The statistical data show that, nearly 700 incidences of coal mine occurred with more than 10 persons a time killed from 1990 to 1999 in China, and the total killed persons had gone beyond 10000 during the 10 years. In such shocking incidents, those methane explosion and water burst ones take over 80% [1]. Mining activity inevitably brings about the rock mass fracturing and collapse, and it's known that, the permeability of fracture is far higher than the pore permeability[2]. In other words, majority of the seepage disasters take place in fractured and overbroken rock mass. By now, the seepage experimental techniques for integrated rock mass and the corresponding theory are both relatively rich. However, the seepage testing method for overbroken rock mass is rare, as undoubtedly influences the further research of the engineering problems involved.

Therefore, a self-designed seepage testing instrument for overbroken rock (STIOR) is suggested in this paper. It can be employed to analyze the permeability of post-broken rock in compression after installed into the MTS815.02 Rock Mechanics Testing System with two independent sets of close loops exerting axial force and porosity pressure.

2 STIOR and its outside equipment

The appearance and constitution of STIOR are shown in figures 1 and 2. Its size is designed just to match the structure of MTS815.02. The following are some detail parameters: 240mm in assembly length(110mm for a piston and 180mm for a cylinder); 160mm and 16.5mm for the outer diameter and thickness of cylinder respectively; No. 45 steel employed for manufacture and thorough quenching process for the whole cylinder; two Φ10mm pores through the centers of piston and base respectively to link the water way in MTS; the threads in such two pores for connecting bars to disassemble the instrument by pulling contrarily; and also the cylinder designed separating from the base for disassembling instrument and changing samples easily. Figure 3 is the matching situation of STIOR and MTS815.02. The designed maximum porosity pressure of STIOR is 78MPa, and maximum axial pressure is 150MPa.

Figure 1. Appearance of STIOR

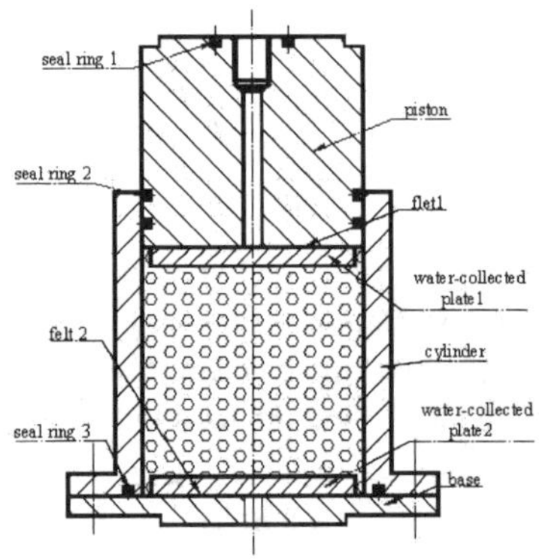

Figure 2. Constitution of STIOR

3 Sample preparation and experiment procedure

The crushed sandstone is taken as an example. Considering the influence of grain size on seepage, by using a set of sieves the crushed sandstone grains are classified into 5 particle-diameter groups: Φ1(2.5~5mm), Φ2(5~10mm), Φ3(10~15mm), Φ4(15~20mm), and Φ5(mixing the above 4 kinds of grain by 1:1:1:1). For every group, three experiments have to be repeated so as to give a true outcome. Before filled into

Figure 3. The matching situation of STIOR and MTS815.02

STIOR, the rock samples have to be watered for the dust around them moved and for slight saturation as well.

The permeability of crushed sandstone is usually high from experience, so a stable testing technique is employed. In this technique, the velocity of trunk piston of water container is controlled, and that is, the rate of flow into the sample is a constant. After that, one collects the pressure difference between two ends of the sample during a stable seepage, and then calculates the permeability *K* from a transformer of Darcy Law, i.e.,

$$K = \frac{QL}{A(h_1 - h_2)} \tag{1}$$

where: Q — rate of flow in sample; $Q = \pi d^2 v / 4$; A, L—cross-section area and length of sample; $h_1 - h_2$ —the pressure difference between two ends of the sample; v — the velocity of trunk piston; d — diameter of the trunk piston, 5.5*cm*.

In STIOR, the length of crushed sandstone sample is set as a constant 11cm (1392.74 *ml* in volume) in every experiment. The sample compressed in STIOR is at three-direction stress situation. Ref. [3] give the ratio of active axial press and passive surrounding pressure 2: 1, and it is known that the singly axial strength of chosen sandstone is 40MPa, so the axial pressures are 10MPa, 20MPa and 30MPa.

In addition, the permeability of the chosen sample sandstone, when integrated, is 10^{-8}cm/s in order of magnitude. And the controlled velocities of trunk piston are 5mm/180s (or 2.78×10^{-3}cm/s), 15mm/180s(or 8.33×10^{-3}cm/s)and 30mm/180s (or 16.67×10^{-3}cm/s), which satisfy the Reynolds condition of Darcy Law.

Table 1 Permeabilities *K* of different particle-diameter crushed sandstone samples (10^{-6}cm/s)

Φ	σ_a	v (mm/s/180)		
(mm)	(MPa)	5.0	15.0	30.0
2.5~5	10.0	4.05	4.05	3.98
	20.0	1.35	1.21	1.22
	30.0	0.28	0.28	0.27
5~10	10.0	10.13	9.95	8.42
	20.0	4.62	3.77	2.99
	30.0	0.74	0.68	0.46
10~15	10.0	19.19	18.24	16.83
	20.0	5.14	4.97	4.47
	30.0	0.83	0.78	0.66
15~20	10.0	21.45	19.89	18.24
	20.0	10.13	9.12	7.55
	30.0	0.99	0.86	0.73
1:1:1:1	10.0	3.04	2.88	2.74
	20.0	0.38	0.32	0.27
	30.0	0.20	0.19	0.19

4 Data collecting and analysis

Data collected in experiments include time, axial load, displacement of trunk piston, axial displacement and pressure difference. Data-collecting durations are set at 3min to omit unstable seepage. Table 1 gives the permeability (average value of three repeated tests) of different particle diameter crushed sandstone under different axial pressures. Figures 4 and 5 illustrate the changes of pressure difference and permeability of crushed sandstone as time ($L = 8cm$; v=8.33x$10^{-3}$$cm/s$).

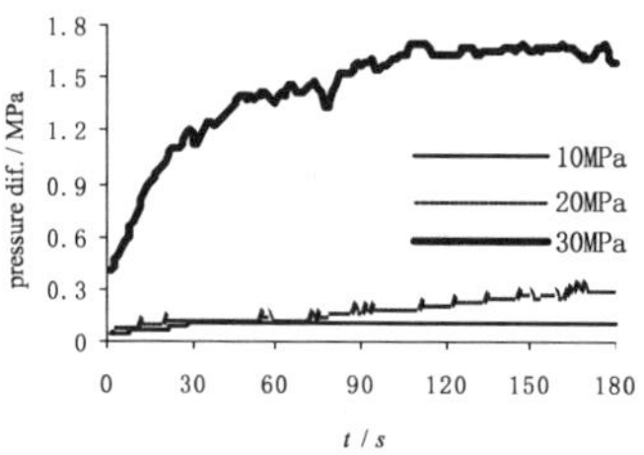

Figure 5. Pressure difference change of Φ 5~10mm crushed sandstone as time

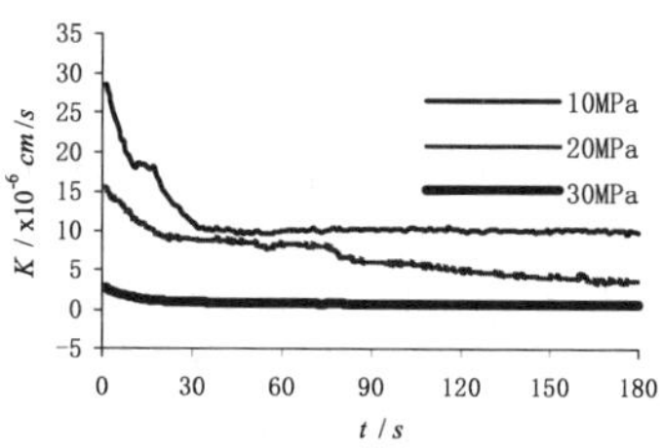

Figure 6. Permeability change Φ 5~10mm crushed sandstone as time

5 Conclusion

STIOR and testing method in this paper can be used to obtain seepage parameters of overbroken rock. Experimental results of crushed sandstone show: the permeability of overbroken rock enlarges obviously as the particle-diameter of crushed rock enlarges, and goes down in order of magnitude as the axial pressure enlarges.

Acknowledgments

This paper is supported by the Key National Natural Science Foundation (50134040) and the PHD Foundation of Educational Ministry of China (20010290003).

References

1. S.N. Zhou et al, *Strategy of Prevention and Control of Coal Methane*. CUMT Press, Xuzhou, China(2001)
2. GI Barenblatt et al, *Prikl. Mat. Mekh.*, Vol **24**, 852~864(1960)
3. X.X. Miao et al, *Experimental Mechanics* Vol **12**, 394~399(1997)

STUDY ON ACOUSTIC EMISSION OF IRREGULAR SINGLE PARTICLE UNDER UNIAXIAL COMPRESSION

P. LIN[1,2] C.A.TANG[2] Y.M.WANG[3] S.H.WANG[2]

[1]*Department of Hydraulic Engineering, Tsinghua University, Beijing, China*

[2]*Center for Rockbursts & Induced Seismicity Research, Northeastern University, China*

[3]*Department of Chemical & Metallurgical Eng., Lulea University of Technology, Sweden*

This paper presents the numerical results for irregular particles compressed under four instant loading conditions. Different shape particles are numerically modeled to investigate the relationship of acoustic emission (AE) and energy under unconfined conditions. Based on the analysis of AE, the less energy can be consumed with increasing the contacting points. Most of the energy released on the breakage of the particles during the failure process is spent in the highly stressed areas, especially in the vicinity of contact points in a crushed zone.

1 Introduction

Comminution is one of the major unit operations in the processing of solids including cement, ceramics, coal, grain, ores and pigments, among other materials. It is estimated that comminution of rock and minerals accounts for about 4 % of all the electrical energy consumed in the world. There is every indication that energy cost of comminution will continue to raise inexorably in the coming years. Thus, it is important to better understand the mechanism of energy consumed in comminution process and to design the machines that enhance energy utilization. Early investigations aimed at the better understanding of the comminution process were concerned with the relationship between the energy consumed by a grinding mill and the amount of size reduction that the consumption of this energy brought about. These investigators created some classical theories of comminution. For example [1-3], Rittinger's "surface" theory (1867), Kick's "volume" theory (1885), Bond created the third theory (1952).

Acoustic emissions are transient elastic waves caused by sudden localized changes in the micro-scale stress state in solids under mechanical load. It is known that final failure zone is not formed immediately, but it is produced as results in coalescence of the local fractures along fracture zone and obtained enough failure energy. It is very important to detect AE signals associated with fracture patterns called "progressive failure" and analysis the energy consumed during the process of communition. In this study, by employing a numerical approach which can be used for analysis and prediction of the breakage behavior of single particle, to investigate the consuming energy of irregular single particle based on change of acoustic emission energy, which was hardly investigated in previous study.

2 Introduction of numerical model

In general, mineral fragment passing through the roller gap is subjected to one of the four instant loading conditions shown in Fig.1, irrespective of whether it is loaded directly by the rollers or by other mineral particles. The first case is termed point-to-point loading (see Fig.1a). The particle is loaded at two points. The second case is termed plane-to-plane loading (see Fig.1b). The piece is loaded between two approximately parallel planes. The third case, point-to-plane loading is a mixture of

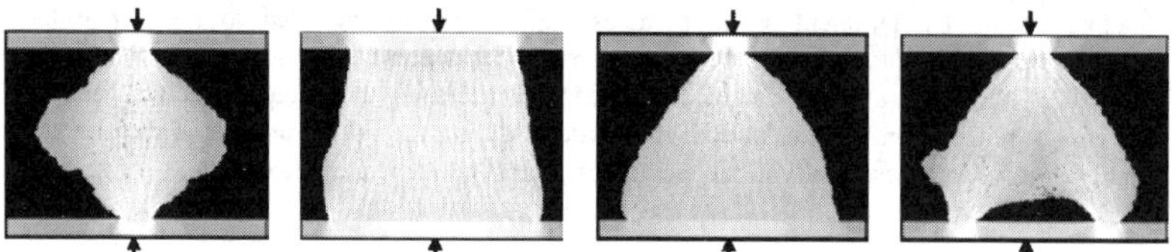

(a) Point-to-point (b) Plane-to-plane (c) Point-to-plane (d) Three-point bending

Figure 1. Stress distribution results of irregular particles under instant loading conditions

cases (a) and (b), as shown in Fig.1c. The fourth loading condition is the multi-point bending (see Fig.1d). This involves at least three or more loading points. Fig.1d only shows an example of three-point-bending. Based on the four instant loading modes and employed the numerical approach [4], It is easy to give the numerical model (see Fig.1) (detailed building model method listed in reference [4,5]

3 Results & Discussion

3.1 General AE observation

Figure 2 shows the AE distribution of irregular particle during the whole failure process under four instant loading conditions. In Fig.2, the ellipse zone represents the tensile zone that leads the tensile crack to produce during the loading process.

Based on the Fig. 2, the first case (see Fig.2a), a main AE signal appeared at the center zone along the diametric direction. The tensile crack propagated through the particle towards to the loading direction, up to the loading platens. The particle is spitted two main parts and formed many small parts around the peak loading. In the second case (see Fig.2b), the AE signals produce continuously along the loading direction, and occur splitting failure in outer zones. In the third case (see Fig.2c), the intensive AE occurs in the area close to the loading points with increasing the load. It leads a few long cracks to start from the upper loading point and extend to the lower loading plane when the loading reaches to the peak loading. In the fourth loading condition, firstly, few AE in the particle appeared a symmetry distribution. Because of the heterogeneity of particle, the AE concentration starts to transfer to another side with increasing the load, and a few long cracks occurred to propagate

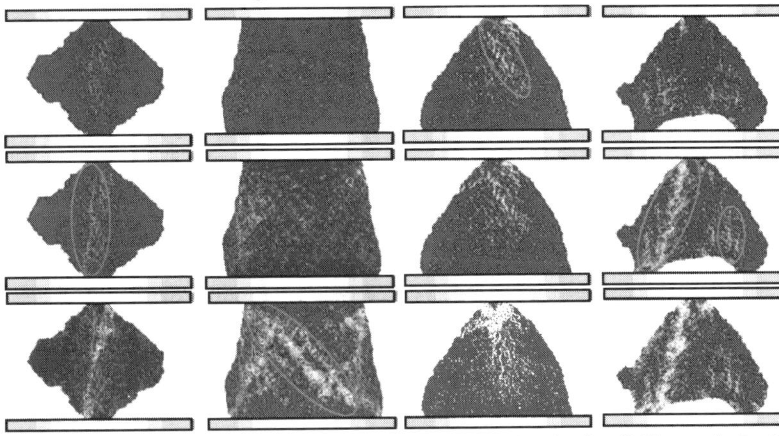

(a) Point-to-point loading (b) Plane-to-plane loading (c) Point-to-plane loading (d) Three-point bending

Figure2. AE distribution of particles under four instant loading conditions (Ellipse means tensile zone)

along the outer surface of particle. The particle is broken into two pieces along two loading points. The particle could be broken into many small fragments in another loading point zone.

3.2 The analysis of acoustic emission energy

Figure 3 shows the relationship between strain and the AE energy. The AE energy was listed in y-axis, and calculated according to following phases:

(a) The energy of each failure element was obtained through the formula:

$$W_i = \frac{V_i\left(\sigma_1^2 + \sigma_3^2 - 2\upsilon\sigma_1\sigma_3\right)}{2E} \tag{1}$$

Where: V_i, ν, σ_1, σ_3, E represents the volume of element, poisson, max principal stress, min principal stress and elastic module, respectively.

(b) The total AE energy is the energy summations of all elements involving in the failure process:

$$W = \sum_{i=1}^{n} W_i \tag{2}$$

Where: n represents the total number of element involving in the failure process.

It can be seen that continuous AE events start at 35% of peak loading (see Fig. 3 (a), (c) & (d)), and AE energy also increases dominantly with loading increasing. The increase in AE energy implies that more and more fractures will occur within the irregular particle. There has large AE energy occurs after the peak loading, shown in Fig.3 (a), (c), & (d). However, the second case, plane-to-plane loading, the few AE occur after the peak loading. Most of the energy required for comminution would be to increase the strain energy. This would have to be done each time when

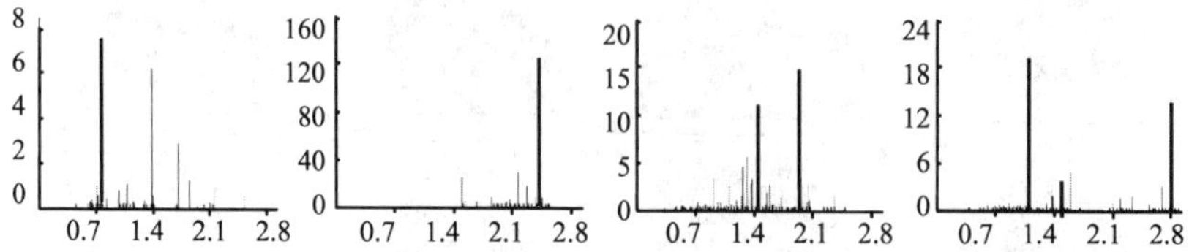

(a) Point-to-point loading (b) Plane-to-plane loading(c) Point-to-plane loading (d) Three-point bending
Figure3. The relationship between the strain and the AE energy of irregular particles under four instant loading conditions [X-axis is Strain($\times 10^{-4}$), Y-axis AE Energy(J) ($\times 10^{-5}$)]

the particles were fractured. The particles can be easily disintegrated later by light impact or other external forces. It is found that the less energy will be consumed with decreasing the area of point contacting.

4 Conclusion

It can be seen that continuous AE events start at 35% of peak loading, and AE energy also increases dominantly with loading increasing. It is also proved that the less energy will be consumed with decreasing the area of point contacting.

Acknowledgements

The primary author is working as a fellow in Tsinghua University, however, the work presented in this paper was mainly funded by the China National Key Fundamental Research “973 Programme” (No. G1998040700) to Prof. C.A. Tang. The primary author would like to particularly thank Prof. W.Y. Zhou, R.Q. Yang & Q. Yang of Tsinghua University for their help.

References

1. Rittinger, P.R., Lehrbuch der aufbereitungskunde, Berlin, 1867
2. Kick, F., Das gesetz der proportionalen widerstande, Arthur Felix Verlag, Leipzig,1885.
3. Bond, F. C., The third theory of comminution, Trans, Aime,193,484-494,1952
4. Tang,C.A., Numerical Simulation on Progressive Failure Leading to Collapse and Associated Seismicity, Int. J. Rock Mech. & Min. Sci., 1997, 34(2), P.249
5. RFPA User’s Guide, CRISR, Northeastern University, P.R.China, 1997.

AN EXPERIMENT INVESTIGATION OF ROCK CRACK FRACTURE UNDER MODE III LOADING

FENGGUANG XU[1], LIYUN LI[1], DA'AN LIU[2]

1.University of Science and Technology Beijing, Beijing, P.R. China
2.Institute of Geology and Geophysics, Chinese Academy of Science, Beijing, P.R. China
E-mail address: ly.li@sohu.com

In this paper, mode III fracture experiments of granite are conducted. Four-point shearing tests of rectangle specimens and three-point shearing tests of circular specimens are used. Regular helicoids fracture surfaces and the relative effect factors have been obtained. Fracture reason has been revealed.

1 Introduction

The crack propagation in rock mass is an important problem in underground engineering. It has realistic meaning to analyze the stability of underground wells, tunnels, holes and slopes, and to illustrate the crack initiation process of earth's crust in earthquake engineering. The investigations among mode I and mode II fracture of rock have made much progress [1-3]. But few researches concerning mode III fracture of rock has been reported [4]. There is no conclusion available to analyses engineering problems up to now.

2 Experiments

In order to study mode III fracture of rocks more deeply, two kinds of specimens are used, which are four-point shearing rectangle specimens and three-point shearing circular specimens (Fig.1 and Fig.2). Rock type used in this study is granite. The mechanical properties of granite are listed in Table.1.

For rectangle specimens, with dimensions $h \times w \times L=30 \times 60 \times 150$mm, three types of crack length 'a' are designed, which are a=24mm, a=30mm and a=36mm. Two types of loading distance 's' are used, which are s=12mm and s=24mm. Three kinds of other loading distance 'd' are also determined, which are d=w (mode A), d=a+10mm (mode B) and d=a (mode C) (Fig.1). The crack h of all these specimens is 0.6mm.

Table 1. Mechanical properties of granite

E (GPa)	v	σ_c (MPa)	σ_t (MPa)	K_{IC}(Mpa*m$^{1/2}$)
57.56	0.22	184.1	19.14	1.215

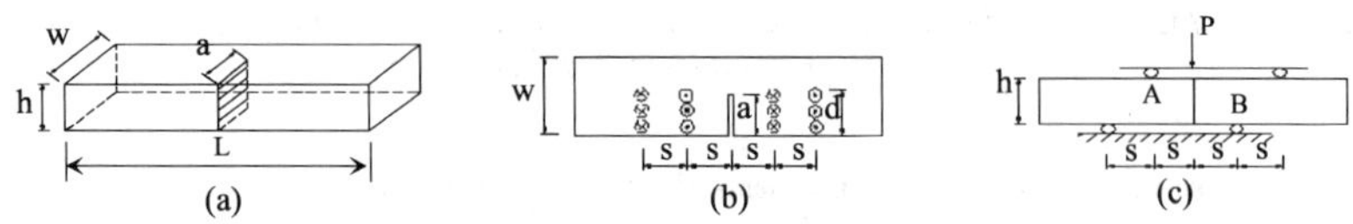

Figure 1. mode III loading of rectangle specimen

For circular specimens, with radius r=30mm and crack lengths a= r=30mm, three kinds of thickness 'h' are chosen, which are h=15mm, h=10mm and h=5mm (Fig.2). The crack width is 0.3mm.

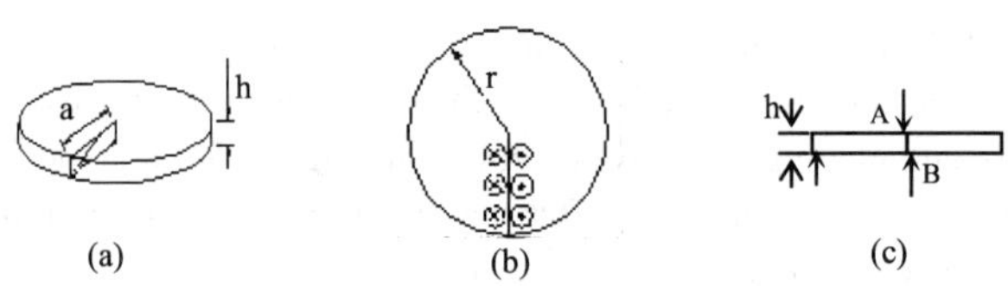

Figure 2. Mode III loading of circular specimen

Table 2. Failure load P_{max} of mode III rack of granite rectangle specimen (unit： KN)

Loading mode	A	B	B	C
a (mm)	s=12mm	s=12mm	s=24mm	s=24mm
24	32.5	10.63	10.62	9.4
30	21.4	9.74	9.7	8.4
36	18.4	9.12	6.76	7.1

Table 3. Failure load P_{max} of mode III rack of granite circular specimen (unit： KN)

h(mm)	15	10	5
Pmax	1130	711	320

All of the tests are carrying out on an RGM-100A omnipotent test machine controled by a computer. The loading velocity in displacement is 0.5mm/min. The final failure loads P_{max} of these two kinds tests are listed in table 2 and 3.

3 Experimental Results and Discussion

The testing results show that the fracture surfaces of all the specimens with different loading modes are similar in shape. The crack propagation initiates at the middle point (h/2) of the crack tip line (Fig. 1(c) and Fig. 2(c)). Then the fracture surface propagates from small smooth helicoids around the point to larger coarse ones with their radii of curvature increasing along propagation direction (Fig.3). Even though the circular specimen is very thin, its fracture surfaces are similar to the rectangle specimen's.

(a) (b)

Figure 3. Some tested specimens

For the rectangle specimen, its failure load decreases with the crack length 'a' increasing if other conditions are the same. Comparing loading mode A with mode B and mode C, the loading capacity of mode A is best and that of mode C is the worst. It should be noticed that the loading distance 's' has direct effect on fracture of specimen. With the loading distance 's' increasing, the helicoids surface area increases, but the failure load decreases.

It is a phenomenon worth to investigate. As shown in Fig.1, when the whole loading 'P' is same, the whole shear force is always one third of P (P/3) on the crack surface with different loading distance 's'. In common sense, the failure load should not change with loading distances 's', but the experimental result is far from it. This shows that different loading distance 's' affects directly stress state of crack tip, thus causing failure load varying with loading distances.

Table 3 shows that the smaller thickness becomes, the smaller failure load is of circular specimens with the same radius.

Whether specimen is rectangle or circular, on material mechanics principle, the shear stress in the crack tip region is null on the specimen surface and is the biggest at the middle point (h/2) of crack tip line. Thus the middle point of crack tip line is under high stress state, and whose stress intensity factor K_{III} must be the biggest one, so it become initial point of crack propagation. The appearance of fracture surface shows that it is tensile stress that leads to crack propagation.

4 Conclusions

Based on above statement the following conclusions can be drawn:

(1) The fracture surfaces are helicoids for both rectangle and circular specimens with edge crack through thickness.

(2) The crack propagation initiates at the middle point (h/2) of the crack tip line. Then the fracture surface propagates from small smooth helicoids around the point to larger coarse ones with their radii of curvature increasing along propagation direction.

(3) Loading distance 's' has significant effect on fracture of specimen. With the loading distance 's' increasing, the helicoids surface area increases, but the failure load decreases.

Acknowledgments

The financial support by National Science Foundation of China (No. 49972090) and Science Developing Foundation of USTB is gratefully appreciated.

References

1. Liyun Li, Zhenci Li, Zongqi Sun. Experiments research and theoretical analyses in Brittle fracture of rock under mixed-mode loads [J]. The Chinese Journal of Rock Mechanics and Engineering, 1994,13(2): 134~140.
2. Zongqi sun et al. Study on determination of shear fracture toughness (K_{IIC}) [J]. The Chinese J. of Rock Mechanics and Eng. 2002,21(2): 199~203.
3. J.Dvies. New geometry for fracture testing. International Journal of Fracture, 60(1993), 33~37.
4. Luong.M.P. Fracture behavior of concrete and rock under mode II and mode III shear loading. International Conference on Recent Development in the Fracture of concrete and rock. Sep 20-22 1990.

3. Numerical Analysis and Computer Simulation

NUMERICAL APPROACH TO STRENGTH HARACTERIZATION OF ROCK UNDER DIRECT UNIAXIAL TENSION

C.A. TANG S.H.WANG
CRISR, Northeastern University, Shenyang, P.R.China

L.G.THAM, P.K.K.LEE, H.LIU
RERC, The University of Hong Kong, Hong Kong, P.R.China

The behavior of a specimen under direct uniaxial tension and the process of microfracture in such a specimen are phenomena of considerable interest from the point of view of understanding strength charaterization of a brittle or quasi-brittle materials such as rock. We present numerical results, based on a Rock Failure Process Analysis code ($RFPA^{2D}$), that show the nucleation and growth of macrocracks in relatively heterogeneous specimen under direct uniaxial tension. A number of features were consistently obtained in the numerical simulation. In the simulation, the diffused AE events or microfractures appeared in the early stage of loading. Macrocrack nucleation starts well before the peak stress is reached and the crack propagation and coalescence can be traced, which can be taken as a precursory to predicate the macrofracture of the specimen.

1 Introduction

Tension failure is one of the most important failure modes in brittle materials such as rock. Even in the situation of compression, it is found that, at a smaller size scale, i.e., the micro- and meso- level, fracture of rock is essentially a tension phenomenon. In addition, since the tensile strength of rock is much lower than its compressive strength, the direct consideration at analyzing the tensile behavior of rock under direct tensile loading conditions is particularly required in designs of underground structures. However, as pointed out by Okubo and Fukui (1996), the current design process fails in this regard and only limited data are available on rock tensile characteristics, especially complete stress-strain curves in uniaxial tension (Okubo and Fukui, 1996; Peng, 1975; Nova and Zaninetti, 1990).

No matter some experiments on the failure modes and the complete stress-strain curves for rocks under uniaxial tension were conducted by Peng (1975) and recently by Okubo and Fukui (1996), due to the difficulty in conducting direct tensile tests, a detailed study on tensile failure behavior of rock is still lack. Stress concentration over the grip location of the specimen is one of the obstacles that are hard to overcome. Another obstacle is the occurrence of a bending moment because of a non-coaxial gripping and some curvature of the specimen. However, even if the apparatus is perfect, slight imperfections in specimen preparation or inhomogeneity in the material itself lead to nonuniform tensile stresses across the section where failure occurs (Pells, 1993).

This lack of fundamental knowledge of tensile failure under uniaxial loading led to the present study. In this paper, uniaxial tensile test with specimen of brittle disordered rock material was numerically studied by using a Rock Failure Process Analysis code ($RFPA^{2D}$), developed recently by CRISR at Northeastern University (Tang, 1997).

2 Numerical Model

2.1 Brief Outline of $RFPA^{2D}$

Numerical simulation is currently the most popular method used for modeling deformation behavior of rock before failed. Even though progress has been made in numerical simulation of failure occurred in rocks, there is clear lacking of satisfied model which can simulate the progressive failure in a more visualized way.

There are two features distinguishing it from other numerical methods:

1. By introducing heterogeneity of rock properties into the model, RFPA code can simulate non-linear deformation of a quasi-brittle behavior with an ideal brittle constitutive law for the local material;

2. By introducing reduction of material parameters after element failure, RFPA can simulate strain-softening and discontinuous mechanics problems in a continuum mechanics method.

In this paper, the disordered material properties (failure-strength σ_c and elastic module e_c) for elements are distributed throughout the specimen by following a Weibull distribution (Tang, Hudson and Xu, 1993; Tang, 1997).

$RFPA^{2D}$ is a Microsoft Windows application package for rock failure analysis of mines/geotechnical structures using finite element techniques. The package can handle heterogeneous material properties of rock and perform nonlinear deformation analysis. Excavation and slope stability analysis can be modeled easily. A user-friendly pre- and post-processor is integrated to generate the finite element mesh and prepare the input data. The result may be displayed graphically as a "movie" to aid in understanding the mechanics of collapse.

2.2 Numerical Specimen and Loading Procedure

The model mesh contains 130×130=16900 elements with a dimensional size of 130×130 mm. The mechanical properties for the specimen are: (1) Homogeneity index , m=1.5; (2) Mean compressive strength (σ_0)= 200 MPa; (3) Mean elastic modules (E_0)=60 GPa; (4) Tension cutoff (λ)=0.1; and (5) Frictional angle (ϕ)= 30 °.

It should be mentioned that, although in meso-scale the specimens are more or less heterogeneous, in macro-scale, however, it is statistically homogeneous since the mechanical property parameters are randomly distributed throughout the whole specimen.

In this study, we use a smeared failure approach that a microfracturing occurs when the stress of an element satisfies a strength criterion. A Coulomb criterion envelope with a tensile cut-off (Brady and Brown, 1993) is used so that the failure of the elements may be either in shear or in tensile.

In all cases, the specimens undergo a plane stress tension, imposed by a relative motion of the upper and bottom rigid plates. We apply an external displacement with a constant rate of 0.0002 mm/step in the vertical direction and compute the stress and deformation in each element. The external displacement in vertical is then slowly increased step by step. In the step the stress in some elements satisfies strength criterion, the element is damaged and becomes weak according to the rules specified by the strength criterion. The stress and deformation distribution throughout the specimen is then adjusted instantaneously after each rupture to reach the equilibrium state. At positions with increased stress due to stress redistribution, the stress may exceed the critical value and further ruptures are caused. The process is repeated until no failure elements are present. Further external displacement is then increased. In this way the system develops a macroscopic fracture. Energy is stored in the elements during the loading process and is released as acoustic emissions through the onset of element failures. Owing to the stress redistribution and the long-range deformation induced interactions, a single important element failure may induce an avalanche of additional failures in neighbor elements leading to a chain reaction releasing more energy.

Figure1. Model geometry and mesh generation of the numerical specimens for direct tension specimen (the gray color represents the relative heterogeneity of the mechanical properties)

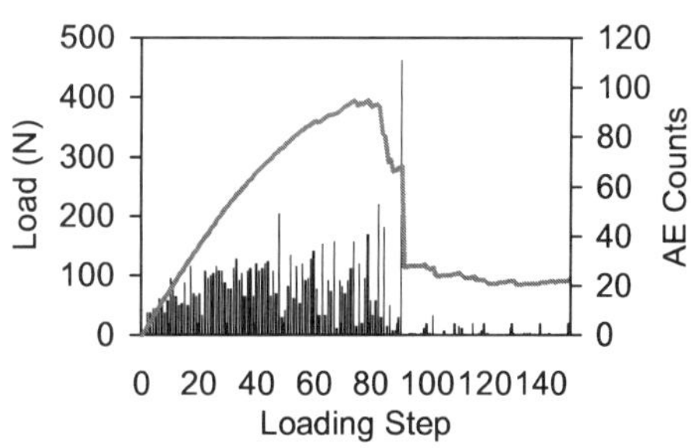

Figure2. Load and AE counts vs loading step (2×10^{-3} mm per step) (RFPA^{2D} simulated)

3 Numerical Simulation Results and Discussions

Fig.2 shows the simulated load-deformation relationship for the specimen (homogeneous index m=1.5). It is clear that the model predicts the non-linear load-deformation curve similar to the typical curve of disordered material which is observed in laboratory tests (Okubo and Fukui, 1996; Peng, 1975), and have essentially the same shape as the compressive stress-strain curves that were shown in a general laboratory test. It is found that, although the constitutive law for the individual element in the numerical model is brittle (with 20% residual strength), a substantial non-linearity exists before the maximum load, and the curve has an obvious post-peak region (strain softening). A detailed study on Fig.2 shows that the load-deformation curves can be divided into three stages. In stage A, the beginning portion of the curve indicates that the randomly distributed micro-fractures have little influence on the overall stress field and the deformation in the specimen is statistically uniform. Decrease of the slope of the load-deformation curve can be observed when the load reaches about 90% peak load. This implies that the specimen tends to deform no-uniformly. Around the maximum load, the specimen undergoes a larger deformation. This is attributed to the crack initiation and propagation. The big stress drop in the curve is found to be in the stage of unstable failure immediately after the peak load is beyond. Then the failure process returned to stable, and the residual strength of specimen is attributed to the failure mode.

Fig.2 also shows the AE counts (event rate) as a function of deformation. It can be clearly seen that for this relative heterogeneous material, lots of AE events occurred before reaching the peak load (macro strength). Fig.2 shows a good relationship between load-deformation curve and event rate. It is found that there are at least two times the AE rate shows an obvious increase. The first time is corresponding to the load drop immediately after the peak load, as shown in Fig.2. The second time is corresponding to the maximum AE count, where the biggest load drop is induced.

In Fig.3 we present the locations of AE events occurred during the loading stages. Each cycle represents one AE event and the diameter of the cycle represents the magnitude of the AE released energy. AE locations for events occurring during loading to 75% peak load are plotted in Fig.3a (step 48). Note that events are distributed throughout the specimen, reflecting the statistically uniformed deformation during this portion of the simulation. It is very difficult to predicate where the macrocrack will initiate in this stage. In step 62, however, some AE events occurred around certain cluster that appears to be the nucleation site for the future fracture plane. While a few events are still occurring throughout the specimen, most events are now clustered near the nucleation zone on the left side of the specimen (step 64 to step 79). From step 79 to step 86, a distinct AE active zone has developed along this left zone of the specimen (as shown in Fig.3e-f) and the events decreases in intensity throughout the rest of the specimen. It is important to find that the maximum load is not reached at step 92 where the highest count of AE events is recorded, but at step 79 where the AE events is less than half of the maximum count. We attribute this phenomenon to the heterogeneity of the specimen. The final AE

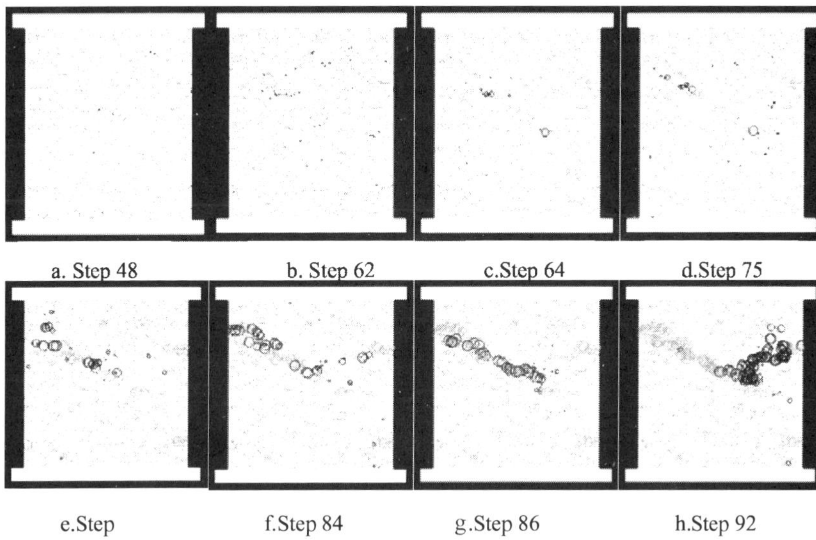

Figure3. Maps of AE source locations in different loading steps for direct tension specimen (RFPA2D simulated)

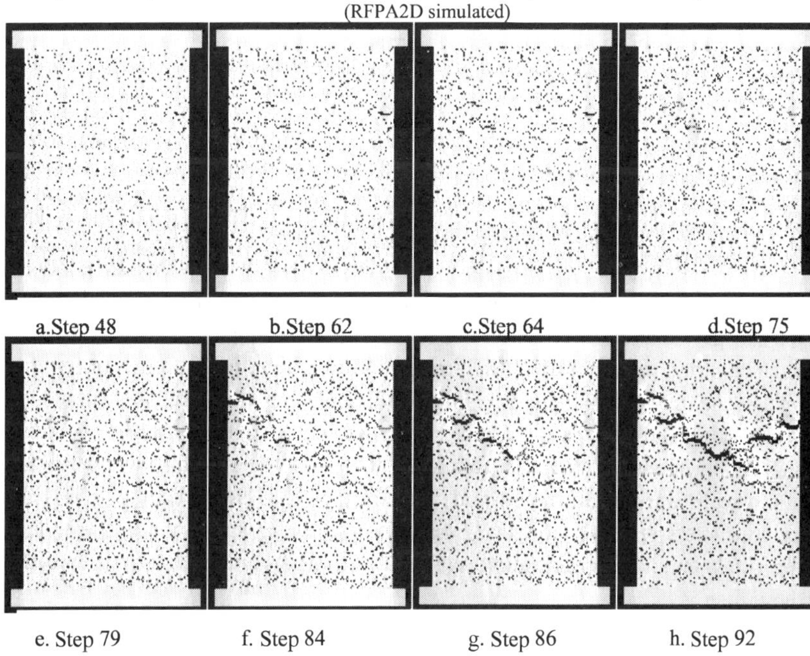

Figure4. Maximum shear stress distribution in different loading steps for un-notched

events (Fig.3h) show the AE active zones are connected together to form a large AE event active zone that is found to be consistent with the site of macrocracks. Again, the AE event maps show that, although most of the AE events occurred prior to the maximum load are distributed throughout the specimen, the AE events along the narrow active zone dissipated most of the acoustic energy (as can be seen by reading the diameters of the AE events).

Fig.4 shows the simulated initiation, propagation and coalescence of the cracks in different loading steps. Generally, the failure process of the specimen can be divided into three stages. Initially, as has been shown in Fig.4a, randomly distributed microfractures are formed due to the heterogeneity of the specimen. The stress distribution shown in Fig.4a shows a totally disordered picture in micro-scale but a statistic uniform in macro-scale. Once the crack nucleation initiated in some sites, mostly in the upper left side of the specimen (Fig.4b-c), they proceed quickly to establish the nascent macrocracks but do not connect (as shown in Fig.4d-f). Following this phase of macrocrack development is the growth phase, in which a band of microfractures developed, indicating a zone of intense stress concentration in the front area of the crack process zone (Fig.4g). It can be seen that the cracks extend from the left side of the specimen into the interior. Different from the beginning stage of loading, the stress field in this stage shows in macro-scale a strongly heterogeneous picture (Fig.4f-h). In the left side of the specimen, a relatively lower stress region forms behind the advancing fracture front. On the other hand, a relatively high stress region forms in the right side, indicating that this portion of the specimen now bears most of the load from the loading frame. Finally, due to the strong interactions between the cracks under this high stress field, the cracks propagate in an unstable manner and the isolated cracks coalesce to form a macrocrack (Fig.4g-h). The crack system shown in the simulation is not as simple as those described by fracture mechanics. It is difficult to define a crack process zone or the crack tip in such a heterogeneous material (Van Mier, 1997).

In a conventional experiment, stress distribution throughout the specimen is difficult, if not impossible, to obtain. The advantage of numerical simulation is that we can get detailed information about the stress distribution including the failure induced stress redistribution. The stress field displayed in Fig.4 shows some new light at the information regarding to the failure mechanism of the specimen. Firstly, these figures show that, although the stress distribution in the beginning of loading is statistically homogeneous in macro-scale, it shows a highly disordered feature in micro-scale due to the micro-scale heterogeneity of the specimen. The failure induced stress redistribution, and, particularly, the high stress concentration due to the crack initiation, propagation and coalescence is demonstrated very clearly in these figures.

From this simulation, we can have a more clear understanding about the concept of material strength. Comparing the load-deformation curve shown in Fig.2 and the failure processes shown in Fig.4, it is found that the maximum strength does not necessarily mean an abrupt failure of the specimen. The load-deformation curve becomes nonlinear when microfractures occur in the specimen (Fig.4a-c). At point e, the specimen reaches its maximum strength. Although the loading bearing capacity drops dramatically from point f to g, this does not indicate the collapse of the specimen. The specimen rapidly lost its loading bearing capacity, yet the crack does not propagate thoroughly until point h. Subsequently, a distinct transition occurs and the loading capacity begins to decrease at a much slower rate until it reaches its residual strength, being about 20% of its maximum strength.

It is generally considered that the microscopic failure mechanisms in tension and compression are considerably different from each other. In compression, a failed portion does not completely lose its loading bearing capacity because of friction or contact with the surroundings. However, in tension the loading capacity of failed portion decreases more rapidly than in compression. This consideration suggests a relative gentle decrease of loading bearing capacity of a specimen in compression. However, as stated early, the shapes of the stress-strain curves do not differ much from each other in tension and compression. It has also been observed by Okubo and Fukui (1996) in their experiments. Though clear explanations for this result are difficult, heterogeneity mentioned above at least can be taken as one of the reasons for the resemblance in stress-strain curves between tension and compression (Okubo and Fukui, 1996).

In experimental study by Okubo and Fukui (1996), one result of particular interest was obtained in uniaxial tension testing: a large amount of residual strength remains in the post-failure region, Peng (1975) conducted uniaxial tension tests and successfully obtained complete stress-strain curves of four rocks. He also found considerable large residual strength for all four rocks. The presence of residual strength in the post-peak region may be attributable to several reasons. Firstly, heterogeneity or strength variation from location to location in a specimen is considered to be one of the reasons. If local strength variation is large, specimen failure starts in weak portions. Even in the post-peak region, strong portions are still intact and some amount of loading capacity remains. Meanwhile, a macrocrack extends rapidly from boundary to boundary in a homogeneous specimen.

4 Concluding Remarks

Although the simulation of tension test here is a two dimensional, and not a quantitative approach, the phenomenological approach shows in great detail the propagation of fracture zone, and the interaction of cracks with the structure of rock material. At least qualitatively, such simulation suggests various mechanisms of rock failure under direct tension. Although many conclusions given here may be obtained by laboratory tests, or even be a common sense, the significance of reproducing these phenomena by numerical simulation is obvious. Firstly, at the moment, almost no convenient experimental method is available to obtain the stress or strain field during the progressive failure process in heterogeneous rocks. Numerical simulation provides supplementary information regarding the stress distribution and the failure induced stress redistribution. Secondly, although AE technique has been used to monitor the microfracture in rocks for many years, very few theoretical methods including numerical tools have been found to be available to simulate the locations of AE event sources. Numerical simulation of AE allows a further check on the results observed with AE techniques. Lastly, and the most importantly, the successful reproduce of the experimentally observed failure phenomena with a numerical method implies that our understanding to the rock failure has reached a reasonable level, which in turn will help us to make further progresses in the field of rock mechanics.

Acknowledgement

The First author is grateful for the support of this work from the Chinese National Key Fundamental Research "973" Program (G1998040700) and the Natural Science Foundation of China (No.49974009 and 50174013) and the partially support from Hong Kong University for a visit of the first author to Civil Engineering Department..

References

1. Brady B.H.G. and Brown E.T., Rock Mechanics (second edition), *Chapman & Hall*, London (1993).
2. Nova R. and Zaninetti A., An investigation into the tensile behavior of a schistose rock, Int. J. Rock Mech. Min. Sci., 27, 231-242(1990)
3. Okubo S. and Fukui K., Complete stress-strain curves for various rock types in uniaxial tension, *Int. J. Rock Mech. Min. Sci.*, 33, pp.549-556(1996).
4. Pells P.J.N., Uniaxial strength testing, Comprehensive Rock Engineering, Ed. J.A. Hudson, Pergamon Press, Oxford (1993).
5. Peng S.S., A note on the fracture propagation and time-dependent behavior of rocks in uniaxial tension, *Int. J. Rock Mech. Min. Sci.*, 12, pp.125-127(1975).
6. Tang C.A., Numerical simulation of progressive rock failure and associated seismicity, *Int. J. Rock Mech. Min. Sci.*, 34, pp.249-262(1997).
7. Van Mier J.G.M, Fracture process of concrete, CRC Press, New York, p.253(1997).

INTEGRATION OF NUMERICAL MODELING AND SEISMIC MONITORING: GENERAL THEORY AND FIRST STEPS

ALEXANDER M. LINKOV
61 Bo'lshoy Prospect V. O., St-Petersburg, 199178, Russia
E-mail: link@mech.ipme.ru

The first steps in integration of numerical modeling and seismic monitoring are discussed. It is noted that although significant progress in joining rock mechanics and seismicity has been reached recently, the existing approaches are not free of shortcomings. Ways to remove them and to gain further progress are suggested on the basis of the general theory. In particular, proper accounting for the combined influence of elasticity, softening and time is strongly advocated. Basic formulae for this are given explicitly. The other suggestion consists in using active and passive elements together when simulating seismicity. The need in developing true dynamic codes by combining ideas of distinct and boundary element methods is emphasized.

1 Introduction

Four years between the 4th and 5th International Symposia on Rockbursts and Seismicity in Mines (RaSiM) have shown progress in joining rock mechanics and seismic monitoring. Recall that at the RaSiM-4 we had mostly theoretical analysis of new geomechanical approaches for developing quantitative seismicity. Although the analysis revealed that bringing together rock mechanics and seismology is feasible [7], practical results were rare (e.g. [15,1-3,10,11,13]). Four years later, at the RaSiM-5 we had quite a different picture: there were many papers on practical integration of numerical modeling and seismicity (e.g. [4,5,12,16-18]). They indicate significant practical achievements. But it looks that now the general theory is somewhat neglected. The paper tends to restore the balance: it focuses on feasible enhancements suggested by the general theory.

2 General considerations

Seismicity is a chain of instabilities occurring in time. Hence, we need to account for instability and also for time effects. To properly simulate seismicity, one must include into a model the three factors [7]: (i) elasticity which provides energy for a jump; (ii) softening which is a source of instability and gives a measure of rock brittleness; and (iii) creep (or other time process) which controls intervals between events. Accounting for these free factors serves for studying dynamic instabilities (seismic events). Furthermore, it reveals fracture acceleration: smooth although accelerating aseismic movements are as common as seismic events.

3 First steps in synthetic seismicity

Numerical simulation of multiple seismic events is termed synthetic seismicity. M. D. G. Salamon was the first to generate it in rock mechanics [15]. The Salamon's scheme looks perfect as an initial step. It allows one to obtain data on location, energy, seismic moment and frequency-magnitude dependence. But we need to remove two its shortcomings. Firstly, the scheme does not account for softening; consequently, it does not contain a measure of rock brittleness. Secondly, the time does not enter constitutive equations. It is accounted only through changes in geometry produced by mining steps. Hence, there are no aseismic deformations.

A natural extension of the scheme is to explicitly include creep into the constitutive equations. This was done by employing the simplest viscous element in the programs DIGS [10,13], MINSIM [12], and MINF [17], what allowed accounting for aseismic effects. In addition to the mentioned dependences, the results include catalogues of synthetic seismicity and chains of events after a blast.

Nevertheless, softening still is not properly accounted for. To the date, softening is used only in a model which includes a softening element in parallel with a viscous element [12]. In accordance with the general theory [7], the latter model exhibits fracture acceleration what was reported in [12], but it ceases to generate instabilities: infinite instant stiffness of the dash-pot prevents instability.

4 Next step: using ESC-model

The Elasticity-Softening-Creep model [7] (Figure 1) is free of this shortcoming. It presents a straightforward extension of the standard linear solid model; it differs from the latter in only one respect: after reaching the strength, the upper element may soften. Since the ESC-model contains the softening element in series with the viscous element (dash-pot), it captures instability as well as fracture acceleration.

The deformation law of the ESC-model is quite simple. For instance, referring it to the shear DD Δu_τ and the shear traction τ and denoting by a dot the derivative with respect to time t, we obtain:

$$\Delta \dot{u}_\tau - \frac{1}{k_c}\dot{\tau} + \frac{1}{t_r}\left[\Delta u_\tau - \left(\frac{1}{E_l} + \frac{1}{k_c}\right)\tau\right] = C_0, \tag{1}$$

where $\Delta u_\tau = u_\tau^- - u_\tau^+$ is the DD in the tangent direction, which is taken to the left from the direction of the normal to the contact; the sign plus (minus) denotes a value on the positive (negative) side of a surface; $t_r = \eta/E_l$ is the retardation time; k_u

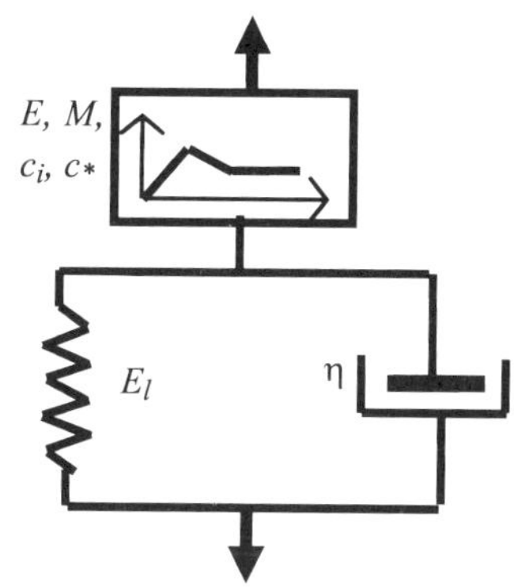

Figure 1. Elasticity-Softening-Creep Model

$= E$ for elastic deformation ($|\tau| < \tau_c$) of the upper element, $k_u = -M$ for its softening and $k_u = 0$ for its residual strength; E is the elastic rigidity of the upper element, M is its softening modulus; the current shear strength τ_c may be defined by the Coulomb's law: $\tau_c = c + (-\sigma_n)\tan\rho_c$ with $c = c_i$ for the initial strength, $c = c_i - |\Delta u_u^p|M/(1+M/E)$ for softening and $c = c_*$ for the residual strength; Δu_u^p is the plastic shear deformation of the upper element; σ_n is the normal traction on the contact ($\sigma_n < 0$ for compression); ρ_c is the angle of contact friction; C_0 is the value of the expression on the l.h.s. of (1) at the moment $t = 0$.

In numerical simulation, equation (1) is applied to each element of a crack, fault or any surface of displacement discontinuity. On the other hand, a boundary integral equation (BIE) for embedding rock provides its own dependence between tractions and DD on an element. By joining these equations, we arrive at a system of ordinary differential equations of the first order in time. The system allows one to distinguish an unstable state by examining its matrix.

In the simplest case, when mutual influence of the elements is neglected, we have for each element from the BIE: $\tau = -k_s\,\Delta u_\tau + \tau_0$, where k_s is so-called 'self-influence' coefficient (for the accepted definition of the DD as $\Delta u_\tau = u_\tau^- - u_\tau^+$, it is positive); τ_0 is a known value. Substitution into (1) gives an elementary differential equation. Its solution is $\Delta u_\tau(t) = A\exp(-\alpha t/t_r) + Ct_r/\alpha$, where $A = \Delta u_\tau(0) - Ct_r/\alpha$, $\alpha = 1 + k_s/(Ea)$, $a = 1 + k_s/k_u$, $C = [(1/E+1/k_u)\tau_0/t_r + C_0]/a$.

From the solution it is easy to see that the most important effects that are: (i) decaying aseismic deformation if $\alpha > 0$; (ii) smooth but accelerating aseismic deformation if $\alpha < 0$ and $M < k_s$; (iii) instant instability if $M < k_s$, are captured by the ESC-model. Being not much more complicated than models used to the date, it provides significant advantages: it allows us to account for rock brittleness; to distinguish between stable and unstable states; to evaluate energy consumption and to follow accelerating aseismic deformations. For these reasons, we strongly commend the model for implanting into the codes used to simulate seismicity.

5 Incorporation of observed seismicity

Generating synthetic seismicity is the *active* way for integrating numerical modeling and seismicity. In this case, we find location, orientation and DD of an element producing an event from the solution of a problem.

Another approach, which may be termed *passive*, uses seismic data on events already occurred (see, e.g. [4,5,18]). From these data, one extracts information on the location, orientation and DD of the elements, which have produced seismic events. In this case, we prescribe location, orientation and DD of elements in advance, as input data. We find numerically stresses induced by these *passive* elements.

The main problem of the passive approach is to transform seismic data into input information on the orientation of elements and values of DD. For events attributed to a surface of discontinuity (a fault, in particular), the elements are taken along the surface. To simplify the problem they are assumed to be circular. The DD on them are prescribed by accepting additional assumptions as described in [4,5,18].

For events not localized on a plane, this scheme is modified. The seismic *strain* distributed in a volume is used instead of the DD. In a numerical code, the strain is accounted for as common additional strain prescribed in advance like, for instance, thermal strain. Again the main problem is to interpret seismic data in terms of values used as input data in a code solving a solid mechanics problem.

The passive approach, by employing seismic data from direct seismic observations in time, allows one to trace changes in the stressed state of rock. Analysis of these changes serves to conclude whether the state is dangerous in respect of violent events. Unfortunately, this approach actually misses aseismic deformations.

6 Next step: joining active and passive approaches

So far active and passive approaches have been used separately. Meanwhile, it is quite feasible to join them. It is sufficient to place active elements outside passive ones. This facilitates new important options: synthetic seismicity generated in this way, being compared with seismic data on following events, may serve for calibration of input data for active elements, for better interpretation of seismic data in mechanical terms and, as a result, for better prediction of hazardous events.

7 Rapid processes and true dynamic modeling

So far, we discussed only static characteristics of dynamic events that are their location, energy and seismic moment. This significantly simplifies numerical modeling by allowing one to use codes for static elasticity problems. This level of integration of numerical modeling and seismic monitoring looks quite reasonable if we take into account that even the mentioned static characteristics are not always available in mines. Meanwhile, there is a problem for which true dynamic codes are indispensable. This is a problem of evaluating the damage effect of strong dynamic events. Inertia of moving particles becomes important in this case; we need to keep dynamic terms in equations and to follow rapid processes in time. Wave amplification, including its extreme case when a small seismic event triggers a strong event with the velocity of order 10 m/s, refers to this problem [9,8].

Simple static evaluations of stresses and energy can be used when calibrating a dynamic model [6]. For having a realistic picture of velocity amplification, the stress drop should be of order of the rock strength σ_0; thus, the energy release per unit volume should be of order $\sigma_0^2/(2E)$, where E is the Young's modulus. Then the velocity v of fragments with the density ρ is of realistic order: $v = \sqrt{\sigma_0^2/(\rho E)}. = c\varepsilon_0$, where ε_0 is the strain corresponding to the strength, c is the elastic wave velocity. For a hard rock with $\sigma_0 = 200$ MPa, $E = 100$ GPa and $c = 6$ km/s we obtain $v \approx 12$ m/s. Even if a half of the energy is lost, we still have $v \approx 8.5$ m/s, what agrees with observations in deep gold mines.

8 Next step: combining distinct and boundary elements

Computational effort needed for a dynamic problem is much greater than for a static one. This explains using simplified models to increase efficiency of dynamic codes such WAVE [3] and a particle flow code (PFC) [14]. We cannot dwell here on discussion of limitations of the existing true dynamic codes. Note only that there is need in developing new codes free of the limitations. In our view, such a code may be developed by combining the ideas of the distinct element method (DEM) with those of BEM, as suggested in [7].

Acknowledgements

The author is grateful to the Russian Fund of Fundamental Investigations (RFFI) for the support of this work (Grant No 05-00-64316).

References

1. P. A. Cundall and J. N. Lemos in *Proc. 2nd Int. Symposium RaSiM-2*, ed. C. Fairhurst (Balkema, Rotterdam, 1990).
2. M. F. Handley et al in *Rock Mechanics*, ed. M Aubertin, F. Hassani & M. Mitri (Balkema: Rotterdam, 1996).
3. M. W. Hildyard et al in *Proc. 35th US Symposium on Rock Mechanics* (Balkema: Rotterdam, 1995).
4. Hofmann G. et al in *Proc. 5th Int. Symposium RaSiM-5*, ed. G. van Aswegen, R. Durrhiem & D. Ortlepp (Balkema: Rotterdam, 2001).
5. R. Lachenicht et al in *Proc. 5th Int. Symposium RaSiM-5*, ed. G. van Aswegen, R. Durrhiem & D. Ortlepp (Balkema: Rotterdam, 2001).
6. A. M. Linkov, *Dynamic Phenomena in Mines and the Problem of Stability* (ISRM, Lisboa, Cedex, Portugal, 1994).
7. A. M. Linkov in *Proc. 4th Int. Symposium RaSiM -4*, ed. S. I. Gibowicz & S. Lasocki (Balkema, Rotterdam, 1997).
8. A. M. Linkov in *Proc. 5th Int. Symposium RaSiM-5*, ed. G. van Aswegen, R. Durrhiem & D. Ortlepp (Balkema: Rotterdam, 2001).
9. A. M. Linkov and R. J. Durrheim in *Proc. 3rd Int. Conf. on Mechanics of Jointed and Faulted Rock MJFR-3*, ed. H.-P. Rossmanith (Balkema, Rotterdam, 1998).
10. D. F. Malan, *Rock Mech. Rock Eng.*, **32**, 173 (1997).
11. D. F. Malan and S. M. Spottiswoode in *Proc. 4th Int. Symposium RaSiM-4*, ed. S. I. Gibowicz & S. Lasocki (Balkema, Rotterdam, 1997).
12. J. A. L. Napier in *Rock Mechanics in National Interests*, ed. J. Tinucci & Heasley (Swets \& Zeitlinger Lisse, 2001).
13. J. A. L. Napier and D. F. Malan, *Int. J. Rock Mech. Mining Sci.*, **34**, 1075 (1997).
14. D. O. Potyondi et al in *Proc. 2nd North American Rock Mechanics Symposium NARMs'96*, ed. M. Aubertin (Balkema, Rotterdam, 1996).
15. M. D. G. Salamon in *Proc. 3rd Int. Symposium RaSiM-3*, ed R. P. Young (Balkema, Rotterdam, 1993).
16. E. J. Sellers and J. A. L. Napier in *Proc. 5th Int. Symposium RaSiM-5*, ed. G. van Aswegen, R. Durrhiem & D. Ortlepp (Balkema: Rotterdam, 2001).
17. S. M. Spottiswoode in *Proc. 5th Int. Symposium RaSiM-5*, ed. G. van Aswegen, R. Durrhiem & D. Ortlepp (Balkema: Rotterdam, 2001).
18. Wiles T. et al in *Proc. 5th Int. Symposium RaSiM-5*, ed. G. van Aswegen, R. Durrhiem & D. Ortlepp (Balkema: Rotterdam, 2001).

NUMERICAL SIMULATION OF SHOUDA MINE IN JAPAN USING UBIQUITOUS JOINT MODEL

GANG LI
Dept. of Civil Engineering, Faculty of Engineering, Yamaguchi University
2-16-1 Tokiwadai, Ube 755-8611, Japan.

HARUMI KATO
DOWA Engineering Co., Ltd., Shinbashi DOWA Bldg.,
5-10-5, Minato-ku, Tokyo 105, Japan

YOSHIAKI MIZUTA
Dept. of Civil Engineering, Faculty of Engineering, Yamaguchi University
2-16-1 Tokiwadai, Ube 755-8611, Japan.

In this paper, the evaluation of methodology and stability for excavation of Shouda Mine in Okayama prefecture in Japan using numerical modeling analysis — i.e., FLAC, is introduced. Ubiquitous-Joint Model is adopted in this analysis, which accounts for the presence of an orientation of weak planes (weathering joints or bedding planes). Yielding may occur in either the solid or along the weak planes or both depending on the stress state, the orientation of the weak planes and the material properties of the solid and weak planes. The numerical modeling analysis provides a reasonable and safer excavation methodology.

1 Introduction

Numerical modeling is used here to investigate the methodology and stability for excavation of Shouda Mine in Okayama prefecture in Japan. It produces insights into the Ubiquitous Joint system under given geometries and boundary conditions. The modeling process is important for evaluation of the excavation process and monitoring of displacement caused by the excavation. Models were run with different geometries and material properties to inspect the influence of aspect ratio of the models, different constitutive models and the complexity of the models. Harmonized excavation was considered to improve the stress distribution during excavation with higher safety factor.

2 Numerical Modeling

Shouda Mine consists of three different kinds of rocks: crystalloid limestone, hornfels and tonalite. The excavation of Shouda Mine starts from level 285m and reaches level 382.5m with 3m intervals each step, which amounts to 26 steps. The space/time history of such a geological system cannot be fully described by any one or even a set of analytical solutions. Thus a numerical model that can follow and step through the history of the geological system of interest is introduced.

FLAC (Fast Lagrangian Analysis of Continua, Cundall & Board 1988) has been used to run the numerical simulation. FLAC is a two-dimensional explicit, finite difference program that performs a Lagrangian analysis with friendly integrated interfaces, which is robust and efficient for modeling elastic, plastic deformation and groundwater flow. Generally in most numerical modeling Mohr-Coulomb model is adopted, which corresponds to a Mohr-Coulomb criterion (shear yield function) with tension cutoff (tensile yield function), while in this case a Ubiquitous-Joint Model is adopted in this analysis, which accounts for the presence of an orientation of weak planes (weathering joints or bedding planes) in a FLAC Mohr-Coulomb model. Yielding may occur in either the solid or along the weak planes or both depending on the stress state, the orientation of the weak planes and the material properties of the solid and weak planes.

FLAC uses a technique named weak plane plastic corrections to represent a Ubiquitous Joint Model, by which general failure is first detected and relevant plastic corrections are applied according to the Mohr-Coulomb criterion. The new stresses are then analyzed for failure on the weak plane and updated accordingly. The criterion for failure on the plane consists in a local form of the Mohr-Coulomb yield condition. In the model presented here a fault zone with orientation of N30E79N has been modeled as the weak plane.

3 Geometry, boundary and initial conditions

First since FLAC is a two-dimensional explicit finite difference program, a plane strain condition assumption was made provided that the geometry of excavation plane and the loading conditions are uniform over a comparatively long distance in the perpendicular direction. Two models containing different numbers of elements (90000 elements and 37500 elements) were run on trial, which gave approximately the same results. Thus the model with fewer elements was adopted in the simulation for the convenience of further amendment since the computer simulation takes rather long time, and is illustrated in Fig. 1.

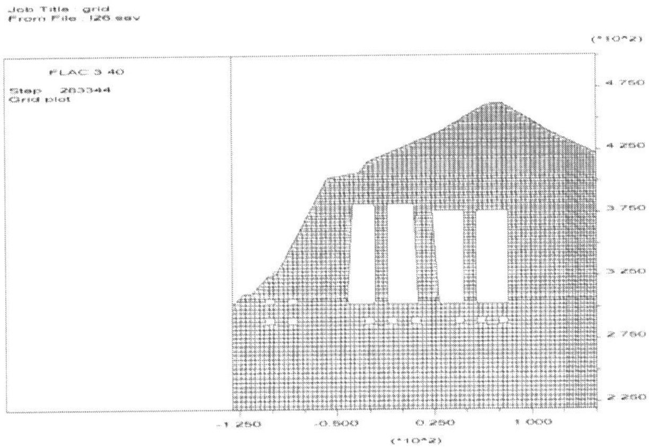

Figure1. Grid plot of the interested part of Shouda Mine

The boundary condition is established with X direction fixed at both sides and X, Y direction fixed at bottom. The initial vertical stress is assigned linearly according to the mass-density of the materials. As for the estimation of in-situ horizontal stress, data obtained by hydraulic fracture measurement method are assigned to the numerical model from depth 37m to 92m, and linear assumption is made for stress distribution along the depth. The simulation then was run to find a suitable equilibrium state, which takes into account the contrast in material properties and layered geometry.

4 Material properties

Table 1. Material properties for the lithologic units

	Area Perc. (%)	density (kg/m^3)	Young's Modulus (Gpa)	Poisson's ratio	Cohesion (Mpa)	Internal Friction angle (°)	Tensile strength (Mpa)
ore (crystalloid limestone)	-	2687	16.1	0.28	6.3	31.1	1.55
hornfels	-	2834	41.2	0.27	19.7	27.5	6.1
tonalite	-	2522	39.8	0.23	30.0	30.4	7.9
first layer pillar	32.6	2687	5.2	0.28	2.0	31.1	0.51
column pillar	24.0	2687	3.9	0.28	1.5	31.1	0.37
column tonalite	24.0	2522	9.6	0.23	30.0	30.4	1.9
joints	-				3.13	20.0	0.78

There are three different kinds of rocks adopted in the numerical simulation, including crystalloid limestone, hornfels and tonalite. The intact rock properties and joint information were gained through field investigation and laboratory experiments, and then extrapolated to field-scale properties as illustrated in table 1.

5 Description and Discussion of Results

According to the principle stress distribution and safety factor illustrated in each step during the simulation, a final elevation of 370.5m can be reached stably. The maximum stress contour and safety factor contour at elevation of 334.5m are shown in Fig. 2. Note that a failure mechanism is indicated only when there is a contiguous line of active plastic zones that joins two surfaces, which does not exist in further study of plastic indicator drawing.

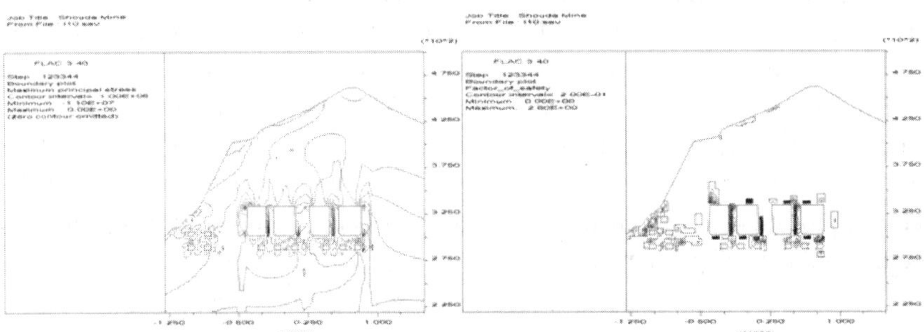

Figure 2. The maximum stress contour and safety factor contour at elevation of 334.5m

To improve the stress distribution condition the harmonized excavation method was considered with higher safety factor. The models were run several times to find out the best excavation sequence in the end. Thus a simple, generic grid is necessary, which has been proved effective at the beginning. Safety factors with different constitutive models, i.e., Mohr-Coulomb model and Ubiquitous-Joint Model are also compared and it has been shown that Ubiquitous-Joint Model has the advantage with the weak plane considered.

References

1. P. Upton and A. Ord, Coupled Mechanical/Fluid Flow Numerical Modelling, (CSIRO, Australia, 1996).
2. Itasca Consulting Company, Inc. FLAC: Fast Langrangian Analysis Continua, User's Guide, Version 3.40 (1999).

A NUMERICAL SIMULATION OF HYDROFRACTURING BY USING HEAVY BENTONITE SLURRY IN COMPRESSED AIR ENERGY STORAGE

HONG LI

Dr., Geoscience Research Laboratory, Inoueasahimon Bldg.3F, 2-2-20, Kouraku Bunkyouku, Tokyo, 112-0004, Japan

YOSHIAKI MIZUTA

Professor, Dr., Faculty of Engineering, Yamaguchi University, Tokiwadai, Ube 755-8611, Japan

When the Compressed Air Energy Storage (CAES) that takes advantage of the underground space in the rock proceeds, using bentonite slurry as hydraulic compensation media is considered to maintain the relatively great air pressure to a certain surrounding stress. In this paper, by two-dimensional boundary element methods (the combination of fictitious stress method and displacement discontinuity method), a numerical simulation of fluid ingress into fractures is carried out in order to study the possibility of natural sealing when bentonite fills and then concretes inside fractures. Nevertheless, mercury is taken as the analogue fluid in this transient attempt due to the known parameters, i.e., surface tension and contact angle.

1 Introduction

The Compressed Air Energy Storage (CAES) is the ongoing technology used for storing electric power by transferring surplus energy from the valley consumption period to the peak time daily. The feasibility of CAES depends on the economic and effective construction of the air storage cavern. Shown as Fig. 1, unlined caverns are used to store the compressed air which is prevented leaking by the subsurface water pressure applied to the surrounding rock.

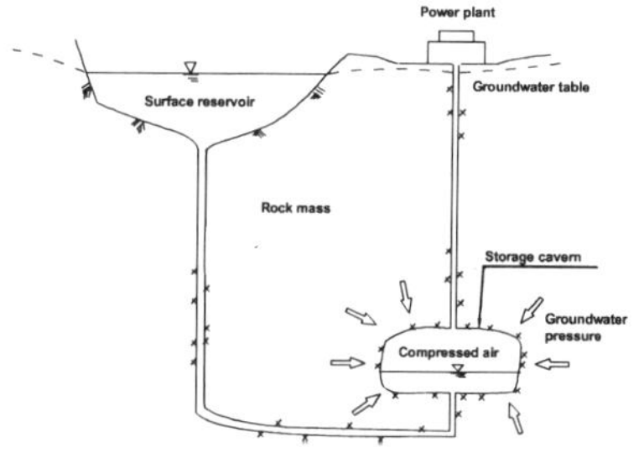

Figure 1. The concept illustration of CAES

If distinct fractures exist surrounding the cavern walls, some water unsaturated areas may be caused due to their extra permeability. As pore pressure dwindles away at these areas, the air tightness of surrounding rock loses its assurance. Heavy bentonite slurry is then considered as filling material in advanced pressed into the fractures surrounding the cavern to keep relatively great air compression and clog up the flows.

2 The model and set conditions of numerical simulation

Shown as Fig. 2, a 2D numerical simulation is done to model the hydrofracturing [1] mechanics of pressing the fluid into the cracks from the wall of borehole.

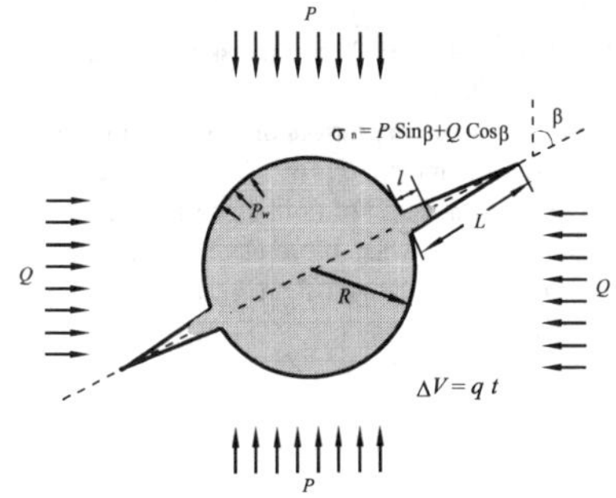

Figure 2. The simulation model and setup conditions

Where R is the radius of borehole (m), P is the maximum principal stress in the surface perpendicular to borehole axis (Pa), Q is the corresponding minimum principal stress (Pa), σ_n is the normal stress component on the surface parallel to the crack (Pa), β is the angle between the surface to where minimum principal stress acts and crack face, L is the length of crack (m), l is the length of fluid ingress (m), ΔV is the incremental of the volume enclosed by the bole hole walls and fracture openings envelope (m^3), q is ingress flow rate (m^3/s), t is time starting from fluid ingress (t), P_w fluid pressure (Pa).

The combination of displacement discontinuity elements for the crack boundaries and fictitious stress elements for the circular wall boundaries is adopted. Furthermore, the surface tension at the intruded fluid fronts and the fracture extension according to that Stress Intensity Factor K_I exceeds toughness K_{IC} (set zero) have also been taken into account. At each crack tip, the accurate square-root elements based on analytical integrations, i.e., closed form expression for the influence coefficients, are realized by taking advantage of computer algebra (symbolic computation) [2] . Bisection method is adopted to gradually approximate the nonlinear interaction between the surface tension of intruded fluid front and the opening surfaces of the extending cracks.

The surface tension effect is expressed as follows

$$P_s = P_w = 2\gamma \cos\theta / d_0 \tag{1}$$

where d_0 is the fracture aperture at the fluid front (m), γ is surface tension of the fluid (N/m), θ is contact angle and P_s is shut-in fluid pressure (Pa). The experimental result about mercury is shown as Table 1.

Table 1 Relation between pressure and ingress limit of mercury

$P(Bar)$	$d_0(\overset{o}{A})$	$P(Bar)$	$d_0(\overset{o}{A})$
0.1	750,000	32	2,400
0.32	240,000	100	750
1.0	75,000	320	240
3.2	24,000	1,000	75
10	7,500	3,200	24

3 Results and analysis

Under condition of stress status $P/Q = 2$ and rock stiffness $G/Q = 417$, where G is the shearing modulus of rock (Pa), the relation between normalized fluid pressure P_s/Q and the spatial variation ratio $\Delta V/V$ driven by fluid at case $\beta = 0$ (when Q is normal to the fracture) and $\beta = \pi/2$ (when P is normal to the fracture) is shown as Fig. 3. Because of $\Delta V = qt$, if ingress rate q keeps constant, the horizontal axis is proportional to the time. Additionally, the rhombus and roundness points in the figure are the calculated values than which if the fluid is pressed at higher pressure and shut-in (fluid rate gets zero), the fluid pressure within the fracture is in equilibrium with the rock stress, which is in accordance with the procedure of the practical hydraulic fracturing stress measurements. The functions of length of crack L/R and portion of fluid ingress l/R respectively along with $\Delta V/V$ are shown as Fig. 4. It can be seen that in case of $\beta = 0$ which means crack is perpendicular to the minimum principal stress and fluid is easy to be pressed, both the length of fracture opening and fluid ingress are greater than the respective ones in case of $\beta = \pi/2$. Fig. 5 can be used as reference.

4 Conclusion

In practice, natural cracks exist with various directions, so the forced fluid inside bore hole will first take the crack with smallest σ_n to fill in. Whilst, after the ingress status is fixed by the concretion of bentonite, the next fluid ingress will then take the crack with newly smallest σ_n to fill for this time. Complying with the iteration of thus a process, all cracks are gradually sealed.

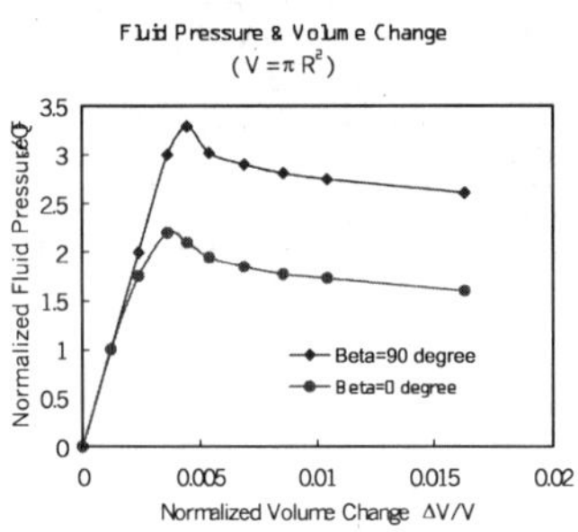

Figure 3 Shut-in pressure at various load process

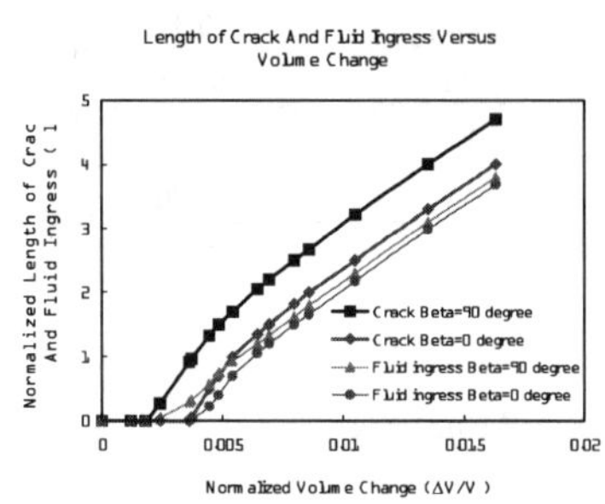

Figure 4 Lengths of crack & ingress fluid at various loads

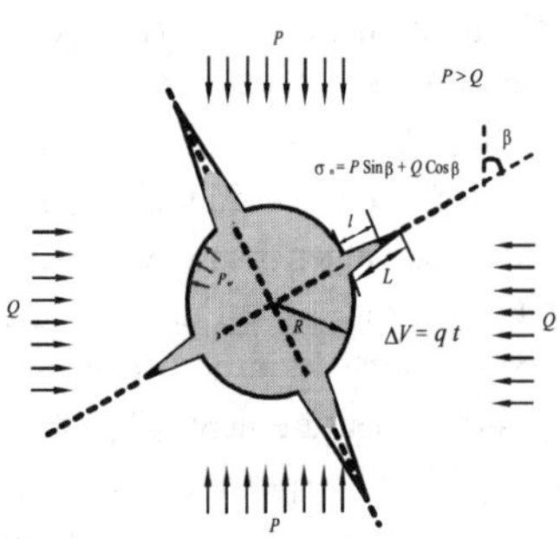

Figure 5 Relation of l and L about two sets of fracture with different σ_n

References

1. Y. Mizuta et al, *Pacific Rocks Workshop at Seattle* (2000).
2. H. Li et al, *Commu. of Numer. Meth. in Engng.* **17**, 365-378(2001).

NUMERICAL ANALYSIS OF MIXED-MODE FRACTURE FOR INTERFACE BETWEEN DAM AND ITS FOUNDATION

XINPU SHEN
[1]Department of Architectural Engineering, Shenyang University of Technology, Shenyang, 110023, China; Email: xinpushen@yahoo.com

LIXIN CHEN
Engineering Center for Coal-bed Methane, Northeastern Bureau of Coal Geology, Shenyang, 110035, China

GUOXIAO SHEN
Department of Applied Sciences, Taiyuan Heavy Machine Institute, Taiyuan, 030024; China

Firstly the focus is put on the formulation of an holonomic constitutive model for a cohesive interface crack at the interface of a concrete structure accounting for the influence of water pressure. Secondly, numerical analysis for mixed-mode cohesive fracture of a concrete-rock interface of a benchmark gravity dam is described.

1 Introduction

Hydraulic pressure exerting at the concrete-rock interface is an unfavourable factor to the safety of gravity dams, and the distribution of hydraulic pressure is hard to be determined experimentally, but the popular opinion in engineering is that the value of hydraulic pressure in the process zone is dependent on the crack mouth opening displacement[1,2,3]. In this work, the formulation of an holonomic constitutive model for cohesive interface crack at the interface of a concrete structure accounting for the influence of water pressure is presented. Numerical example is presented for the purpose of checking the capability of the proposed model.

2 Definitions of Variables For a Dam-Foundation Interface

At the interface between dam and rock-foundation the following vectors and relationships are defined as:

$$\mathbf{p} = \mathbf{p}^- = -\mathbf{p}^+ = \begin{Bmatrix} p_n \\ p_t \end{Bmatrix}, \quad \mathbf{w} = \mathbf{u}^+ - \mathbf{u}^- = \begin{Bmatrix} w_n \\ w_t \end{Bmatrix} \tag{1}$$

where $\mathbf{p}^-$ is the traction vector calculated from the lower part of the solid and $\mathbf{p}^+$ is the traction vector calculated from the upper part of the solid by FEM, and $\mathbf{p}$ is the traction vector acting on the interface; $\mathbf{w}$ is the vector of displacement discontinuity across the interface crack; subscripts "n" and "t" denote the components in the normal direction and tangential direction respectively.

3 Constitutive Law

The holonomic elasto-plastic constitutive law of the interface variables is expressed with effective traction as follows:

$$\mathbf{p}' = \mathbf{K}(\mathbf{w} - \mathbf{V}\boldsymbol{\lambda}) \tag{2}$$

where $\mathbf{K}$ is the elastic stiffness matrix of the interfacial material and

$$\mathbf{K} = \begin{bmatrix} k_n & 0 \\ 0 & k_t \end{bmatrix}, k_n = E/t, \quad k_t = G/t \tag{3}$$

where E is the Young's modulus and G is the shear modulus, and t is the thickness of the interface layer. If t=0, then $k_n=k_t=\infty$. The matrix $\mathbf{V}$ in Eq. (2) is the collection of plastic flow vectors of all the 5 yielding surfaces, and is expressed by Shen[4] as

$$\mathbf{V} = \begin{bmatrix} 1 & \mu & \mu & 0 & -\mu \\ 0 & 1 & -1 & 0 & 0 \end{bmatrix} \tag{4}$$

The yielding surface for concrete-like materials presented by Shen[4] is adopted. Its matrix form for a given traction vector is

$$\varphi = \mathbf{N}^{\mathrm{T}}\mathbf{p} - \mathbf{H}\boldsymbol{\lambda} - \mathbf{Y} \leq \mathbf{0} \tag{5}$$

The detailed expressions for the matrices and vectors in Eq. (5) can be found in Shen[4]. Since the hydraulic pressure in the process zone being accounted for, , Terzaghi's assumption is adopted and there is

$$p_n = p_n' - p_n^{(f)} \qquad p_t = p'_t \tag{6}$$

where p_n' is the normal component of effective traction of the skeleton and p_n is the normal component of total traction. Water pressure in the process zone, i.e. $p_n^{(f)}$, doesn't influence the distribution of traction component in tangential direction. According to the linear-assumption of water pressure inside the cohesive zone[3], we have for the mixed-mode crack the following expression for $p_n^{(f)}$:

$$p_n^{(f)} = \frac{w_n - \lambda_4}{w_{nc}} \hat{p}_n \tag{7}$$

where $\hat{p}_n$ is the value of the hydraulic pressure at the mouth of the process zone, i.e. the tip of the macro-crack; w_{nc} is the critical value of normal crack opening and w_n is the normal displacement at a point in the interface process zone; λ_4 is the 4th component of λ and its function here is to keep the value of $p_n^{(f)}$ at a point in the process zone to be not larger than $\hat{p}_n$. The matrix expression of yielding surfaces is:

$$\varphi = \mathbf{N}^T\mathbf{p} - \mathbf{H}_f\boldsymbol{\lambda} - \mathbf{Y} \le \mathbf{0} \tag{8}$$

where

$$\mathbf{H}_f = \begin{bmatrix} H_{nf} & \left(\frac{\chi_0}{c_0}H_t - \frac{\mu\hat{p}_n}{w_{nc}}\right) & \left(\frac{\chi_0}{c_0}H_t - \frac{\mu\hat{p}_n}{w_{nc}}\right) & -H_{nf} & -\left(\frac{\chi_0}{c_0}H_t - \frac{\mu\hat{p}_n}{w_{nc}}\right) \\ \left(\alpha_n\frac{c_0}{\chi_0}H_n - \frac{\mu p_n}{w_{nc}}\right) & \left(H_t - \frac{\mu^2\hat{p}_n}{w_{nc}}\right) & \left(H_t - \frac{\mu^2\hat{p}_n}{w_{nc}}\right) & -\left(\alpha_n\frac{c_0}{\chi_0}H_n - \frac{\mu p_n}{w_{nc}}\right) & -\left(H_t - \frac{\mu^2\hat{p}_n}{w_{nc}}\right) \\ \left(\alpha_n\frac{c_0}{\chi_0}H_n - \frac{\mu p_n}{w_{nc}}\right) & \left(H_t - \frac{\mu^2\hat{p}_n}{w_{nc}}\right) & \left(H_t - \frac{\mu^2\hat{p}_n}{w_{nc}}\right) & -\left(\alpha_n\frac{c_0}{\chi_0}H_n - \frac{\mu p_n}{w_{nc}}\right) & -\left(H_t - \frac{\mu^2\hat{p}_n}{w_{nc}}\right) \\ H_n & \frac{\chi_0}{c_0}H_t & \frac{\chi_0}{c_0}H_t & -H_n & -\frac{\chi_0}{c_0}H_t \\ \alpha_n\frac{c_0}{\chi_0}H_n & H_t & H_t & -\alpha_n\frac{c_0}{\chi_0}H_n & -H_t \end{bmatrix} \tag{9}$$

$$H_{nf} = H_n - \frac{\mu\hat{p}_n}{w_{nc}} \tag{10}$$

In Eq. (9), the off-diagonal items of **H** are for the coupled influence of the softening in the normal direction to properties in the tangential direction of the interface crack and vice-versa.

4 Numerical Application

With the above cohesive crack model, numerical calculation has been carried out. The geometry of the Benchmark gravity dam is defined to be of 80-meter height and 60-meter width at bottom. The elastic modulus, Poisson's ratio and the mass per cubic meter for concrete are $E = 24000MPa, \upsilon = 0.15, \gamma = 2400kg/m^3$ respectively. A total of 61 interfacial nodes are uniformly distributed along the interface, and it is clear that the distance from a node to the crack mouth is X=(nodal number - 1) meters.

Material parameters and model parameters at interface are chosen as follows: Frictional angle $\phi = \arctan(0.7)$, $\mu = 0.7$; tensile strength and shear strength are $\chi_0 = 0.3MPa, c_0 = 1MPa$ respectively; softening modulus in the normal and tangential directions are $H_n = -0.5N/mm^3$ and $H_t = -1.0N/mm^3$, which correspond to fracture energy for mode I crack $G_f^I = 0.09N/mm$ and for mode II fracture energy $G_f^{II} = 0.5N/mm$. The value of the coupling parameter is α=0.8.

The resultant figure for the full hydraulic pressure together with a 5.15-meter overtopping water-head are shown in Figure 1.It is seen from Figure **1** that, after the initiation of a cohesive crack at the concrete-rock interface in the presence of water pressure, there is a stable stage of crack propagation with increase of the overtopping pressure.

5 Conclusions and Ending Remarks

The formulation of an holonomic constitutive model for a cohesive interface crack at the interface of a concrete structure accounting for the influence of water pressure has been presented. Numerical analysis has been carried out for the mixed-mode quasi-brittle fracture of the concrete-rock interface between a benchmark dam and its foundation. The numerical results has indicated that shearing fracture is the dominant factor for propagation of cohesive zone under the conditions adopted in this calculation.

References

1. E. Bruhwiler, E. Saouma, *ACI Mater. J.*, **92**(3), 296-303 (1995).
2. E. Bruhwiler, E. Saouma, *ACI Mater. J.*, **92**(4), 383-390 (1995).
3. G. Cocchetti PhD Thesis, Politecnico di Milano, Italy (1998).
4. Xinpu Shen, *Chinese J. Geotech Engng*(English article), **23**(2), 250-256 (2001).
5. C. Comi, G. Maier, U. Perego, *Comput. Meth. Appl. Mech. Engng.*, **96**(1), 133-171 (1992).

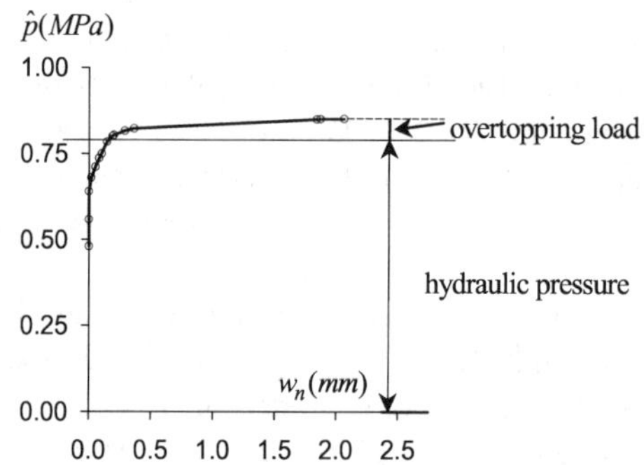

Figure 1. Diagram of load *vs* crack mouth opening in normal direction.

NUMERICAL SIMULATION ON TRANSVERSE EFFECT OF BOLT IN DISCONTINUOUS ROCK MASS

HONGWEI SONG[1] JIAN ZHAO[1,2] YINGXUAN JIA[1]
College of Arch. & Civil Eng., CUMT, Xuzhou, Jiangsu, 221008, China [1]
School of Civil and Environmental Eng., NTU, Singapore, 639798 [2]

By use of finite element of 3D models, the transverse-acting mechanism of a bolt in discontinuous rock mass is analyzed. The deformation process of discontinuous rock mass restricted with a bolt is simulated. The interaction law of between a bolt and rocks, the law of displacement and loading of discontinuous rock mass and bolt are gained. An unusual characteristic phenomenon has been discovered, which is called 'the railway function of bolts'. Mechanical action of 'the railway function of bolts' is analyzed.

1 Introduction

Rock mass is a discontinuous material that has joints or fractures in it. When using bolt as support, the bolt, the rock blocks and the joints form a structural system. Seeing about the function of a bolt in stabilizing the system is of great importance for effectively design bolt support.

Using physical model, the practical stress states of bolt and rocks near-by cannot be fully acquired because of lack of effective means. Using continuous numerical method, the interaction between bolt and the rock blocks cannot be clearly described. In recent years, the discontinuous numerical simulation begins to be used. It is used to study the interaction between bolt and the rock blocks in this paper.

2 Model and the simulation

A 3D model was used. The 3D model is cut out from the place that a bolt threads the joint with a size of $0.4\times0.4\times0.4$ m. A joint (as a contact face) cuts through the middle part of the model. A bolt of 0.02 m in diameter threads through the joint. The 4 sides of x and z direction of the model are restricted. A pair of shearing stresses is put on the model in y direction. Different angles between the bolt and the joint are used in the simulation.

The inner stress and displacement in the model and the stress and deformation in the bolt under different loads and angles are studied in the simulation. The material parameters are listed in the table 1. The arrangement of the simulation is listed in the table 2.

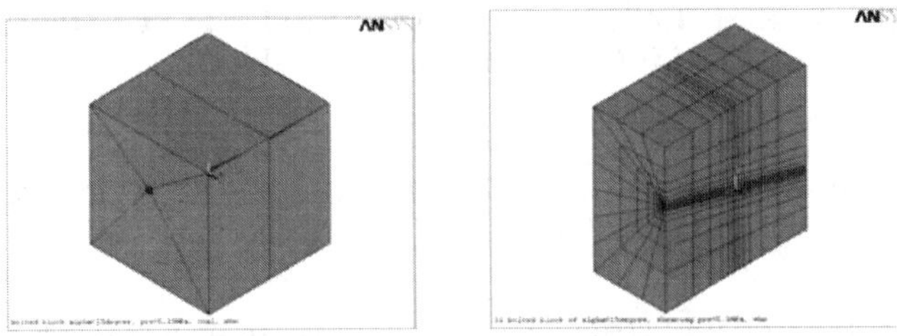

The 3D model The meshing of the model

Figure1. The analyzing model

Table1. The material parameters

Materials	Blocks(Coal)	Bolt (A3 Steel)
Properties of materials	E: 3000MPa V: 0.28 R_l: 0.9MPa R_a: 9Mpa	E: 200Gpa V: 0.25

Table 2. The arrangement of the simulation

Items / Models	Angle (°)	Material	Friction Coef.	Load (MPa)
1	0	Block And bolt	0.6	0.020—0.197
2	15			
3	-15			

3 The analysis of simulation results

3.1 The Law of Interaction between Rock Blocks and Bolt

Under shearing loads, a displacement occurs along discontinuous face between rock blocks. The bolt inflects under this displacement. The bolt stresses counteract on the rock blocks. A reaction stress field is established in the rock blocks. This arrests further displacement of rock blocks and reinforces discontinuous of rock mass. (Figure 2). The rock block put the force on the section of bolt near by and makes the local deformation of the bolt or even failure.

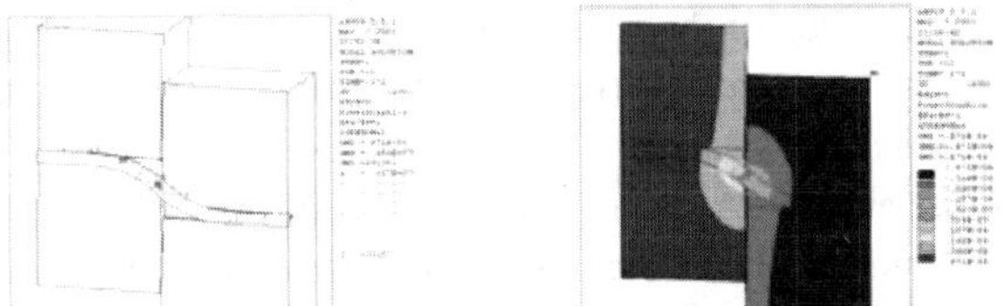

Sy value when Po=0.171MPa Uy value when Po=0.171MPa

Figure 2. Interactions between rock blocks and bolt

3.2 The Law of loading and deformation of a bolt

Under shear loading, the bolt is deformed along its axes in an "S"form (Figure 2). This make the bolt have an outward incline angle. The bolt acts like a rail and make the rock blocks to move outwards along it. This phenomenon is called "the rail function of a bolt".

The simulation shows that the stresses in the bolt are very complicated. But the law is obvious that the deformation and loading in the bolt concentrate at small part near the joint. The bolt curves remarkably.

3.3 The Law of loading and deformation of rock

Under shear loading, the blocks move outwards along the curved bolt, that is the phenomenon of "the rail function of bolts". This phenomenon shows that the blocks have a trend to split at the joint. And the normal contact force between the two blocks becomes small or even zero. The simulation shows that even under different angles between the bolt and the joint, this phenomenon occurs constantly (Figure 3).

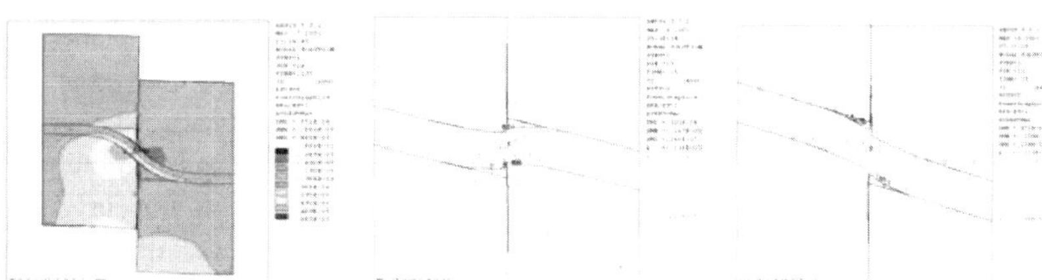

Angle= 0 ° Angle=—15 ° Angle=15 °

Figure 3. Uz value in the model

4 The analysis of "the Rail function of bolts"

According to above discussion, the rail function of a bolt can be analyzed with Figure 4. Shearing force acts on the bolt, two components of force occur. T is along the bolt axis, N is perpendicular to the bolt axis. The relationship is shown as:

$$T = PCos\ \theta \qquad (1)$$
$$N = PSin\ \theta$$

As the loading increases, at the crossing point between bolt and joint, the angle between bolt axis and the joint plane, θ decreases. Then T increases and N decreases.

The rock block will slide along the curved bolt axis and split with other block when the component force T is larger than the friction force between bolt and the block.

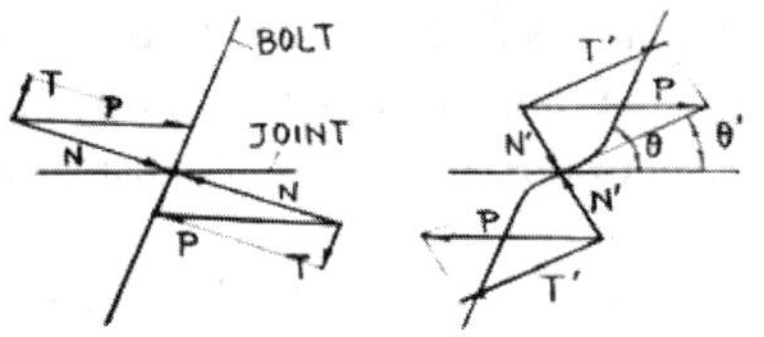

Figure 4. The analysis of rail function of bolt

From eq.(1), in projection direction normal to joint plane, T and N counteract each other. So contact force between two rock blocks cannot be changed.

Therefore, it is valuable to deliberate the point of view [1,2,3] that the contact force will increase when a bolt through a joint under shearing loading.

5 Conclusions

(1) Under shearing loading, the rock blocks slide along the joint plane. Then larger deformation and higher stresses are produced in the bolt body near joint. These deformation and stresses retroact on the blocks and produce stress fields in the blocks. Then the further slide move is prevented. This is the mechanism of bolts.

(2) Under shear loading, the rock blocks move outwards along the curved bolt, this phenomenon is called "the rail function of bolts". This function of bolts makes the rock blocks have a trend to split at the joint, and results the normal contact force between the two blocks becomes small or even zero. The function also decreases the friction force between the blocks and causes a negative bolting effect.

(3) In projection direction normal to joint plane, T and N counteract each other. So the normal contact force between two rock blocks cannot be changed.

References

1. Shiliang LU, Lei TANG, Xin'an YANG. The bolting force and bolting techniques. Beijing: Coal Industrial Publishing House, 1998.11, pp39-44
2. G. E. Bannes. Soil Mechanics : Principles and Practice. London: MACMILLAN PRESS LTD. 1995, pp263-265
3. Bo LIU. Studies on the bolt-shearing effects and comprehensive supporting force. Doctoral thesis of CUMT, 1998, pp100.

THE FEM RESEARCH OF TERRANE CONTROLLING OF THE GALLERY EXCAVATED IN "THREE SOFTS" STRATA WITH A LARGE OBLIQUITY

SHUGANG CAO, JIN BIAN, PENG LI
The resource and environment college, Chongqing University 400044

In this article, the author conducts an elastic-plastic FEM analysis in terrane controlling of K_b^3 coal seam, Songzao colliery, which has a large obliquity and is overlaid by "three softs" terrane. The following acknowledgements are made: the incidence of gallery excavation is nearly symmetrical along the perpendicular direction of the terrane; there is a high stress area in the coal pillar; the maximum stress is found in the side of coal pillar along the goaf; the stress grads drop as the coal column becomes wide; the distribution of the plastic area is determined by the gallery section, but whether the goaf is packed or not has a little influence on it; and so on. These conclusions have some instruction- al significance on the engineering practices. With the increase of the mining depth in recent years, it becomes more prominent that the surrounding rock owns the characteristics of large deformation, high stress and difficult supplying[2][3]. The movement and destruction of the overlaid terrane is a dynamic and rheological procedure[4][1]. Therefore, knowing the stress distribution and the deformation of rocks surrounding the gallery in "three softs" terrane has an important significance on terrane controlling.

1 The Establishment of the Calculation Model

For the research of the influence of the excavation of the gallery and the pillar width on the deformation of surrounding rocks, a calculation model is designed on the geological conditions of $1^{\#}$ mine, Songzao Colliery. The model is 120m wide and 150m high .The dip angle of terrane is 30^0. The coal seam has a thickness of 2.5m and the obliquity of 30^0. It is assumed to be a problem of plane strain .The horizontal restriction is applied in the left verge of the model; the horizontal pressure is done in the right verge which is calculated by Janenick formula .The gravity of overlying stratum in the upper verge; the restriction in the direction of the plumb line in the verge. There is a layer of the terrane at the roof of the gallery and is 50m at the floor .The walls of the anomalistic gallery are parallel to the plumb line, and its floor is perpendicular to the walls, and its roof is parallel to the terrane. The distance of the midpoints of the bottom and roof of the gallery is 2.4m.The width of the floor is 2.7m.

2 The FEM analysis of the gallery deformation

With the FEM, the stress distribution figures and plastic areas of the following items are computed: when galleries of the anomalistic section, trapezoid one and anti-trapezoid one are excavated; when the goaf is or is not stuffed, and the width of coal pillar is 5m, 10m, 15m, 20m, and 30m, respectively, with the gallery of the anomalistic section; when the goaf is or is not stuffed, and the width of coal pillar is 5m and 10m,respectively with galleries of trapezoid.

There are too many rock strata, for the convenience of calculation and reasonable calculating error, some rock strata are merged. The program ANSYS is used. Calculating the value of first principal stress in the figures, the value of third principal stress and the disturbing plastic area of each combination are obtained.

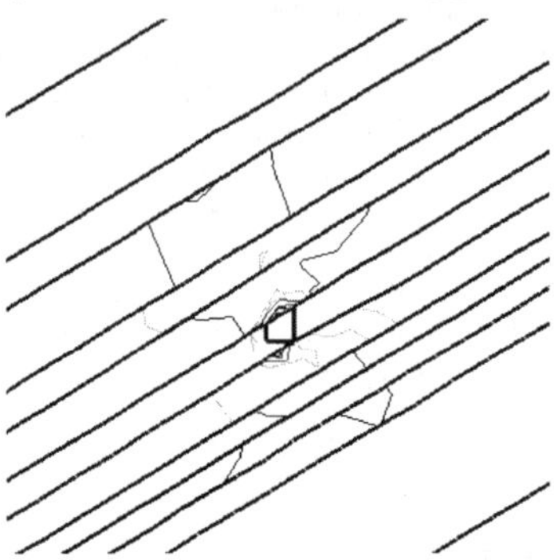

Figure1. the first principal stress of the anomalistic Gallery

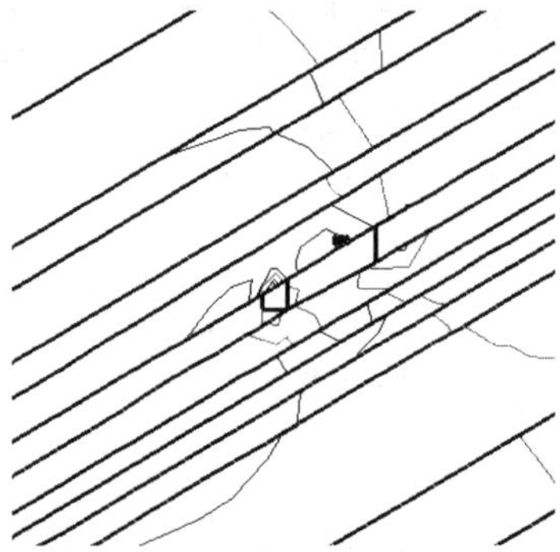

Figure2. the first principal stress of the anomalistic gallery of the pillar width L=10m when the goaf not being packed

According to the values of principal stress and relative plastic area, the following conclusions have been obtained: In Fig.1 and 2, the incidence of the gallery excavation is basically symmetric along the perpendicular direction to the terrane. But its excavation exerts a greater effect on the upper stratum than on the lower strata. The distribution of the stress isoclines is similar in different kind of section. But the stress gradients in the surrounding rocks around the anti-trapezoid gallery are the largest.

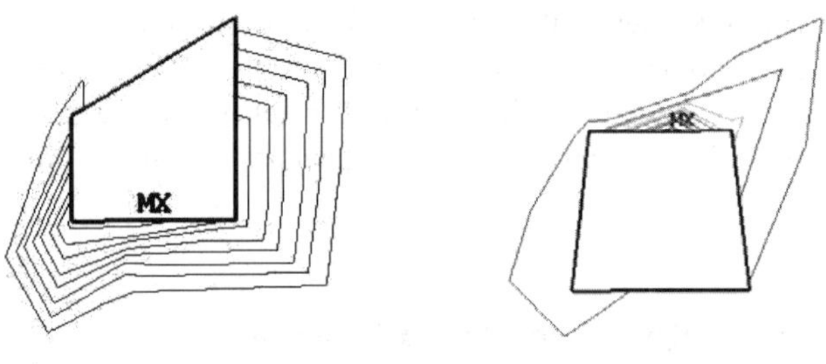

(a) anomalistic (b) trapezoid

Figure 3. the plastic area of the gallery with different sections after its excavation

If the goaf is not stuffed, the plastic area spreads around the gallery in all kinds of the combination of the gallery section and the width of coal pillars.
No matter how the mining vacancy is stuffed or not, the stress isocline's distribution is similar along the vertical direction in all calculation things. But the stress gradients become large with the width increase of the coal pillar .In the coal pillar there is a stress concentration area, of which the change of stress gradients follows the same regulation as above. The stress field is similar with the fluid one, in which the streamline becomes dense in the narrow sharp curve, especially the inner side.
When the mining vacancy is stuffed, the stress values of nodes decrease .The degree of stress concentration descends. When the gallery section is trapezoid or anti-trapezoid, the fluctuation is smaller and stress values are larger.

Generally, the maximum pressure area appears in the side of coal pillar that approaches the goaf and the tensile stress field lies in the bottom and the roof of the gallery.

The wider the coal pillar is, the better the stress distribution is. However, too wide coal pillar wastes resources. Analyzed from the stress figures, the values of the third principal stress of nodes and stress gradients drops quickly while the width of coal pillar increase from 5m to 15m, and drops slowly from 30m to 15m. And in consideration of production safety, the optimum combination is the anomalistic gallery and the 15m width of coal pillar.

3 The Forecast of the rheological principles

During the calculation process, the stress values in midpoints of two walls of the gallery with the anomalistic section are obtained while the goaf is not stuffed and the width of coal pillar is 10m and 20m respectively. In the roof of the gallery, the tensile stress appears. Because rheological characteristics in compression are largely different from ones in tension, the strain of the roof and the floor of the gallery are not calculated. Fig.4 is the observation displacement curve of the 1# well, Songzao colliery. Fig.5 is the displacement curve calculated in term of the formula[5] :

$$E^{'} = \frac{S_0^{'}}{2G_1} + \frac{S_0^{'}}{2G_2}[1 - \exp(-\frac{G_2}{\eta_2}t)]$$

In the beginning of the curve, the values of strain rises quickly and then becomes Slowly when t =10d. The strain increase regularity is similar with the displacement rise one. In consideration of surveying errors and the difference of stress and rheological parameters .So the equation can be used to predict the rheological properties of rocks and coal.

4 Conclusions

1. No matter how the goaf is stuffed or not, the stress isocline's distribution is similar along the vertical direction when the pillar width is invariable.

2. When the pillar width changes, the stress isocline's distribution is similar along the vertical direction .The stress grads become large with the width increase of the coal pillar.

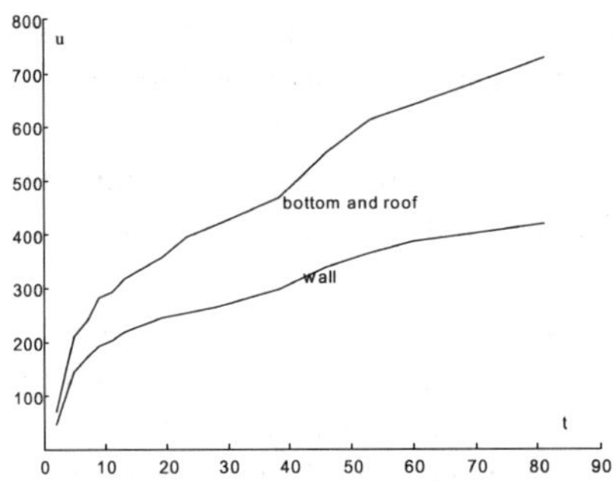

Figure 4. the displacement-time curve of bottom, roof and walls in 1#survey station

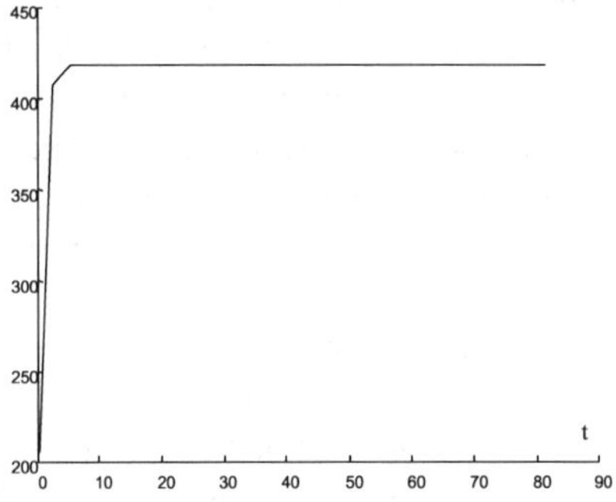

Figure 5. the strain-time forecast curve of the walls with the pillar width L=10

3. When the goaf is stuffed, the stress values of nodes decrease and the degree of tress concentration descends.

4. In the coal pillar there is a stress concentration area, of which the change of stress grads drops with the increase of the coal pillar. The stress grads rise in the side of the coal pillar that approaches the goaf.

5. The plastic area varies mainly with the gallery section. There is a little influence on the distribution of the plastic area, no matter the goaf is stuffed or not .

THE SIMULATING PROCESS OF ULTIMATE BEARING CAPACITY IN LARGE DIAMETER PILE

XIAOLI QIAN (1,2)   LI ZHOU (2)

1. Department of Resource Exploitation Engineering ,CUMT ,Beijing 100083, P.R.China;
2. Department of Civil Engineering, Heilongjiang Institute of Science & Technology, Jixi, Heilongjiang, 158105, P.R.China

To solve the difficulty of measuring ultimate bearing capacity in large diameter pile , according to the theory of displacement method in structural mechanics and finite element method, the paper infers a simulated nonlinear analysis method to forecast the relation of load -- displacement in large pile, and work out a calculating program which contrast with the result of engineering. The results indicate that for all kinds of large piles, such as diameter, length, parameter of soil, and so on, the load--displacement curve of program and engineering is similar.

1 Calculating theory [1-2]

1.1 General Assumption and division of cell

1. Pile is vertical and the surface of ground is horizontal ;
2. The surrounding soil is nonlinear ideal elasticity、homogeneous and isotropic ;
3. The physics and mechanics character of surrounding soil is unchangeable ;
4. The interface of pile and soil system use non--thickness cell，shearing force and shearing deformation is nonlinear, and parameter is changed from soil ;
5. The system of pile and soil is axis-symmetry ;
6. The interface is non-slippage.

The division of cell is shown in figure 1.

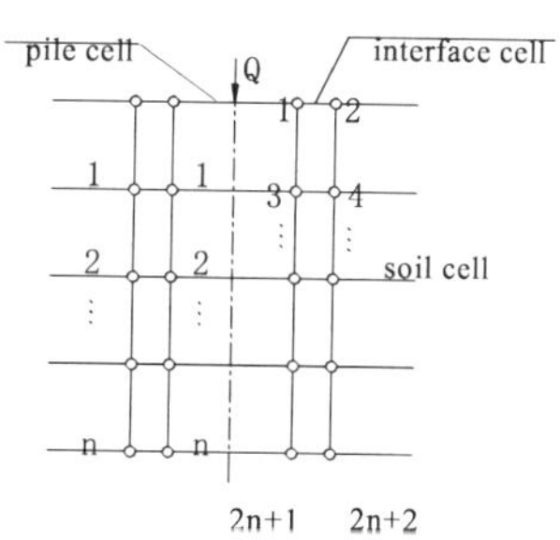

Fig.1 Finite cell partition

1.2 The Establishment of Cell Stiffness Matrix

1. The stiffness matrix of soil around the pile

Suppose $G(\gamma)$、$E(\gamma)$ are modulus relate to shear strain.

Order function $$\beta(r)=\frac{u(r,z)}{u(r_0,z)} \qquad (1)$$

Suppose $\beta(r)=e^{\alpha(r-r_0)}$ then $\alpha=\dfrac{\ln\beta(r_1)}{r_1-r_0}$

$$\therefore u(r,z)=e^{\ln\beta(r_1)\frac{r-r_0}{r_1-r_0}}u(r_0,z) \qquad (2)$$

Suppose $u(r_0,z)$ is linear function of coordinate Z, then

$$u(r,z)=e^{\ln\beta(r_1)\frac{r-r_0}{r_1-r_0}}\left[\frac{z(i+1)-z}{z(i+1)-z(i)},\frac{z-z(i)}{z(i+1)-z(i)}\right]\cdot\begin{Bmatrix}u(i)\\u(i+1)\end{Bmatrix} \qquad (3)$$

Shear strain γ and normal strain ε of any point in soil can be developed by (3)equation, and the shearing and compress cell stiffness of soil around the pile can be developed by idle work principle .

$$[K(\gamma)]=2\pi\left[\frac{\ln\beta(r_1)}{r_1-r_0}\right]^2\int_{r_0}^{r_1}G(\gamma)\,e^{2\ln\beta(r_1)\frac{r-r_0}{r_1-r_0}}\cdot rdr\cdot\frac{z(i+1)-z(i)}{6}\begin{bmatrix}2&1\\1&2\end{bmatrix}$$

$$[K(\varepsilon)]=2\pi\int_{r_0}^{r_1}E(\gamma)\,e^{2\ln\beta(r_1)\frac{r-r_0}{r_1-r_0}}\cdot rdr\cdot\frac{1}{z(i+1)-z(i)}\begin{bmatrix}1&-1\\-1&1\end{bmatrix}$$

∴The cell stiffness matrix of soil around the pile must be:

$$[K]_c=[K(\gamma)]+[K(\varepsilon)]$$

2. The cell stiffness matrix of soil under the pile

The relation between force and displacement of junction in soil is:

$$\{\delta\}^e=[I]_{2\times2}\cdot\{P\}^e$$

Suppose soil under the pile is half infinite nonlinear elestomer, namely request any vertical displacement of half extensity on level bordering, then the flexibility matrix $[I]$ of soil under the pile can be requested by Boussinesq formula.

Then the stiffness matrix $[K]_d$ of soil under the pile can be requested by revert the flexibility matrix.

3. Mechanics model and cell stiffness matrix of interface cell between pile –soil(fig.2).

Interface cell has a column surface ,whose radius equal to pile. Suppose the interface cell only have contact on junction with apposition cells on the ahead or on the below .At the same time, suppose the displacement model of cell is a most simple apposition function .namely

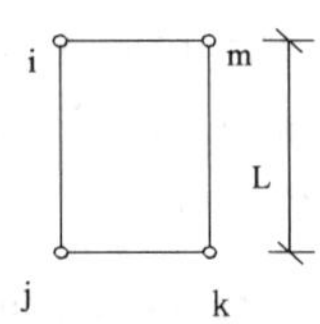

Fig.2 Interface cell

$$\begin{cases} N_i = 1 - \dfrac{x}{L} = N_m = A \\ N_j = \dfrac{x}{L} = N_k = B \end{cases}$$

For any displacement in the cell

$$\begin{cases} u_{\times 6} = A\rho_i + B\rho_j \\ u \quad = A\rho_m + B\rho_k \end{cases}$$

Any displacement difference between two surface in the cell is:

$$\Delta\rho = (u_{\times 6} - u \quad) = \{AB - B - A\}\begin{Bmatrix} \rho_i \\ \rho_j \\ \rho_k \\ \rho_m \end{Bmatrix} = \{B\}\cdot\{\rho\}_e$$

$\{B\}$ is shape function of the cell

Suppose K_s, shear stiffness of the cell , which is equal to the force which cell produce single displacement， suppose K_s is secant modulus， then

$$\tau_S = K_s \cdot \Delta\rho = K_s \cdot \{B\}\cdot\{\rho\}_e$$

Following idle work, interface cell stiffness matrix can be developed:

$$[K]_j = 2\pi r_0 \int_0^L \{B\}^T K_s \{B\} dx = \frac{\pi r_0 L}{3} K_s \begin{bmatrix} 2 & 1 & -1 & -2 \\ 1 & 2 & -2 & -1 \\ -1 & -2 & 2 & 1 \\ -2 & -1 & 1 & 2 \end{bmatrix}$$

4. The stiffness matrix of pile

Suppose the pile is equal in section and material is averaging，elasticity modulus of pile is E， length of pile is L(fig. 3).

Suppose displacement pattern is linear function

Order $\begin{cases} N_i = 1 - \dfrac{x}{L} \\ N_j = \dfrac{x}{L} \end{cases}$

Then the common expression of any displacement of junction in the cell is:

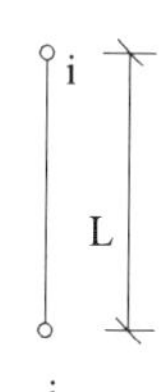

Fig.3 Pile cell

$$u = [N_i, N_j]\cdot[\rho_i, \rho_j]^T = \left(1 - \frac{x}{L}\right)\rho_i + \frac{x}{L}\rho_j$$

According to geometry equation, any strain of junction in the pill cell is :

$$\varepsilon_x = [B]\{\rho\}_e \quad [B] = \left[-\frac{1}{L}, \frac{1}{L} \right]$$

[B] is strain matrix of the cell

The internal force and deformation of any junction in pile cell meet Huke' s principle.

Following idle work principle, pile cell stiffness matrix $[K]_z$ can be developed :

$$[K]_z = A_o \int_0^L E[B]^T [B] dx = \frac{EA_0}{L} \begin{bmatrix} 1 & -1 \\ -1 & 1 \end{bmatrix}$$

1.3 Holistic Analysis and the flowchart of program

According to the cell stiffness matrix $[K_c]$、the cell stiffness matrix $[K]_d$ 、the cell stiffness matrix $[K]_j$ and the cell stiffness matrix $[K]_Z$, the general stiffness matrix $[K]$ can be created 。

This paper considers that the stiffness of soil around the pile、the cell stiffness of soil under the pile and the cell stiffness of interface between pile –soil are all gradual intenerate to follow accreting displacement ,namely stiffness are relate to displacement . So, this paper must suppose the initial displacement .

The flowchart of program is shown at figure 4

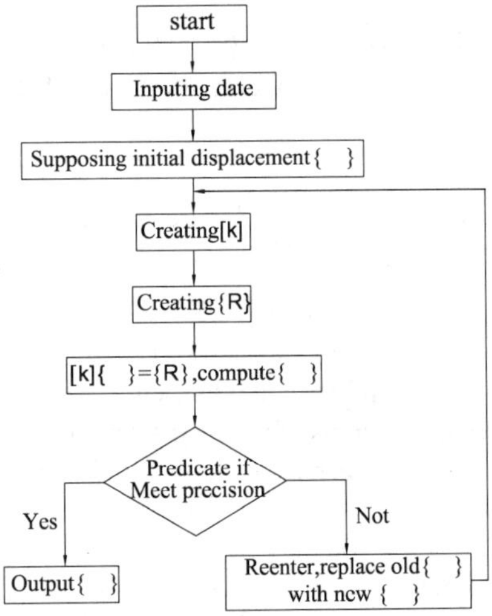

Fig.4 Program flowchart

1.4 Considering of Nonlinear

Carrying through nonlinear modify on the cell stiffness of interface between pile –soil、the cell stiffness of soil around and under the pile.

Suppose the relationship between shearing stress τ and shearing displacement μ of the interface is hyperbola，the curve is shown in figure 5.

a is the reciprocal of initial shearing modulus in interface ,namely

$$a = \frac{1}{K_1} \qquad (4)$$

b is the reciprocal of ultimate shearing stress ,namely

$$b = \frac{R_f}{P_f} \qquad (5)$$

R_f —failure ratio, allowable resistance to ultimate frictional resistance .

P_f —the most frictional resistance of any cell .

According to figure 5， then the secant module between origin and any point $c(\tau, \mu)$ in hyperbola is :

$$K_s = \frac{\tau}{\mu} = \frac{\mu}{a + b\mu} / \mu = \frac{1}{a + b\mu} \qquad (6)$$

Substitute equation(4)(5)to (6),then

$$K_s = \frac{1}{\frac{1}{K_1} + \frac{R_f \mu}{P_f}} = \frac{K_1}{1 + \frac{K_1 \cdot R_f \mu}{P_f}}$$

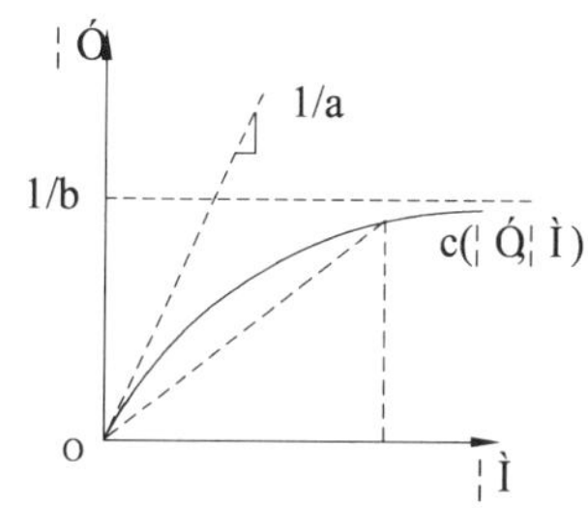

Fig.5 ¦Ó¦Ì relation curve

Amelioration of soil under the pile cell is equal to interface cell .

The cell stiffness of soil around the pile varies from junction displacement , so shearing module and compress module of stiffness matrix vary from shear strain.

Order $G(\gamma) = G_{\max} \dfrac{1}{1 + \gamma / \gamma_r}$ $\qquad E(\gamma) = 2(1 + \mu)G(\gamma)$

$G_{\max}$: most shearing module of soil , γ_r : reference shear strain

2 Program Validation and Example Analysis

Basing on the analysis method and program of antecedent, this paper fitting load –displacement curve for large diameter impregnate piles beyond 20， meanwhile contrast the program with the result of engineering.

2.1 Calculation parameter

Parameter index of soil in the simulation calculation are as follows:

Shearing modulus G_{max}=1~3.2MP$_a$; Poisson's Ratio μ=0.3~0.35 ; reference shear strain γ_{γ}=0.008 ; initial shearing stiffness K_1 =100~150MP$_a$/m ; most elasticity modulus of soil under pile E_{max}=6~9.1MP_a; ultimate resistance of soil around pile q_{su}=10~90KP_a; ultimate resistance of soil under pile q_{su}=1 MP_a; failure ratio R_f=0.85~0.9; neglected ratio around pile $\beta(\gamma_1)$=0.002. Elasticity modulus of pile $E_p=2.5\times10^5$N/mm.

2.2 Analysis fitting effect

Contrast effect is shown in figure 6. Fitting curve is very approach to actual measurement curve, especially two ultimate bearing capacities close to each other. Therefore this method can reflect load – precipitation specificity basically.

3 Conclusion

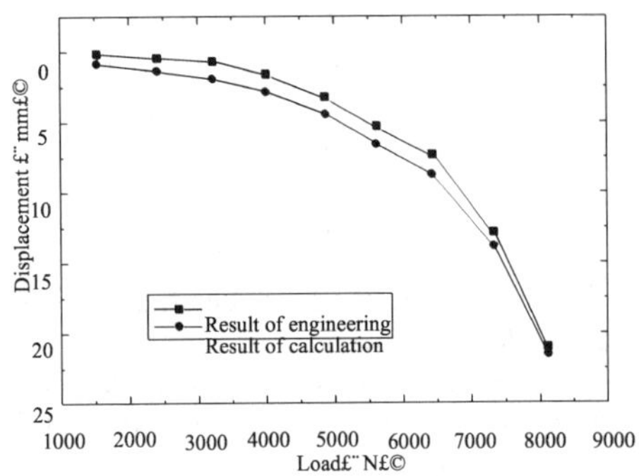

Fig.6 Contrast of ZS-2 Q-S Curve

This paper creates a REM to simulate the interaction between pile-soil system and introduce program. Adopting exponent function to simulate shearing deformation of soil around the pile , modifying soil under the pile rigidity to simulate the nonlinear of soil under the pile resistance ,the nonlinear of interface cell ,and so on. Fitting result indicates that for all kinds of large piles, such as diameter, length, parameter of soil, and so on, the load--displacement curve of program and engineering is similar, ultimate bearing capacity is very approximately. Therefore, the simulate method and program of pile--soil system are correct in theory and feasible in engineering.

References

1. Peigong Shi. Practical engineering handbook of pile. Beijing :1999
2. Xiaoli Qian. The Simulate and Statistics of Ultimate Bearing Capacity of Large Diameter Pile. [Thesis for Master Degree][D].Fuxin : Liaoning Technical University , 2001, 17~38

Biograph: Xiaoli Qian (1970_), female, lecturer, doctoral student, engaged in the research of geo-technical engineering .

PARAMETRIC STUDIES ON BEHAVIORS OF SOIL-NAILED STRUCTURE BY THREE-DIMENSIONAL (3-D) FINITE ELEMENT METHOD (FEM)

MINGJU ZHANG
Graduate Department, Shijazhuang Railway Institute, Shijazhuang, 050043, China
E-mail: zhangmingju@tsinghua.org.cn

QIAN YANG
Graduate Department, Shijazhuang Railway Institute, Shijazhuang ,050043, China
E-mail: yangqian3555@sina.com

TIECHENG SUN
Civil Engineering Department, Shijazhuang Railway Institute, Shijazhuang, 050043, China

A discrete three-dimensional (3-D) finite element model has been adopted for the analysis of soil-nailed structures. The soil/nail interactions are modeled by two-node double spring elements, the staged excavation and retaining are simulated and the soil constitutive law is chosen to analogize the behaviors of soil undergoing excavation. On the basis of an engineering case to be analyzed, the working principles and strain-stress performances of soil nailing system are researched. The effects of several parameters on the soil-nailed structures are examined. Conclusions may be of very high practical value in design and construction of soil-nailed structures.

1 3-D FEM of soil nailing

A discrete 3-D finite element model, considering the soil/nail interactions, the staged excavation and retaining and the soil constitutive law which is chosen to analogize the behaviors of soil undergoing excavation, has been adopted for the analysis of soil-nailed structures[1, 2]. The width of the 3-D model is equal to half the horizontal spacing of nails and the cut width is supposed to be the same magnitude as the excavated depth H. The FE mesh boundaries are($2\sim3$)H far from the excavated surface and 0.5 H deep below the final excavated bottom. Simple, clear and definite restrains are added to the boundaries of the FE mesh. The nodes of the vertical faces are restrained by rollers and those of the bottom are fixed.

The 20-node isoparametric elements were employed for the soil. The shotcrete facing was modeled by the same elements surrounded by thin soil elements. Corresponding to the surrounding soil elements, a nail was discretized into a number of 3-node bar elements. The grout nail is composed of the reinforced bar and the wrapped cement grout, so its tension –deformation relationship can be considered to be linear elasticity approximately [3].

Since a strong adhesion is existed between the nail and the surrounding soil, the nail can bear a very big pull force in its axial. A double-spring connect element was applied to simulate the stick or slip mode between soil and nail. The soil/nail interface may be in either stick or slip mode depending on the magnitude of the shear stress compared to the ultimate shear strength.

The analysis of a soil-nailed structure was carried out in steps that simulated approximately its actual staged construction. The construction process was modeled by successive excavation of soil followed by the placement of nails and shotcrete. In programming, switches were established for all kinds of elements. Unless the switch for an element was on, the element stiffness matrix would not be added to the global stiffness matrix. The excavation was simulated by substituting excavation loads for the soil to be excavated and the removed elements were turned off. The nail installation was modeled by turning on the nail elements. The shotcrete placement was modeled by switching the corresponding soil element properties into those of shotcrete.

The initial state of stress in the soil was first set up for the original soil layers. When the first layer of soil elements was excavated, the excavated surface was made stress free by applying equivalent nodal forces from the removed elements to the excavated boundary. After a non-linear analysis, in which the load was divided into a number of increments, was carried out, the displacement and stress after the first excavation was obtained by accumulating the increments. A similar process was repeated in the successive construction simulations until the excavation was completed.

2 Soil constitutive model in excavation case

Since the soil near excavation area is in a successive unloading or reloading state, it is necessary to choose a constitutive law which can model the non-linear stress-strain relationship of soil undergoing excavation. It is believed that the non-linear *E–B* model, proposed by Duncan et al. [4], is one of comparatively simple and practical model. It embodies the principal behavior of soil deformation, that is, non-linearity. It also takes the modulus of unloading and reloading into consideration, so the inelasticity and influence of stress history is reflected. The model is, therefore, employed in the present analysis after consideration of soil behavior undergoing excavation. As a few modifications have been made here, a brief description is given in the following.

(1) It is essential to give a criterion which can distinguish primary loading and unloading or reloading, so that the value of E_{ur} is adopted in unloading or reloading cases. Since σ_3 keeps constant in laboratory test, unloading will occur if $\sigma_1 - \sigma_3$ decreases. However, because σ_3 varies in the field, it is possible that the stress level will increase and plastic deformation will take place when both σ_3 and $\sigma_1 - \sigma_3$

decrease. Duncan et al., at first, employed such criterion as

$$f_1 = \frac{(\sigma_1 - \sigma_3)}{(\sigma_1 - \sigma_3)_f} \qquad (1)$$

However, it was unreasonable that the soil tangent Young's modulus took E_{ur} instead of E_t in so many soil elements because of adopting this criterion. Based on the researches and applications of the criterion over a long period of time, Duncan et al. made some improvements and modified the criterion as

$$f_1 = \frac{(\sigma_1 - \sigma_3)}{(\sigma_1 - \sigma_3)_f} \sqrt[4]{\sigma_3} \qquad (2)$$

which could give reasonable results. In the present analysis, the modified criterion was adopted.

(2) Since the vertical pressures exerted on the soil elements below the excavated bottom decrease after excavation and the horizontal surrounding pressures are correspondingly reduced, the soil tangent Young's modulus, which is calculated by the method mentioned above, is comparatively low. The soil has actually accepted so high earth pressure that it has higher bearing capacity and larger modulus than those calculated. However, the formulation of the modulus in this case was not properly given in the soil model. Excavation exhibits successive unloading and the reduction of vertical pressure on the soil element introduces the decrease of horizontal surrounding pressure. In such case, the soil is situated the condition corresponding to consolidation pressure reduction, so the soil tangent Young's modulus depends on the horizontal surrounding pressure before excavation. In programming, the soil stress took σ_{3max}, which is the greatest consolidation stress in the stress history, instead of σ_3 to calculate the initial modulus E_i, the bulk modulus B_t and the unloading modulus E_{ur}.

3 Verification of the model and parametric studies

3.1 Comparison with the measured

To verify the model, a field prototype of soil nailing system, conducted by Shen et al.[5], was analyzed. The soil deposit in the test site was primarily composed of

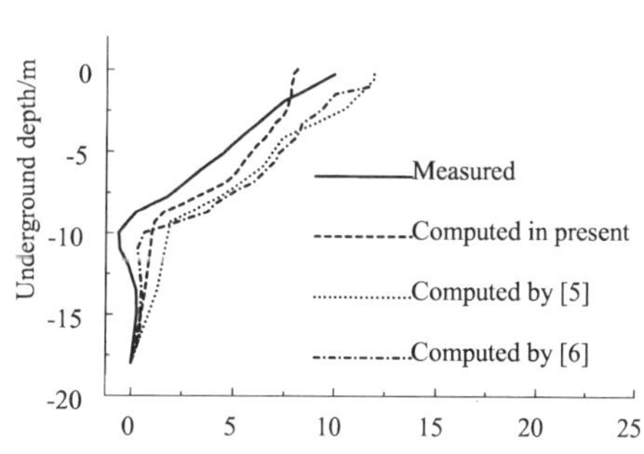

Figure 1. Distribution of deflections

sandy silts and silty clays. The 9.2m deep excavation was constructed in five lifts with each lift approximately 1.83m thick. The nails were 6.1m in length, inclined at 20° , and at a spacing of 1.8 × 1.8m in a gird pattern. According to the survey readings, a horizontal deflection profile and a nail tension profile about the underground depth are shown in Fig. 1 and Fig. 2 respectively.

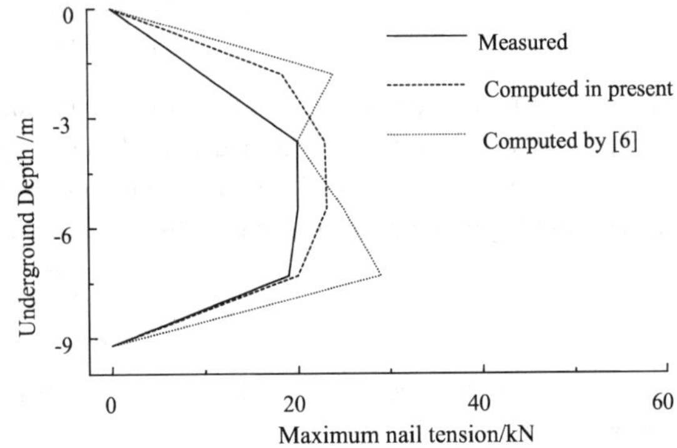

Figure 2. Comparison of maximum nail tensions

The horizontal deflection profile and the nail tension profile about the underground depth, computed by present analysis, are also shown in Fig. 1 and Fig. 2 respectively. As may be seen, the computed results are in close agreement with the measurements.

3.2 Comparison with the two-dimensional FE analysis results

As shown in Fig. 1 and Fig. 2, the present analysis has a better agreement results than the two-dimensional FE analysis compared with the measurements.

3.3 Parametric studies

The primary parameters in the present analysis were adopted as follows. The 9m deep excavation was constructed in five lifts with each lift approximately 1.8m thick. The nails were 6m in length, installed horizontally, and at a spacing of 1.8×1.8m in a gird pattern. A set of soil parameters is, c=3kP_a, Φ_1=28° , $\Delta\ \Phi$=5° , R_f=0.8, K=800, K_{ur}=1600, n= 0.6,

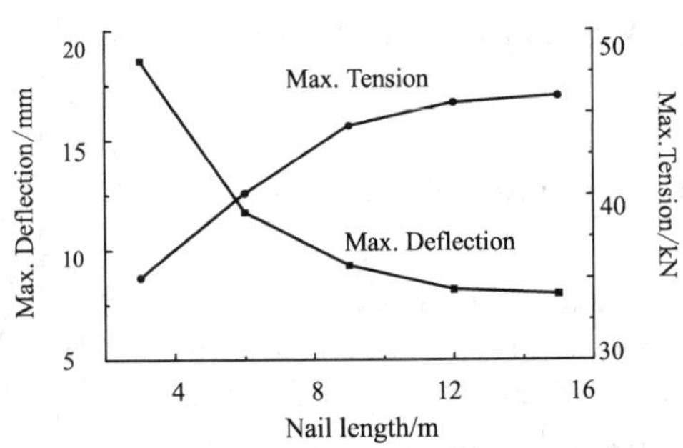

Figure 3. Influence of nail length

K_b=500, m=0.5, γ=19 kN/m^3, K_0=0.67 and $_{ult}$=80kP_a.

Based on the information, the effects of the change of one of the parameters on the stress and strain of soil nailing system are examined.

(a) Influence of nail length

Different cases, in which the nail length was changed from 3m to 15m, were analyzed. The maximum deflection and the maximum nail tension corresponding to the nail length are shown in Fig. 3. In order to control the deflection of excavation, it is essential to increase the nail length properly. However, the deflection is not direct proportion to nail length, so the deflection controlling is not very effective and the strength of the nail can not be brought into full play because of a little increasing of the nail tension.

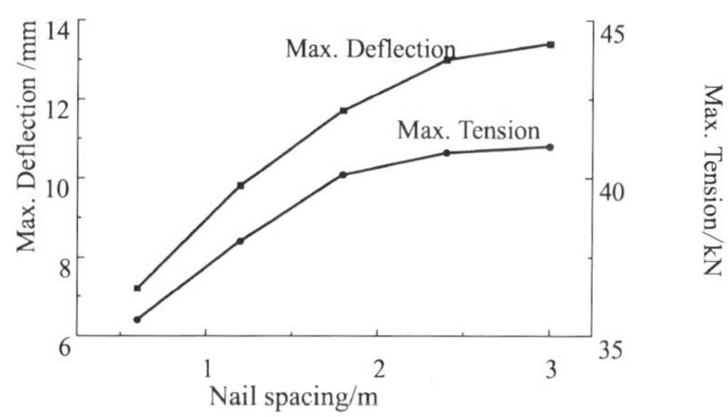

Figure 4. Influence of nail spacing

(b) Influence of nail spacing

Different cases, in which the spacing was changed from 0.6m to 3.0m, were then analyzed. The relationship of the maximum deflection and the maximum nail tension with the nail spacing are shown in Fig. 4. From the view of soil/nail interactions, if the nails are in a fine gird pattern, they will bear the excavation loading together, which may be effective in restrained the deflections.

(c) Influence of nail inclination

Different cases with different inclinations from 0° to 30° were also analyzed. The record of the maximum deflection and the maximum nail tension about the nail inclination are shown in Fig. 5. As may be observed, the nails should be installed horizontally or at a little inclination to reduce the deflections.

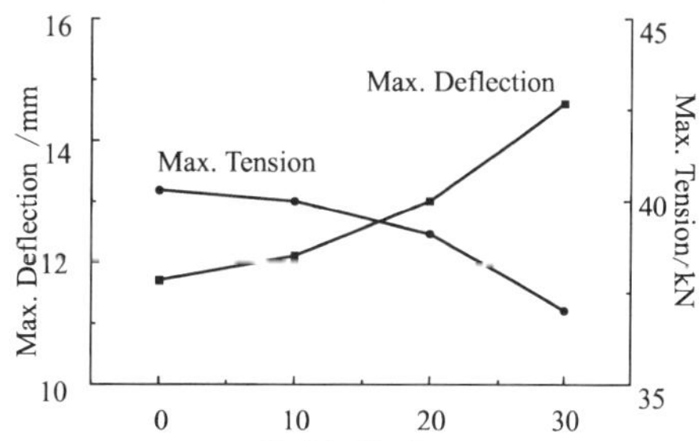

Figure 5. Influence of nail inclination

4 Conclusions

The present analysis has a better agreement results than the two-dimensional FE analysis compared with the measurements. Although it is essential to increase the nail length properly to control the deflections, the deflection controlling may not be very effective and the strength of the nail may not be brought into full play. If the nails are in a fine gird pattern, they will bear the excavation loads together, which may be effective in restrained the deflections. The nails should be installed horizontally or at a little inclinations to reduce the deflections.

References

1. Zhang, M. J., Song, E. X., Chen, Z. Y. Ground movement analysis of soil nailing construction by three-dimensional(3–D)finite element modeling (FEM). Computers and Geotechnics. 1999(25):191~204
2. Zhang, M. J., Song, E. X. Analysis of deflection behaviors of soil nailing by FEM. China Civil Engineering Journal. 1999, 32(6): 59~63 (in Chinese)
3. Schlosser, F. & Unterreiner, P. French Research Program CLOUTERRE on Soil Nailing. Geotechnical Special Publication, ASCE, 1992, Vol. 2, No.30: 739~749
4. Zhu, B. L., Shen, Z. J. Computational soil mechanics. Shanghai: Press of Science and Technology of Shanghai,1990 (in Chinese)
5. Shen, C. K., Bang, S., Romstad, K. M., & etc. Field measurements of an earth support system. J. Geotech. Eng. Division, ASCE, 1981, 107(12): 1625～1642
6. Song, E. X., Chen, Z. Y. Soil nailing technology and its analysis by FEM. Geotechnical Investigation and Surveying, 1996, 139(2): 1~5 (in Chinese)

FINITE ELEMENT MODE TO SIMULATE THE UNDERGROUND OPENING

QINGKAI CHEN XINGMIN ZHANG
Northeastern University Shenyang, P. R. China
E-mail:Cqk1228@sohu.com

XIANHONG MENG
314 WBB Building, 135S 1460E,SLC, UT 84112,USA
E-mail: mengxh@hotmail.com

The stress field around an opening under ground is calculated by using finite element method. The result shows that tangential stress is a maximum at the horizontal boundary of the opening, whereas radial stress is only around 500 psi. The maximum shear stress also occurs at the boundary about 45 degrees from the horizontal direction. The radial stress is small around the opening than far away from it. The FEM result has almost the same result as that from the theory.

1 Introduction

The problem of designing an underground opening in massive rock is one of determining the matrix stress that exists in the rock around the opening. Only when the maximum stress is less than the ultimate strength of the rock can the opening be stable and remain open without artificial support. The objective of this paper is to compute the stress fields about a circular hole excavated in a homogenous, isotropic, linearly elastic material by FEM method. Then the results from FEM and from theory will be compared in order to make some analysis about the discrepancy of the FEM method from theory

2 Basic assumption

1). The massive rock obeys Hook's law and is isotropic and homogeneous with respect to its elastic properties.
2). The opening is an infinite medium.
3). The cross section of the opening is represented by simple geometric shape, such as circle.
4). The axes of the cross sections of the openings are horizontal.
5). The vertical stress is equal to the weight of the rock above the section; thus, $\sigma_{yy}= \gamma * h$

6). The horizontal stress is given by σ_{xx}= M*σ_{yy}. There are three types of stress field represented by M=0, M=1/3 and M=1. We choose M=0 to represent stress field occurring at shallow depths or near vertical free surfaces.

3 Initial and boundary condition

1). Use an initial stress state of σ_{yy}=-1000psi and all other stresses equal to zero.
2). Use boundary condition which prevent displacement perpendicular to the sides of the mesh at the mesh sides.
3). Use a Young's modulus of 2.4(10^6) psi and a Poisson's ratio of 0.20.

4 Solving steps

1). mesh set up. Arrange the finite element around the circular of the excavated hole. Use 18 elements for each ¼ ring around the hole from horizontal to vertical direction. The total element for the ¼ area around the hole is 360. Also use the same scheme to arrange the node with the number of totally 399. The mesh is shown in figure 1.

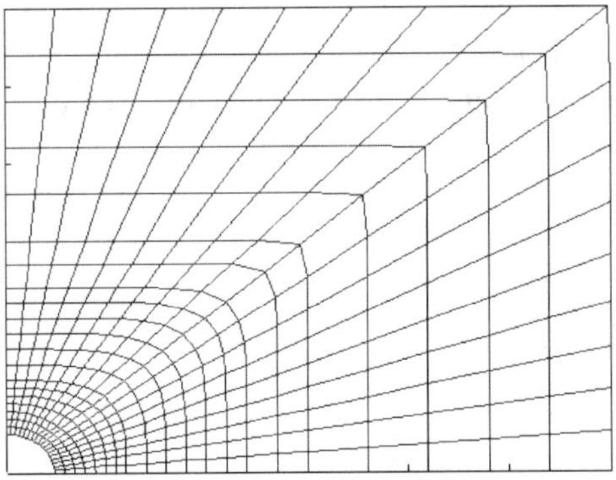

Figure 1. Finite element mesh

2). programming. Finite element programs are compiled with Gauss-seidel integration as the solver.
3). input files set up. The files to input node coordinate, element node, material properties, initial conditional, boundary condition and main guideline are set up.
4). output files produced. The displacement, stress, strain and principal stress file are produced when the program is executed.
5). theory calculation of stress by using:

- $\sigma_{yy}= \gamma h$; $\sigma_{xx}= M\sigma_{yy}$; M=0.

- $\sigma_r=(\sigma_{xx}+\sigma_{yy})(1-a^2/r^2)+ (\sigma_{xx}-\sigma_{yy})(1-4a^2/r^2+3a^4/r^4)\cos 2\theta$
- $\sigma_\theta=(\sigma_{xx}+\sigma_{yy})(1+a^2/r^2)- (\sigma_{xx}-\sigma_{yy})(1+3a^4/r^4)\cos 2\theta$
- $\tau_{r\theta}=(\sigma_{yy}-\sigma_{xx})\ (1+2a^2/r2-3a4/r^4)\sin 2\theta$

where a=hole radius, r=radial distance from center of hole.

6). Stress transference by the following formula:

- $\sigma_r=\sigma_{xx}\cos^2\theta+\sigma yy\sin 2\theta+\tau_{xy}\sin 2\theta$
- $\sigma_\theta=\sigma_{xx}\sin^2\theta+\sigma_{yy}\cos 2\theta-\tau_{xy}\sin 2\theta$
- $\tau_{r\theta}=((\sigma_{yy}-\sigma_{xx})\sin 2\theta)/2+\tau_{xy}\cos 2\theta$

5 Computing result and analysis

1). Stress field. The stress of elements around the hole is computed with both theoretical and FEM method. While computing all the stress components, both of the methods use the center coordinates of each element as reference. So all the calculating points are located on the circle with radius of 3.25 and with 5 degrees increase from 2.5 degree to 87.5 degree. The principal stresses (SIG1 & GIG3), tangentialσ_θ(S_tht), and radial stressσ_r(S_rr) field are shown in figure 2.

2). Radial and tangential normal stress as function of angel from the horizontal x-axis to the vertical y-axis by FEM and theoretical methods are shown in figure 3.

3). Shear stress as function of angel from the horizontal x-axis to the vertical y-axis by FEM and theoretical methods are shown in figure 4.

4). circumferential normal stress concentration from the first ring of elements over the plot of the analytical solution for M=0 is shown on the figure 5. Because the FEM and analytical solution both use the element centers of the first ring element as the reference points to calculate the circumferential stress of each element, there is little discrepancy observed between the two curves from the two methods. The reason to produce the discrepancy may be the input data error, program mistakes and the computing error of the computer.

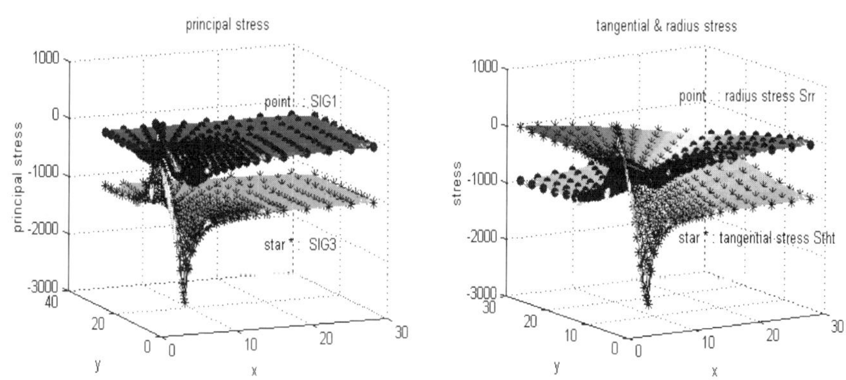

Figure 2. Stress field around the opening

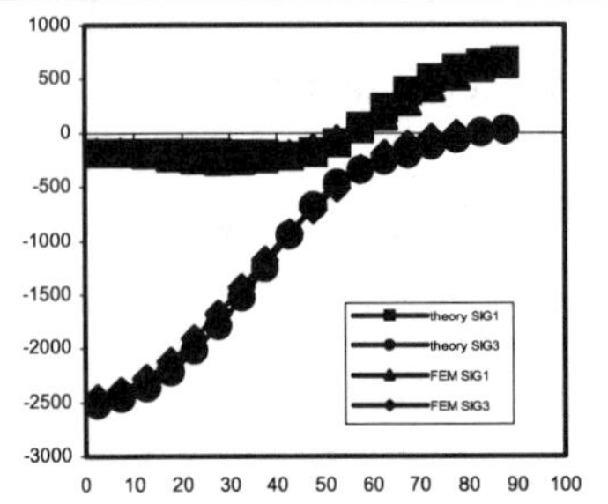

Figure 3. Relationship between normal stress and angle

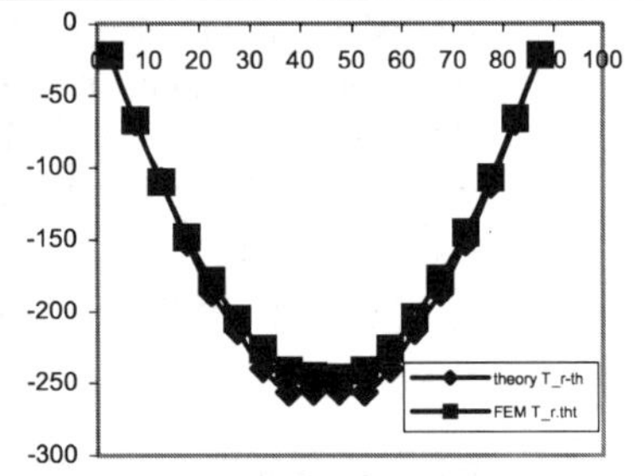

Figure 4 Relationship between stress and angle

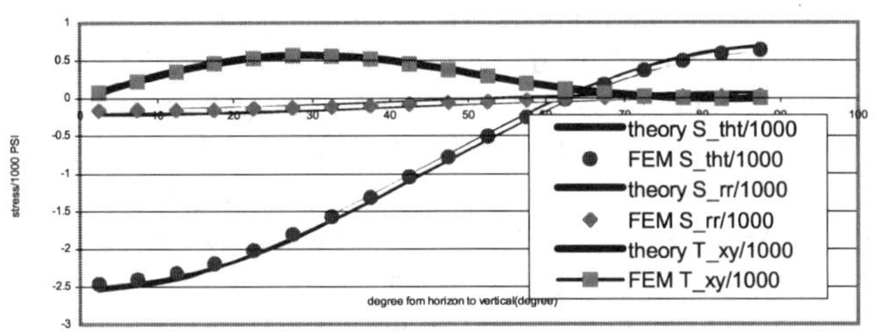

Figure 5. Stress concentration around the opening

6 Conclusion

The result and analysis show that the finite element method has little error and is reliable to calculate stress and stress concentration of the excavated hole. tangential stress is a maximum at the horizontal boundary of the opening, whereas radial stress is small around the opening. The maximum shear stress also occurs at the boundary about 30 degrees from the horizontal direction

References

1. Alfreds R.Jumikis. 1983. Rock Mechanics, Gulf Publishing Company, Texas, USA, pp.312-335
2. W.G.Pariseau. 2002. Introduction To Finite Element Modeling In Geomechanics.

STABILITY CRITERION OF JINPING ARCH DAM BY FINITE ELEMENT ANALYSIS

XIN CHEN
Department of Hydraulic Engineering, Tsinghua University, Beijing, 100084, China
E-mail: chenxin99@mails.tsinghua.edu.cn

WEIYUAN ZHOU
Department of Hydraulic Engineering, Tsinghua University, Beijing, 100084, China
E-mail: zhouwy@mail.tsinghua.edu.cn

In this article, the stability and bifurcation criterions is studied for arch dam structure. The bifurcation model in failure status is derived from the uniqueness of the equilibrium and the relation between positive definiteness of the structural tangent stiffness and bifurcation is described. Then the structural collapse criteria are discussed in the energy mode. A stability criterion based on the connection of yield region is proposed according to the implementation of elasto-plastic FEM (finite element method) numerical calculations. In accordance with this judging method, the nonlinear analysis of FEM is used for Jinping arch dam analysis. The yield region is sought and the failure process of the dam is simulated by the overloading analysis.

1 Introduction

Overloading factor of arch dam is considered in engineering science. In this article, bifurcation and collapse criteria of overall stability is studied by using the uniqueness of bifurcation, and stability of equilibrium and bifurcation criterions is discussed based on the energy mode. Jinping arch dam is studied as engineering sample on yield region development of dam.

2 Stability of Equilibrium and Bifurcation Criterions

Define Π as potential function of a structure. It can be expressed as [1]

$$\Pi = \int_V W dV - \int_V F_i u_i dV - \int_S T_i u_i dS \tag{1}$$

Right parts of the equal mark show the external work done by total strain energy, body load and surface load. From the virtual work principle, the sufficient and necessary condition is

$$\delta\Pi = \int_V \sigma_{ij} \delta\varepsilon_{ij} dV - \int_V F_i \delta u_i dV - \int_S T_i \delta u_i dS = 0 \tag{2}$$

The sufficient condition can be altered into $\delta^2\Pi > 0$ considering about Taylor series. They both establish the minimum potential energy principle. If the condition is $\delta^2\Pi = 0$, the system is reaching the critical point of stability. In such status, two possibilities including stability and instability exist in the system.

3 Bifurcation Criterions of Localization Failure

Noncontinuous bifurcation of rate of displacement $\dot{u}_i$ and its gradient $\dot{u}_{i,j}$ appears in internal singularity surface Ω when keeping on loading. In both sides of singularity surface, displacement field keeps continuous, whereas displacement gradient $[\dot{u}]$ will appear noncontinuous skip. Defining tangent stiffness tensor $T_{il} = n_j D_{ijkl} n_k$, appearance of localized bifurcation means that nonzero result exists in expression $T_{il} c_l = 0$ in deformation period. So the problem can be transferred to finding zero eigenvalue of tensor T_{il}.

$$\det T_{il} = 0 \tag{3}$$

4 Judgment of Overall Stability in Numerical Calculation

Evaluation of stability in civil engineering, which may be disturbed by many external factors, is quite complex and undependable. Judgment of overall stability can be realized through the points below.

4.1 Judgment of Iteration Convergence in Numerical Calculation

If the qualification satisfies $\Delta F > 0, \Delta\delta \to \infty$, elasto-plastic iteration appears divergence, the structure is considered as failure. In other words, an equilibrium status cannot achieve in the iterative process.

4.2 Judgment of Overall Stiffness Singularity Based on Energy Criterion

If external work Π_R is not more than increment of internal energy Π [2], civil engineering system is regarded as stable. That is to say, if there exists a set of virtual displacement to satisfy the condition that virtual work done by external load is greater than increment of internal energy, excess parts are transferred to kinetic energy of the system. The system is unstable. So failure criterion of the system is

$$\int_v \delta [U]^T [K] \delta [U] dV < 0 \tag{4}$$

Where K is overall stiffness matrix in FEM analysis. Stability of a system can be judged directly by positive definiteness of K. Defining that $\lambda_k (k=1, 2, \ldots, n)$ is n eigenvalues of K, since $\det K = \lambda_1 \lambda_2 \ldots \lambda_n$, stability criterion is

$$\det K = 0 \tag{5}$$

4.3 Connection of Yield Region

In proceeding of arch dam analysis, yield region will expand and connect, which can result in losing of structure capability of dam abutment. If yield region is not completely connected, the structure can still considered of supporting load.

5 Bifurcation and Failure Used in FEM [3]

In the case of calculating overloading capability of high arch dam, stability criterion can be applied to judge positive definiteness of sub-stiffness matrix $[k]$. If $[k]$ is not positive, relevant element is considered of destroyed. Judgment of local stability should carry out by searching direction $\boldsymbol{n}$ of local cracks based on tangent stiffness tensor D_{ijkl}. If $\boldsymbol{n}$ satisfies requirement that tangent stiffness tensor $T_{il} = n_j D_{ijkl} n_k$ loses its positive definiteness localization failure occurs.

6 Case Study of Jinping Arch Dam

In this article, stress and displacement of Jinping high arch dam is researched based on FEM model. Overloading analysis [4] is also applied to failure development of dam in limit status. In overloading process, one water pressure will be put in every loading step to consider about point safety factor (P.S.F.) and yield elements.

Table 1. Limit Load of the Dam in Failure Simulation Process

	Upstream surface	Downstream surface
Normal water pressure	Surface stress is in normal level and P.S.F.>1.0	P.S.F.>1.0
2nd water pressure	P.S.F.>1.2	P.S.F.>1.0， yield region appears around the dam
3rd water pressure	General P.S.F.>1.2 and yield region appears around the dam	Yield region fills in the dam boundary, P.S.F.=1.2 and shear band comes into being
4th water pressure	Yield region of first stress σ_1 becomes continuous	Yield region of third stress σ_3 becomes continuous

From the table above, yield region of the dam appears at about 4th water pressure. So this pressure can be regarded as the limit load.

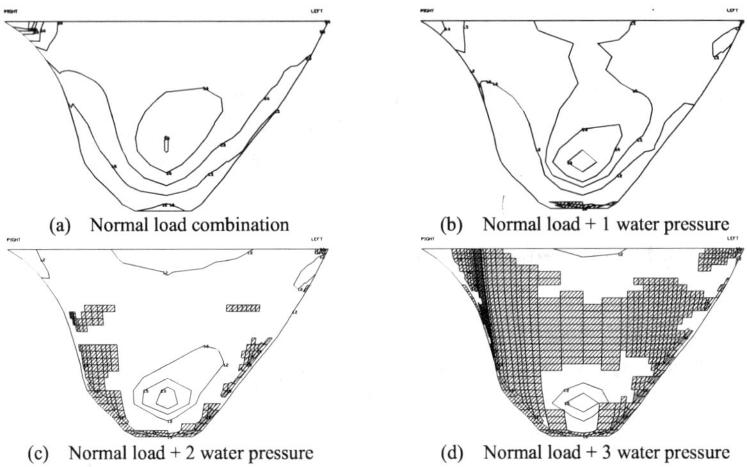

(a) Normal load combination

(b) Normal load + 1 water pressure

(c) Normal load + 2 water pressure

(d) Normal load + 3 water pressure

Figure 1. P.S.F. and yield region of downstream surface

7 Conclusion

In this article, failure mode of structure is discussed from the point of energy with application on Jinping arch dam. The scope of safety factor and yield region calculated by overloading analysis satisfies rules of dam failure. The result indicates the feasibility of this method applied for arch dam research.

References

1 Roclfstra F.E. And Wittmann F.H. (1986). Numerical method to Link Strain Softening with Failure of Concrete. *Fracture Toughness and Fracture Energy of Concrete*, edited by Wittmann F. H., Elsevier

2 Rots J.G. (1986). Strain Softening Analysis of Concrete Fracture Specimens. Fracture *Toughness and Fracture Energy of Concrete*, edited by Wittmann F.H., Elsevier

3 Tianchi Qiang (2001). A Study on Stability Analysis System for Large Arch Dams Based on Nonlinear Failure Simulation and Feedback Analysis. *Ph. D. Thesis*, Dept. of Hydr. Engng, Tsinghua Univ. in Chinese

4 Weiyuan Zhou, Ruoqiong Yang (1988). Overall Three Dimension Nonlinear FEM Analysis in Concrete Arch Dam (TFINE FEM Program Series). Dept. of Hydr. Engng, Tsinghua Univ. in Chinese

NUMERICAL ANALYSIS OF DEFORMATION AND FAILURE OF SOIL-NAILED SLOPE

ZHENGZHAO LIANG CHUN'AN TANG SHUHONG WANG MINGLI HUANG
Center for Rockburst & Induced Seismicity Research, Shenyang, 110004, China

Based on the mechanics characteristics of soil nailing, a FEM simulative code SNDFAS with friendly pre-process and post-process interface is developed. To introduce the excavation and construction of soil nailing, mechanics and simulation method of soil nailing are investigated. By simulating the deformation and failure in the excavation process with and without soil nailing respectively, work performance of soil nailing in the composite soil is discussed, and it provides a reference for design.

1 Introduction

Mechanics of soil nailing is influenced by many factors such as soil nail length, soil nail spatial distribution, and physical properties of soil etc, so it is not realistic to take into account all of them. Different ultimate equilibrium methods are commonly used to analyze slope stability with soil-nailed slope, each of them belong to either internal instability analysis or external instability analysis[1-3]. According to the safety factor derived from ultimate equilibrium method, foundation slope may be prevented from slipping in some degree, but it does not mean the analysis method is reliable and reasonable to reflect soil nailing mechanics. Moreover, it is very important to finish every step of excavation in time and little delay may lead to fatal damage of bracing system, so ultimate equilibrium methods should be used with much caution. Finite element method becomes so useful to investigate complicated problems in rock and soil engineering for many years, especially in dealing with complex boundaries in geometry, which is much difficult to be resolved by analytic method. Chen Z.Y. and Song E.X. etc. studied composite soil nailing in deep foundation pit excavation[1]. Shen C K and Shafiee investigated influence of some parameters on the work performance of soil- nailed wall[2]. Even though sufficient researches have been conducted to study the soil nailing, the mechanics and the parameters related to deformation and failure still remains unknown due to the complicated interaction between its components. Moreover, FEM has many disadvantages when applied. On one hand, it is not easy to prepare all data for calculation and errors when initial data input are difficult to avoid and any of them can lead to wrong results. On the other hand, instead of numerical results obtained directly from FEM, visual results are needed to analyze the deformation and failure state clearly and directly, such as figures, pictures and tables. In this paper, by comparing displacement and failure mode with soil nailing with those without soil nailing, a numerical code SNDFAS(Soil Nailing Deformation and Failure Analysis System)is introduced to study the soil nailing performance in excavation process. In SNDFAS, not only does data input become more convenient and simple, but also grids are plotted automatically. Fluid underground, surface loading, foundation and interactions between soil and soil nails can all be taken into account. Different from other FEM code, SNDFAS can simulation both displacement and stress field and failure state in each excavation step.

2 Model Description and simulation process

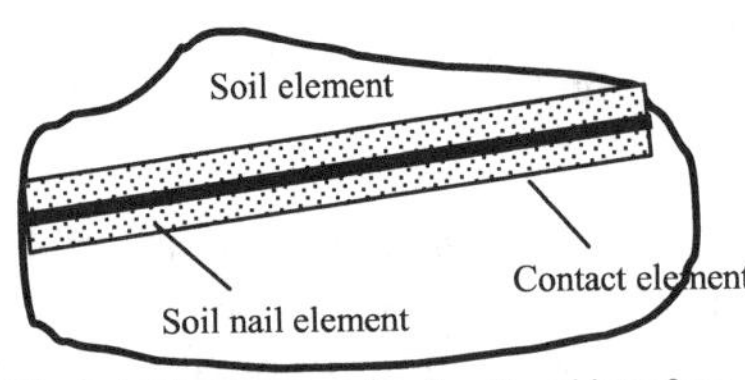

Figure1. Sketch map of soil nail and interface

There are four types of element models: soil element, topping element, soil nail element and contact element. Topping element is used to describe the support function of reinforced concrete on the top of pit slope. With reinforcing steel bar, concrete's strength gets too high to fracture and failure, and in practice topping concrete is scarcely ever found to collapse, so only strength verification is needed. Soil nail element can be looked on as bolt element in other finite element methods. Each soil nail element is composed of two nodes and stress bond between them, and each soil nail is composed of many soil nail elements. By comparing the soil nail strength with the tens stress between two nodes, we can get the safe factor of it. Concrete mortar injected into nail hole becomes solidity, which is different both from soil nail and soil. This type of element that we call contact element is a quadrangle. Due to its thinness, it is regarded that the stress on the contact element's surface acts on the soil nail element immediately, and this provides boundary conditions for calculating stress of adjacent soil nail element.
Other researchers presented many constitutive laws on soil-like materials in the last decades, and these constitutive laws can all explain one or more constitutive characteristics of soil. Users can select either nonlinear elastic model or elasto-plastic model in SNDFAS. Duncan-Chang model and Drucker—Prager model are used in nonlinear and plastic-elastic constitutive law respectively. As to failure criteria, Mohr-Coulomb model with tension cut-off is taken into consideration in SNDFAS.

Usually grids and nodes' generation is resolved by virtue of other powerful tools, for example ANSYS. In SNDFAS, grid plotting and node generation become an automatic thing. Elements in the current excavation zone are excluded from the total matrix, and as the excavation process, new elements of topping concrete and soil nails should be added to the current elastic modulus matrix for FEM calculation. Thus the simulation process is accomplished. For example, if in i-1 step the results is $\{\sigma\}i\text{-}1$, $\{\varepsilon\}i\text{-}1$, and $\{\delta\}i\text{-}1$, respectively presenting the current stress matrix, strain matrix and displacement matrix, in the n step, regarding the results of n-1 step as the initial value, calculating program can start when excavation elements are got rid of and newly generated elements are added into. Then the result in the n step is obtained, $\{\sigma^*\}i=\{\sigma\}i\text{-}1+\{\Delta\sigma\}$, $\{\varepsilon^*\}i=\{\varepsilon\}i\text{-}1+\{\Delta\varepsilon\}i$, $\{\delta^*\}i=\{\delta\}i\text{-}1+\{\Delta\delta\}i$. $\{\Delta\sigma\}i$, $\{\Delta\varepsilon\}i$, and $\{\Delta\delta\}I$ are not the final results of the i step, and stress adjustment is necessary when any of the elements satisfies the failure threshold described by failure criteria. Equivalent load of excavation soil in each step is obtained the function below:

$$\{f\}=\sum_{i-1}^{n}\iint_{e}[B]^{T}\{\sigma\}dxdy$$

where n is the total step, [B] is strain matrix, and $\{\sigma\}$ is stress matrix. Equivalent load, with direction reversed, is put on the nodes adjacent to the elements excavated.

3 Simulation analyses

A great number of experiments have confirmed that strength of soil-nailed wall is double of without natural soil wall. Due to the restriction and control feature of the composite soil of soil, soil nail and concrete, the soil-nailed slope has a stronger capability to resist slide and fall down.

Deformation and failure state of slope with soil nailing compared with those without soil nailing are analyzed. The depth in the first excavation step is 2.0m, and in the following steps it is 1.5m. Horizontal displacement of almost all points on the left side border of slope is much more smaller when the slope is soil nailed. The final horizontal displacement curve is shown in Fig.2, and the maximum displacement decreases from about 27mm to 18mm. At the bottom of the slope, the displacement decreases about 4mm. In the vertical direction, the decrease on the top of the slope and along the slope is not so much. So we can arrive at a conclusion that soil nailing has much more effective

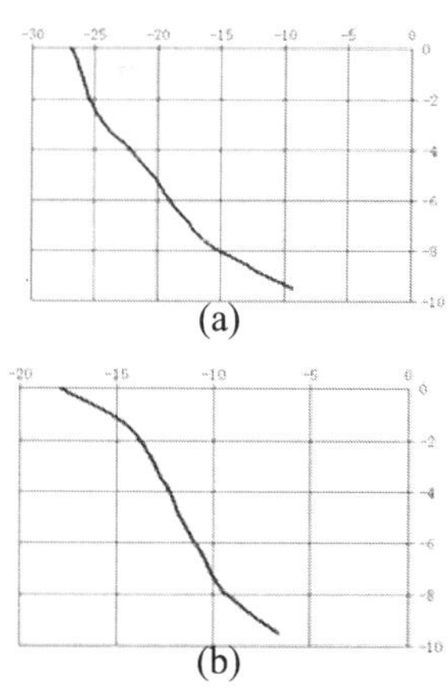

Figure2. Horizontal displacement curve of slope
(a) Without soil nail
(b) With soil nailing

restriction in the horizontal displacement than vertical displacement, as is shown in Fig.3. Without soil nailing, the failure area is more widely distributed than with soil nailing, and it can be found that the slide plan has formed and traverses the first four lines of soil nails.

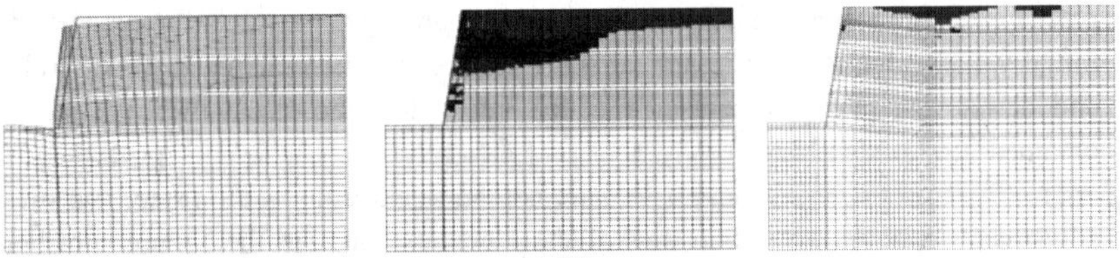

(a) (b) (c)

Figure 3. Slope deformation and failure state
(a) Slope deformation without soil nailing (b) Failure state of slope without soil nailing (c) Failure state with soil nailing

4 Conclusion

Based on FEM and visual technique, mechanics and simulation method of soil nailing are investigated. Soil nailed slope has a stronger capability to prevent from sliding and soil nailing has a preference to control the displacement in the horizontal direction. With SNDFAS code, further simulations in other papers have explained which and how parameters related to soil nail, soil properties, and fluid underground influence the deformation and failure state of soil nailed slope. These have more close relationship with soil nailing design.

5 Acknowledgements

The authors are grateful for the support of this work from the Chinese National Key Fundamental Research "973" Program (G1998040700) and the Natural Science Foundation of China (No.49974009, 50134040 and 50174013).

Reference

1. Song E X and Chen Z Y, Soil nailing bracing and FEM analysis[J], Engineering Reconnaissance, 1996(2).
2. Shen C K, et al. Ground Movement Analysis of Earth Support System, J. Geotech. Eng. ASCE, Dec, 1981.

THREE-DIMENSIONAL THERMAL STRESS NUMERICAL SIMULATION SYSTEM FOR GEOTHERMAL DEVELOPMENT

CHENGLUN LIU[1], ZHONGCHENG QIN[1], YOSHIAKI MIZUTA[2], KEN KURIYAMA[2]
[1] *Shandong University of Science and Technology, Taian, 271019, China;*
[2] *Faculty of Engineering ,Yamaguchi University,Ube,755,Japan*

In order to solve the problem of rock thermal stress in geothermal development of high temperature rock mass, according to the characteristic of high temperature rock mass and manual geothermal reservoir, the temperature field was simulated by using the boundary element numerical simulation system of 3D non-steady heat conduction; As an equivalent static elasticity problem, the stress field was simulated by using 3-D FSM (Fictitious Stress Method) and DDM (Displacement Discontinuity Method). The simulation system is reliable and effective

1 Introduction

Before excavation of high temperature rock mass geothermal, the feasibility must be carried through. First, the rational excavated extent (i.e. output power) must be determined and its available life must be predicted. So, the boundary element numerical simulation system of 3-D non-steady heat conduction problem[1][2] and the FSM·DDM[3][4][5] of 3-D elastic problem have been used here. This paper mainly discussed the establishment of the numerical simulation system of uncoupled thermo-elastic problem. Namely, the displacement field and the temperature field are uncoupled, the temperature distribution of each time element can be calculated by BEM numerical simulation system of 3D non-steady heat conduction; and then carry through the numerical simulation of thermal stress and thermal deformation originated to it. The disposal method is same as the static elasticity problem.

2 Navier's equation of thermoelastic problem

Eq. (1) is Navier's equation for quasi-static uncoupled thermoelastic problem When there is no volume force. Let n represents the normal direction cosine on its boundary surface, and then the corresponded boundary constraints are as Eq. (2).

$$Gu_{i,kk} + (\lambda + G)u_{k,ki} + b_i - \beta\theta_{,i} = 0 \tag{1}$$

$$\left.\begin{array}{ll} X_i + \beta\theta n_i = \sigma_{ij} n_j & \text{on} \ \Gamma_p \\ u_i = \overline{u}_i & \text{on} \ \Gamma_u \end{array}\right\} \tag{2}$$

Where, λ, G are Lame′ elastic constants; $\beta = \alpha E/(1-2\nu)$, α is linear expanding coefficient, E is elastic module of surrounding rock, ν is Poisson's ratio of surrounding rock; $\theta(t,P) = T(t,P) - T_0$, $T(t,P)$ is temperature at point P at t

moment[1], T_0 is benchmark temperature; $u_i(u_i(P))$ is the displacement in i direction at point P in the region, $\bar{u}_i$ is the forced displacement; b_i is the body force in i direction; X_i is the surface force in i direction; Γ_p is the stress boundary, Γ_u is the displacement boundary; σ_{ij} is stress.

3 Establishment of integration equations

According to Duhamel's similar theory, and when the volume force $b_i = 0$, Equation (1) can be regarded as equal temperature static-elastic problem. Thus, the interior field integration equations of the induced displacement and induced stress corresponded to FSM·DDM are as follows:

$$u_l(P) = \int_{\Gamma_1} p_k(s')u^*_{lk}(P,s')ds' + \int_{\Gamma_2} D_k(s')v^*_{lk}(P,s')ds' - \int_{\Omega} \beta\theta_{,k}(\bar{t},P')u^*_{lk}(P,P')dP' \quad (3)$$

$$\sigma_{lm}(P) = \int_{\Gamma_1} p_k(s')\sigma^*_{lmk}(P,s')ds' + \int_{\Gamma_2} D_k(s')\eta^*_{lmk}(P,s')ds' - \int_{\Omega} \beta\theta_{,k}(\bar{t},P')\sigma^*_{lmk}(P,P')dP' \quad (4)$$

The boundary integration equation is

$$u_l(s) = \int_{\Gamma_1} p_k(s')u^*_{lk}(s,s')ds' + \int_{\Gamma_2} D_k(s')v^*_{lk}(s,s')ds' - \int_{\Omega} \beta\theta_{,k}(\bar{t},P')u^*_{lk}(s,P')dP' \quad (5)$$

$$\sigma_{lm}(s) = \int_{\Gamma_1} p_k(s')\sigma^*_{lmk}(s,s')ds' + \int_{\Gamma_2} D_k(s')\eta^*_{lmk}(s,s')ds' - \int_{\Omega} \beta\theta_{,k}(\bar{t},P')\sigma^*_{lmk}(s,P')dP' \quad (6)$$

Where, $\Gamma = \Gamma_1 + \Gamma_2$, Γ_1 is the dummy stress boundary, Γ_2 is the dummy displacement boundary; s' is the source point, s,P are the field points on the boundary and interior respectively; $u_l(P), \sigma_{lm}(P)$ and $u_l(s), \sigma_{lm}(s)$ are the displacement and stress of field point P and s respectively; u^*_{lk}, σ^*_{lmk}, v^*_{lk} and η^*_{lmk} are the fundamental solution of displacement and stress respectively[3][4]; $p_k(s')$, $D_k(s')$ are the dummy stress and displacement of s' respectively.

4 Establish the discretizatioin equation of integral equation

Suppose the boundary Γ of resolved object Ω is divided into N triangle elements (Δ_i, $i = 1 \sim N$), the interior is divided into n_v tetrahedron elements, then the discretization equation of corresponded interior and boundary integration equations (3),(4),(5),(6) can be combined as follows:

$$\left[P_x^j \; P_{yz}^j \; P_z^j \; P_{xy}^j \; P_{yz}^j \; P_{zx}^j\right]^T = \sum_{i=1}^{N}[T_{ji}][S_{ji}]\left[X_x^i \; X_y^i \; X_z^i\right]^T + \sum_{l=1}^{n_v}[R_{jl}][\sigma_T^l] \quad (j = 1, N) \quad (7)$$

$$\left[u_z^j \; u_y^j \; u_x^j\right]^T = \sum_{i=1}^{N}[E_{ji}][F_{ji}]\left[X_x^i \; X_y^i \; X_z^i\right]^T + \sum_{l=1}^{n_v}[L_{jl}][U_T^l] \quad (j = 1 \sim N) \quad (8)$$

Where, $[S_{ji}], [F_{ji}]$ are the influence coefficient matrixes in i local coordinate system respectively; $[T_{ji}], [E_{ji}]$ are the transform matrixes of stress and strain in i to j local coordinate system respectively; $[R_{jl}], [L_{jl}]$ are the transform matrixes of stress and strain in l to j local coordinate system respectively; [σ_T^l]and[U_T^l] are the stress and displacement matrixes caused by temperature variation of the volume element in l local coordinate system respectively, and can be determined by Gauss numerical integral method.

5 Solving procedure and example

(1) solving of the dummy stress and displacement of boundary elements

For the geothermal development problem, the boundary conditions are mainly stress boundary conditions, the solving equation group is

$$\begin{bmatrix} P_0 - P_z^{jo} - \beta\theta^j \\ -P_{yz}^{jo} \\ -P_{zx}^{jo} \end{bmatrix} = \sum_{i=1}^{N}[T_{ji}][S_{ji}]\begin{bmatrix} X_x^i \\ X_y^i \\ X_z^i \end{bmatrix} + \sum_{l=1}^{n_v}[R_{jl}][\sigma_T^l] \quad (j=1,N) \tag{9}$$

Where, P_0 is the water pressure inside the artificial geothermal reservoir; $P_z^{jo}, P_{yz}^{jo}, P_{zx}^{jo}$ are the original ground stresses of field point (barycenter of element j) in j local coordinate system respectively.

(2) solving of the redistribution stress and induced displacement of arbitrary point $j(x,y,z)$ inside the region and on the boundary

By using Eqs (10) and (11), the redistribution stress and induced displacement of arbitraries point can be solved.

$$\begin{aligned}\left[P_x^j\ Py_z^j\ P_z^j\ P_{xy}^j\ P_{yz}^j\ P_{zx}^j\right]^T &= \left[P_x^{jo}+\beta\theta\ \ Py_z^{jo}+\beta\theta\ \ P_z^{jo}+\beta\theta\ \ P_{xy}^{jo}\ \ P_{yz}^{jo}\ \ P_{zx}^{jo}\right]^T \\ &+ \sum_{i=1}^{N}[T_{ji}][S_{ji}]\left[X_x^i\ X_y^i\ X_z^i\right]^T + \sum_{l=1}^{n_v}[R_{jl}][\sigma_T^l]\end{aligned} \tag{10}$$

$$\left[u_z^j\ u_y^j\ u_x^j\right]^T = \sum_{i=1}^{N}[T_{ji}][S_{ji}]\left[X_x^i\ X_y^i\ X_z^i\right]^T + \sum_{l=1}^{n_v}[L_{jl}][U_T^l] \tag{11}$$

An example is shown here. There is a granite solid sphere with radius b(2.0m) and temperature T_0(100℃), put it into water with temperature Ts(0℃), then to calculate the interior temperature distribution and radial displacement after Δt (20) hours.

Using the boundary element thermal stress numerical simulation system of 3D non-steady heat conduction developed by authors to simulate the above problem, and then compares the simulated results with its analytical solution.

The basic parameters for numerical simulation are shown in table 1.

Table 1 the basic parameters for numerical simulation

Rock physical and mechanical property parameters				numerical analytic parameters	
Heat conduct coefficient K(m²/h)	Linear heat expansion coefficient α(1/ □)	Elastic modulus E(MPa)	Poisson's ratio ν	Numbers of boundary element	Numbers of volume element
0.004	0.000011	14100.0	0.104	816	15504

The comparing results of the stress simulation with its analytical solution are shown in Fig.1; Fig.2 shows the displacement numerical simulation solution.

6 Existing problems and improvement direction

Although the simulation system has high calculating precision, it still has some problem. On one hand, the boundary elements are too much and decrease the calculation speed; On other hand, the temperature, also the thermal stress and displacement varies obviously near the boundary, but the calculation errors on there are more large. Therefore, the following two aspects should be probed in the future.

(1) The volume element should be divided rationally.

(2) The Gauss numerical integration point and difference parameter should be selected in reason.

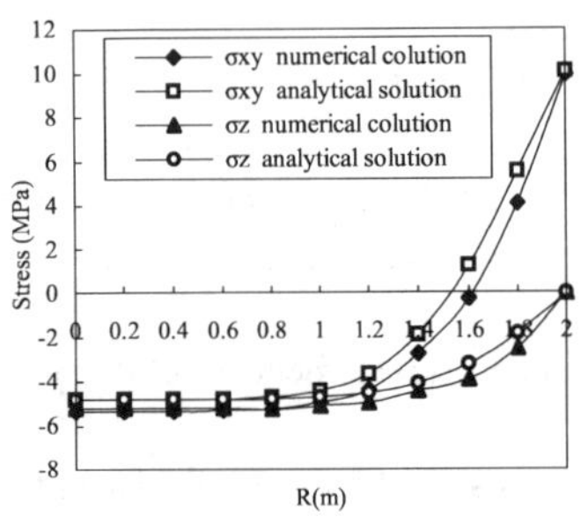

Figure1. Stress distribution on sphere surface

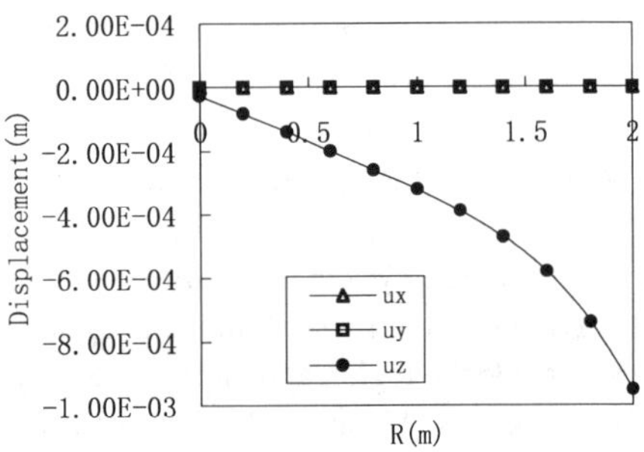

Figure2. Displacement distribution on sphere surface

References

1.Chenglun Liu, Ken Kuriyama, Yoshiaki Mizuta: Development of Numerical Codes by Direct Boundary Integral Method for Three-dimensional Non-steady Heat Conduction. Journal of the Mining and Materials Processing Institute of Japan. Vol.114, pp.225~228, 1998

2.Chenglun Liu, Ken Kuriyama, Yoshiaki Mizuta: Predictive Calculation of Three-dimensional Heat Conduction in Medium Including Fracture Boundary. Journal of the Mining and Materials Processing Institute of Japan. Vol.115,pp.73~76,1999.

3.Kuriyama.K and Mizuta.Y: Three-dimensional Elastic Analysis by the BEM with Analytical Integrations over Triangular Leaf Elements, Int. J. Rock Mech. Min. Sci. & Geomech. Abstr. Vol. 32, No.1 pp.77-83,1995.

4.Chenglun Liu, Yoshiaki Mizuta, Ken Kuriyama,: Analytical Integrations for Three-dimensional Fictitious Stress Method Based on Kelvin Solution. Journal of the Mining and Materials Processing Institute of Japan. Vol.115,pp. 719~724,1999.

5.Yoshiaki Mizuta, Ken Kuriyama, etc.. Three-dimensional Numerical Analysis by Boundary Element Method for Geo-science. JASCOME , Vol.14,pp.105~110,1997

APPROACH TO PILLAR ROCKBURST AND ASSOCIATED SEISMIC BEHAVIOUR

SHANYONG WANG CHUNAN TANG SHUHONG WANG TIANHONG YANG MINGLI HUANG YONGBIN ZHANG

Center for Rock Instability and Seismicity Research, Northeastern University, Shenyang.P.R.China 110004

Based on the assumption that there exists a proportional relation between the number of microseismic events and microfractured elements, By using a developed numerical code, RFPA2D (Rock Failure Process Analysis), the process of the pillar rock burst is simulated and the simulated results reflect macroscopic failure evolution process induced by microscopic fracture and the spatial as well as the temporal distribution characteristics of acoustic emissions event hypocenter. Especially the heterogeneity of rock is considered which has a good influence on the AE resulted from the rupture of the rock.

1 Introduction

Rock bursts inside any mining excavation are essentially rock failures due to alteration of the virgin state of stress by mining and creating an opening through the rock mass. As mining depth increases progressively, so do the ambient stresses, depth increases progressively, so do the ambient stresses, which proportionately increases the severity of mine seismicity. Commonly recognized as a seismic event, a rock burst usually results in damage to underground workings caused by the uncontrolled disruption of the rock equilibrium and is associated with biolent release of strain energy stored in the rock body[1]

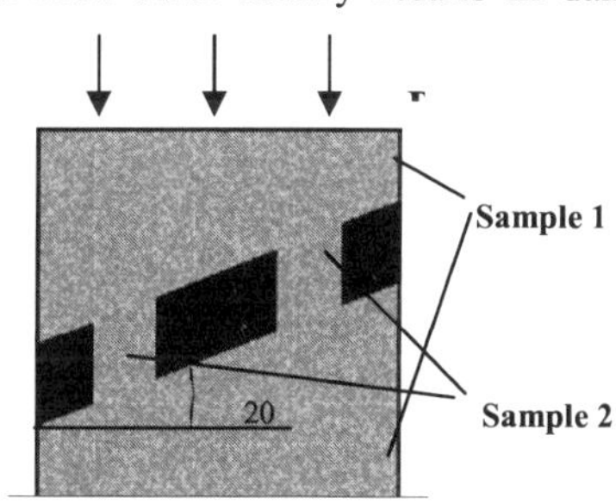

Fig.1 Numerical model and the boundary conditions

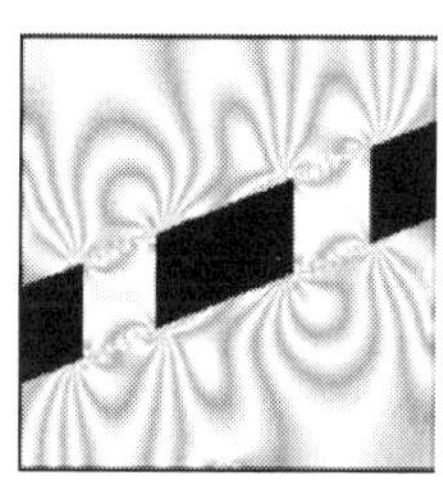

Fig.2 Photo elastic description of shear

In this paper the progressive failure process and associated sesmicity in rock samples is simulated with a newly developed numerical code, RFPA2D[2-3].

2 Numerical Simulation

Mathematically, the RFPA^{2D} code is a linear finite element method (FEM) and continuum mechanics method for numerically processing non-linear and discontinuum problems. In RFPA, the heterogeneity of material properties is introduced into the model by assigning the element strengths and elastic modulus as statistically distributed values within the model mesh using a Weibull distribution as defined in equation (1), and the distributed state of the elastic modulus and the shear stress can be shown from Fig.1 and Fig.2.Especially, Fig.2 represent virtually the shear distribution with the photo elastic description, the stress field like water-waves due to the heterogeneity of the rock and the loading pressure. For the double rock sample model, it is assumed that the two samples have two different Weibull probability density distributions .The rock properties are chosen so as to simulate the brittle behavior of rock material. For sample 1 as shown in Fig.1 to simulate the surrounding body of the rock mass, the mean compressive strength, δ_0 is equal to 200 MPa; the mean elastic constant E is equal to 60 GPa and Poisson's ratio υ =0.25。 The homogeneity index, m for the strength is 2.5 and for the elastic constant, 10. For sample 2 to simulate the failed body of the rock mass, the mean strength is 25% lower than sample 1 and the elastic constant is the same as sample 1. The size of the mesh was 200×200 with 40000 elements. The simulation was limited to a two-dimensional problem.

$$f(u)=\frac{m}{u_0}\left(\frac{u}{u_0}\right)^{m-1}\exp\left(-\frac{u}{u_0}\right)^{m} \tag{1}$$

Once the mechanical properties are assigned to the elements, the double specimen is compressed axially in displacement control mode to simulate a collapse process leading to rock bursts. The total boundary displacement 0.15 is divided into 100 steps, i.e. 0.15% for every step. Each increment of the displacement is regarded as a disturbance induced by mining. The numerical simulation is the same as for the elastic finite element method. When the twin sample is loaded step by step, the positions of the elements change, and so do the stresses and strains of the elements. Figure 3 shows the collapse process obtained in the simulation. Also presented in these plots are the severity stresses in Fig.3(c) which are calculated as the following equation:

$$\delta_s = \delta_1 - \gamma\delta_3 \tag{2}$$

where δ_s represents the severity extent of the elements to failure, which is shown by the brightness of the gray, γ is the ratio of the compression strength to the tensile and δ_1, δ_3 are two principal stresses, respectively. Figure 3(b) shows the locations of micro-seismics and their magnitude White circles represent the micro-seismics of the current step and black represent all previous steps. The stress-strain, seismic event and seismic energy release curve for a double sample is plotted in Fig.3 (a).

As can be seen from Fig.3, we simulate numerically the evolution process of the progressive failure leading to sudden fracture by using the Rock Failure Process Analysis code (RFPA) which allows us to examine the progressive failure process visually on a continuous basis during the loading process, when the applied boundary displacement increases, the micro-seismic events occurred as shown in Fig.3 (b) before step 19. Several micro-seismic events occur randomly and non-interacting fractures are observed. Some small clusters of micro-seismic form, extend or coalescence. Sequences of related micro-seismic events occur during this deformation stage. However, these sequences involve a limited number of events and would be isolated from each other. The principal characteristic of this stage would be associated with disorder and heterogeneity in the entire rock mass. As the boundary displacement continues to increase, the micro seismic events will become dense enough to begin interacting in steps 46. More damage occurs, forming a few larger clusters before fault nucleation. At this stage, the micro seismic activity in the failed body of the rock mass increases rapidly and forms several larger clusters. For a strong roof and floor, seismicity is almost exclusively confined to the pillars and failure occurs in shear, As soon as the subvertical fracture propagates (Fig.3 (c), step (46), shear failure, creating a wedge-shaped punch-failure mode, causes the right pillar to failure almost instantaneously (step 52). After some further shearing, the left pillar fails in the same failure mode during a single loading step (step 58), This second pillar failure is associated with more energy release because a more extensive shear zone forms than during the failure of the first pillar. Fig.3 (a) shows two seismicity peaks related to the failure of the two pillars. It is apparent from the Fig.3 (a) that before the two big rock bursts, there is always little micro-seismic events occurring, which is similar to the quiescence of seismic activity, and it is also found that following the big rock bursts, there is a relatively long time quiescence of seismic activity for the large energy release.

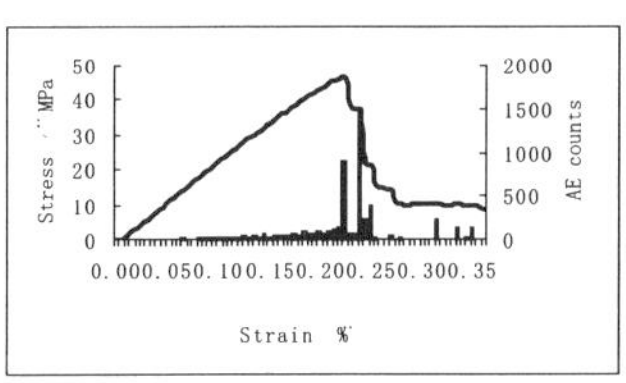

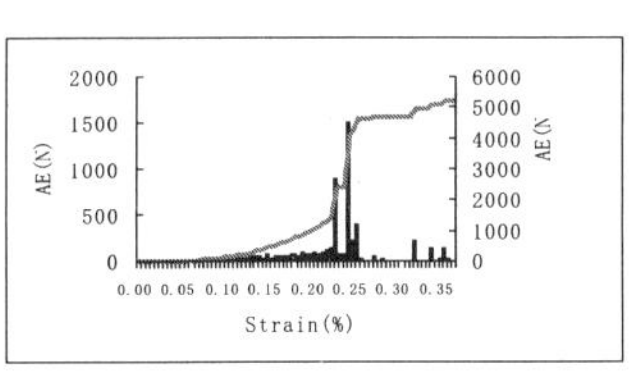

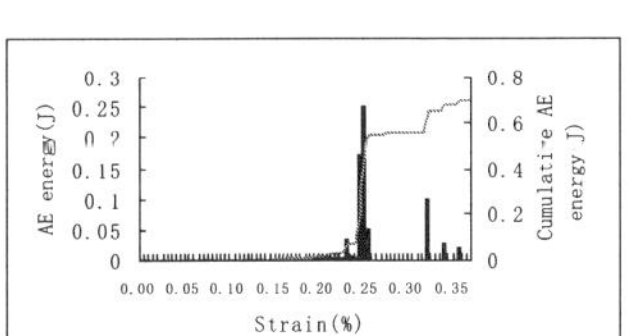

Fig.3a Stress-strain, seismic event and seismic energy release during pillar

3 Conclusion

The Rock Failure Process Analysis code (RFPA^{2D}) was applied to simulate the damage initiation and propagation causing seismic energy release during unstable failure (rock burst) of brittle rock. The progressive failure process leading to the development of a shear or fault zone in the pillar, and the eventually collapse of a heterogeneous pillar is modeled. Since the constitutive law for each element in the model is elastic-brittle, seismic energy is radiated whenever an element fails. Due to the heterogeneity of rock properties, seismic quiescence may occasionally occur within the nucleation zone. It was found that significant decreases followed by sudden increases in radiated seismic energy may be an indicator of potentially unstable nucleation.

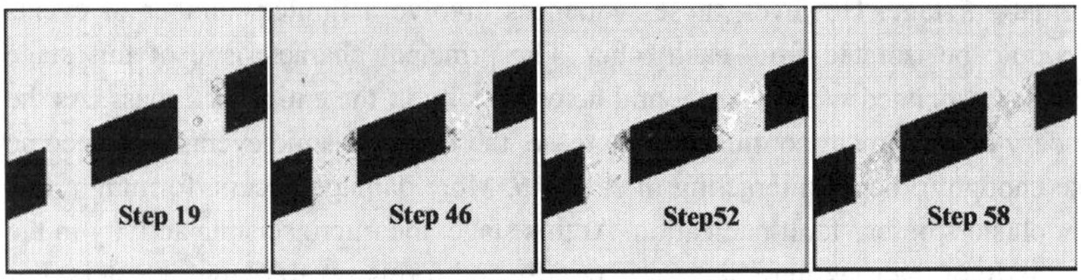

Figure 3b. Seismic source location during pillar failure with hanging wall

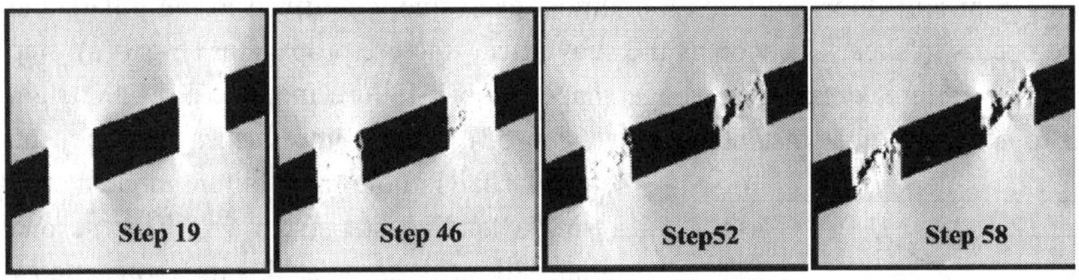

Figure3c. Pillar failure mode with hanging wall shear

Acknowledgements

The study was supported by the China National Key Fundamental Research "973 Programme" (No.G1998040700) and the China National Natural Science Foundation (No.50134040, No.50174013, No.49974009)。

References

1.Cook, N.G.W., Seismicity associated with mining. Eng.Geol.,1976,10,99-122

2. Tang, C., Hudson, J.A. and Xu, x., rock Failure Instability and Related Aspects of Earthquake Mechanisms. China Coal Industry, Beijing, 1993

3.Tang, C., Hudson, J.A. and Xu,x.,rock Failure Instability and Related Aspects of Earthquake Mechanisms. China Coal Industry, Beijing, 1993

NUMERICAL SIMULATION ON MIXED MODE FRACTURE PROCESS OF ROCK

W.C. Zhu, C.A. Tang
1Center for Rock Instability and Seismicity Research, Box 138, Northeastern University
Shenyang, 110004, P.R. China
M.L. Huang
2Institute of Rock and Soil Mechanics, The Chinese Academy of Science
Wuhan, 430071, P.R. China

A numerical code named RFPA2D is used to simulate the mixed fracture of rock specimen. The numerical simulation predicts well the crack trajectories of rock specimen when the orientation angle of notch are different, which prove that the mixed fracture as well as model II fracture observed at the macroscopic level is actually caused by tensile damage at mesoscopic level.

1 Introduction

In recent years, there has been growing interest in numerically modeling the fracture process of rock. Efforts in this direction are very necessary for knowing the real fracture mechanism of rock [1]. In the past, most material characterizations had been based on the approximation of representing the rock by one of the classical mathematical models such as elasticity, plasticity and viscoelasticity. This kind of approximation can't capture the complete fracture process of rock as characterized by initiation, propagation and coalescence of microcracks. The macroscopic features can be explained in terms of the material structures at a low level, i.e. mesoscopic level, at which the heterogeneous material properties should be incorporated in the numerical model. In this investigation, a numerical tool named RFPA^{2D} (abbreviated from Rock Failure Process Analysis) is used to simulate the mixed-mode fracture of rock specimen. In the past years, the RFPA^{2D} had been successfully used to the numerical simulation of the rock failure and induced seismicity and crack propagation in rock, and the numerical results compared well with the experimental results [2-3].

2 Numerical mode

A disc-type notched specimen as shown in Fig.1 is used widely to determine the fracture toughness of rock [4-5]. For this specimen, it is possible to study the mode of fracture from tensile (mode I for $\beta=0^\circ$) to mixed mode (tension-shear for $\beta<30^\circ$ and compression- shear for $\beta>30^\circ$) by changing the notch inclination angle β with respect to the loading direction. The following material data for this rock are used: Young's modulus E = 52GPa; Poisson's ratio ν = 0.276; uniaxial compressive strength f_c = 105MPa; tensile strength f_t = 2.31MPa. The diameter of disc is 98mm, and a/R = 0.3 is used. Rock specimen is denoted with 22500 mesoscopic elements and simulated as plane stress problem. Between the loading plates and specimen a cushion material (such as cardboard) is inserted to distribute the high contact stress. In this study b/R = 0.12. The mechanical properties (including Young's modulus and strength) of these elements are assumed to be heterogeneous and be conformed to Weibull distribution.

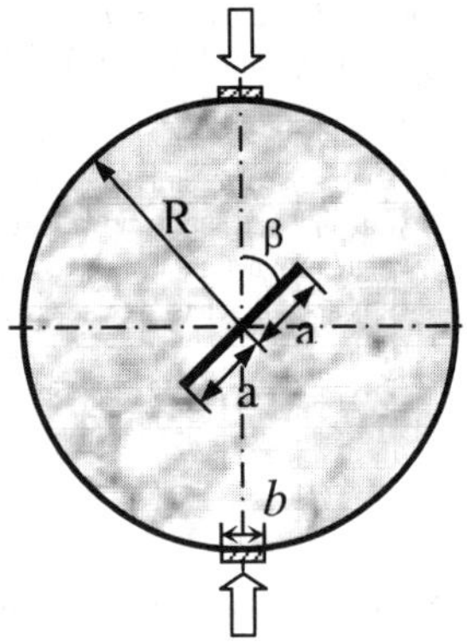

Figure 1. Numerical model of disc-type rock specimen

3 Numerical results

The distribution of maximum shear stress and minor principal stress before crack initiation in the specimen is shown in Fig.2. In this figure, the complex stress field can be observed. The cracking patterns of rock specimen when the values of β are 30, 45 and 60 are shown in Figure 3. For all the specimens, the crack (it is generally called primary crack) initiates at the crack tip, and propagates toward the loading plates. Then the secondary cracks are also observed for all specimens. Due to different notch inclinations, the final patterns are also different, which agree with the observation obtained by Jia et al. (1996) on concrete. In Fig.4 several typical the failure processes of specimen for β = 30° are given Fig.3. From the distribution of failure elements as given in Fig.3(b), It can be found that the so-called mixed mode fracture of rock specimens are dominantly caused by tensile failure of mesoscopic elements (which are indicated with red color). Due to the heterogeneity, the crack propagation trajectories for each specimen are not symmetrical at the crack initiates at two notch tips does not start at the same loading step. These failure phenomena can be simulated when the heterogeneity of rock are considered in the numerical model.

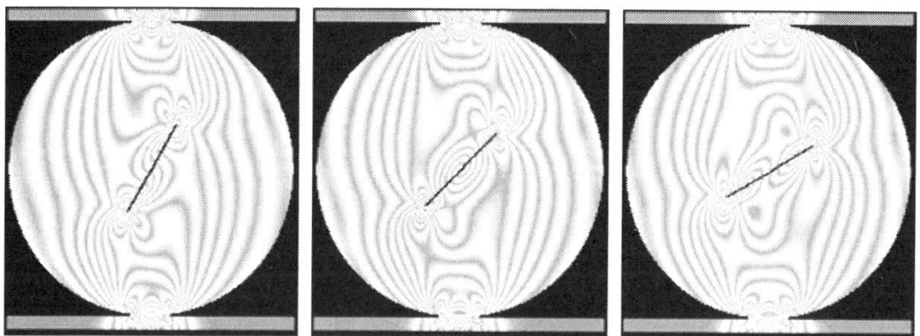

β = 30°　　β = 45°　　β = 60°

(a) Maximum shear stress

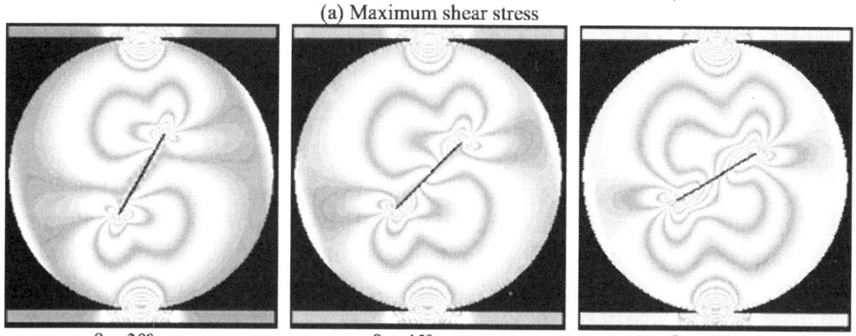

β = 30°　　β = 45°　　β = 60°

(b) Mimor principal stress (maximum tensile stress)

Fig2. Distribution of initial stress in three specimens for different notch inclination

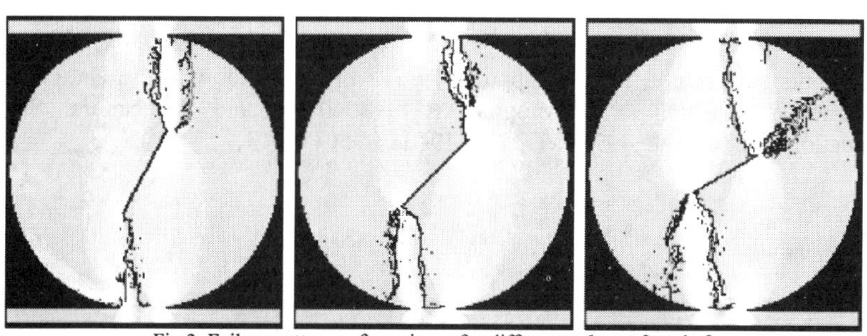

Fig.3. Failure patterns of specimen for different values of angle β

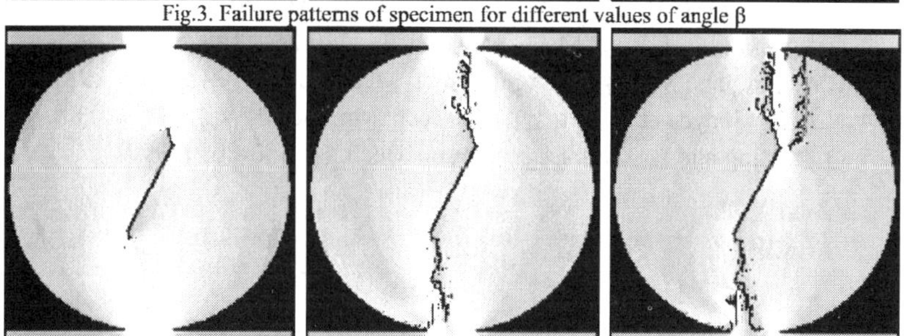

(a) Distribution of maximum shear stress

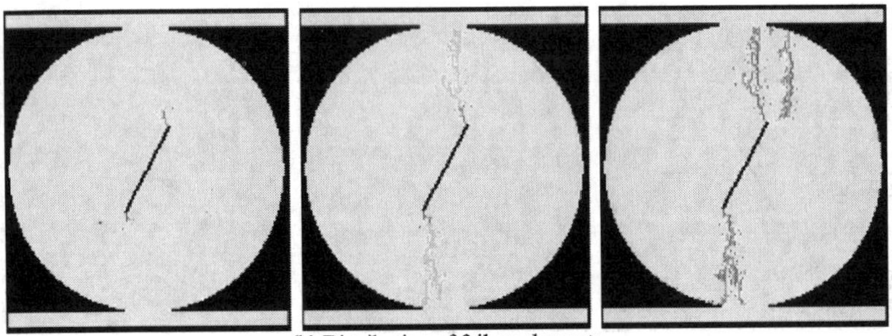

(b) Distribution of failure elements

Fig2. Failure process of specimen for $\beta = 30^{o}$

4 Conclusions

The crack propagation trajectories and final fracture patterns in cracked disc specimen has been obtained from numerical simulation of RFPA2D and show a good agreement with the experimental results. Due to the heterogeneity of rock, the local crack propagation path is curvilinear and the mixed-mode fracture of rock is predominantly caused by tensile damage at the mesoscopic level.

Acknowledgements

The authors are grateful for the support of this work from the Chinese National Key Fundamental Research "973" Program (G1998040700) and the Natural Science Foundation of China (No.49974009, 50134040 and 50174013).

References

1. A. Bobet and H.H. Einstein, Int. J. Rock Mech. Min. Sci. 35, 863-888 (1998).
2. C.A. Tang, Int. J. Rock Mech. Min. Sci. 34, 249-261 (1997).
3. C.A. Tang and S.Q. Kou, Engng. Fract. Mech. 61(3/4), 311-324 (1998).
4. N.A. Al-Shayea et al., Int. J. Rock Mech. Min. Sci. 37, 629-643 (2000).
5. Q.Z. Wang and L. Xing, Engng. Fract. Mech. 64, 193-201 (1999).

THE EFFECTS OF CONFINING COMPRESSION ON PROPAGATION AND COALESCENCE MECHANISM OF CRACKS IN BRITTLE ROCK MATERIAL

MINGLI HUANG [1] XIATING FENG [1] SHANYONG WANG[2] CHUN'AN TANG[2]
1 Key lab of Rock and soil mechanics, institute of rock and soil mechanics, the Chinese academy of Sciences, wuhan 430071 china
2 CRISR, Northeastern University, Shenyang, 110004, People's Republic of China.

A series of numerical simulations on biaxial compression tests were performed on model specimens with two parallel and offset cracks. The main objective is to study the propagation and coalescence mechanism of cracks in fractured rock masses. The numerical simulations are implemented with a Rock Failure Process Analysis code, RFPA^{2D}, and it shows that different confining pressures leads to different stress field in the rock bridge significantly, and different coalescence path and failure modes. The numerical results are in good agreement with the related experimental conclusion.

1 Introduction

Mechanism of propagation, interaction and coalescence of multi-crack in rock-like materials is one of the important research fields in rock fracture mechanics. Studies show that propagation and coalescence of cracks are due to the interaction between each other; we can gain further insight into propagation and coalescence modes and their relations with stress field and acoustic emission characteristics. Generally rock masses exist at three dimensional stress state in real engineering, but they are often simplified as two dimensional plane problem for the purpose of convenience. When confining pressure exists, crack propagation, interaction and coalescence mechanism are greatly different from those under the axial load. So more studies focused on it.

In this paper, with a 2-D rock failure process analysis code, RFPA^{2D} (Tang, 1997), a series of numerical simulations on rock samples with two parallel crack re-existing under the biaxial compression are carried out. The main purpose of the research is to gain further insight into the behaviors of coalescence of micro-cracks in rock masses under different confining pressure. Not only does such numerical studies satisfy the curiosity of the academically inclined, but also provide a basis for meaningful extrapolation from available experimental results.

2 Numerical Model

The scale of model is 100 mm × 100 mm, and divided into 100 × 100 = 10000 elements. Two pre-existing open cracks with lengths $2c$=24 mm, and inclination α=45°, are located in the center of the model (Fig.1). The rock bridge between the two cracks is also 24mm in length. The rock bridges are inclined at β=90° to the horizontal direction (shown in Fig.1). Mechanical properties of two cracks, greatly smaller than other elements are approximately equal to those of hole element. The following material data for this rock are used: Young's modulus E=6GPA; Poisson's ratio ν=0.30; uniaxial compressive strength σ_c= 60MPa;tensile strength σ_t=6MPa; friction angle Φ=30° .

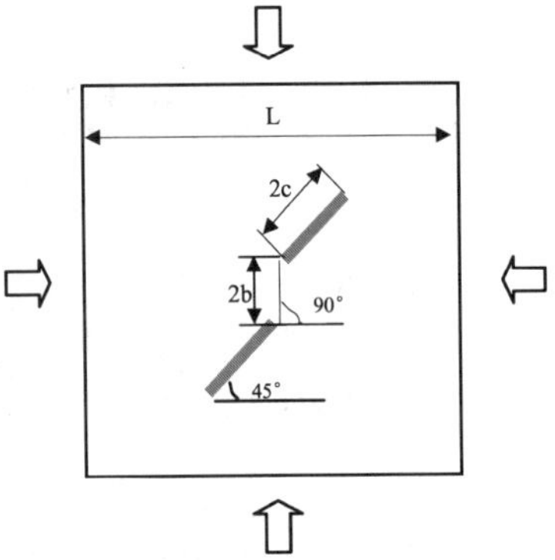

Figure1. model of specimen

Load was controlled by displacement load. The total displacement was 0.2mm, and divided into 100 steps. Confining pressures are 0Mpa, 2.5Mpa, 5Mpa, 10Mpa respectively. All the cases are assumed as plane strain condition.

3 Numerical simulation results and discussion

Many researches have shown that new cracks initiating from pre-existing ones can be classified as wing crack induced by tensile stress and secondary crack induced by shear stress on mechanism (Bobet et.al., 1998; Wong et.al., 1998; Shen, et. al., 1995). Because resisting tension strength is smaller than resisting compression strength, wing cracks, which are extensive and propagate in the bigger angle direction from pre-existing fracture, always grow from the tip of pre-existing fractures where stress concentrates at first. The outer wing cracks propagate rapidly in general. With the new crack occurring and propagating continuously, stress field modifies by itself, and tensile stress concentration at the tip of the cracks is becoming weak. When thc tensile stress is perpendicular to main stress, compressive stress concentration occurs at the tip and cracks will not grow any more along the direction of main compressive stress. But secondary cracks are generally found only when crack approaches to coalescence , and initiate from the tip of crack and propagate stably along pre-existed surface , which is the main cause of coalescence.

Based on rock-like materials with two pre-existed cracks under uniaxial and biaxial compression experiment, A. Bobet and H.H.einsein studied the influence of confining pressure on crack propagation and coalescence characteristics systematically, and the results showed that coalescence mechanism is affected by stress state of crack pronouncedly. After loading the specimen with two pre-existed crack, stress concentration with different characteristics occurs at the tip of crack, and the obvious tensile stress concentration near the direction of main compressive stress in general. With cracks propagating, stress field evolutes at the tip of crack and

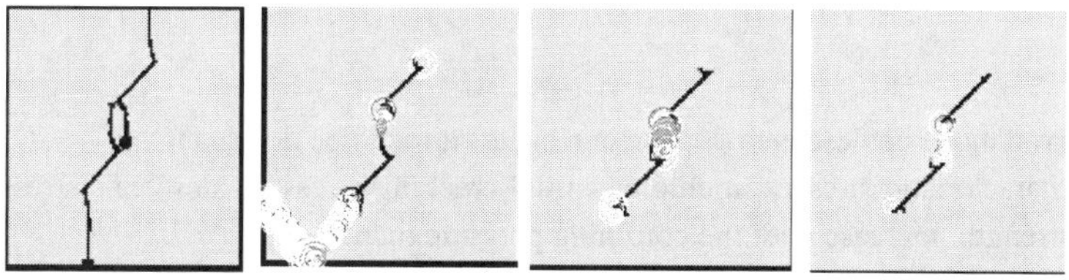

a) σ_3=0MPa b) σ_3=2.5MPa d) σ_3=5MPa e) σ_3=10MPa

Figure 2. AE distribution when crack coalesces

induces new crack occurrence, then interaction of cracks with different characteristics results in different coalescence mechanism.

It can be observed from Fig.2 that the crack coalescence mode will change with the confining pressure increasing. When σ_3 =0, it is the tension or wing mode; and σ_3=2.5Mpa, it becomes the mixed tension/shear mode, where wing crack is main; when confining pressure increases to 5Mpa, it is still TS mode, but secondary cracks are dominating. So with the increasing of the confining pressure, although wing crack occurs, its propagation at the inner and outer tip of pre-existing crack are suppressed by the confining pressure and stop soon, and the coalescence mode changes from tensile to shear failure gradually. When confining pressure increases to 10Mpa, it turns to be the shear failure mode that secondary cracks connect each other. This supports conclusions from reference (Bobet, et.al., 1998).

The influence of confining pressure on characteristic stresses is described in Table 1. It is shown that strength increase with the increase of confining pressure, and it is because coalescence mode changes from tensile to shear failure gradually with confining pressure increasing, so only axial load leading to crack failure increases, it could reach the characteristic stress of crack initiate and failure. The results have a good agreement with relative experiment.

Table1. Characteristic stress under different confining pressures.

Confining pressure （Mpa）	Coalescence mode	Initiating stress （MPa）	Coalescence stress （MPa）	Peak strength （MPa）
0	Tensile mode(T)	5.06	13.64	21.54
2.5	Tensile shear mode(TS)	8.47	24.19	24.19
5	Shear tensile mode(ST)	12.53	25.64	27.54
10	Shear model (S)	17.88	27.79	27.79

4 Conclusion

The effects of confining compression on propagation and coalescence mechanism of cracks brittle rock material are studied by means of numerical simulation and the results show:

1. With confining pressure increasing, growth of wing crack will be suppressed, and

the mixed mode coalescence (tensile and shear) turns to be dominant.

2. Characteristic stresses, including initial cracking stress, coalescence stress and peak strength, increase with the confining pressure increasing.

Acknowledgements

The work presented in this paper was supported by the Chinese National Key Fundamental Research "973 Program ' (No.95-13-07-01), the National Natural Science Foundation of China (No. 59939190) and the Administrative Committee of CAS, K.C. Wong Post-doctoral Research Award Fund.

References

1. Bobet.A.,Einstein.H.H.1998.Fracture coalescence in rock-type materials under uniaxial and biaxial compression , International Journal of Rock Mechanics and Mining Science.35(7):863-888
2. Wong R.H.C., Lin P., Chau K.T., Tang C.A.,2000.the effects of confining compression on fracture coalesence in Rock-like material. Key engineering materials vols,183-187pp.857-862
3. Shen B., Stephansson O., Einstein H.H., 1995. Ghahreman B., Coalescence of fractures under shear stresses in experiments. J. Geophys. Res., 100(B4): 5975-5990.
4. Tang C.A. 1997. Numerical simulation of rock failure and associated seismicity. Int. J. Rock Mech. Min. Sci., 34: pp.249-262.

A NUMERICAL SIMULATION OF PERMEABILITY EVOLUTION IN FAILURE PROCESS OF ROCK WITH PRE-EXISTING CRACK

T.H. YANG S.H.WANG C.A.TANG L.C. LI Z.ZANG
CRISR, Northeastern University, Shenyang 110006, P.R.China

Experimental results provide strong evidence that rock permeability is not a constant but a function of stresses and stress-induced damage. In this paper, A newly developed numerical code, the flow-coupled Rock Failure Process Analysis (F-RFPA2D), is applied to account the growth of the pre-existing crack and formation of new fractures for heterogeneous and permeable rocks . Implemented with this code (F-RFPA2D), we investigated the behavior of fluid flow, damage evolution and their coupling action in samples subjected to both hydraulic and biaxial compressive loading. The modeling results suggest that dramatic permeability increase occurs as soon as the macro fracture forms in the rock.

1 Introduction

Among the current problems that rock or civil engineers face, perhaps there is none as challenging as the characterization of fluid flow though fracturing rocks, especially when they are highly stressed. In mining and civil engineering projects, the re-distribution of the stress field during the excavation of tunnels and underground chambers lead to the formation of new fractures. These damages may cause dramatic changes in the permeability of the rock masses. As a result, the rate of water flowing into the tunnels and chambers will increase [1-2].

2 Numerical Model

A number of 2D analyses were done using the flow-coupled Rock Failure Process Analysis code (F-RFPA2D) described in literature[3]. The program is written specifically for modeling rock failure including crack initiation, propagation and coalescence in heterogeneous and permeable materials. The analysis presented here is concentrated on the effects of heterogeneity on hydraulic fracture initiation, propagation and breakdown in permeable rock samples. For a given hydraulic pressure, the stress and strain in individual element is solved at every loading step. At any step, if either shear or tension strength of a element is exceeded, that element fails irreversibly to initiate a new fracture or extend an old one and bears only a residual stress. For the failed element, its Young's modulus is considerably decreased and its permeability is considerably increased [1]. Then the stress and strain in individual element is solved again, which results in the stress redistribution, until a new equilibrium is obtained. Thus, the evolution of this hydraulic system is determined, and the fracture initiation and propagation process are simulated.

Due to the high stress concentration at the crack tip or low strength elements, some weaker elements that are not on the fracture surface may fail to form micro-fractures that distribute in the vicinity of the fracture zone. Some of them may grow and join the main fracture, while others remain isolated beyond the main fracture. Unlike in the model such as developed by Kim and Yao [2] in which the fluid pressure in the fracture is achieved by adding full pressure to the new main fracture surface instantaneously and no water pressure exists inside the isolated micro-fractures, in our model, the fluid pressure in individual element is calculated by solving coupling equation between stress and flow .

3 Modeling and discussion

The numerical sample is a two dimensional rectangle block with 40000 elements ,the sample width is 50mm, and the height is 80mm. .In these numerical experiments, samples are tested fully saturated. The detailed structure and the sample installation is illustrated in Fig.1. Constant water pressure difference, 1.5 MPa. The elements are characterized by their Young's modulus, E, uniaxial compressive, σc, Poisson's ratio, v, and permeability, k. The elements provide resistance against compressive or tensile deformations which are governed by constitutive law shown in literature[3]. The input parameters used in the model are selected to simulate a heterogeneous and brittle rock with pre-existing crack and they are listed in the Table 1 below.

Table 1. Input material mechanical properties – mean values

Young's modulus	$E0$	10 GPa
Poisson's ratio	v	0.25
Compression strength	σc	100 MPa
Tension strength	σt	10 MPa
Coefficient of residual strength	λ	0.2
internal friction angle	ϕ	30°
Permeability	k	1 m/s
Water pressure	ΔP	1.5 Mpa
Water pressure coefficient	α	0.6
Homogeneity index	m	3
Crack length	l	14mm
Crack angle	a	53°

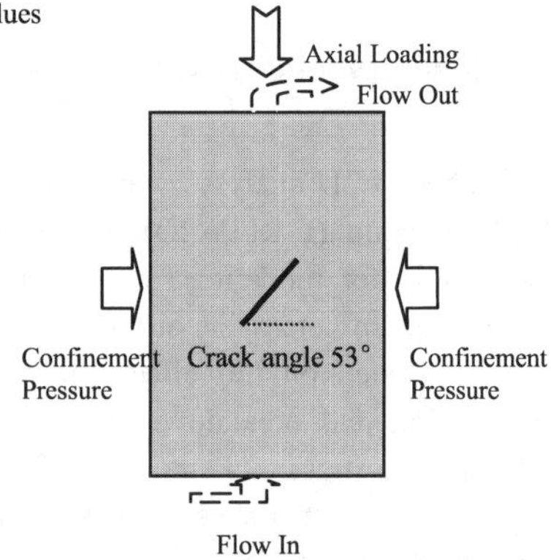

Figure 1. Configuration for numarucal tests

Numerical simulations have also been performed to investigate the axial permeability of numerical rock samples subjected to biaxial compressive stresses.

Fig.2 shows the numerically obtained relationship among load, permeability and loading step of the sample with the homogeneity index m=3. The overall permeability is obtained by the following equation:

$$K = L \quad Q / \Delta P = L \Sigma q / \Delta P \tag{1}$$

where $Q = \Sigma q$ is the total flux flowing through the sample, L is the width of sample, and ΔP is the hydraulic pressure difference between the two ends of the sample. During the elastic deformations, rock permeability decreases. As the sample is loaded to the nonlinear deformation stage, the permeability slowly increases until yielding where one can observe a significant increase in the permeability. The numerical result reveals that every permeability increase corresponds to a stress drop. Since the sample is loaded in a displacement control manner, the sudden failure with large number of acoustic emissions results in a rapid stress drop. Therefore, it can be concluded that the permeability jump is due to the sudden damage (micro-fractures) within the sample. It is obvious from Fig.2 that changes in permeability and damage are coherent in trend, and this phenomenon illustrates that rock damage and permeability are related by some kind of functional relationships.

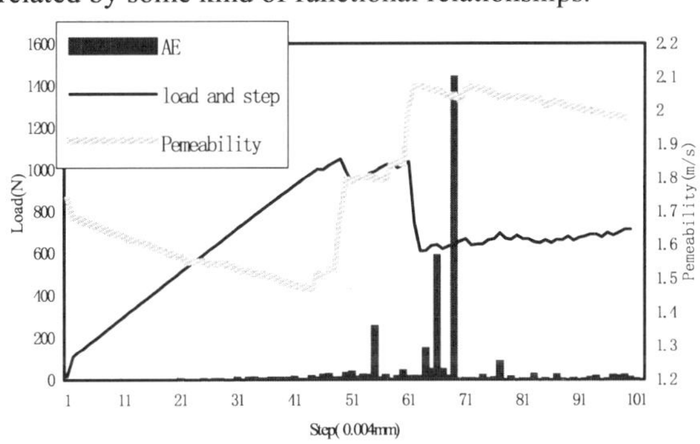

Figure 2. Numerically obtained relationship among load, permeability and loading step and the associated acoustic emissions (AE events, relative counts)

Fig.3a-d show the numerical simulated progressive failure process and the formation of macro-fractures during hydraulic and boundary loading, with the load The numerically obtained stress field during the failure process in the sample is shown in Fig.3. The stress level is represented by the brightness of the gray scale: a higher stress level is indicated by a brighter scale.

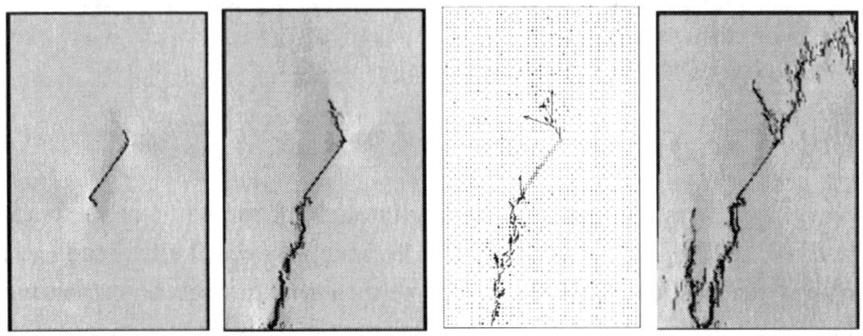

a. Step34 b. Step 49 c. Step49 (flow velocity field) d step61

Figure 3. Numerically obtained stress and flow velocity evolution during failure

4 Conclusions

Numerical results demonstrate that the trend and magnitude of permeability variations are controlled by the stress and the damage evolution developed in rocks. with pre-existing crack. In the elastic deformation region, rock permeability decreases when the rock compacts. The decrease rate of the permeability starts to slow down or gradually increases again when micro fractures begin to nucleate. Dramatic permeability increase occurs as soon as the macro fracture forms in the rock.

Acknowledgement

The authors are grateful for the support of this work from the Chinese National Key Fundamental Research "973" Programme (G1998040700).

References

1. Zhang, J.C., Bai, M. et al, Experimental determination of stress-permeability relationship, *Paciffic Rock 2000*, Balkema, Rotterdam, 2000, 817-822.
2. Kim, K., Yao, C.Y., The in fluence of constitutive behavior on the fracture process zone and stress field evolution during hydraulic fracturing, *Rock Mechanics,* Balkema, Rotterdam, 1994, 193-199.
3. Tang, C.A., Yang, T.H. et al, Coupled analysis of flow, stress and damage (FSD) in rock failure, *Int. J. Rock Mech. Min. Sci.* (in press).

NUMERICAL STUDY OF EFFECT OF MESOSCOPIC HETEROGENEITY ON SEISMIC SEQUENCE PATTERNS

MINGRUO JIAO CHUNAN TANG MINGLI HUANG

Center for Rockburst and Induced Seismisity research, Northeastern University, Shenyang, Liangning, 110006

A newly-developed numerical method RFPA^{2D} is applied to study the effect of mesoscopic heterogeneity on macroscopic behavior of rock failure as well as AE sequence types. The failure procedures of rock sample with 5 different kinds of heterogeneity are simulated, showing that rock failure with the different heterogeneity will produce different seismic sequence types, mainly including (1)main shock-aftershock type, (2) foreshock-main shock-aftershock type, and (3)swarm type.

1 Introduction

If the occurrence of an earthquake is considered to be a result of unstable failure of rock, much of the experimental research on fractures in brittle rock material will provide a powerful lead to elucidating the mechanism by which earthquakes occur within the earth's crust. The most interesting fields of the laboratory studies is focus on seismic activities. The most important way of observing damage during rock deformation experiments is to monitor the acoustic emissions (AE). It shows that three types of acoustic emission sequences have been found by Mogi [1] (1962), namely (1)main shock-aftershock type, (2) foreshock-main shock-aftershock type, and (3)swarm type. It concluded that these types are the result of differences in the heterogeneity of the medium. Tang[2] (1993) has given the analytical explanations based on a simple mechanical model for the testing machine-specimen system. However, the theoretical study concerning the AE response of geologic materials, especially the seismic activities or acoustic emission sequences is quite a shortage in the existing literatures. We know the analytical models have to be simplified and inevitably ignore some key factors influencing the material behavior, such as the heterogeneity. In many cases, the analytical models cannot deal with the rocks with some discontinuities. For this, it seems that the numerical method can become a useful tool to provide an insight into seismic activity of a rock specimen.

2 Model description

Based on the linear finite finite element method, RFPA^{2D}(Rock Failure Process Analysis code) is a two-dimensional package for the analysis of rock failure process. The program allows to simulating the progressive failure of brittle rock from initial micro- failure to collapse by virtue of a simple approximation that gets rid of the numerical complexities of non-linear and discontinuum codes. By choosing a new Coulomb criterion with a tensile cut-off as the failure criterion or threshold, it allows the failure of an element in either shear or tension. Detailed introductions of the RFPA code are given in published literature [3].

Five specimens with different heterogeneity indices m = 1.1, 1.5, 2, 3, 5, representing materials from relative heterogeneity to relative homogeneity under uniaxial compression are simulated. The mesh for the model is 260 x 130=33,800 elements representing a sample geometry of 260 x130mm in scale(Table 1). The element property is selected by setting the Weibull's distribution. The mean Young's modulus of specimens is 60,000MPa with mean compressive strength 200MPa. In the following simulations a plain strain condition is assumed. In this paper, smeared failure approach is used to predict the occurrence of microfracturing when the stress state of an element satisfies a strength criterion. The samples are loaded in a displacement manner at a constant rate of 0.002 mm/step.

Table 1 Material properties for specimens in uniaxial loading condition

Specimens for testing influence of heterogeneity on failure mode and strength characterization	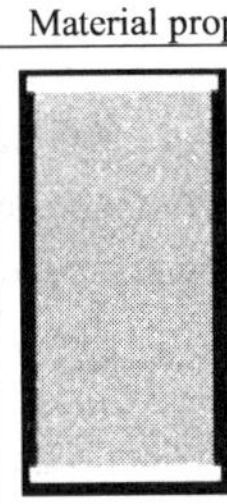	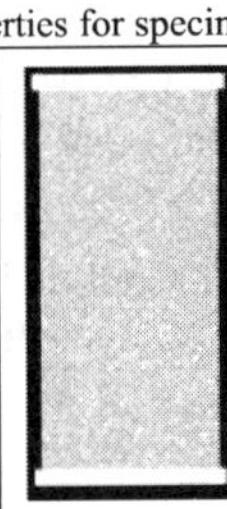	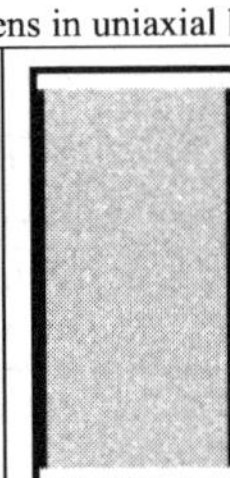		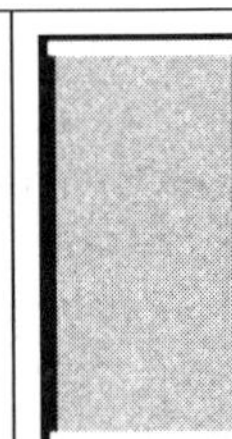
Specimen name	s-h-1	s-h-2	s-h-3	s-h-4	s-h-5
Homogeneity index	m=1.1	m=1.5	m=2	m=3	m=5

3 Numerical simulation results and discuss

Stress-strain curve is divided into three stages. In stage I (up to about 80% strength): Linear-elastic behavior of rock is typical for this phase, showing little elastic energy was released, though some AE events occurred, and no crack development is expected (see Fig.2a). At the onset of stage II (from about 80% to maximum strength): The stress-strain curve is no more linear, and evident fractures can be expected. An increasing rate of AE activity assumption that the AE were generated by discontinuous dynamic micro-deformation events such as cracks. A large number of AE events occurred before reaching the peak load (maximum strength) for this relatively heterogeneous specimen. Besides, up to the maximum strength, it is symbolized by the formation of isolated fracture zones. Local fracturing characterizes the relief of stress accompanied the inelastic phase, consistent with the produced by heterogeneity of rocks. Besides, most of fractures are oriented parallel to the applied stress (see Fig.2 b, c, d and e). In stage III: It is post-peak phase. Strain-softening characterized this phase. Local fracturing becomes extremely prominent, predominating

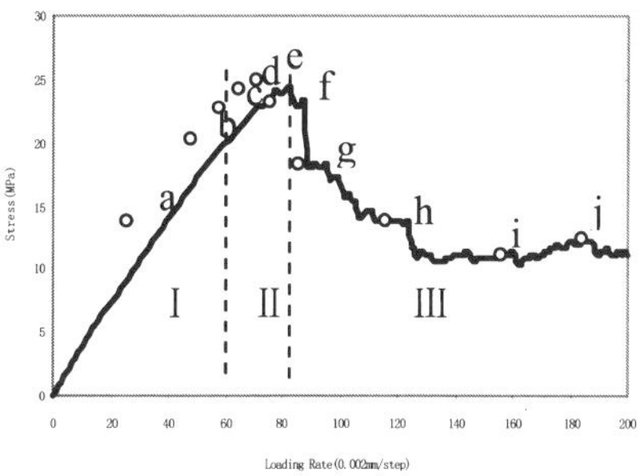

Fig.1 stress-strain curve for specimen s-h-2

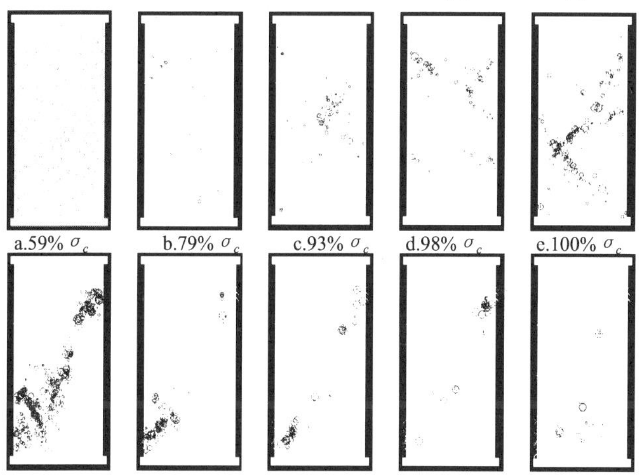

f.95% σ_c g.70% σ_c h.63% σ_c i.56% σ_c j.49% σ_c

Figure2. Plots of simulated AE locations for specimen s-h-2.

along certain narrow zones. Besides, the specimen has still remained some loads. The numbers of cracks remains practically the same, and almost no new cracks are generated (see Fig.2 f, g, h, i and j). The AE seismic patterns for specimens with different homogeneity index m under uniaxial loading are shown in Fig.3. The numerical results produced three types of AE event sequence patterns: (1)main-shock-aftershock type, (2) foreshock-main shock-aftershock type, and (3) swarm type. That is, type (1) occurs in relatively homogeneous cases (s-h-4 and s-h-5), (2) in moderately heterogeneous cases (s-h-2 and s-h-3), and (3) in extremely heterogeneous cases (s-h-1). The simulative results show a good coincidence with the natural observations according to the research based on the data of China, Japan and these types do not occur randomly in spatial terms but show a obvious regionalism. Consequently, it let to the conclusion that these types are the results of differences in the heterogeneity of the medium.

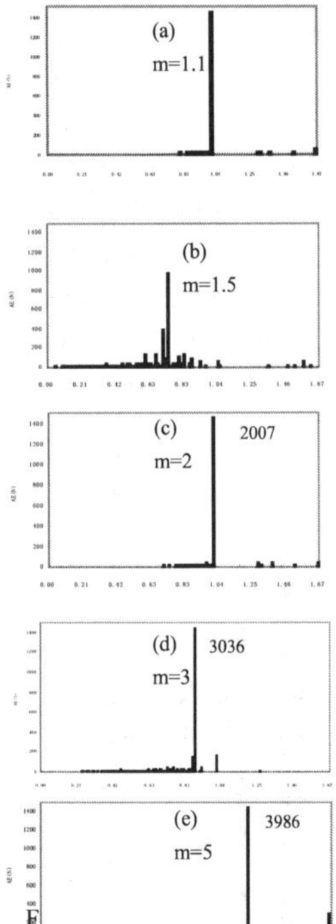

F for specimens with different homogeneity

4 Conclusions

1. The complete stress-strain curve for a heterogeneous specimen(taking the specimen s-h-2 as an example with m=1.5) can be modeled containing three characteristic stages.
2. The numerical results demonstrate that the AE event patterns are influenced greatly by the degree of heterogeneity of the materials. There are three types of seismic consequence patterns: (1)main-shock-aftershock type, (2) foreshock-main shock-aftershock type, and (3) swarm type, accordingly corresponding to type (1) occurs in homogeneous cases, (2) in moderately heterogeneous cases, and (3) in extremely heterogeneous cases.

Acknowledgements

The study was supported by the China National Key Fundamental Research “973 Programme” (No.G1998040700) and the China National Natural Science Foundation (No.50134040, No.50174013, No.49974009)。

References

1. Mogi k.. On the time distribution of aftershocks accompanying the recent major earthquakes in and near Japan, Bull. Earthquake Res. Inst., 1962a, 40, 107-124.
2. Tang Chun’an, John A. Hudson, Xiaohe Xu. Rock Failure Instability and Related Aspects of Earthquake Mechanisms. Beijing: China Coal Industry Publishing House. 1993, 50-60.
3. Tang CA. Numerical simulation on progressive failure leading to collapse and associated seismicity. Int. J. Rock Mech. Min. Sci. , 1997; 34: 249-262

NUMERICAL SIMULATION AND COMPARISON STUDY OF ROCK SPECIMEN'S ELASTIC-PLASTIC FAILURE PROCESS

LI SONG
Department of Civil Engineering, Dalian University, Dalian,116622, P.R.China

LIPING XIAO
Department of Civil Engineering, Dalian University, Dalian, 116622, P.R.China

YUNMEI LIN
Civil Engineering Institute, Northeast University ,Shenyang, 110006, P.R.China

A series of comparison study is conducted on the rock specimen's elastic-plastic failure process with the help of REPFPA. This not only permits us to reproduce the rock specimen's elastic-plastic failure process on the computer, but also enables us to draw many constructive conclusions.

1 Rock Characteristic and Its Mathematical Description

Rock medium displays a quality of uneven distribution and discontinuity within the scope of human eyes; whereas in the computations of mechanics problems of mining and surface projects with the help of the traditional FEM, a hypothesis is established that it is evenly distributed, continuous and isotropic. In fact, this hypothesis is out of conformity with the micro mechanics property of rock medium. Not taking into account the uneven distribution and discontinuity of rock medium and the dispersion and random of distribution of material parameters will lead to a relatively big deviation between the numerical computation and the practice situation.

As a matter of fact, the various material parameters of rock medium in building projects, such as young's modulus and compression strength, do possess the quality of dispersion and random. We hence adopt Weibull distribution law to describe the distribution pattern of material parameters:

$$P(\sigma)=cm(\sigma/\sigma_0)^{m-1}\exp[-(\sigma/\sigma_0)^m]$$

where c stands for normalization coefficient, m the pattern parameter of statistical distribution function, which determines the material parameters' distribution pattern within materials, of which the larger the value is, the more evenly distributed the material is. σ_0 stands for a certain reference value of σ.

The author of this paper has developed a program of the rock elastic-plastic failure process based on FPS4.0 platform by means of Monte Carlo Method. Using this program, numerical simulation and comparison study is herein conducted with rock specimen.

2 The Numerical Simulation Analysis of Plane Stress Specimen's Elastic-Plastic Failure Process

The specimen's size is 300mm×200mm, the thickness of which is one unit, given automatically by the programme. The specimen is divided into 600 8-node Serendipity elements. The order of Gaussian quadrature is 3, thus there are 9 Gaussian Points in each element. There are altogether 1901 nodes in the mesh.

The material parameters of the specimen are given in tab.1, the values of which are statistical mean values. We here presuppose that the material parameters conform to Weibull distribution, and that, in this project, the pattern parameter of the statistical distribution function is m_i=2.5.

Tab.1. Rock specimen's material parameters

E	μ	ρ	C	H'	S_c	S_t	φ
50000.0	0.25	2.5	30.0	30000.0	100.0	20.0	45.0

The displacement "load" is imposed on the specimen, the total is –0.45mm, the imposition is divided into 12 steps. Tab.2 gives the every "load" step's coefficient and increment.

Tab.2. "load" step's coefficient and increment

Load step	1	2	3	4	5	6	7	8	9	10	11	12
Load coefficient	0.4	0.5	0.55	0.6	0.62	0.63	0.635	0.64	0.645	0.65	0.7	0.75
Load increment	0.4	0.1	0.05	0.05	0.02	0.01	0.005	0.005	0.005	0.005	0.05	0.05

The failure process can be divided into three stages according to the distribution of the rupture element. Ⅰ. Steps 1 to 6 are the random distribution of the rupture element: Though the rupture element increase step by step, the distribution of it in the specimen is basically random. The stage load is 0.84 times the total toad when reaching the sixth step. In this stage, the rupture is not obvious, basically with those whose strength is weak breaking first. Ⅱ .Steps 7 to 9 is transition stage, in which the rupture element is focusing on some strength-weak ribbon zone. The stage load is 0.86 times the total load when reaching this stage. It is obvious that the rupture element increase greatly with little increase in load. The detailed statistical results are listed in Tab.3. Ⅲ. This is the taking-shape stage of the macro crack. Because of the decreased value of the ruptured element's material parameters, the material capability of resisting deformation and destruction is becoming weaker and weaker, which in turn causes the neighbouring element's stress to centralize. This will lead to the increasing rupture probability of the neighbouring elements and thus a macro rupture belt is formed. This phenomenon can be clearly seen through the changing process of the destruction analysis diagram.

Tab.3. Statistics in rupture element

Load step	1	2	3	4	5	6	7	8	9	10	11	12
Number of ruptured element	0	32	56	134	99	84	91	132	191	210	280	176
Total number of ruptured element	0	32	88	222	321	405	196	628	819	1029	1309	1485

Since we did not load at regular intervals, tab.3 didn't reflect the true nature, that is, the changing situation of ruptured elements with the change in load. The following table gives the number of ruptured elements when loading is imposed at regular intervals.

Tab.4. The change in number of rupture element with the change of loading

Even loading intervals	1-80	81-100	101-110	111-120	121-124	125-126	127	128	129	130	131-140	141-150
Number of rupture element	0	1.6	5.6	13.4	24.75	42	91	132	191	210	28	17.6

3 The Comparison Study of the Pattern Parameter m of Statistical Distribution Function

The material parameters here are the same as those in tab.1, so a comparison can be conveniently made between the conclusion of this Section and that of Section 2.

The pattern parameter of statistical distribution function m_i=5.5. Tab.5 gives the number of rupture elements with the regular change of load steps.

Tab.5. The change in number of rupture element with the change of loading

Even loading intervals	1-120	121-140	141-150	151-160	161-164	165-166	167	168	169	170	171-180	181-190
Number of rupture element	0	1.2	7.3	33.7	149	345	233	67	25	27	7.8	3.2

Through the comparison we can see that the specimen's medium is becoming more evenly-distributed with the value increase of pattern parameters. Thus the peak number of the rupture elements appears later and more concentrated than that in section 2.

The peak number of rupture elements appears between step 127 and step 130,while in this section, it appears between step 161 and step 167.

4 Other Comparisons and Their Conclusions

The comparison study of random assignment method of material parameters:
The random assignment method of material parameters in this section is different from that in section2. In section2, the random assignment of the material parameters

is conducted within the “Small element” of all Gaussian points; hence 5400 random number between (0,1) are concluded. The random assignment of material parameters is despatched to each “small element”. While in this section, this is done in the 600 8-node elements.Through this comparison we conclude: The specimen’s whole strength becomes weaker and a macro rupture belt is more likely formed when assigning random data of material parameters in elements than in Gaussian point “small element”. In fact, this conclusion is based on the fact that the former’s distribution evenness is higher than the latter’s.

The comparison and its conclusion between plane strain and plane stress: When comparing the two stress state, we find that, on the condition that other factors are the same, the number of the specimen’s rupture elements under the state of plane strain is much greater than that under the state of plane stress when the same displacement “load” is imposed. This is mainly because the displacement in the Z direction is restrained under the state of plane strain.

The comparison and its conclusion between the specimens with weak zone and without weak zone in the centre of the specimen: When conducting the simulation analysis of the specimen with weak zone, we find that, because of the phenomenon of “rigid body” transtational motion, the diagonal line zone is sheared hardly, thus the ruptured elements appears on the diagonal line rather than randomly.

The comparison and its conclusion of the number of elements in the mesh: When conducting the comparison of the different division number of element, we conclude that, regardless of the element number, the specimen’s total strain is approximately the same when out-of-stability destruction appears. The main difference exists between the size of the macro crack belt and the number of ruptured element.

The comparison and its conclusion of the seed for the random number generator: When conducting the comparison of different effect of the random material parameters produced by different seed for the random number generator on the rock specimen’s failure process, we find that, except for the different position of the macro crack belts, other results are approximately the same, especially in the peak and number of rupture elements.

The comparison and its conclusion of displacement boundary condition and stress boundary condition: In comparing the different effect of stress boundary condition on the specimen’s rupture process, we find that, the specimen’s load-bearing capacity under the condition of displacement boundary is much higher than that under the condition of stress boundary and its stability of keeping balance is much better.

MODELLING OF MIXED-CRACK PROPAGATION AND COALSCENCE IN ROCK

SHUHONG WANG CHUN'AN TANG WANCHENG ZHU TAO XU
Center for Rock Instability and Seismicity Research , Northeastern University, Shenyang 110004, China

By using Rock Failure and Process Analysis code, i.e., RFPA^{2D} , numerical and similar materials simulations on rock samples with two pre-existing cracks in uniaxial compression were conducted to investigate the initiation, propagation, coalescence of crack and failure mechanism of rock. Numerical simulations visually reproduce the process of crack initiation, propagation and coalescence in rock, which are well tallied with experiments in laboratory.

1 Introduction

Failure of rock with pre-existing flaws, such as cracks, pores and other defects is a very important topic in rock mechanics research. A number of studies have been done on crack initiation, propagation and coalescence in solids in the past decades. For example, the works of Ingraffea and Heuze on the limestone[1], Hoek and Bieniawski on glass[2], Petit and Barquins on standstone[3], etc., are all on preexisting flaws. More recently, Shen et al. [4], Bobet and Einstein [5,6], Wong and Chau [7] and Vassarhelyi and Bobet [8] have investigated crack propagation and coalescence in rocklike materials containing two inclined flaws. It is recognized that under the compressive loading, both tensile and shear stress concentrations can develop at preexisting cracks in rock.

Great process has been achieved on crack initiation, propagation and coalescence in rock; however, the mechanism of mixed crack propagation and coalescence has been a puzzle. To a better understanding of the failure mechanism of rock with pre-existing cracks, a series of numerical analyses and corresponding similar materials simulations were carried out to investigate the behavior of crack propagation and coalescence in rock samples. Our investigation should provide the fundamental understanding of crack propagation, interaction and coalescence in rock with pre-existing cracks under uniaxial compression.

2 Brief outline of RFPA^{2D}

Numerical simulation is currently a popular and effective method used for modeling deformation behavior of heterogeneous materials before failure, so does it for rock. Even though the process has been made in numerical simulation of failure occurring in rocks [9], there is a lack of satisfactory models, which can simulate the progressive failure in a more visual way, including simulation of the failure process, failure induced seismic events and failure induced stress redistribution.

The demand for new tools, which may contribute to a better understanding of the failure mechanisms of brittle rocks, initiated the development of RFPA^{2D} (Rock Failure Process Analysis code). Details of the RFPA^{2D} code has been described in published literatures [10-11], and so only a brief outline will be given here.

RFPA^{2D} is a progressive elastic damage model. There are two features distinguishing RFPA^{2D} from other numerical methods: (1) By introducing heterogeneity of rock properties into the model, the RFPA code can simulate nonlinear deformation of a quasi-brittle behavior with an ideal brittle constitutive law for the local material; (2) By introducing a reduction of material parameters after element failure, the RFPA code can simulate strain-softening and discontinuum mechanics problems in a continuum mechanics mode.

3 Numerical and Similar Materials Simulation

3.1 Numerical Model and Simulated Results

In this paper, numerical experiments on rock containing two inclined pre-existing parallel cracks are conducted to investigate crack propagation and coalescence in uniaxial compression. The material parameters for model specimens are Φ_E (2,6000MPa)and Φ_o(2, 50MPa), receptively. Fig. 1 (a) and (b) show the configuration of pre-existing cracks in model specimens and the numerically simulated results, respectively.

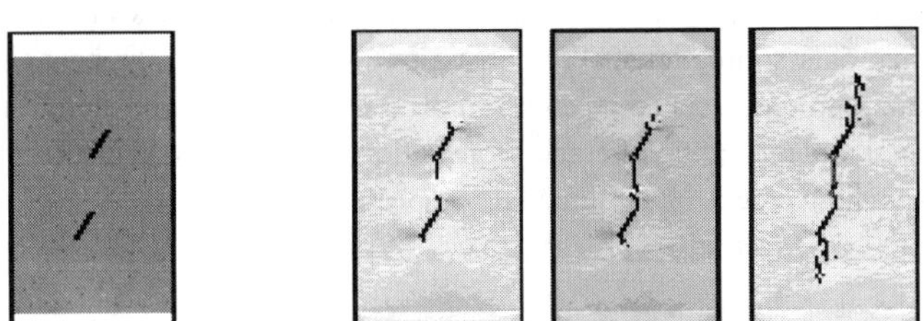

(a) (b)

Figure 1. (a) Configuration of model specimens; (b) Numerically simulated results

3.2 Similar Materials Simulation

Similar materials simulation tests on rock specimens with pre-existing mixed cracks under compressive loading were carried out in laboratory to investigate the mechanism of crack initiation, propagation and coalescence. The materials used in the experiment are fine sands, calcium carbonate, gypsum and borax. The materials were mingled at an appropriate mixing ratio to make model specimens with two pre-existing parallel cracks. The similar materials simulated testing results are shown in Fig. 2

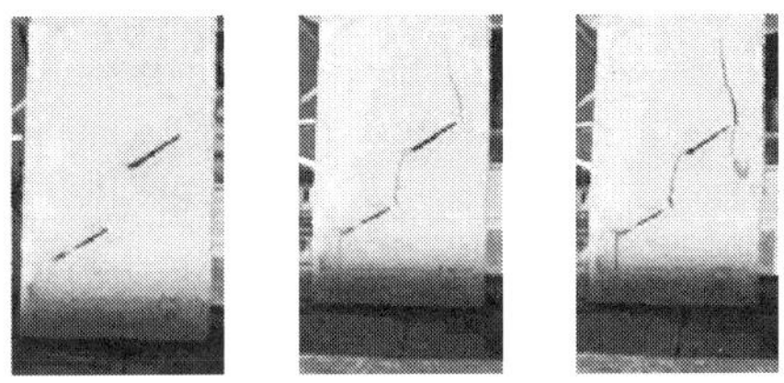

Figure 2. Similar material simulated results

4 Discussion and conclusions

As shown in Fig. 1 and Fig. 2, the mode of crack initiation, propagation and coalescence in numerical simulation is good agreement with that of similar materials simulation in laboratory. The variations of stress field around crack tips and the complete process of crack initiation, propagation and coalescence are both visually duplicated in numerical simulation and similar materials simulation in laboratory. It can be seen from Fig. 1 that stress concentrations occur between cracks tips at the beginning of loading, and with the increase of loading, the stress concentrations at crack tips gradually become sharp enough to lead to the crack propagation and the initiation of branch cracks. The internal cracks between the two preexisting cracks gradually propagate and coalesce in the end under tensile stress. In the meanwhile, the branch cracks also gradually propagate during the coalescence of internal cracks and form serrated cracks. It is noticeable from Fig. 1 that stress concentrations between internal cracks tips are extremely sharp prior to the point of coalescence and the cracks coalesce under the interactions of compressive stress and shear stress in the end. Moreover, the branch cracks initiate from the tips of pre-existing cracks and propagate in a curvilinear path at first and gradually align with the direction of compressive stress, which is known as wing cracks. The wing crack surfaces are characterized by crushed rock materials and material powder.

In addition, compared with the numerically simulated results, the similar materials simulation shows good agreement with the numerical simulation results stated above, as shown in Fig. 2. As a result, numerical simulation on rock-like materials with pre-existing cracks using $RFPA^{2D}$ is a simple and effective method to investigate the mechanism of rock failure due to the initiation, propagation and coalescence of crack.

Acknowledgement

The authors are grateful to The Chinese National Key Fundamental Research "973" Program (G1998040700) ,The Natural Science Foundation of China (No.49974009 and 50174013) and The Post-doctor Natural Science Foundation of China (No. 2001) for their support.

References

1. Ingraffea A.R., Heuze F.E. Finite element models for rock fracture mechanics. Int. J. Numer. Anal. Mech. Geomech. ,1980,4:25-43
2. Hoek E., Bieniawski Z.T. Brittle fracture propagation in rock under compression. Int. J. Fracture, 1984,26:276-294
3. Petit J., Barquins M. Can natural faults propagate under II mode conditions? Tectonics, 1988, 7(6):1243-1256
4. Shen B., Stephansson O., Einstein H.H., et al. Coalescence of fracture under shear stresses in experiments. J. Geophys. Res., 1995,100(B4): 725-729
5. Bobet A., Einstein H.H. Fracture coalescence in rock-type materials under uniaxial and biaxial compression. Int. J. Rock Mech. Min. Sci., 1998,35(7):863-889
6. Bobet A., Einstein H.H. Numerical modeling of fracture coalescence in rock materials. Int. J. Fracture, 1998,45:263-282
7. Wong R.H.C., Chau R.T. Crack coalescence in a rock-like material containing two cracks. Int. J. Rock Mech. Min. Sci., 1998,35(2): 147-164
8. Vasarhelyi B., Bobet A. Modeling of crack initiation, propagation and coalescence in uniaxial comp. Rock Mech. Rock Eng., 2000,33(2): 119-139
9. Hart R. Cundall PA. Lemos J. Formulation of a 3D distinct element model. Part II : mechanical calculations for motion and interaction of a system composed of many polyhedral blocks. Int. J. Rock Mech. Min. Sci. 1988,25:117-125
10. Tang CA. Numerical simulation on progessive failure leading to collapse and associated seismicity. Int. J. Rock Mech. Min. Sci. 1997,34:249-262
11. Tang CA. A new approach to numerical method of modeling geological processes and rock eng. problems. Engineering Geology 1998,49:207-214

NUMERICAL SIMUALTION TO EXCAVATION OF THICK COAL SEAM

JIANXIN LIU CHUN'AN TANG SHUHONG WANG
Center of Rock Instability & Seismicity Research, Northeastern University, Shenyang, 110004

SFPA is utilized to simulate the process of thick coal seam mining. The simulated results reproduce the dynamic evolution process, and point out the influence to the rock burst of the existence of internal and external stress fields.

1 Introduction

The rock burst is the result of mining, and it is caused by that the in-situ stress field is disturbed by excavation. When the coal (rock) comes to its critical condition, the rock burst will be induced by some external factors. Recently, CRISR (Center of Rock Instability & Seismicity Research) from NEU (Northeastern University) made systemic researches on rock failure process and rockburst, and pointed out that the occurrence of rock burst is dynamic, however its pregnancy is static. Therefore the rock burst is the mechanics process from static deformation to dynamic failure, and it is of dynamic-static double characters. One theory and its relevant research method which are static and dynamic simultaneously are need to establish, and the research emphases is the formability process of rock burst, not its occurrence process, that is to say, how the coal (rock) structure converts from stable equilibrium state to instable equilibrium state.

2 SFPA simulation applications

SFPA^{2D} system utilized in this paper is one of numerical simulation tools, which can simulate the dynamic progressive process of vertical discontinuous displacement, bending, sagging, fracturing and caving of overlying rock strata. The code has been developed in considering the deformation of an elastic material containing an initial random distribution of micro-features [2][3]. The stress distribution and its curve in the coal seam simulated by SFPA are shown in Fig.1. It is clear that there are the internal stress field and the external stress field in the coal seam. The formability of the two stress fields is mainly due to that after the upper coal seam ruptures, the stress field near the rupture line decreases and transfers to the flanks, and with the working face advancing, the pinnacle of internal stress field moves towards the coal wall and the stress of external stress field expands outwards. Song [1] pointed out that when mining in the internal stress field, the stress value of the internal stress field will not make the occurrence of rock burst. However, the excavation continues forwards to the external stress field, the stress value comes to the peak value immediately and leads to the occurrence of rock burst. A visible stress arch is shown in Fig. 1 in the upper coal. Based on the literature [4], after the development workings excavated, the elements composing of the adjoining rocks move towards the drift under the gravity. And during the process of moving downwards, the elements are forced to interactively squeeze. On the other hand, the elements of roof will contact with each other for bulk when moving downwards, and which make themselves slim rotary motion that increases the wedge-caulking force as well. The wedge-caulking degree of every point in roof changes with coordination, therefore during the rocks slump along the contact face and rotary slimly, a limit equilibrium arch will appear in upper roofs—natural self-supporting arch.

3 Numerical simulation results and Discussions

It is simulated and discussed that the coal seam breakage caused by excavation and the formability and distribution of internal and external stress fields in the coal internal. When the middle of the discovered rock seam is not drawn, the occurrence of breakage is determined by the working face height. Generally, the breakage height is 60 percent of worked-out area, and when the working face advances 10m, the breakage height is about 15m. In this paper, the worked-out area is simulated by 15m heights so that the movement of coal seam can be simulated conveniently.

The advantage of $SFPA^{2D}$ is that it can reproduce the failure process of rock and easily obtain the curve of surface movement. Fig.2 shows the representative simulated results and the gray scale reflects the rock mass mechanic parameters distribution. The value of rock mass mechanic parameter is larger if gray scale is higher.

In virtue of the existence of joint and bedding, the coal body will rupture and breakage with the working face advancing, not the structure failure. Before the working face advances 30m, the discovered coals will not rupture and breakage under the gravity for the strength of themselves (See Fig. 2(a)). When the working face comes to 30m, the discovered coals will rupture under the gravity, and breakage (See Fig. (b)(c)). With the working face advancing, natural self-supporting arch forms gradually, and it will expand upwards until comes to the rock strata upper the coal seam (See Fig.2 (d)~(h)). When the working face comes to about 100m, the breakage arrives to the top of the coal seam (See Fig.2 (i)(j)), that is to say, the equilibrium stress arch comes to the rock strata upper the coal seam, and the rock strata is discovered, the sagging and caving will appear gradually.

4 Conclusion

SFPA is utilized to simulate the process of thick coal seam mining. The simulated results reproduce the dynamic evolution process, and point out the influence to the rock burst of the existence of internal and external stress fields.

Acknowledgements

The study was supported by the China National Key Fundamental Research "973 Program" (No.G1998040700) and the China National Natural Science Foundation (No.50134040, No.50174013, No.49974009)。

References

1. Song Z.Q., Practical support stress control in mining engineering, Beijing, Coal Industry Published House, 1992
2. Tang C.A., Numerical Simulation of AE in Rock Failure [J], Journal of Rock Mechanics and Engineering, 1997,16(4): 368～374
3. Tang C. A., Numerical simulation of rock failure and associated seismicity[J]. Int J. Rock Mech Min Sci, 1997, 34(2): 249~262
4. Lin Y. M., Ground pressure tutorial, Coal Industry Published House, 1981.12

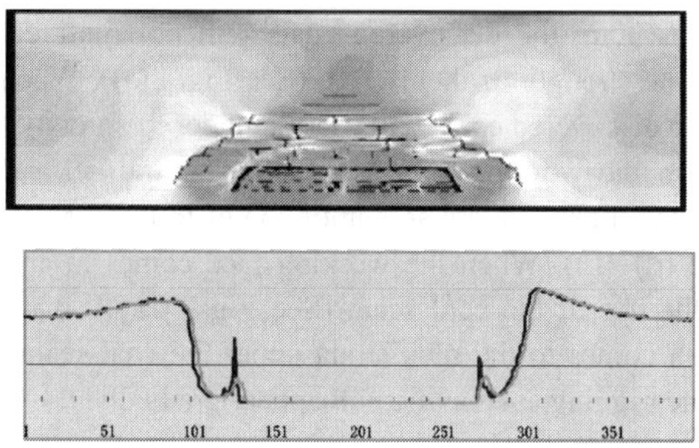

Figure1. the simulated stress distribution and its curve (the red line is the stress fitting curve)

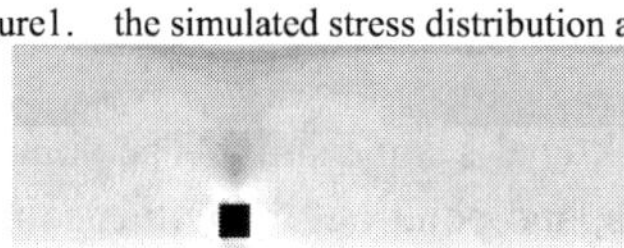

(a) Working face advancing 15m

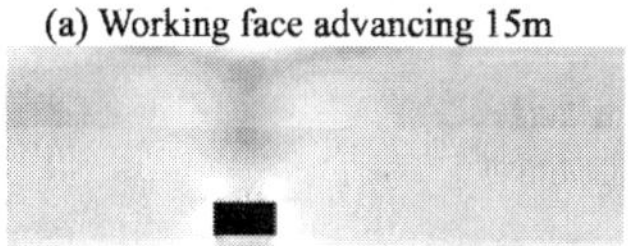

(b) Working face advancing 30m

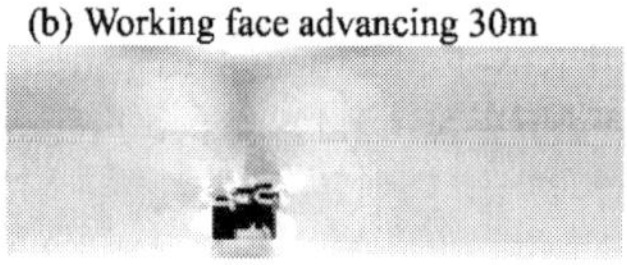

(c) Working face advancing 30m

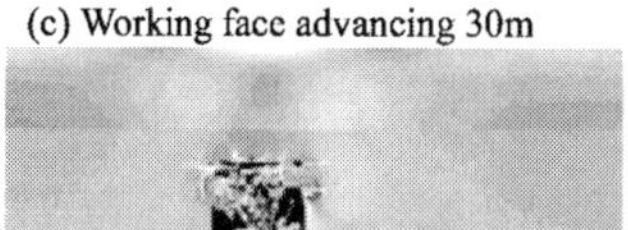

(d) Working face advancing 45m

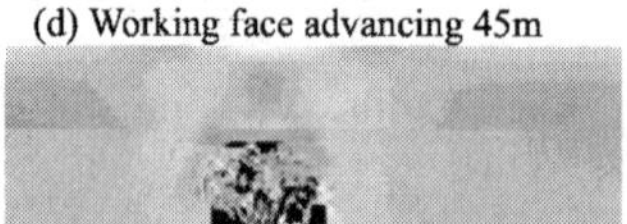

(e) Working face advancing 55m

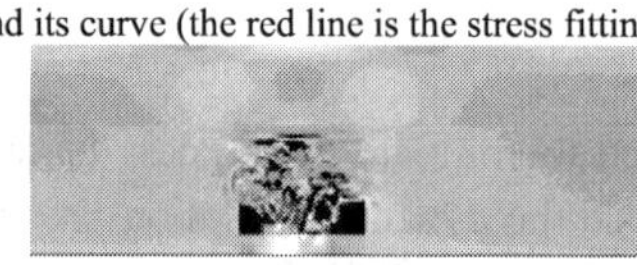

(f) Working face advancing 60m

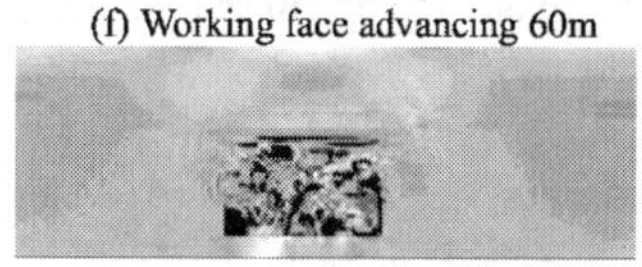

(g) Working face advancing 75m

(h) Working face advancing 90m

(i) Working face advancing 95m

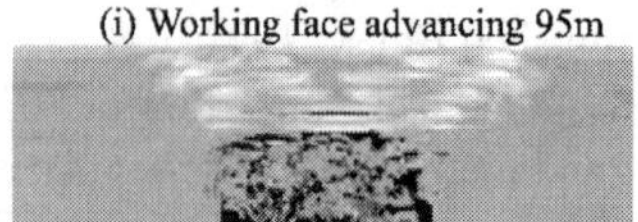

(j) Working face advancing100m

Figure 2. the simulated stress distribution during breakage process

4. Artificial Intelligence and System Science

HEURISTIC ANT COLONY ALGORITHM FOR ROCK SLOPE ENGINEERING*

CHANGFU CHEN, YU YANG
Institute of Geotechnical Engineering, Hunan University, 410082, Changsha, China

Based on modifying the basic ant colony system's structure and the transfer probability of ant, a new heuristic ant colony algorithm (HACA) is proposed to analyze the stability of rock slope engineering. The application of HACA to search for the critical slip surface of rock slope is discussed. The results show that the presented method has obvious advantages compared with traditional methods.

1 Introduction

The research shows that ants communicate with each other through a material called pheromone during searching for foods. When ants move, they will release and feel the pheromone on the route. Ants can decide which direction to take according to the intensity of the material. Thus, the more ants pass through one route, the more pheromone trails will be left and the greater the probability will be that ants will choose this route. According to this, M. Dorigo *et al* have developed a new kind of bionic algorithm—Ant Colony Algorithm [1] (ACA) and it is successfully applied to solve the combination optimization problems such as TSP [2]. As ACA has many excellent features, for example parallelism, robustness and strong applicability, it has been given increasing attention in industrial engineering world.

The stability analysis of slope aims to find out the minimal factor of safety and the corresponding critical slip surface. It is a nonlinear programming problem with multi extreme values and usually, the object functions could not be expressed as the explicit expressions. Therefore, the critical slip surface of complex slopes is hard to be found out by conventional optimizing methods. In this paper, a new heuristic ant colony algorithm (HACA) is proposed based on modifying the basic ant colony system's structure and the transfer probability of ant and it is used to search for the critical slip surface and the minimal safety coefficient of rock slope.

2 HACA for Stability Analysis of Rock Slope

For ACA being fit for solving the problem of slope stability, a slope is discrete by the

* Supported by the Science Fund of Hunan Province in China (No. X99P51059)

separating lines and nodes shown in Figure 1. An ant starts from the point START, and then passes the nodes on the separating line one by one up to the point END. Thus, a circulation is finished and a slip surface formed. It is noticed that the dotted lines in Figure 1 are only used to guide the ants to search for slip surfaces.

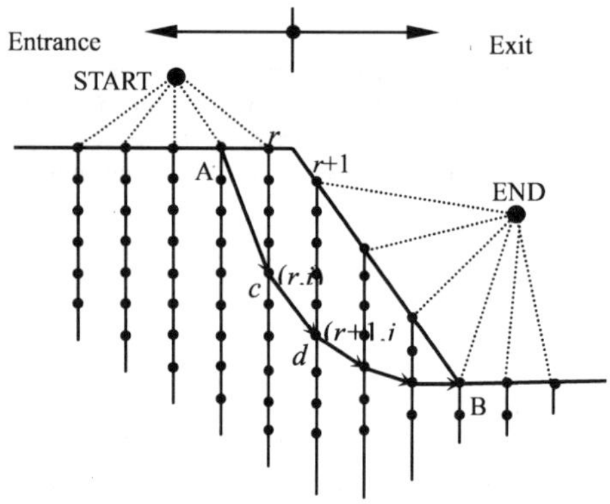

Figure 1. Search techniques of critical slip surface of slope

Suppose that (r, i) represents the ith node on the rth separating line; $(r+1, j)$ represents the jth node on the $(r+1)$th separating line; $[(r, i), (r+1, j)]$ represents the line from (r, i) to $(r+1, j)$ and the number of ants in ant colony is m. During the movement of ants, ant k (k=1, 2, ⋯, m) decides its transfer direction according to the pheromone quantity on the routes. At moment t, the probability $p^k_{[(r,i),(r+1,j)]}(t)$ of ant k transferring from (r, i) to $(r+1, j)$ can be expressed as

$$p^k_{[(r,i),(r+1,j)]}(t) = \frac{\left[\tau_{[(r,i),(r+1,j)]}(t)\right]^{\alpha}\left[\eta_{[(r,i),(r+1,j)]}(t)\right]^{\beta}}{\sum\limits_{j=1,2,\cdots M_{r+1}}\left[\tau_{[(r,i),(r+1,j)]}(t)\right]^{\alpha}\left[\eta_{[(r,i),(r+1,j)]}(t)\right]^{\beta}} \tag{1}$$

in which, $\tau_{[(r,i),(r+1,j)]}(t)$ is the intensity of the remained pheromone on the line $[(r, i), (r+1, j)]$ at moment t. At initial moment, $\tau_{[(r,i),(r+1,j)]}(0)=C$ (C is a constant). α and β are respectively the importance coefficients ($\alpha \geq 0$, $\beta \geq 0$). $\eta_{[(r,i),(r+1,j)]}(t)$ is the visibility of the pheromone on the route $[(r, i), (r+1, j)]$ at moment t. Here, we have

$$\eta_{[(r,i),(r+1,j)]}(t) = \begin{cases} \dfrac{M_{r+1}+1-f_{r+1}}{(M_{r+1}+1-f_{r+1})f_{r+1}+\sum\limits_{s=f_{r+1}+1}^{M_{r+1}+1}(M_{r+1}+1-s)}, & 1 \leq j \leq f_{r+1}\quad f_{r+1}=\left[\dfrac{M_{r+1}}{2}\right]+1 \\ \dfrac{M_{r+1}+1-j}{(M_{r+1}+1-f_{r+1})f_{r+1}+\sum\limits_{s=f_{r+1}+1}^{M_{r+1}+1}(M_{r+1}+1-s)}, & f_{r+1} < j \leq M_{r+1} \end{cases} \tag{2}$$

in which, M_{r+1} is the number of nodes on the separating line (r+1).

With time going on, the intensity of the remained pheromone on the routes ants have passed will gradually abate. Parameter ρ ($0 \leq \rho \leq 1$) represents the lasting ability of pheromone, (1−ρ) is regarded as the evaporation degree of pheromone. During the interval Δt, each of the m artificial ants finds out a feasible solution, that is, finishes a circulation; the intensity of pheromone on each route [(r, i), (r+1, j)] can be adjusted according to the following formula

$$\begin{cases} \tau_{[(r,i),(r+1,j)]}(t+\Delta t) = \rho\tau_{[(r,i),(r+1,j)]}(t) + \Delta\tau_{[(r,i),(r+1,j)]} \\ \Delta\tau_{[(r,i),(r+1,j)]} = \sum_{k=1}^{m} \Delta\tau^{k}_{[(r,i),(r+1,j)]} \end{cases} \quad (3)$$

in which, $\Delta\tau^{k}_{[(r,i),(r+1,j)]}$ represents the pheromone quantity ant k releases on the route [(r, i), (r+1, j)] in this circulation.

The different computation methods of $\Delta\tau^{k}_{ij}$ can constitute different ACA. For slope stability analysis, a special heuristic operator is presented as follows:

$$\Delta\tau^{k}_{[(r,i),(r+1,j)]} = \begin{cases} Q/F^{k}_{s}, & \text{if route } [(r,i),(r+1,j)] \text{ is on the slip surface} \\ & \text{searched by ant } k; \\ 0, & \text{others.} \end{cases} \quad (4)$$

in which, Q is a constant; F^{k}_{s} is the safety factor corresponding to the slip surface searched by ant k in this circulation and it can be computed through Janbu's method.

The HACA for searching for critical slip surface can be described as follows:

STEP 1: $t := 0$; $\tau_{[(r,i),(r+1,j)]}(0) := C$; $\Delta\tau_{[(r,i),(r+1,j)]} := 0$;

STEP 2: m ants search for the effective slip surfaces from point START to point END according to Eq. (1).

STEP 3: compute the F^{k}_{s} of each effective slip surface and save the best one;

STEP 4: compute $\Delta\tau^{k}_{[(r,i),(r+1,j)]}$ and $\tau_{[(r,i),(r+1,j)]}(t)$ by Eq. (3) and (4) ;

STEP 5: $t := t+1$; If $t \geq t_{\max}$ (the maximum circulation number), output the best route (the critical slip surface) and the minimal safety factor; else turn to step 2.

3 Example

An open-pit slope is 99.3m high and its inclination angel is 53° . The profile of slope is shown in Fig.2 and the computation parameters are γ =27.2kPa, C_m=0.1125Mpa, φ_m=36.121 °.

Based on Janbu's method, the presented heuristic ant colony algorithm is applied

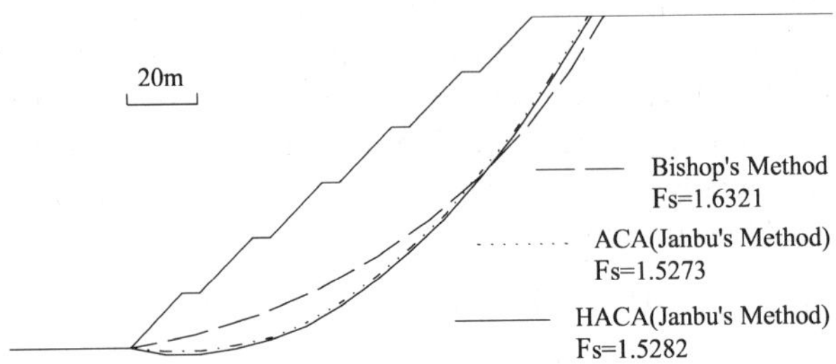

Figure 2. The computing results

to search for the critical slip surface and the result is shown in Figure 2. The results of ACA method and Bishop's method are also shown in Figure 2. From Figure 2, the results of HACA are very consistent with that of ACA, but the HACA consumes lesser computing time.

4 Conclusions

The stability analysis of complicated slope is actually a nonlinear programming problem with multiple extreme values and the objective function usually is not explicit. The traditional optimization methods are difficulty to find out critical slip surface. For the ant colony algorithm being fit for solving the stability of rock slope, the basic ACA's structure is modified and a heuristic operator is introduced to determine transfer probability of ant. And a heuristic ant colony algorithm is presented. It is used to search for the critical slip surface and evaluate the stability of open-pit slope. The computation results show that the proposed method (HACA) is efficient and feasible. It can also evaluate various slopes with complicated soil layers.

References

1 Dorigo M, *et al.* Ant system: optimization by a colony of cooperation agents. IEEE Trans. on SMC-Part B, 1996, 26(1): 29-41.

2 Dorigo M, Gambardella L M. Ant colony system: a cooperative learning approach to the travelling salesman problem. IEEE Trans. on Evolutionary Computation, 1997, 1(1): 53-56.

PARAMETER DETERMINATION FOR CONCRETE DAM AND ROCK FOUNDATION USING GENETIC ALGORITHM

SHOUJU LI, YINGXI LIU, XIANG HE, YUANPAI ZHOU
Dept. of Engrg. Mechanics, Dalian Univ. of Technology, Dalian116024, P. R. China

The inverse problem of material constants identification of concrete dam and rock masses is formulated as an optimization problem, which is then solved by using genetic algorithm(GA). Based on the measured horizontal displacements of dam, the inversion procedure was proposed for determining the material constants of rock masses and concrete dam. To speed up the rate of convergence, the paper proposes a hybrid GA that combines the gradient descent technique. The practical application to Fengman concrete dam shows that the parameter inversion approach proposed by the paper can obviously enhance the forecasting precision of dam deformation.

1 Introduction

The most difficult aspect in analyzing the deformation and stress of a dam is the estimation of the material constants of concrete dam and rock masses. There is no doubt that the laboratory investigation and in-situ testing are the important measures, but have their limitations. In order to compensate for such deficiency, the revision of the material constants is very important according to the observation of dam deformation, so that the deformation of the dam may be better predicted and agree with the observed ones. The material constants of rock masses and concrete dam can be identified according to measured deformations of the dam by means of back analysis. The goal of the paper is to build the parameter identification procedure to estimate the material constants in the rock masses and concrete dam according to the observing data of deformation.

2 Genetic Algorithms for Parameter Inversion of Concrete Dam

Based on the concept of genetics, GA simulates the evolutionary process numerically[1]. Analogous to genes in genetics, GA represents the parameters in a given problem by encoding them in a single point in traditional mathematical optimization methods, in GA a set of points, that is, a population of coded strings, is used to search for the optimal solution. The main processes of genetic algorithm are illustrated as follows:

1) Establishment of objective function and fitness function[2]

$$J_j = \|u_m - u_c(X)\|_2^2 \qquad (j = 1,2,..., M) \tag{1}$$

Where X is the vector of identified parameters, $X=\{E_c,E_r\}^T$; E_c and E_r are the elastic moduli of concrete dam and rock masses, respectively; U_c is the vector of calculated displacements at observation points based on some estimated values of material constants, and U_m is vector of observed displacements. The fitness function F_j is shown as follows:

$$F_j = C - J_j \tag{2}$$

2) Representation of solution[3]. An essential characteristic of the genetic algorithm is the coding of the variables that describe the problem. For several reasons, the binary or bit string representation of solutions has dominated genetic algorithm research. Bit string representations are still the most commonly used encoding techniques and have been used in many applications of GA.

3) Reproduction operation. This operator produces a new population, extracting with repetition individuals from the old population. The extraction can be carried out in several ways. One of the most commonly used is the roulette wheel selection, where individuals are extracted in probability following a Monte Carlo procedure. The extraction probability of each individual is proportional to its fitness as a ratio to the average fitness of all the individuals in the old population.

4) Crossover operation[4]. This is a process by which information contained in tow candidate solutions is combined. The crossover process allows for an exchange of design characteristics among members of the population pool with the intent of improving the fitness of the next generation. The crossover operator combines the features of two parent structures to form two similar offspring.

5) Mutation operation. Mutation safeguards the genetic search process from a premature loss of valuable genetic material during the selection and crossover. The process of mutation is simply to choose few members from the population pool according to the probability of mutation, and to switch a 0 to 1 or vice versa at a randomly selected mutation site on the selected string.

6) Local search strategy. A hybrid genetic algorithm is one in which a GA is combined with a different optimization algorithm. Local search is usually a simple gradient descent heuristic that carries each solution to a local optimum.

3 Application

The Fengman concrete dam is located in the middle reaches of Songhua river, Jilin province, China. The concrete dam is built in 1937. The dam is a concrete gravity one with a total length of 1080m and a maximum height of 91.7m. There are total 60 blocks, each of which has a length of 18m. The crest horizontal displacement of concrete dam can be expressed as follows:

$$d = d_h + d_T + d_\theta \qquad (3)$$

Where d is the crest horizontal displacement of concrete dam, mm; d_h is elastic component caused by water level variations; d_T is elastic component caused by temperature variations; d_θ is an elastic component. By means of the multi-elements regression techniques, the elastic component of dam displacement caused by water level variations can be separated from formula (3). The elastic component of dam displacement caused by water level variations at block 17 is expressed in the follows:

$$d_h = 0.361 - 17.184H + 21.339H^2 \qquad (4)$$

Where $H=(h-200)/600$, h is the water level at the upstream of dam. Taking d_h as measured displacement vector u_m at different water level, we can do parameter inversion research for rock masses and concrete dam by using genetic algorithm.

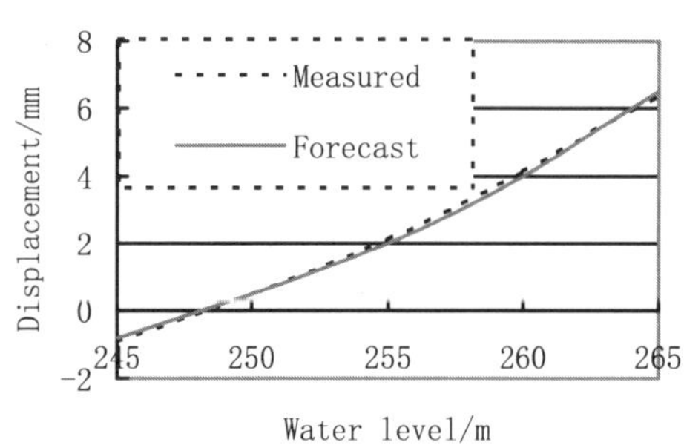

Figure 1. Comparison of measured horizontal displacements with forecasted ones at different water level at block 17

The observed data of crest displacements has been recorded since 1954 in the Fengman concrete dam. And the laser monitoring system was built in 1984, its observing precision can be less than 0.06mm. Based upon the improved genetic algorithm, the elastic moduli of the concrete dam and the rock foundation are identified according to the horizontal displacements of the Fengman concrete dam. The identified values of elastic moduli of concrete dam and rock masses are shown as follows respectively: Ec=31.458GPa, E_r=10.379Gpa. Figure 1 shows the comparison of measured horizontal displacements with forecasted ones at different water level at block 17.

4 Conclusion

In order to monitor the operation state of concrete dam, it is necessary to set up mathematic model of monitoring index for the large dam safety. The estimation of elastic parameters for concrete dam and rock foundation, in turn, is the basement of setting up mathematic model of large dam safety. The feasibility of applying hybrid GA to the inverse problem was researched in the paper. The practical application to Fengman concrete dam shows that the proposed inversion method can speed up the convergence of the inverse problem solution in searching for the global optimization.

References

1. M I. Friswell, Computers & structures , 69,547(1998).
2. K N. Sankar, Geophys. J. Int. 138,108(1999).
3. G. S. Maifeld, Electric Power Systems Research, 31,147(1994).
4. J. H. Chou, Computers & structures, 79,1335(2001).

STUDY ON BOREHOLE STABILITY PREDICTION BEFORE DRILLING BY NEURAL NETWORK

MIAN CHEN YAN JIN
University of Petroleum, China

Borehole stability analysis after drilling is easy and practical for the drilled well, but the result may be impractical for the well to be drilled because of the change in tectonic structure. In addition, the stability of wildcat well is generally determined by experience, so the disaster at complex formations is unable to be avoided. In order to prevent from borehole instability problems, a new method for predicting borehole stability before drilling is studied closely in this paper. According to neural network theory, a new model for predicting borehole stability by seismic data is established, and successful in field application.

1 Introduction

Neural network theory and method have been widely applied in many oil fields in recent years, such as lithology and permeability analysis by using logging data, reservoir parameters prediction and formation analysis by using seismic data, petroleum production prediction, petroleum pipelines design, bit type selection in petroleum engineering. The main principles of these applications are identical. Message input layer, hidden layers and message output layer constitute a structure similar to that of human brain. This kind of structure can be of an effective prediction and identification system. By properly picking up input message and topology of the network, neural network can solve highly non-linear problems.

During drilling process, the borehole stability of various types of formation relies on controllable factors and uncontrollable factors [1]. The uncontrollable factors include the in-situ stress, rock elasticity and strength, anisotropy induced by geometric occurrence and joints as a result of tectonic activities. The controllable factors include density and chemical properties of the mud, hydraulic parameters and operation program, etc. As a result, borehole stability research should be focused on the effects of controllable factors to select optimal mud weight and performance, and operation parameters, based on understanding the regularities of uncontrollable factors. It is an urgent problem for drilling at foot-hill to determine the safe mud density range according to the characteristics of foot-hill structures so that borehole mechanic instability can control successfully.

It is much easier for borhole stability analysis after well completion than borehole stability prediction before drilling, because drilling log, logging, and core data are available for stability analysis while only seismic data are available for stability prediction. For pre-drilling prediction, the only data that can be used are seismic data and borehole stability of drilled wells, and no one has researched this problem until now. based on borehole stability analysis of the drilled wells in adjacent structures, a neural network model for predicting borehole stability is established, and is applied to predict borehole stability for a new well to be drilled.

2 BP Model For Prediction Borehole Stability

Maintaining a stable borehole is of primary importance for the successful drilling of a well, which relies on the correct choice of mud weight after consideration of the break-out and fracture pressure in each hole section. Borehole stability requires a proper balance between the uncontrollable factors of in-situ stresses, rock strength, and pore pressure, and controllable factors of borehole fluid pressure and mud chemical composition. As for the formation to drilled, it is important to understand the relationships between the uncontrollable factors and seismic wave velocity, which are the basis of the neural net model of borehole stability prediction .

Horizontal in-situ stresses at great depth induced by overburden pressure and tectonic stress field can be expressed with the overburden pressure. So horizontal in-situ stresses can be described as shown

$$\begin{aligned} \sigma_H &= f_H(V_n)\sigma_v = f_H(V_n)\int \rho g dH \\ \sigma_h &= f_h(V_n)\sigma_v = f_h(V_n)\int \rho g dH \end{aligned} \qquad (1)$$

where σ_H is the maximum horizontal in-situ stress, σ_h is the minimum horizontal in-situ stress, σ_v is overburden pressure, V_n is seismic wave velocity, $f_H(V_n)$ is function of in-situ stress for σ_H, $f_h(V_n)$ is function of in-situ stress for σ_h, ρ is density of formation, H is well depth and $g = 9.8N/Kg$.

If the overburden pressure in the structure to be drilled is obtained from the statistic result of drilled region, and can be expressed by a function $f_g(H)$, formula (1) becomes

$$\begin{aligned} \sigma_H &= f_H(V_n)f_g(H) \\ \sigma_h &= f_h(V_n)f_g(H) \end{aligned} \qquad (2)$$

It is found that there are some relationships between rock strength and sonic velocity. The faster the sonic velocity through rock is, the higher rock strength, elastic module and density are. At the same time, rock strength and sonic velocity are both affected by in-situ stresses. So rock strength can be written as

$$Strength = f_C(V_n)f_g(H) \qquad (3)$$

where $f_C(V_n)$ is function of rock strength

In addition, some discrepancy exists between the seismic wave velocity and the sonic velocity from logging data in the same well. The later is more precise and can reflect real formation information. The in-situ stress and strength prediction models are the statistic results of logging data, so the discrepancy should be corrected if the seismic wave velocity is used for borehole stability prediction.

Finally, based on BP neural network theory [2,3], the borehole stability prediction model can be shown as Figure 1.

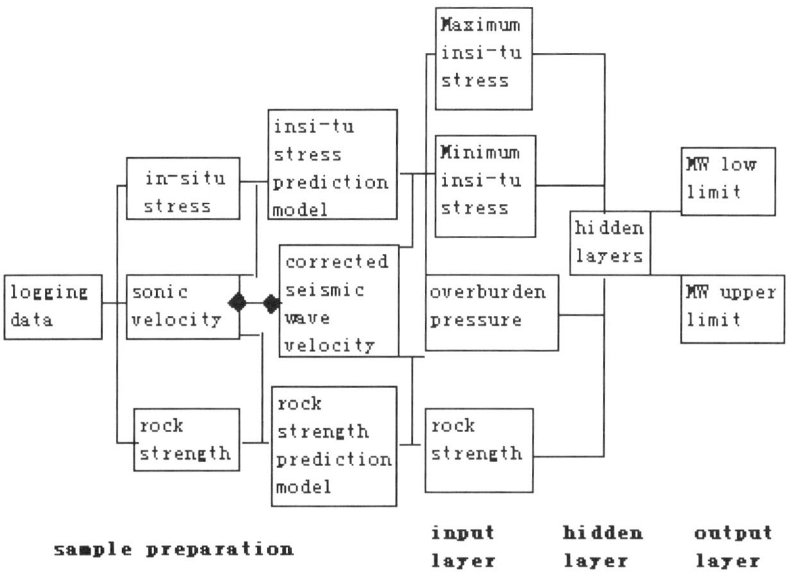

Figure 1. Neural network model for borehole stability prediction

3 Field Application

Yiqikelike is main rich gas region in Tarim, and important for the national project transferring gas from west to east China. Yn2 well was drilled and experienced fatal borehole instability which wasted nearly 2 million dollars. Considerable effort was directed toward understanding the mechanism of borehole instability, and the conclusion was arrived that the mud weight program was unsuitable for controlling borehole instability. In order to avoid borehole instability for Ys4 well to be drilled, it was emphasized on studying the safe mud range for special formations of Yn2 well. But the results can not be applied to guide Ys4 drilling operation directly, because of a fault between Yn2 well region and Ys4 well region. Fortunately, borhole stability prediction model in this paper can challenge the problems. By comparing with the predicted safe mud range before drilling and analyzed safe mud range after drilling, the prediction model can guide drilling operation. The result is shown as Figure 2.

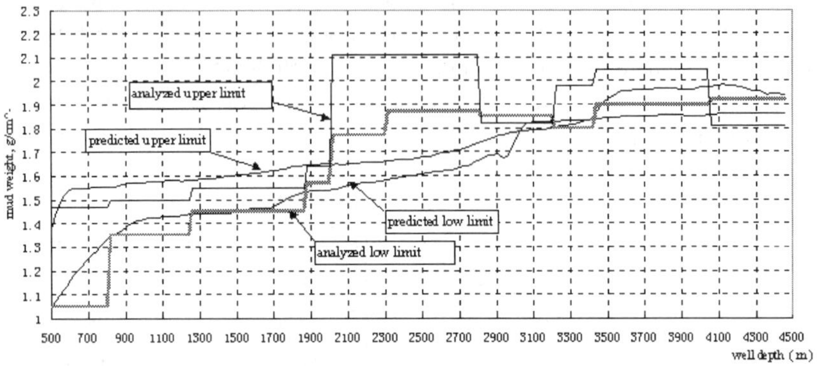

Figure 2. Comparison between the prediction and the practical mud weight range

4 Conclusions

The model of borhole stability prediction overcomes the difficulty that the safe mud weight range can not be predicted before drilling using seismic data. The model is successful in field application, however, the precision of prediction is affected by the accuracy of seismic wave velocity and the formation thickness. Generally, seismic wave velocity could not precisely reflect thinner intermediate layer, so the prediction is not reasonable. The model is still to be improved.

References

1. Gao Deli (1999), The instability and control of borehole system for oil and gas drilling, China science and technology Press,China, 1-9
2. Jiao Licheng (1993), Computation of artificial neural networks, Xi'an electronics science and technology university Press, China,268-279
3. Zhang Liming (1994), Models and applications of artificial neural networks, Fudan University Press, China, 25-34

OVERBURDEN FAILURE PREDICTION BASED UPON ARTIFICIAL NEURAL NETWORK MODEL

DONG QINGHONG
School of Resources and Earth Sciences, China University of Mining and Technology, Xuzhou, Jiangsu Province, P.R.C, 221008

Usually, mechanics property of rock mass and many other factors play important roles in deciding its overburden failure during mining. The paper first analyses several pivotal factors' influencing on the height of overburden failure zone, such as thickness of overburden rock mass, mechanics property of rock mass, clay rock content ratio in overburden rock mass, mining system, obliquity of coal-measures, size of working face, distributing of water-bearing zone in soil mass and etc. Easily, each position in mining workaround is corresponding to a set of such factors and the actual height of failure. And that is considered as a sample. Then, 56 samples of a certain zone are obtained and filtered and then imported to an artificial neural network model. After about 95110 times training, the model gives the result with a relativity error below 0.1 m^2. Then an artificial neural network model for overburden failure prediction is made, and which is a back propagation networks model with an input layer of 11 elements, two hidden layers each has a suitable elements number 45 and 8, and an output layer of failure height include 2 elements. Finally, 8304 working face, another zone of Taiping coal mine is selected and from which 7 samples are imported to the model. In this area, the thickness of overburden rock mass above No.3 coal seam ranges from 20m to 80m, and the result of failure height, each site according to its position, varies from 12.68 m to 22.32 m. And this prediction proved to be successful in production.

1 Introduction

Overburden failure during mining relates with mine method and physic- mechanical properties of rockmass etc. This failure zone is vital to decide the rational mining up limit under thin overlay rock mass and with water bearing Quaternary system above. Taiping coalmine, which lies in the west of Yanzhou digging of Shandong, China, is a typical one of such geology conditions. NO.3 coal seam, with the thickness of 9 meters, embeds in Permo-carboniferous system and the thickness of overburden rock mass above No.3 coal seam ranges from 20 to 80m. On the above is Quaternary system, which is composed of 3 aquifers and 2 water- resisting layers interlacing. Aquafer 3 is the very danger of No. 3 coal seam mining. An artificial neural network model with back propagation analysis is founded to solve this multi-factors problem.

2 learning arithmetic of BP network model

Back propagation network model is commonly abbreviated to BP network model. In a BP mode, Input vectors and the corresponding target vectors are used to train a network until it can approximate a function in an appropriate way as one defined. Standard back propagation model is a gradient descent algorithm, in which the network weights are moved along the negative of the gradient of the performance function. Such multiple layers of neurons allow the network to learn nonlinear relationships between input and output vectors. As with momentum, the model error is controlled and self optimized to the lowest level but lower. Figure 2 presents the structure of the network model. The input and output vectors are shown in the graph and two hidden layers with 45 elements and 8 elements reveals respectively. And transfer functions of log-sigmode, tan-sigmod, linear are picked for each layer correspondingly in the paper.

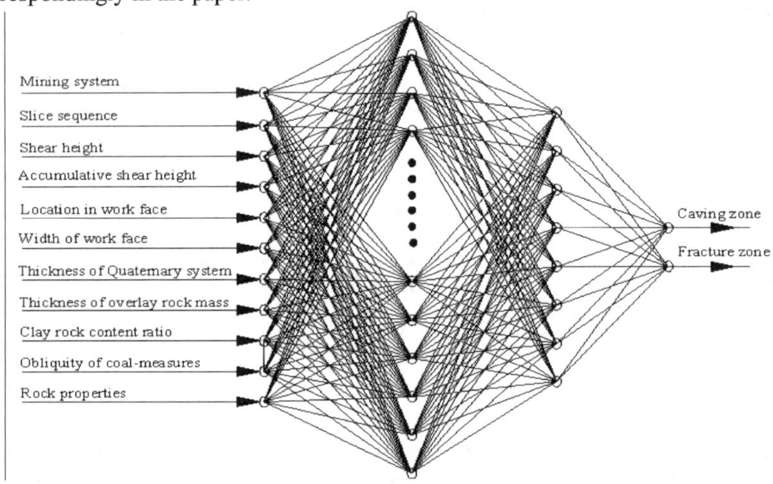

Figure 1. Overburden failure prediction model

3 Selection of pivotal factors

It is well known in the past researches that mechanics property of rock mass; mining system; dimension of working face; obliquity of coal-measures and etc. controlled the process of overburden failure. What's more, there are several other agents listing below play important roles. Firstly, thickness of overburden rock mass is the maximum dimension of failure zone, and it is proved that mining under thinner overlay rock mass always with lower failure extension. Secondly, clay rock content ratio in overburden rock mass and weathering zone are restraints of rock cracking because of gravity closing effect. Thirdly, water-bearing zone in soil mass is the direct water source that hang over the work face. If the failure zone extends to this zone without any resistance then a crisis may happen. The fourth factor is roof clay, embedding at the bottom of Quaternary system, which reinforces protective layer effectively. Then the penman and colleagues take all such influential and advisable factors into consideration as follows: thickness of overburden rock mass, mechanics property of rock mass, clay rock content ratio in overburden rock mass, mining system, obliquity of coal-measures, size of working face, distribution of water-bearing zone in soil mass and distribution of roof clay. Eleven factors deducing from the former 6 ones constitute the input vectors and the latter two are conditions aided design the rational up limit. And the failure dimension include cave zone and fractured zone make up the target vectors. The total number of samples gained is 56 and 27 are filtered out because of inconsequence and unavailability. And 10 samples of which are listed here in table 1 for explanation. As a matter of convenience, input vectors and target vectors are abbreviated to M_s, S_s, S_h, A_h, L, W_w, T_q, T_r, C_r, O_c, R_p, C_a, C_p, F_a, F_p according to column 1 of the table, and quantification of such vectors are made.

Table 1. Main factors quantification (10 of 56 samples)

Sample No.	1	2	3	4	5	6	7	8	9	10
Mining system	2	0	0	2	0	0	0	1	1	0
Slice sequence	0	2	2	0	2	1	1	0	0	1
Shear height	9.4	2.7	2.7	7.5	2.6	2.8	2.8	2.5	2.8	2.2
Accumulative shear height	9.4	5.2	5.4	7.5	5.4	2.8	2.8	7.3	7.6	2.2
Location in work face	0	0	0	1	0	0	1	0	0	1
Width of work face	1.2	1.4	1.4	1.4	1.4	1.4	1.4	1.4	1.4	1.1
Thickness of Quaternary system	1.61	1.85	1.8	1.75	1.8	1.7	1.75	1.95	1.95	1.87
Thickness of overlay rock mass	0.8	0.5	0.6	0.8	0.9	1.1	1.2	0.90	0.8	0.33
Clay rock content ratio	0.42	0.35	0.3	0.25	0.25	0.2	0.25	0.25	0.3	0.13
Obliquity of coal-measures	10.0	5.0	8.0	8.0	8.0	5.0	5.0	8.0	8.0	6.0
Rock properties	69.4	95.0	100	100	105	105	105	105	105	78.5
Actual caving zone (m)	24.8	19.1	17.2	31.2	17.4	11.6	9.23	20.0	26.0	5.98
Prediction caving zone (m)	24.9	18.6	18.7	27.2	19.5	12.3	10.5	21.2	26.5	6.13
Actual fracture zone (m)	56.4	39.7	41.5	61.5	43.4	31.2	19.9	36.2	43.5	10.7
Prediction Fracture zone (m)	56.5	39.9	40.6	62.9	42.3	31.2	19.4	37.1	43.4	12.9

4 Training and Prediction

Program is compiled to fulfill this model. The system error is restricted to E<0.1m^2 and takes the program runs for 95110 times. The

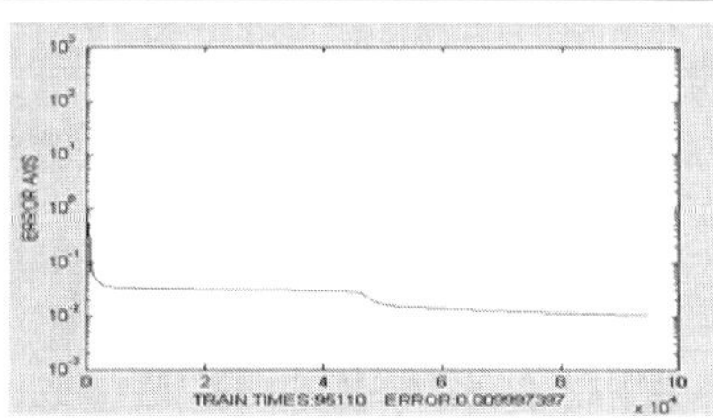

Figure 2. Convergence and Error adjustment

process of error droop and model convergence see figure 3 and results reveal in table 1. As a result the weight, bias of each element and transfer functions are reserved for model application. Finally, factors of 8304 work face in Taiping coalmine is imported in the model and then the failure dimension of slice mining for the first sequence is given, see table 2. Here C_p^* and F_p^* are caving zone and fractured zone gained from finite element model.

Table 2. Overburden failure prediction of 8304work face in Taiping coalmine

L	S_h	W_w	T_q	T_r	C_r	O_c	R_p	C_p	F_p	C_p^*	F_p^*
3	1.4	65	150.90	17.27	18.93	2.00	39.79	6.57	13.14	5.2	13.5
1	1.8	65	155.23	34.24	21.47	3.00	49.38	8.87	13.90	7.5	12.9
2	1.8	65	158.10	28.94	18.69	5.00	43.07	6.85	15.75	7.5	17.6
5	1.4	80	150.90	17.27	18.93	2.00	39.79	6.40	12.68	5.2	13.5
7	1.4	80	151.96	17.55	6.38	2.00	40.60	5.89	13.74	5.3	14.2
4	1.8	80	158.10	28.94	18.69	5.00	43.07	6.76	22.32	7.5	17.8
6	1.8	80	150.90	54.50	23.6	9.00	58.25	6.57	17.17	7.5	17.4

And the fact of production proves that the failure zone and up limit educed by BP network model is rational and credible.

Conclusions

It is easy to gain from the paper that multi-factor controls the overburden failure under thin overlay rock mass. And BP network model is a reasonable method applied in researching on up limit of coalmine under water bearing Quaternary system. Research indicates that proper elements number leads to the fast convergence and lower system error, and the model in the paper includes two hidden layers each with 45 and 8 elements is perfectly. And it is particular important for one to select reasonable samples and train the model. The BP network model used cooperating with finite element analysis produce the appropriate outcome.

references

1. Jiao Licheng. Theory of Neural Network System. Publishing house of electronics industry, 1990.
2. Stornefta W,Huberman B. . An improved three-layer back propagation algorithm. Proceedings of the IEEE First International Conference on Neural Network, 1987, (II): 637-644.
3. Shi Zhizhong. Neural Computation. Publishing house of electronics industry,1993 : 85-90.

DISPLACEMENT PREDICTION OF GEOTECHNICAL ENGINEERING WITH ANN METHOD

LEI XUEWEN

Wuhan University of Science & Technology ,Wuhan430070, China

SUN JIZHU

Institute of Rock & Soil Mechanics, Chinese Academy of Sciences,Wuhan430071, China

TANG BIN

Wuhan University of Science & Technology ,Wuhan430070, China

In this paper, the artificial neural network theory and method is adopted to predict the displacement of geo-technical engineering, built the feed-forward network model and presented the new learning algorithm. The forecasting result shows the feasibility and effectiveness of the neural networks method

1 Introduction

The information changing in geo-technical engineering system is implicated in the surveyed displacement time series for the system. The rules of system evolution are expected to be discovered from these displacement data. Therefore, the displacement prediction and forecast become a great bearing problem to the benefit of the design and construction in geo-technical engineering. The process of rock and soil mass displacement pregnancy and generation is a complicated and physical one, and influenced by the factors of many aspects. The internal factors are mainly the high non-homogeneity of rock and soil mass structure, the complex nature of topography, terrain and ground stress. The external factors are coupled and compacted each other, causing the rock and soil mass displacement and very strong nonlinear characteristics, and so making the geo-technical engineering system become a complex nonlinear system and the prediction of rock and soil mass displacement become a very complicated research problem.

Generally speaking, in the prediction of rock and soil mass displacement, there are four kinds of model: mechanism models, statistical models, information models and gray models. These models present some feasible methods for the displacement prediction. It is indeed that if the nonlinear dynamic mechanism model of the complicated relation between the displacement and the internal-external factors can be actually built up, the displacement prediction will not be complicated again. But, it is quite difficult to do this. So the model of statistic like, taking the time series analysis as representation, such as AR, ARMA, ARIMA, SETAR etc., is more effective for the linear and smooth time series or the non-stationary and, nonlinear time series which can be simply transferred into a smooth and linear one. Theoretically speaking, this kind of statistic model still exists shortcomings for the rock and soil mass displacement such a complicated nonlinear system. Moreover the statistics, information and gray models are not yet convenient to analyze the inner essential rules of the geotechnical system. From this idea, it is necessary for us to study the nonlinear theory and method for the prediction of rock and soil mass displacement.

The artificial neural network is a broad scale nonlinear dynamic system composed of a great deal of neural cells. It has the functions of a broad scale parallel treatment, distributed storage, self-organization, self-adaptation, self-learning, etc.. It is successfully applied in the fields of predicating and forecast, module identification, optimizing computation, intellectual control, etc.. It is one of the front fields of international researches at present time and begins to be applied to geotechnical engineering. In view of the above-mentioned facts, the artificial neural network theory and method is adopted to predict the displacement of rock and soil mass, build the feed-forward network model and develop the prediction software of corresponding neural network. The prediction results show the feasibility and effectiveness of the neural network method.

2 BP Artificial Neural Network Model

2.1 Network Model and Learning Algorithm

A three-layer feed-forward network is adopted for prediction in our research. The advance BP algorithm is used to train the feed-forward network. The momentum method and self-adaptive adjusting maneuver to the learning rate are used for the advance BP algorithm to raise learning speed and to increase the reliability of the algorithm.

$$w_{ij}(t+1) = w_{ij}(t) + \alpha(t)[(1-\eta)D(t) + \eta D(t-1)] \tag{1}$$

$$\theta_j(t+1) = \eta\theta_j(t) + \alpha(t)(1-\eta)\delta_j(t) \tag{2}$$

In Eg.(1) and Eg.(2), $W_{ij}(t)$ represents the weight coefficient connecting input layer joint i and output layer joint j in t times iteration. D(t) represents the minus gradient in t times iteration, $D(t)=\delta_j(t)x_{i,}$ $\delta_j(t)$ is the error value of joint j in t times iteration, x_i is the output of joint i. η represents momentum constant, $0\leq\eta<1$. $\alpha(t)$ represents learning rate in t times iteration, $\alpha(t)>0$. $\theta_j(t)$ represents the taking value of the threshold of joint j in the implicit layer or the output layer of the network in t times iteration.

2.2 Training steps of the network

According to the above-mentioned analysis, the advance BP algorithm of multi-layer feed-forward network can be conduced in accordance with the following steps:

Step 1. The produced random number is assigned to the weight coefficient and threshold value and used as their initial value. The parameters such as the error range ε, momentum constant η, α_0 and β are supposed.

Step 2. The learning samples (the input x_i and expectative output d_i) are supplied.

Step 3. The output value in the implicit layer is calculated with sigmoid function. The output value in the output layer are calculated with pure-liner function .

Step 4.The weight coefficient and threshold value are renovated according to Egs. (1) -- (2).

Step 5. If the sum of the errors E is less than ε, then the calculation is come to a conclusion, otherwise, transferred into step 3.

Table 1. Predicting Results of the Settlement of Soft Clay Ground

Time/d	15	33	49	73	94	115
Measured Settlements/cm	5.2	8.5	9.9	11.2	11.7	
Gray Forecasting Settlements /cm	5.2	9.0	9.7	11.0	12.2	13.5
GF error /%	0.0	-5.88	2.02	1.79	-4.27	
ANN method prediction/cm	5.1883	8.9925	9.7113	10.9820	12.0106	12.2110
ANN error /%	0.23	-5.79	1.91	1.95	-2.65	

3 Prediction of Rock and Soil Mass Displacement

Using the above-mentioned network model and algorithm, the space displacements of a part monitoring points of two engineering actual examples are predicted. One of the two examples is the settlement of the soft clay ground under a large commercial warehouse in Southern China, and another one is the slope displacement of some open iron pit. The original data are cited from actual measure in the site. The parameters are taken as η=0.9, ε=0.0007, α_0=0.1. The gray forecasting results and ANN prediction results of the settlement of soft clay ground are shown in Tab. 1. It is shown that ANN method is in advance of gray forecasting method. The major results of simulation and prediction of slope displacement with ANN method are shown in Tab. 2. For the limited space, the results of only two monitoring points are listed here.

Table 2. ANN Predicting Results of Slope Displacement

Time step	No.1 Measured Displacement (mm)	Predicting Displacement (mm)	Relative Error (%)	No.2 Measured Displacement (mm)	Predicting Displacement (mm)	Relative Error (%)
73	764	761.8395	0.28	451	452.2453	-0.27
74	767	768.0195	-0.13	453	453.8218	-0.18
...	....	...	...	...	...	...
87	1038	1037.7765	0.02	718	727.0818	-1.26
88	1050	1047.510*	0.24	726	729.4491*	-0.48
89	1055	1052.9175*	0.20	729	731.0756*	-0.28

*external deduced predicting displacement

4 Conclusion

From the computing results and Tab.1-2, the following cognition and conclusion can be preliminarily obtained: The prediction method of feed-forward network is a method based on the nonlinear system and has a very strong mapping capability. This method is an effective and feasible one for the prediction of rock and soil mass displacement. In comparison with the information and gray model, ANN method is more convenient to describe the mechanism relation between the rock and soil mass displacement and the influential factors. Under the allowable circumstances of the predicted data, the sample number should be added as fully as possible.

References

1. Xia-Ting Feng. Guide on Intelligent Rock Mechanics[M]. Beijing: Science Press (2000)
2. Chang Laiyao, Yan Pingfan. Neural Network and Fuzzy Control[M]. Beijing: Qinghua Univ. Press (1998)
3. Lei Xuewen, Bai Shiwei, Meng Qingshan. The application of gray forecasting to analyzing soft foundation settlements, Chinese J. of Rock and Soil Mechanics, Vol.21, 145-147(2000)

MODELING WASTE DISPOSAL USING A COMPREHENSIVE COUPLED APPROACH

MAO BAI, SIDNEY GREEN, ROBERTO SUAREZ-RIVERA
TerraTek, Inc., Salt Lake City, Utah 84108, USA

JEAN-CLAUDE ROEGIERS
University of Oklahoma, Norman, Oklahoma 73019, USA

BAOPING ZHANG
Branch of Scientific Research Institute, CNPC, Langfang, China

Waste disposal involves complicated multiple processes of solid deformation, fluid flow, thermal transfer, and particle transport. This paper presents a unified formulation of a comprehensive coupling among the mechanical, transport, thermal, and flow processes. The application of the general formulation focuses on the transport process where the critical parameter that affects the injection rate decline has been identified.

1 Introduction

In oil production particularly in later stage reservoir production, significant amounts of contaminated water, frequently mixed with oil residue, must be disposed. Disposal of oily water is an issue of environmental concern, and is subjected to numerous restrictions by different environmental protection agencies. Also, substantial incidental costs are incurred for this disposal. Injecting the contaminated water back into the ground is considered an economically and environmentally viable means for disposal. In addition, in some cases the injected water may be used to flood the reservoir and increase oil production. One of the problems associated with re-injection is that the residual oil contaminated water frequently also contains other chemical solutes together with fine grains, minerals, clays, and other solid particles. This may significantly reduce the available pore spaces, change the reservoir wettability, and therefore degrade the formation permeability. Thus, the injectivity is reduced, and in the case of the water flooding, reduced flood efficiencies also occur. Such a degradation is reflected as difficulties in maintaining a stable injection rate.

The migration of suspended solids within the water is a complex transport problem. Using the concept of random walk method, Bai et al.[1] derived the solution of particle migration through gravel packs that matched the experimental results. The impact from particle transport phenomena on injectivity decline can be critical, if not primary. Particle deposition should not be viewed as a pure transport process, but a combination of force equilibrium (i.e., dragging and gravitational forces, and surface tension), material balance (i.e., flow continuity), energy balance (i.e., temperature variations), and mass conservation (i.e., solute transport due to concentration gradients).[2-5] The purpose of the present paper is to present a general formulation of the comprehensive coupled processes that has not been presented before. Attention is focused on identifying the dominant factor resulting in injectivity decline during waste injection operations. Using a simplified case of waste injection in a two-layer medium with certain contrasting properties, the study reveals the critical factor resulting in injectivity decline.

2 General Formulation of the Coupled Process

Incorporating the effects of mechanical, fluid flow, thermal flow and solute transport to a two-layer medium, the governing equations can be described as follows:

$$Gu_{i,jj} + (\lambda + G)u_{k,ki} = \sum_{m=1}^{2} (\alpha_m p_{m,i} + \beta_m T_{m,i}) \tag{1}$$

$$\frac{k_m}{\mu} p_{m,kk} = \alpha_m \dot{\varepsilon}_{kk} + c_m^* \dot{p}_m - \alpha_{mh} \dot{T}_m \pm \Gamma_p (p_1 - p_2) \tag{2}$$

$$K_m^* T_{m,kk} = s_m^\# \dot{T}_m \pm \Gamma_T (T_1 - T_2) \tag{3}$$

$$-(v_{mp} + l_h v_{mT}) c_{m,k} = (\phi_m + K_{mp}) \dot{c}_m \pm \Gamma_c (c_1 - c_2) \tag{4}$$

where u, p, T and c are the solid displacement, pore pressure, temperature and solute concentration, respectively; λ and G are Lame's constants; the subscript m represents a layer number ($m=1$ or $m=2$); α is the Biot coefficient, β is the thermal expansion factor; k is the permeability; μ is the fluid dynamic viscosity; c_m^* is the lumped compressibility; α_h is the thermal expansion coefficient; Γ_p is the inter-layer fluid leakage factor; K_m^* is the thermal conductivity; $s_m^\#$ is the lumped intrinsic thermal capacity; Γ_T is the inter-layer thermal leakage factor; v_{mp} is the convective fluid flow velocity; v_{mT} is the convective thermal flow velocity; l_h is the compatibility factor; ϕ is the porosity; K_p is the sorption intensive factor; and Γ_c is the inter-layer transport leakage factor. The superscript *dot* denotes the time derivative.

An analytical solution equations (1) through (4) was obtained for plane radial flow conditions, based on the following procedures:

- Using Goodier's displacement potential[6] to relate u directly to p_1, p_2, T_1 and T_2 in equation (1), thus representing ε_{kk} with p_1, p_2, T_1 and T_2 in equation (2).
- Solving T_1 and T_2 from equation (3) and substituting them into equation (2) to obtain p_1 and p_2.
- Calculating u from equation (1) once p_1, p_2, T_1 and T_2 are known.
- Calculating c_1 and c_2 from equations (4) once the rates of fluid flow and thermal flow are derived from p_1, p_2, T_1 and T_2.

3 Identifying Critical Factors

A case of well injection into a two-layer medium is shown in Figure 1. Primary modeling parameters that are different between the two layers are: (a) layer 1: porosity - 0.3, permeability - 5000 md, inter-layer transport leakage factor - 0.001m^{-2}; and (b) layer 2: porosity - 0.2 permeability - 500. md, inter-layer transport leakage factor - 0.01m^{-2}. All other parameters are identical between two layers. While seeking the sound theoretical basis in evaluating the sources of injectivity decline, it is imperative to identify the critical factors that may show controlling influences in the process. The attention is focused on the impact from sorption intensity factor (K_p). Calculating results are shown in Figure 2.

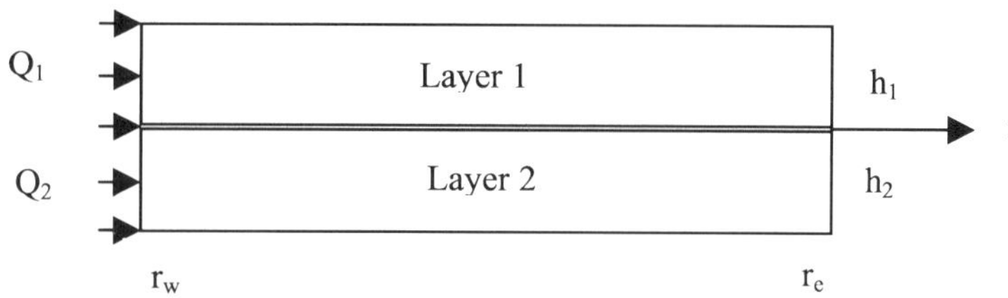

Figure 1. Schematics of well injection in a two-layer radial flow system. Q_1 and Q_2 can be either flow rates or fluid pressures. h1 and h2 is layer thickness for layer 1 and 2.

The insertions in Figure 2 show the spatial concentrations of total suspended solids (TSS) of the injected fluid at various times (e.g., day 1 and days 101, 201, 301, and 401) from the initial injection for two respective layers with two different sorption intensity factors (i.e., K_{1p}=5 and K_{2p}=10). Insert A indicates that the particles tend to depose at the close vicinity from the well mainly at early time (i.e., day 1). Particle plugging is more severe for larger K_p. For example, layer 2 is completely blocked by the particle deposition at day 1 when K_p is equal to 10 (i.e., no concentration change at day 1 for layer 2, see Insert B).

With the results from the spatial concentration versus distance for a wide range of sorption intensity factors (i.e., 0.05 to 100), the relationship between solids travel distance and time from initial injection can be derived (see Figure 2). As shown in Figure 2, the solid particles travel quite short distances at the initial stage of injection (i.e., less than 100 days), depending on the magnitude of the sorption intensity factors. The larger the intensity factor, the smaller the travel distance. The deposition becomes more uniform (i.e., change is less radical) when the time is greater than 200 days from the initial injection. Considering the long duration of the transport process, it seems that particle deposition occurs in extremely near-field regions of the injection well. As K_p becomes larger, more solid particles would be trapped in the near-well region. The asymptotical trend as K_p approaches to infinity implies that no solids could be further injected if they fill up the pore spaces. In reality, this would represent a complete failure to sustain long well injectivity due to particle plugging.

4 Concluding Remarks

This paper provides a unified approach for the analysis of waste injection in deep geologic formations that has not been presented before. For the radial flow in a two-layer medium with contrasting porosity, permeability and inter-layer transport leakage factors, the transport process is identified as a dominant process. This comprehensive and coupled approach includes all important parameters in the analysis. In the transport process with assumed conditions, the critical parameter

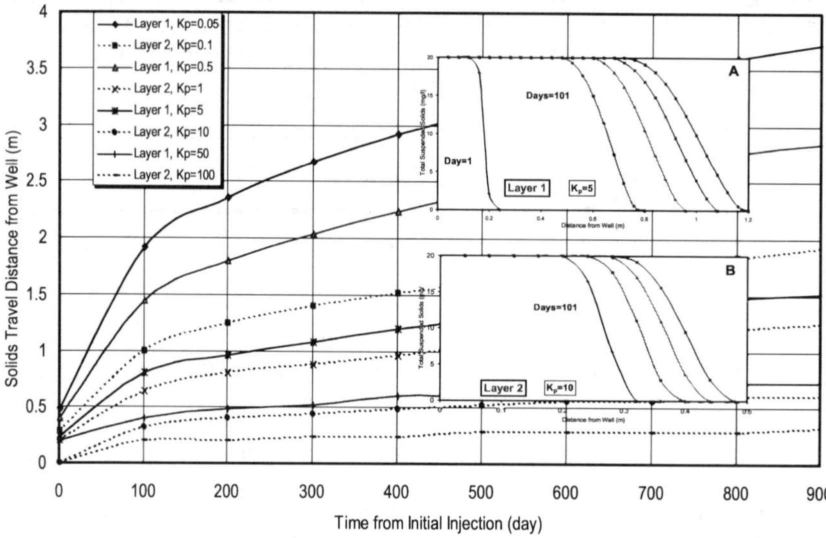

Figure 2. Injected solids travel distance versus time from initial injection for two layers with various sorption intensity factors. Total suspended solids versus distances from injection well for layers 1 (Insert A) and 2 (Insert B).

leading to injectivity decline is the sorption intensity factor (K_p). This factor characterizes the solid clogging mechanism in the near-well region, particularly within the early stage of waste injection.

Acknowledgments

Discussion in the early developing stage with Dr. John McLennan is greatly appreciated.

References

1. M. Bai, H.I. Inyang,, F. Meng, Y. Abousleiman, and J.-C. Roegiers, *Proc. 4th Int. Symp. Environmental Geotechnology and Global Sustainable Development*, Boston, Massachusetts, (1998).
2. M.B. Dusseault and F.J. Santarelli, Rock at Great Depth, V. Maury and D. Fourmaintraux (eds.), *Balkema,* Rotterdam, 789-797 (1989),.
3. K.E. Wennberg, Doctoral Dissertation, The Norwegian Univ. of Sci. and Tech. (1998).
4. I.C. Walton, *SPE 63108*, SPE Annual Tech. Conf. and Exhib., Dallas, TX (2000).
5. T.B. Jensen, Computer Methods and Advances in Geomechanics, Siriwardane & Zaman (eds.), *Balkema*, Rotterdam, 2077-2082 (1994),.
6. J,N. Goodier, *Physics, 7, 156* (1936).

PREDICTING THE AMOUNT OF WATER INGRESS IN TUNNEL BY STATIONARY RANDOM MODEL

XINGGAO LI WEINING LIU
Northern Jiaotong University, School of Civil Engineering and Architecture, Beijing, 100044, P.R.China.
E-mail:lxg_njtu@163.com

Prediction of water ingress in tunnel is important for tunnel design and the safety in transportation. This paper sets up a stationary random model to forecast the water ingress per day in the next year and a numerical analysis is performed to testify the model. The prediction agrees well with in site measurements and therefore proves the validity of this model.

1 Introduction

In China, water infiltrations problem is very common in tunnels and underground works. For example, there are more than 5000 railway tunnels now in China, but there exists water leakage in more than 1000 tunnels. Water leakage is the most common maintenance hazard, causing problems during the working life of the tunnel, not only to the tunnel lining, but also to the fittings within the tunnel, especially it threaten the safety in transportations. Therefore, a accurate forecast of water ingress in tunnel is important to tunnel design and the safety of transportations. Accordingly, this paper sets up a stationary random model to forecast the water ingress per day in the next year.

2 Geological Condition of Dayaoshan Tunnel

Dayaoshan tunnel is 14.925 km in length, and located at 700-900m in depth, whose surrounding rock is sandstone, sandstone-slate inter-bedding and limestone. The tunnel is made through 11 faults, and the ground water is plentiful nearby the No. 8 and the No. 9 fault. Especially near the No. 9 fault and karst area, the surface water and ground water join together, thus leading to disaster to the tunnel. In order to ensure safe transportation , many observations concerning water table, amount of water ingress, flow rate, soil and sand content etc. are carried out in tunnel and near the entrance and exit. Especially, the detailed data on water ingress per day through the central water conduct at the location of K1980+435 of Dayaoshan tunnel are recorded, which are the base of the prediction model in the paper.

3 Stationary Random Model Predicting Amount of Water Ingress in Tunnel

3.1 Discrete Stationary Random Model

Suppose x_t ($t = 1, 2, \ldots, N$) to be a stationary time series, then the sample mean and standard deviation are

$$\bar{x} = \frac{1}{N}\sum_{t=1}^{N} x_t \quad \text{and} \quad \sigma = \sqrt{\frac{1}{N}\sum_{t=1}^{N}\left(x_t - \bar{x}\right)^2} \tag{1}$$

If the deviation of the sequence meets the requirement of confidence level, the mean of the known sequence will be considered as the predication value; if not, a predication value of another new consequence $x_t' = x_t - \bar{x}$ will be added to the mean value to get the final result.

Predicting the new consequence x_t' is done according to the following steps[*WANG*, 1986].

Step 1.Prediction model of the new sequence can be expressed as the following linear equation

$$\bar{x}'_{N+\tau} = a_1 x'_N + a_2 x'_{N-1} + a_3 x'_{N-2} + \cdots + a_k x'_{N-k+1} + \cdots + a_N x'_1 \tag{2}$$

Find the undetermined coefficients $a_1, a_2, a_3, \ldots, a_N$ `to minimize the standard deviation

$$\min \sigma^2 \tau = \sum_N E\left(x'_{N+\tau} - \bar{x}'_{N+\tau}\right)^2 \tag{3}$$

where the integer of τ is the extrapolation step.

Step 2.The differentiation method is adopted to determine the extremum

$$\frac{\partial \sigma^2 \tau}{\partial a_k} = \frac{\partial}{\partial a_k} E\left(x'_{N+\tau} - \bar{x}'_{N+\tau}\right)^2 = 0 \tag{4}$$

yielding

$$-E\left(x'_{N+\tau} \cdot x'_{N-k+1}\right) + E\left(a_1 x'_N \cdot x'_{N-k+1}\right) + E\left(a_2 x'_{N-1} \cdot x'_{N-k+1}\right) + \cdots$$

$$+ E\left(a_N x'_1 \cdot x'_{N-k+1}\right) = 0 \tag{5}$$

By the ergodic theorem of stationary random time series, we can draw the equalities

$$R(\tau + k - 1) = E\left(x'_{N+\tau} \cdot x'_{N-k+1}\right) \quad (6.1)$$

$$R(k - N) = E\left(x'_1 \cdot x'_{N-k+1}\right) \quad (6.N+\tau)$$

The substitution of Eq.(6.1)-(6. N+ τ) into Eq.(5) gives linear equations about the undetermined coefficients a_1, a_2, ..., a_N

$$a_1 R(k-1) + a_2 R(k-2) + \cdots + a_k R(0) + \cdots$$

$$+ a_N R(k-N) = R(\tau + k - 1) \quad (7)$$

where $k = 1, 2, \ldots, N$. Finding the coefficients a_1, a_2, ..., a_N and substitution them into Eq.(2) can compute the prediction value $x'_{N+\tau}$ of the new time series x'_t.

3.2 Model Application

If the amount of water ingress per day in the same month of two successive years is taken as a time series, the water ingress in the next year can be predicted by way of the known observation data. Based on the observation data of 1997, we perform the forecast of water ingress in 1998 and the results are presented in Fig. 1, and only the figures of from January to March are listed for simplicity. As is shown in Fig. 1, the prediction agrees well with in-situ measurements to prove the validity of the model.

4 Conclusions

A stationary random model is set up to predict the amount of water ingress in tunnel in this paper. Its application into Dayaoshan tunnel shows the forecast agrees well with in-situ measurements and thus can meet the requirements of tunnel engineering. Also the calculations of the model are not complex and easy to do in the computer, therefore it has a high reliability and can be adopted as a short-term method to forecast the water ingress in the tunnel.

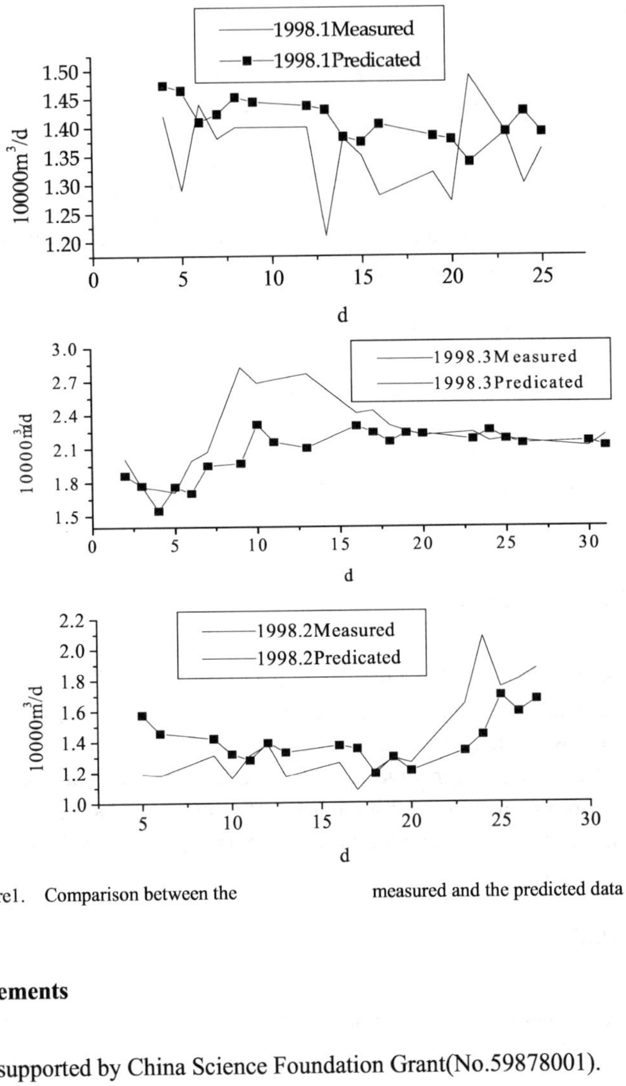

Figure1. Comparison between the measured and the predicted data

Acknowledgements

This work is supported by China Science Foundation Grant(No.59878001).

Reference

Forecast Calculation Method, ed. Yongling Wang. (Science Publishing House, Beijing, P. R. China, 1986)

RESEARCH OF GEOMETRICAL CHARACTERISTICS AND VISULIZATION FOR ROCK JOINTS BY USING GIS

HUA LIU, HUILI WENG, SHOUJI DU
School of Civil Engineering & Mechanics, Shanghai Jiao Tong University, Shanghai 200030

In the natural rock mass there exist aperture of joints whose mechanical characteristics are greatly important to the stability of rock mass slope and underground works. The shear strength, deformation behavior and flow properties of the rough surface of rock joint are dependent on the geometrical characteristics (such as surface roughness and aperture). GIS technology is implanted in the research of rock joints geometrical characteristics. After getting the precise measurements about surface roughness of granite and sandstone rock joints, GIS is used to realize the visualization of the three-dimensional rock joint rough surface, analyze the geometrical characteristics (surface roughness, slope and aspect) of the rough surface and damage characteristics of rough surface under different constant normal stresses before and after shearing.

1 Introduction

It's not sufficient to draw only the rock as research object when the research is about the mechanical stability of rock engineering such as mass rock slope and underground works and the hydromechanical behavior of mass rock relevant to radioactive waste disposal sites. The mechanical and hydromechanical characteristics of the rock joint that existed in the mass rock play an dominant role on the design and construction of the rock engineering. The surface roughness should be studied thoroughly in order to represent the mechanical and hydromechanical characteristics of the rock joint surface essentially. At the present time the common methods used to describe the surface roughness of rock joint are: 1). Barton's JRC method[1]; 2). Fractal joint method which is based on fractal theory[2]. The two methods are based on the analysis of two-dimensional rough lines. However there are obvious limitation and one-sideness for using two-dimensional rough parameters to describe the three-dimensional surface roughness. Therefore GIS technology is firstly implanted in this paper. Three-dimensional data of the rough surface precisely obtained by laser scanning system are inputted into GIS and the visualization of rough surface is realized. Then statistic analysis is preceded on the geometrical parameters of the rock joint. At last the damage characteristics of the rough surface under different constant normal stress before and after the shearing are studied.

2 GIS technology and establishment of the joint model

ArcView is one of the GIS technology software that can support many types of data. Besides the ability to operate and manage spatial data, several powerful extensions are available for ArcView. ArcView has complete programming language and developments environments. Raw spatial data of the rough surface are saved by Microsoft Excel as dbf files. Then, by a series of GIS commands and data transformation, the view of the three-dimensional rock joint rough surface is created. The surface analysis model TIN is created and is also converted to GRID mesh analysis model by the mesh size of 1mm. The various factor quantities, data and parameters can be stored at same mesh grid. The Histogram tool in ArcView provides a univariate (one-variable) description of the data. The tool displays the frequency distribution for the dataset of interest and calculates summary statistics.

3 Visualization and analysis on characteristics of the joint surface by using GIS

Three-dimensional rough surface data of rock joints used in this paper are gained from references[3]. According to the method in the above paragraph, three-dimensional views of rock joints can be created (Fig.1, 2). In general, Fig.1 and Fig.2 clearly show that the two opposing surfaces are similar to each other.

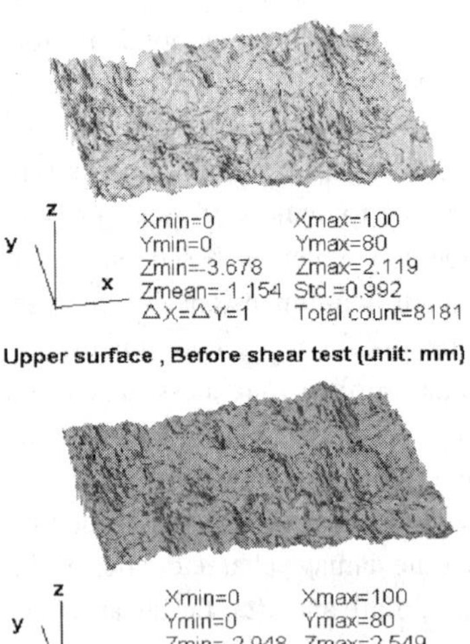

Fig.1Two opposing surfaces of joint in Granite

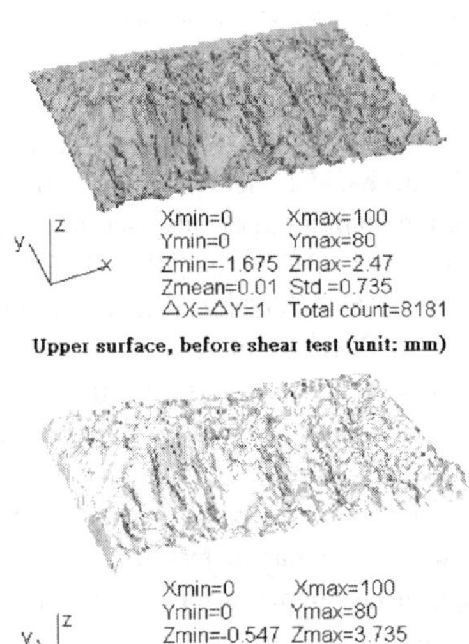

Fig.2 Two opposing surfaces of joint in sandstone

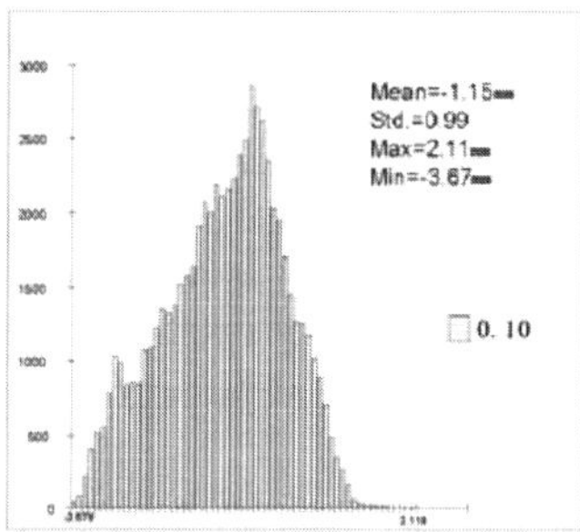

Fig.3 Histogram plot for asperity height of granite joint

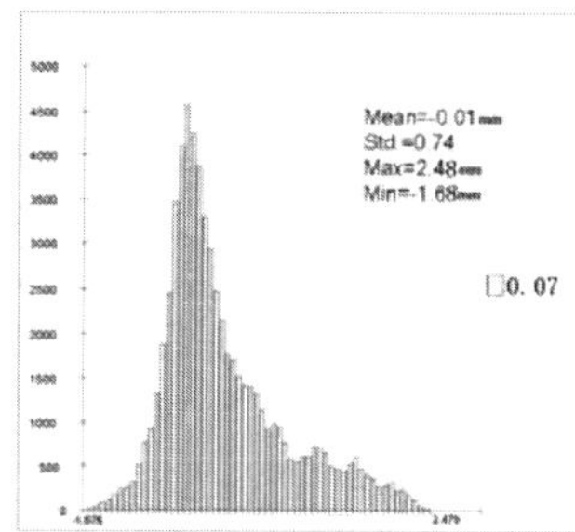

Fig.4 Histogram plot for asperity height of sandstone joint

Fig.3 and Fig.4 show the histograms of the asperity heights for the granite and sandstone joints. All the two histogram plots of the asperity heights can be approximated by the normal distribution. However, the histogram of the granite joint has a relatively wide-spread distribution ranging from –3.67mm to 2.11mm and the histogram of the sandstone joint has a relatively sharp distribution ranging from –1.68mm to 2.48mm. This means the granite joint has a more undulant rough surface than sandstone joint.

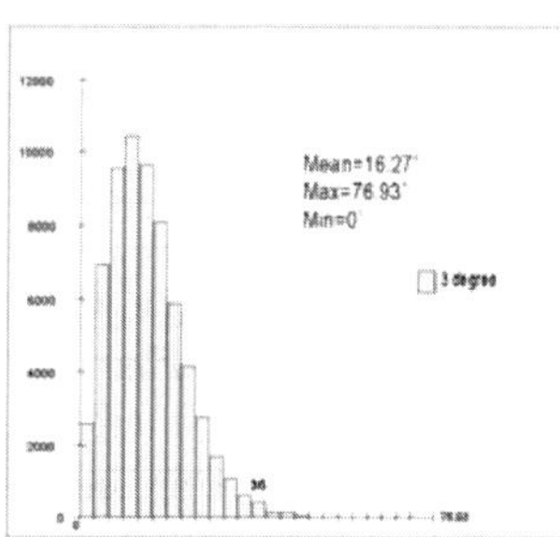

and san
e joint ha
nges fro
and 36^0.

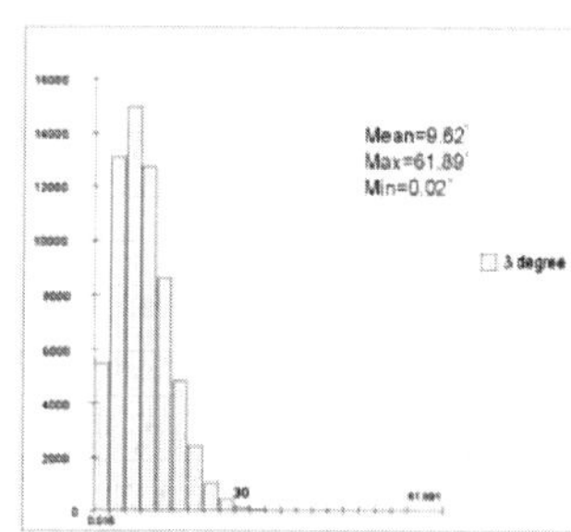

Fig.5 Histogram plot for slope of granite joint

Fig.6 Histogram plot for slope of sandstone joint

The histograms of slope for granite and sandstone joints using ArcView tools are shown in Fig.5, 6. The slope of granite joint has the mean of 16.27^0 in upper surface and 16.45^0 in lower surface and ranges from 0^0 to 76.93^0. However, the slope distribution has the 98.5% between 0^0 and 36^0. The slope of sandstone joint has the mean of 9.62^0 in upper surface and 10.02^0 in lower surface and ranges from 0^0 to 61.89^0. However, the slope distribution has only 0.7% over 31^0. Therefore, the sandstone joint has the smaller slope than the granite joint. The above two histogram plots of the slope can be approximated by the logarithmic normal distribution.

The histograms of the aspect for granite and sandstone joints are shown in Fig.7 and Fig.8. The aspect distribution of both granite and sandstone joint is similar with uniform distribution. The aspect of granite joint has the mean of 178.56^0 in upper surface and 178.29^0 in lower surface. The aspect of sandstone joint has the mean of 171.26^0 in upper surface and 170.19^0 in lower surface.

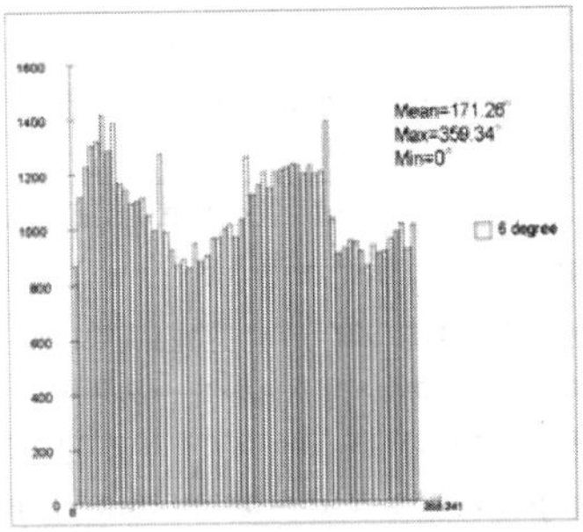

Fig.7 Histogram plot for slope of granite joint

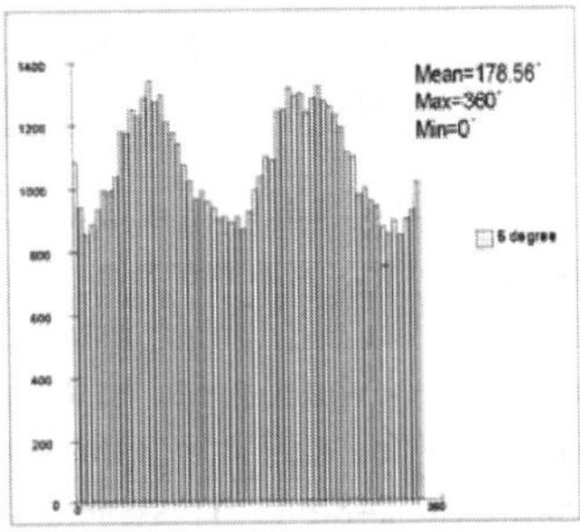

Fig.8 Histogram plot for slope of sandstone joint

4 Effect of different normal stress history on geometric characteristics of rock joints

The shear tests results show that surface damage is relevant to the normal stress history. In order to describe the damage characteristics of joint surface, based on the measurements of surface roughness that conducted before and after shearing (reference [3]), the asperity heights before and after the shear tests are compared. The relationship between the mean slope and the shear history is obtained by using GIS method.

Fig.9 shows the relational curves of asperity heights of granite and sandstone joint surface and normal stress history. With the increasing of normal stress the rougher asperities are damaged, the rough surfaces become more flat, the distance between the upper and lower surface decreases and the contact area increases. As it refers to granite joint, the distance between the upper and lower surface is 0.99mm without normal stress history and the distance decreases to 0.54mm when the normal stress is 30MPa. The distance between the upper and lower surface of the sandstone joint changes from 1.13mm when there is no normal stress to 0.91 when the normal stress is 20MPa. Fig.10 shows the relational curves of slope of joint surface and normal stress and shear displacement history. As the normal stress increases the mean slope of granite joint decreases obviously and the curves of the upper and lower surface are almost uniform. As the normal stress increases change rate of the sandstone slope is great at initial stage and drives to a stable value at last. It can be drawn from

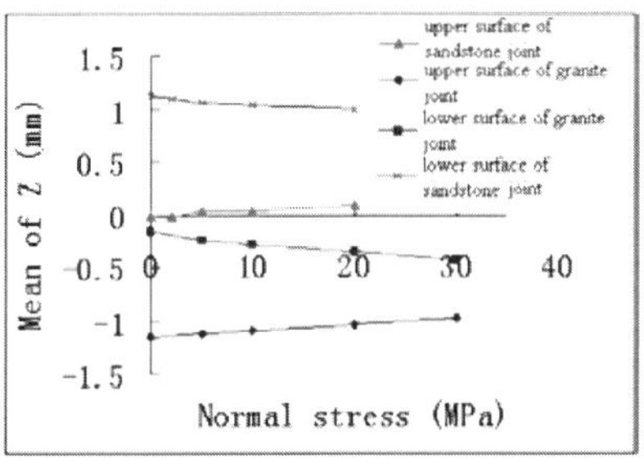

Fig.9 Relation between normal stress and mean of asperity height

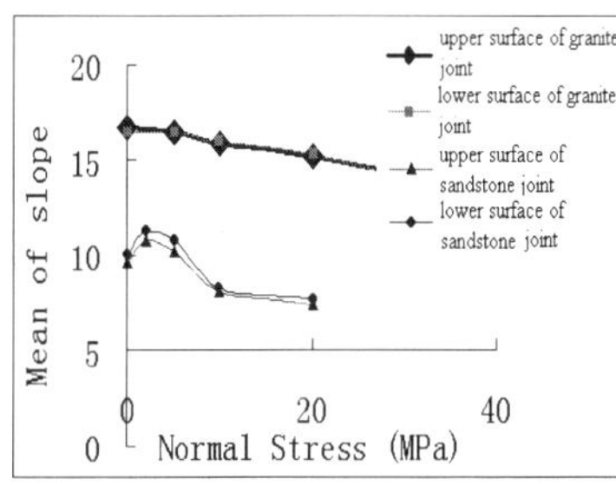

Fig.10 Relation between normal stress and mean of slope

Fig.10 that effect of normal stress on the slope sandstone joint is more sever than that of the granite.

5 Conclusion

GIS technology combined with precise measurements of the rough surface is used in this paper to create three-dimensional views of joint surface and to analyze the geometrical statistic such as asperity, slope and aspect. Regularities of distribution of the geometrical characteristics are obtained from the analysis result. Qualitative analysis is carried out on the relationship between surface damage and normal stress. The result indicates that the greater the normal stress is, the greater the surface damage is. At the same normal stress level, surface damage of granite joints is less than that of sandstone joints. This study shows that GIS technology provides an effective and practical way for analysis on the geometrical characteristics and three-dimensional visualization for rock joints.

Acknowledgments

The authors wish to express sincere thanks and gratitude to the Committee of NSFC for its support in this research (NSFC Reference Number: 50079012).

References

1. Barton N, and Choubey V., Rock Mechanics, 10, 1-65(1977);
2. Heping Xie, Introduction of Fractal-Rock Mechanics, Science Press, 1996
3. Shouji Du, Experimental and Theoretical Research on Geometrical-Mechanical-Hydromechanical Characteristics of Rock Joints, Dissertation, June 1997.
4. Qiming Qin, ArcView GIS Practical Tutorial, Beijing University Press, 2000

RESEARCH ABOUT SEMI-RIGID PAVEMENT STRESS INTENSITY FACTOR OF CREVICES BY NEURAL NETWORK

BIN ZHANG[1], CHUNDONG JIANG[2], XIANFENG LI

[1]*Institute of Civil and Architecture Engineering, Liaoning Technical University, Fuxin, 123000,China*

[2]*Constrution engineering head office of Qitaihe coal group, Qitaihe, China*

Many means are used to assay propagation regularity of semi-rigid base course asphalt pavement. This paper investigates the feasibility of using neural Networks to do this work. Neural networks attempt to calculate clinking tip stress intensity factor that are influenced by traffic load and temperature load after the base course's cracking. With test material coming from spot, cracking mechanism is researched. Results indicate that neural network's algorithm is efficacious and feasible.

1 Introduction

With the development of national economy, road traffic is playing a more and more important role and asphalt concrete pavement has been used widely in road construction. But asphalt pavement has a main defect that is clinking. The clinking of asphalt pavement is influenced by many complicated factors .The destroying mechanics is very complex. Moreover, the structural parameter and contributory factor of stress intensity factors in different parts belong to nonlinear strain, which leads to the complicated calculation. The artificial neural network is a front field, which has been developed. Since the middle of 80s in the world. At present, it has been successfully used in artificial intelligence and cognitive science, such as the cognition of complicated model, picture disposal, etc, some typical applications of it in civil engineering works are structural analysis of pock pillar. This paper try to calculate and analyze cracking stress intensity of asphalt concrete pavement by using artificial neural network model according to the specific property of neural network informative disposal.。

2 Neural network method

Artificial neural network is complicated network system, which is composed by a great deal of formal neural connecting with each other. Their connection and degree of influence can be presented by weight. The development of neural network goes through two-research climax. In this paper we use reversed error propagation network, which was put forward by Rumelhart et al in 1986. According to pictured 1, this network is a feed back network which is composed of output layer, input layer and hidden layer. Each layer has many units. Input units correspond independent variable of practical problems and output units correspond dependent variable.

At present, BP network model is one of the most widely used neural network model. It consists of positive propagation and reversed propagation. After stochastically giring initial value to connective weight's coefficient, the positive propagation begin to choose any model in model group as input layer. After hidden layer disposes the input layer, the output value will be calculated. Output value can get in touch with any nonlinear function. The related knowledge of non-function can be found in reference [3], [4]

Input layer $I_i=X_i$ (1)

Hidden layer $Y_j=F(\Sigma W_{ji}X_i-\theta_j)$ (2)

Output layer $Y_k=F(\Sigma W_{kj}Y_j-H_k)$ (3)

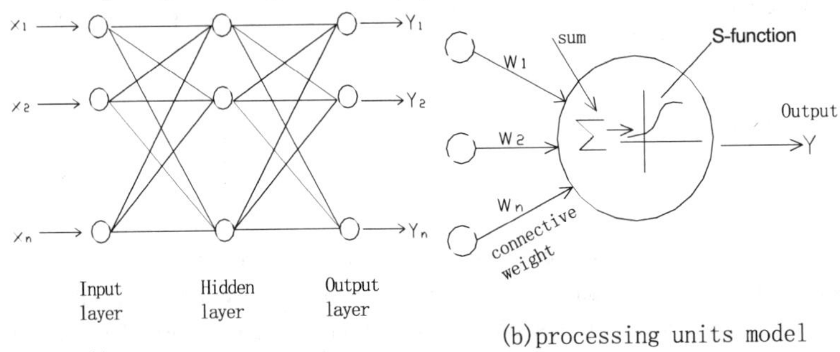

Figure1. typical BP network with three layers

Among them, X_i presents the input of network; Y_j presents the output value of No.j node in secret unit ; Y_k presents the actual output of No. k node in output layer ; F presents active function ; W_{ji} 、 W_{kj} present connective weight from input layer to hidden layer and from hidden layer to output layer; θ_j H_k present unit threshold of secret layer and output layer. $F=1/(1+e^{-x})$ (4)

The definition of square error function is : $E_p=0.5\Sigma(O_{pk}-Y_{pk})$ (5)

In the formula, O_{pk} is the expected output of k of the No.p input model; Y_{pk} is the output of network. The threshold of node is adjusted as following :

$\Delta W_{ij}(t+1)=\eta\delta_j Y_i+\alpha\Delta_{ij}(t)$ (6)

$\Delta\theta_j(t+1)=\eta\delta_j Y_i+\alpha\Delta\theta_j(t)$ (7)

In the above formulas, Y_i neuron is the actual output of i at t, η is studing rate , a is inertia coefficient, δ_j is the error of node j , According to the form of node j, we can make the following calculation.

When j is the node of output layer : $\delta_j=Y_j(1-Y_j)(O_j-Y_j)$ (8)

When j is node of secret layer : $\delta_j=Y_j(1-Y_j)\Sigma\delta_k W_{Kj}$ (9)

In the above formulas, Y_j is the actual output of neuron j at t , k is the number of next layer neuron of neuron j. To train the network by using BP study algorithmic, after the data chain (composed of training samples) is input into the network, the network calculates the value of each neuron by reversed error propagation process and adjusts all the threshold which connects weight coefficient to each neuron. This process will be repeated until it has finished all the calculation of models. After finishing the study of whole network, we can get a group of weight, which can describe corresponding connection between input and output.

3 The neural network model of stress intensity calculation

3.1 The structural design of BP network

Usually the structural design of BP network is composed of :(1)design of input layer and output layer .(2)choosing the hidden layer quantity.(3)choosing the quantity of secret unit. This paper uses 3-layer front BP network structure to establish the calculation model of stress intensity factor. BP network consists of three parts input layer, hidden layer and output layer. In this paper, we choose 11 parameters as input units, which present thickness, Poisson ration, elastic modulus and crack dimension of each layer. The quantity of hidden unit is adjusted according to the principle of the least final error until it reaches the optimum stage.

3.2 The study of network

We take to groups of data, which are collected from the lab as studying simples to carry out network training and take 10 groups of data as testing samples to carry out network test (part of the data, please see table 1 and 2). After studying training we can demand value of average error. The optimum quantity of network-hidden unit is 15. Accordingly , the studying rate is η =0.9 and inertia coefficient is α =0.7。

Table 1. Training example

序号	X1	X2	X3	X4	X5	X6	X7	X8	X9	X10	X11	Y1	Y2
1	16	12	32	144	30000	10000	5000	500	0.25	0.35	2	6.36	0.93
2	12	12	32	144	20000	10000	5000	500	0.25	0.35	4	8.55	1.87
3	16	12	32	144	30000	10000	5000	500	0.25	0.35	6	10.11	2.05
4	12	12	32	144	25000	10000	5000	500	0.25	0.35	8	10.60	2.27
5	16	12	32	144	15000	10000	5000	500	0.25	0.35	10	11.50	2.74
6	12	12	32	144	25000	10000	5000	500	0.25	0.35	12	14.90	3.01

3.3 The function of network:

We adopt the above network structure and take optimized studying coefficient as samples to carry out network training. From the result we can see that the network training will appear little turbulance when it reaches 500 times, But when it reaches 10000 times. the network will be tending towards stability. Moreover, the network convergence is more rapid when the training reaches 18000 times, So it demonstrates that the network is reliable and practical

Table2. Text example

序号	X1	X2	X3	X4	X5	X6	X7	X8	X9	X10x	X11	Y1	Y2
1	16	12	32	144	30000	10000	5000	500	0.25	0.35	4	8.89	1.82
2	12	12	32	144	30000	10000	5000	500	0.25	0.35	6	10.12	1.76
3	12	12	32	144	30000	10000	5000	500	0.25	0.35	8	11.13	1.92
4	16	12	32	144	25000	10000	5000	500	0.25	0.35	2	6.25	0.93
5	12	12	32	144	20000	10000	5000	500	0.25	0.35	12	13.05	2.61
6	8	12	32	144	30000	10000	5000	500	0.25	0.35	10	11.43	1.62

4 Conclusion

The research purpose of this paper is to effectively establish knowledge system based on groups of sample by multi-layer front neural network model. Connecting the different conditions which influence stress intensity factors with it, we can overcome the malpractice of adopting mathematical model calculation which has a large aberration between theatrical data and practical date. Moreover, with the increase of practical data, the research method based on neural network of calculating stress intensity factor can continually absorb new knowledge and has a bright future.

Reference

1. Sha-Qinglin, The semi-rigid base course asplalt pavement of high grade road, Beijing: The people communicant press, 1998
2. Wang-shitong, The neural fuzzy system and application, Beijing: Beijing aviation and airspace university press,1998
3. Chen-ming, The neural network model, Dalian: Dalian science and engineering university press, 1995
4. Han-liqun, The artificial neural network theory, design and application. Beijing: The chemical industry press, 1954

5. Rock Mass Engineering Stability Analysis, Design, Support and Construction

CONSTRUCTION TECHNOLOGY FOR LARGE-SCALE CAVERNS IN JAPAN

SATOSHI HIBINO
Central Research Institute of Electric Power Industry
1646 Abiko, Abiko-shi, Chiba 270-1194 JAPAN

The author has revealed characteristic behavior of the rock mass through analyzing the measurement results of deformation and stress. Measures taken to reinforce rock masses and troubles that occurred during excavation were also investigated. The main findings were drawn as follows: (1) Structures in rock masses should be designed based upon the displacements of surrounding rock masses, (2) Deformation of a rock mass consists of two parts, i.e., "strain displacement" and "opening displacement", (3) Troubles were mainly arisen by the geological conditions which were difficult to predict and the "information-aided construction" is needed.

1 Introduction

More than 20 large-scale caverns were constructed for pumped storage power stations in the last 30 years in Japan. The scale of these caverns is about 50m high, 25m wide, and 150m long. The cross-section area of these caverns is larger than 1000m^2, i.e., 10 times greater than that of ordinary scale tunnels for motorways or railways, which generally have an area ranging from 100 to 200m^2. Drs. Hayashi and author had pioneered the "excavation analysis method" to predict the deformations and stresses in rock masses for safety and economic construction works. The method was applied to sites and has been greatly contributing to the construction of a great number of large-scale caverns. We also revealed the behavior of rock masses using the measurement results of rock masses during excavation. The above is described together with the reinforcements measures and troubles encountered during excavation work.

2 Conditions of caverns and development of excavation analysis method

The dimensions and rock mass conditions of large-scale caverns constructed are shown in Table 1. The sites with favorable rock mass conditions were generally selected for caverns, as the cavern scales were large. Some caverns had to be constructed at sites with poor rock conditions. Since the Japanese Archipelago had suffered intensive crustal movement in the past, poor geological conditions such as faults, crushed zones and local concentrations of joints formed. The rock masses are inhomogeneous and the extent of weathering varies widely depending on locations. All these make the excavation work not only difficult but troublesome (see Chapter 4).

Using the finite element method, the author developed the "excavation analysis method[1]" with the following described features in 1968. Main items taken into consideration for the method include: 1) Dependency of deformability on stress level and deterioration of strength caused by stress concentration under excavation, 2) Proposal of the concept of "excavation equivalent external force", a method to express mechanical effects of "excavation" in the numerical analysis, 3) Excavation process. The latest research result about this has been listed in the reference[2].

Table 1 Dimensions of the caverns and rock conditions

	Site	Kind of rock mass	Size of cavern(m)			Ground pressure(MPa)				Properties of rock mass[1)]		
			Height	Width	Length	σ_h	σ_v	τ	h(m)[2)]	E_d(GPa)	τ_R(MPa)	α
1	Kisenyama	(1968) Shale, sandstone, chert	51.0	25.6	60.4	1.3	3.9	–	250	6~12	1.5~2.9	0.16
2	Niikappu	(1972) Schalstein	43.8	19.8	50.8	3.3	4.4	0.7	110	24	2.4	0.16
3	Okutataragi	(1972) Rhyolite, diabase	49.2	24.9	133.4	5.8	6.5	–	240	35~10	3.9~4.9	0.05
4	Oohira	(1973) Sandstone, slate	45.4	22.0	82.8	5.7	7.8	1.6	280	10~29	1.0~2.5	0.17
5	Nabara	(1974) Granite	47.7	25.0	85.6	7.2	6.3	1.1	180	3~9	0.5~1.5	0.2
6	Shintakase	(1975) Granodiorite, diorite	54.5	27.0	165	2.0	5.9	2.2	250	14/7[3)]	3.1/1.3[3)]	1
7	Okuyoshino	(1976) Shale, sandstone	41.6	20.1	157.8	6.6	6.9	2.3	180	13/6	2.0/0.8	0.3
8	Okuyahagi	(1978) Granite	47.8	22.4	103.3	7.4	10.8	1.0	340	15/7	2.9/1.2	0.8
9	Numazawa No.2	(1979) Rhyolite	47.6	26.0	96.5	4.5	3.4	0.8	160	10	1.4	0.1
10	Tanbara	(1979) Conglomerate	49.5	26.6	116.3	4.5	7.0	1.9	240	16~20	2.4~2.9	0.4
11	Arimine	(1979) Granite	20.8	14.6	30	1.3	1.9	–	63	4/2	1.7/1.3	0.4
12	Honkawa	(1980) Black schist	47.4	26.3	98	5.4	7.1	0.9	270	12/8	2.5/1.3	0.7
13	Takami	(1981) Schalstein	43.3	21.5	55	7.1	5.9	0.2	220	3~8	0.14	0.8
14	Matano	(1981) Granite, porphyrite	46.2	23.5	155.5	18.5	12.5	6.6	350	15~20	2.9	0.5
15	Tenzan	(1982) Granodiorite	48.0	24.0	89.0	15.0	11.0	1.2	500	25	7.4	0.5
16	Imaichi	(1982) Sandstone, breccia	51.0	33.5	160.0	7.6	9.1	0	400	18	1.9	0.4
17	Shiobara	(1986) Rhyolite	51.0	29.0	165	3.0	4.7	0.4	200	3~5	1.0	0.05
18	Kazunogawa	(1993) Mudstone, sandstone	54.0	34.0	210	11.1	12.3	0.3	500	12/10 6/5	1.5~0.8	0.1
19	Okutataragi Add.	(1994) Rhyolite	47	25	130	7.6	6.6	1.8	250	16	1.3	0.07
20	Kannagawa	(1998) Mixed rock	52	33	216	6.0	11.1	3.2	480	30/15	2~5	–
21	Omarugawa	(1999) Granodiorite	46.8	20	186	5.9	9.8	1.0	410	20~45	5.9	–

1) E_d: Deformability, τ_R: Cohesive strength, α : Creep coefficient 2) Depth of overburden 3) Anisotropy

3 Characteristics of rock mass behavior – findings from measurement results

Displacements of rock masses and stresses in arch concrete linings were measured during excavation at a number of sites as shown in Table 1. Characteristics of rock mass behavior were revealed by the investigation on the measurement results. The effects of discontinuities, such as joints of rock masses, were more remarkably demonstrated in the excavation of these large-scale caverns than in the excavation of ordinary scale tunnels with a cross-section area 100・200m^2 for motorways and so on, as the cross section area of the former is far greater.

Major findings are the followings (refer to the details[3]):

1) The values of stress in the arch concrete linings are proportional to the displacements of surrounding rock masses. The stress is generated by the horizontal displacements of the cavern walls toward inside of the cavern during excavation (Figure 1). Numbers assigned to the plots in figure correspond to the site names (Table 1). Each plot indicates the average of all measurements for the individual site.

2) By comparing the measurement results of the rock mass displacements with the observation data by using the borehole TV, it was found that the displacement of rock mass consists of two parts: the continuous "strain displacement" caused by changes in strains of the rock mass and the discontinuous "opening displacement" generated by openings of discontinuities such as joints. That is,

"Total displacement of rock mass" = "Strain displacement" + "Opening displacement". (1)
Although occurrence of "opening displacement" has been observed at the time of rock mass breakage even in ordinary scale tunnels, few case are reported during the stable excavation of those tunnels. It is likely that the "opening displacement" is apt to be generated when the scale of a cavern goes large.

3) The rate of opening displacement *k* ("opening displacement"/"total displacement of rock mass") varies with the type of rock mass. The value of *k* for rock mass with joint-forming characteristics (igneous rock origin, for example) was about 0.6, whereas rock mass with non-joint-forming characteristics (sedimentary rock origin, for example) had a low value of only about 0.2.

The values of stress induced in the arch concrete lining vary by 2 to 3 times, depending upon the type of rock mass (Figure 2). This is supposed to be caused by the types of rock mass with different values of *k*.

4 Measures to reinforce rock masses and confronted troubles during excavation

The examples of reinforcing measures under poor geological conditions and troubles occurred are given below.

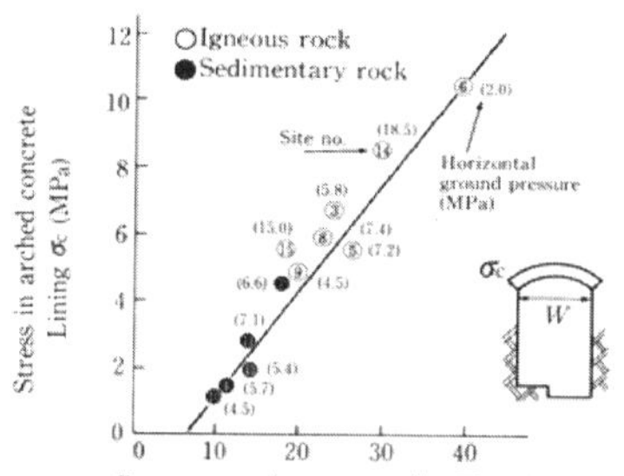

Figure 1. Relationship between stresses in the linings and horizontal displacements of walls

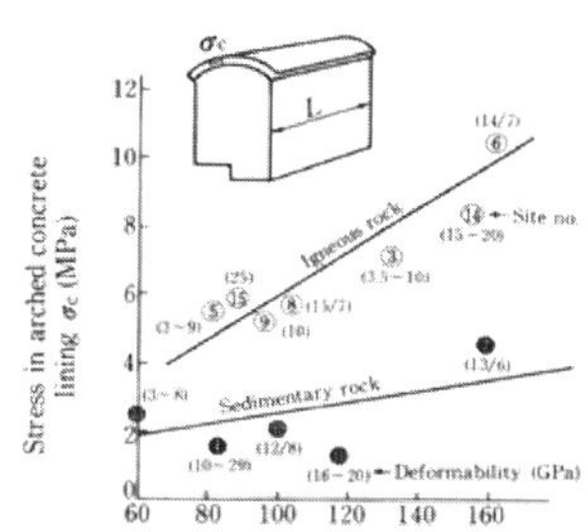

Figure 2. Relationship between stresses in the linings and the lengths of caverns

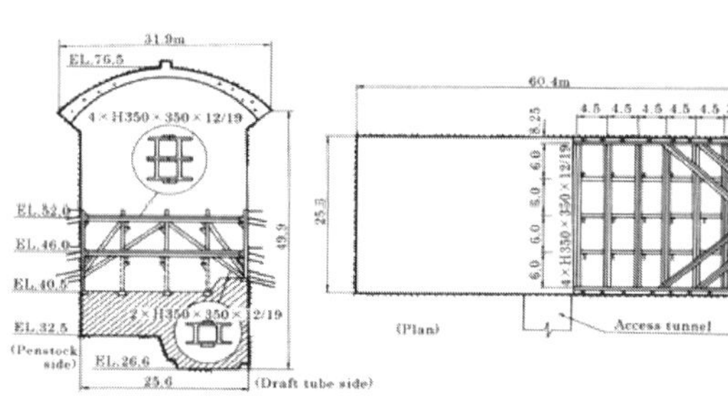

Figure 3. Support with steel struts[4] (Kisenyama 1968)

1) Reinforcement by steel struts (Kisenyama[4] 1968)

A 35mm horizontal displacement took place on the wall of the draft tunnel side with soft slate rock when the cavern was excavated down to EL.40.5m. As the increasing of the displacement slowly continued even after the blasting, the excavation work was suspended and the measures to control it were discussed (Figure 3). Two horizontal sets of steel struts with a 4.5m interval were installed (a unit of strut was formed with four H steels of 350mm x 350mm x 12/19) to control the horizontal displacement. Prior to resuming the excavation, we estimated values of stress in the struts induced by the coming excavation, using the "excavation analysis method". The comparison of results between measured values and estimated ones has shown an agreement to a certain extent, which indicates the effectiveness of this method.

2) Reinforcement by rock struts (Okutataragi[5] 1972)

Reinforcement by steel struts is considerably hard work, as it takes high cost and much time. As the experience of Kisenyama we proposed a concept of rock strut in stead of steel struts in poor rock conditions. One block of a rock mass at a longitudinal midpoint of the cavern is left unexcavated as a strut. This type of rock strut functions well to suppress the horizontal displacements of cavern walls and its top surface can be used as a place for assembling generators. We have applied rock struts for Okutataragi, Okuyoshino and some other sites.

3) Reinforcement of arch abutment and cavern walls (Takami[6] 1981)

The kind of rock mass at Takami is green tuff. Its elastic modulus is 3 to 8 GPa, which is smaller than that of other sites (Table 1). There was ever a worry that localized sliding might break out at the arch abutment area of the cavern and remarkable deformation might occur in a wide region of the cavern walls during the excavation. Accordingly, the rock mass at the arch abutment was improved by consolidation grouting. The reinforcement was performed by installing SEEE cables in the rock masses between drain galleries (EL.88m and EL.72m) and cavern walls so as to apply tension (686 to 882kN) to those rock masses (Figure 4). A number of advantages exist in this case, for example, fastening and tensioning the SEEE cable can be performed

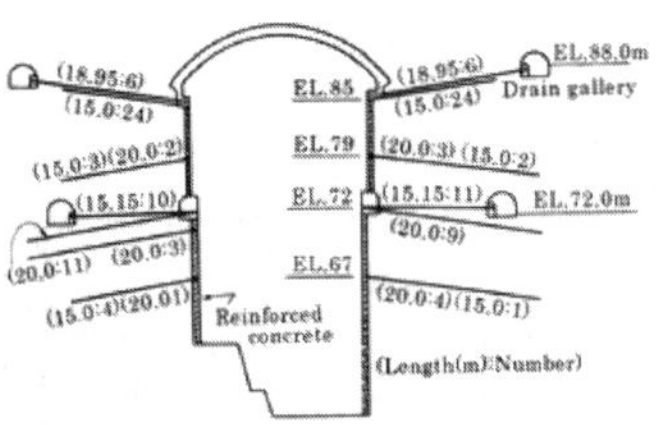

Figure 4. Reinforcement with SEEE cable[6] (Takami 1981)

with certainty and the adjustment of tension is performed with ease because the fastened part of SEEE cable is located in the drain galleries. The engineering of integrated rock mass is realized by the following cycles: (1) cavern excavation of 2.7m depth by one blasting, (2) placing reinforced concrete to form walls of 50cm thick by inverted construction method, (3) applying tension (196 to 329 kN) to rock mass by PS cables from the surface of the concrete walls.

4) Utilization of cavern shape effects: Application of egg-shape cavern (Imaichi[7] 1982)

In the case of Imaichi, cracks appeared on the surface of shotcrete wall over an aria of about 5m× 20m when the excavation reached EL.473m. The construction work was suspended for half a year while the causes were investigated and adequate measures were examined using the "excavation analysis method". It was found that the tangential stress in the cavern wall had increased to an extremely great value until it had caused cracks in the shotcrete wall because of the presence of seams parallel to the wall, which were not perceived before the event. Subsequently, the shotcrete around the cracks was chiseled out, while new shotcrete was placed over that area. After the area was well reinforced by rock bolts and PS cables, the excavation work was resumed. The cavern was successfully completed eventually.

5) More rational support design (Addition of Okutataragi[8] 1994)

Experience in large-scale cavern excavation has been steadily accumulated during 1980's and 1990's, resulting in more rational and economic construction. As an example, the addition cavern of Okutataragi power station can be related (Figure 5). The amount of reinforcement work was successfully reduced to about half that required in other existing caverns by employing the slot drilling process for the excavation of the arch part to reduce relaxed zones as small as possible, and at the same time, by applying reinforcements on the relaxed zones only (Figure 6).

6) Summary

Even if detailed investigations, rock mass tests, numerical analyses and others were carried out in advance of excavation work, troubles may still occur when unexpected poor geological conditions such as faults are encountered during excavation. Therefore, so-called "information-aided construction" is essential for construction works, with which effective information is consistently collected from surrounding rock masses by various measurements to detect the changes of geological conditions and abnormality of rock masses at earliest.

7) Tasks for the future

How large caverns can be excavated in rock masses? It is not easy, to assert if a cubic cavern with a volume of 1 million m^3 can be constructed or not. What is needed to make an answer to this question? Necessary items to be solved are listed below due to the limited pages of this report.

(1) how to detect unexpected geological poor conditions such as hidden weak zones or discontinuities, (2) how to evaluate macroscopic deformation modulus and strength of rock masses[9], (3) what relation exists between cavern scale and effects of discontinuities, (4) effects of macroscopic an-isotropic properties on the stability of

caverns[10], (5) what is the state of stress and strength in relaxed zones, (6) "information-aided construction" is essential and what-like management standard should be established.

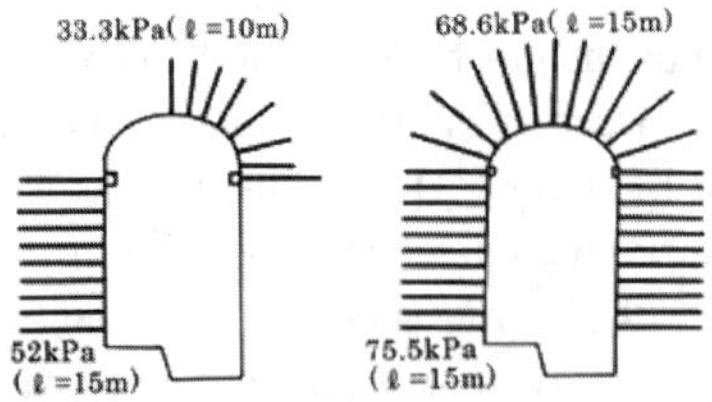

Figure 5. Comparison of reinforcement layouts[8]

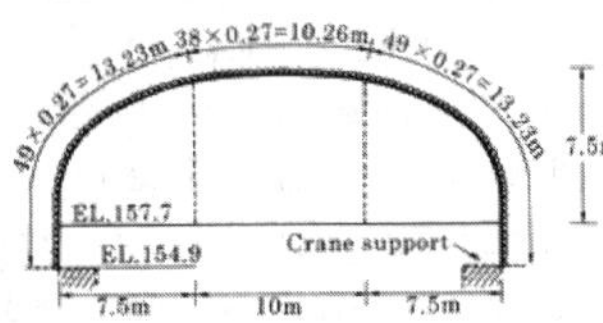

Figure 6. Slot drilling method[8]

5 Conclusions

1. Structures in rock mass should be designed based on the displacements of surrounding rock masses.
2. The "total displacement of rock mass" consists of two parts: "strain displacement" and "opening displacement".
3. As unexpected poor geological conditions are main source of trouble occurrences during excavation, "information-aided construction" is essential to avoid them.

References

1. M. Hayashi and S. Hibino, *Int. Symp. Rock Mech.* Madrid, 343-349 (1968)
2. S. Hibino et al, *GeoEng 2000*, Melbourne, CD-ROM (2000)
3. S. Hibino and M. Motojima, *8th Cong. ISRM,* Balkema, **2**, 583-586 (1995) or *Compr. Rock Eng.*, Pergamon, **4**, 631-651, (1993)
4. M. Hayashi and S. Hibino, *CRIEPI Report*, No. 68123 (in Japanese) (1969)
5. M. Hayashi et al, *CRIEPI Report*, No. 73526 (in Japanese) (1973)
6. Hokkaido Elec. Power Comp. *Const. Report of Takami Power Station* (in Japanese) (1984)
7. Y. Mimaki et al, *Elect. Power Civil Eng.*, No. 185, 47-63 (in Japanese) (1983)
8. H. Yanase and K. Yoshimura, *Elect. Power Civil Eng.*, No.277, 45-50 (in Japanese) (1998)
9. S.Hibino and M.Kamijo, *NARS'98*, CD-ROM, Cancun, (1998)
10. S.Hibino and M.Motojima, *9th Cong.* ISRM, **1**, 385-388(1999)

Mailing address:Central Research Institute of Electric Power Industry
1646 Abiko, Abiko-shi, Chiba-ken 270-1194 JAPAN
E-mail address: hibino@criepi.denken.or.jp
Fax number: +81 471 82 9417

ON THE ENGINEERING POSER AND COUNTERMEASURES OF CROSSING THE PEARL RIVER WITH TWO USED SHIELD MACHINES(TBMS) IN GUANGZHOU METRO LINE 2

WEIBIN ZHU, SHIJIAN JU, BAOFENG XIE, WEIXIONG LUO
Guangzhou Metro corporation

The construction method and related countermeasures while TBM was driving under the Peal River are introduced in this paper.

1 General situation of the work

Guangzhou Metro Line 2 is totally 23.3km with 20 stations, in which the section of Haizhu square station to Shiergong station crossed under the pearl River. (fig.1).At the beginning three constructive methods were proposed: mining method, submerged tube method and shield tunneling method which is finally choose because cost can be saved.

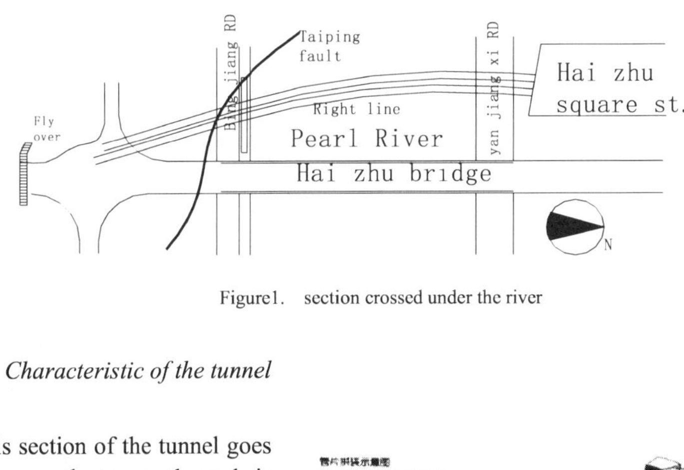

Figure1. section crossed under the river

1.1 Characteristic of the tunnel

This section of the tunnel goes from north to south and is perpendicular to the Pearl River, which total length 840m, section of crossing river 185m, max. gradient 2.5%, min. vertical curve radius 3000m, min. buried cover in the river 5.5m. Internal diameter of tunnel is 5.4m, outer diameter of tunnel is 6.0m, width of

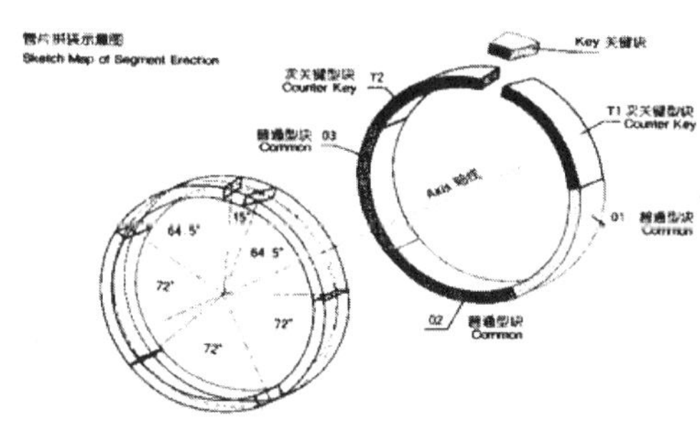

Figure 2. structure of segments each ring

segment is 1.2m, No. of segments each ring is 6 (pieces), 24 pieces of bolt with M24 and M27 each ring, the material of gasket is the swelled rubber with water, the cavity grouting is through the center hole of the segment. (fig.2)

1.2 Reformation of the used shield machines

Reformed mainly for the head face (fig. 3 and 4) and disposal system of spoil (detail see Table 1).

2 Geology

The surface of the river bed is plain ,the max. slope is 7% and the depth of the river is 8—12 m.

Table 1. Comparison of shield characteristic

Item	Shield's size(m)		Thrust KN	Head face							Spoil disposal
	D	L		Torque KN.m	opening ratio	Rpm	roller cutter	cutters	copy cutter	center cutters	
Original	6.17	8.185	32340	2420	20%	0.95	12	135	2	Yes	Pipe system
After	6.21	8.185	32340	3430	28%	2.9	41	86	0	No	Screw conveyer

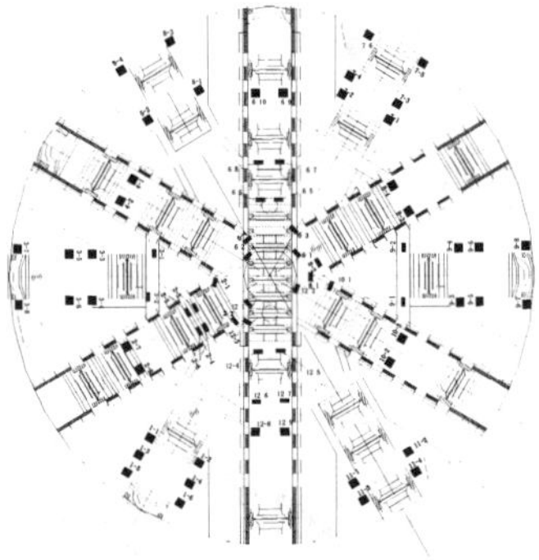

Figure 3. head face before reformed

Figure 4. head face after reformed

2.1 Formation

Mainly consists of two sets, the upper part of which is Quaternary soft earth of sand and sandy gravel layers, and the lower part is Cretaceous silty mud rock weathered in different degree with the max. strength of 15Mpa. (Fig.5)

2.2 Structure and hydrology

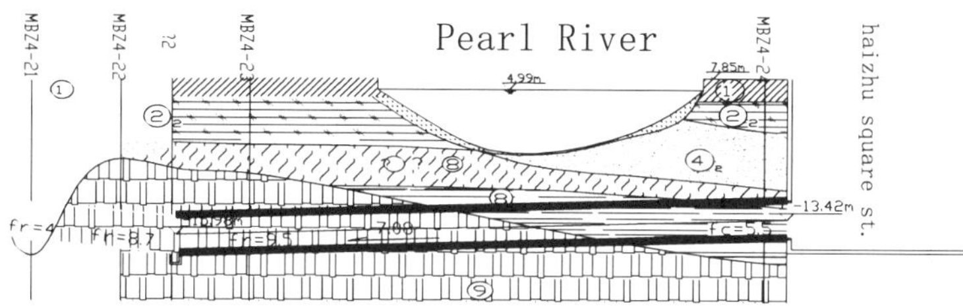

Fig.5 geological section crossed under the river

The formation is simply inclined to the south, the Taiping fault [F2] crossed with the Line 2, so the cracks and the fissures are developed well with high permeability.

2.3 Engineering geological evaluation

There are 6 points where may cause shield tunneling posers related to the geological conditions.

1) Absence of residual clay layer, so the river water will be linked with rock bed directly.

2) The ground condition of the section of tunnel are quite different, the top part of nearly 50% of tunnel length is seriously weathered rock and lower part is medium weathered rock..

3) Rock's thickness of buried cover is less than 3m.

There is a water pipe with Φ 1.6m which foundation is only 1.5m away from the left tunnel on vertical direction. (see fig.6)

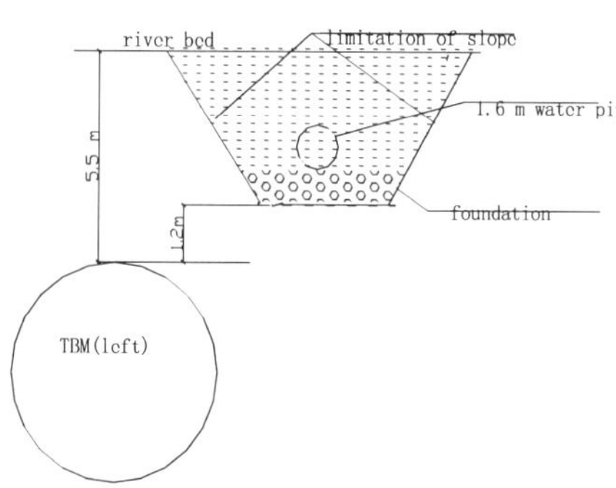

Figure 6. location of water pipe and tunnel

4) The characteristics of Cretaceous silty mudstone can be found in table 2 compared with the other kinds of rocks.

5) The cracks are developed in the bed rock which are the passage of the water.

6) Some boreholes which are not sealed well will be passage of the water.

3 The engineering poser and countermeasures

The problems mainly are the stability of the ground in front of the launching shaft , the suitability of the shield machine reformed and the countermeasures for the construction problems etc.

Table 2. Comparison of physical characteristics of rock

Rock / Test item	Weathered silty mudstone			Siltstone	Sandston e	Conglom erate	limestone
	seriously	medium	slightly				
Softing radio	0.16	0.32	0.48	——	0.65~ 0.97	0.5×0.96	0.70~ 0.94
Elas modu. mpa	0.8~1.5 1.×10³	6.4×10³	7.9×10³	10~32×10³	17~45×10³	——	35~39×10³

3.1 The improvement of the ground condition in the area of TBM launching shaft.

After the demonstration repeatedly, it was decided to do the ground improvement with freezing method which is the first time used in Guangzhou.

The ground was consolidated very well after one mouth freezing with the temperature of –5℃.

3.2 Tthe suitability of the shield machine reformed for the ground condition

The shield machines (TBM) are not working well in the first 43 days, for instance, the average speed is only 0.65m/d or 2~5mm/min. The problem are mainly in two aspects: firstly, the roller disk is only 20mm higher than the cutter bits which is difficult to break the rock into pieces, especially in soft silty mudstone; Secondly, the torque is not strong enough.

After second time reformation, especially optimizing the head pattern of the roller disks and cutter bits, the driving speed can be reached to 10~24mm/min.

3.3 Dealing with the construction accident

1. Tunneling collapse under the river. The reasons of collapse are as follows:
   1) The buried cover of rock is too thin and the top of the tunnel section is

very soft compared with the bottom of the section, so the ground materials of top part will easier go into the chamber than the bottom part.

2) The total thrust is too high (>1500t), the formation in front of the TBM will be lifted, and then deformed and broken.

Several countermeasures had being taken because of protecting the pipes in the river and the shield machines. such as Stop driving temporarily. Close the screw conveyer and make the working face earth balance. Protect the tail seal with grouting a kind of polymer around the TBM. Grout grease to the tail seal. Re-driving the TBM and as fast as possible to pas the collapse area.

2. Prevent the formation of clay cake in the chamber.

1) Adding some foam to change the workability of the spoil.

2) Mount more agitating steel bar on the back face of the head or on the plate of chamber.

3) Turn the head face without thrust, force the clay cake breaking into pieces.

4) Drive in the high pressure air if the permeability is not too high in the bed rocks.

5) Go into the chamber to clean the clay cake if the working face is stable.

3. Spurting from the gate of the screw conveyer (fig.7)

The spurting from the screw conveyer happened more than 20 times when TBM was driving under the Pearl river. The countermeasures are as follows:

1) Close the gate of screw conveyer and push the water out of the chamber while TBM driving forward, but take care of the pressure not too high.

2) Drive with high pressure air, but prevent the air leakage.

3) Pump slurry with high density or foam to change the workability of the spoil.

4) Choose center axial type of screw conveyer with earth plug design.

5) Mount another pump system on some place of screw conveyer, turn EPB mode to slurry mode.

6) Change opening radio and replace the cutters.

4. Striking through the tail seal

The accident of striking through the tail seal had happened while two TBMs drove under the river, the contractor did second compensation grouting to the outside tunnel 10 meter after TBM passed, which way stopped the water flow into the tunnel from the tail.

3.4 Quality problem of tunnels

The problems mentioned above not only delayed the construction time, but also made quality of the tunnels worse.

1. Offset of segments and spalling

Because of the soft ground on the upper part and hard lower part and the accident of spurting and clay cake, the TBM's attitude is difficultly controlled; on the other hand, the grouting result is not always good, so the offset of segments happened again and again which is max. 4~5cm. And the offset will be continuing to lead to other kind of accident, for instance, spalling of segments.

2. Crack of segments

The earth pressure under the river is more than $1.8kg/cm^2$, so the ordinary driving thrust of the TBM is always higher than 1500t. Sometime it is more than 2000t. If the section face of ring is not exactly plain, it will result in the crack where the water has leaked out, water also leaked out from joints between the segments.

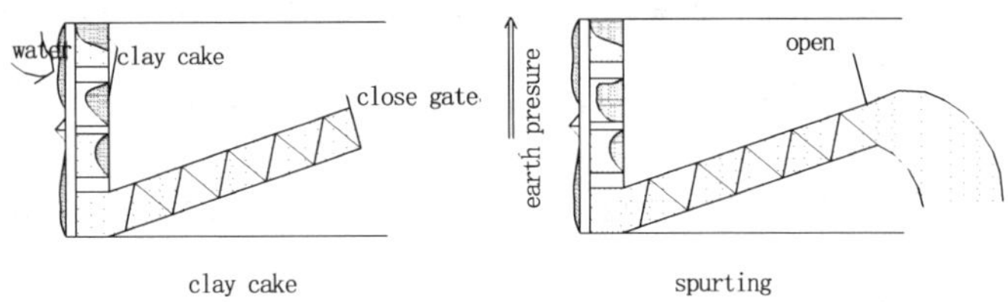

Figure7. Principle of clay cake-spurting

4 Summarization and suggestion

It is the best choice with shield tunneling machine to drive under the river according to the economic or safe reasons. If the used machine is reformed reasonably, it can be used second times and even more times. Before the reformation of the machine, the ground condition must be studied. There were so many accidents because of mainly the lack of construction experience in the similar geology condition. While TBM drive through the buried cover where water pressure is higher than 1D (D means TBM diameter). We had better choose screw convey with earth plug system for the EPBM. If the spoil is mainly consisted of sticky clay, the preventing clay cake formed should be under the consideration. If the spoil is mainly consisted of quartz or quartzite materials which hardness is higher than the steel, so the special treatment for the cutter and head face should be considered. Strengthen the management between man and the machine. Man is the first and the machine is the second, if the man depends on machine, the accidents must happen. The key of preventing accidents is to control the balance of the working face always.

Present this paper to the project team of shanghai tunneling Engineering Co. who drove the two shield machines crossing the Pearl River successfully first time in Guangzhou, as well as to academician Zhongheng Shi and former chief engineer Mr. Zhenxin Wang who are the first generation experts of shield tunneling and gave counsel how to cross the river.

EFFECT OF SHAPES ON STABILITY OF SURROUNDING SOIL USING SLURRY PIPE-JACKING

HIDEKI SHIMADA, TAKASHI SASAOKA, HITOMI YAMAKAWA, SHIRO KUBOTA, KIKUO MATSUI
Dept. of Earth Resources and Mining Engineering, Graduate School of Engineering, Kyushu University, 6-10-1 Hakozaki, Higashi-ku, Fukuoka, Japan
E-mail: shimada@mine.kyushu-u.ac.jp

Slurry pipe-jacking is firmly established as a special method for the non-disruptive construction of the underground pipelines of sewage systems. Pipe jacking, in its traditional form, has occasionally been used for short railways, roads, rivers, and other projects. Basically the system involves the pushing or thrusting of a drivage machine through concrete pipes ahead of jacks. This method utilizes mud slurry that is formed around the pipes in order to stabilize the surrounding soil. However, the behavior of the soil and mud slurry around the pipes is not clearly understood. However, it is not well understood the behavior of the soils and mud slurry around the pipes. From this point of view, this paper discusses the effect of shapes on stability of surrounding soil by means of the numerical analysis.

1 Introduction

In order to protect workers' safety during construction, as well as for environmental and cost reasons, efficient small-diameter shallow tunneling methods have become increasingly important in regards to outside plant engineering such as for water supplies, electricity, telecommunications and gas recently. The effects of the above projects, in overcrowded urban areas are significant and often result in substantial impact and traffic delays with associated loss of travel time. Clearly the solution to these utility placement problems, if the full impact of trench excavation is to be avoided, is trenchless technology. In particular, for construction work near existing facilities, underground tunnels that are excavated by slurry pipe-jacking are being increasingly employed in order to avoid problems.

Slurry pipe-jacking is firmly established as a special method for the non-disruptive construction of the underground pipelines of sewage systems. Pipe jacking, in its traditional form, has occasionally been used for short railways, roads, rivers, and other projects. Basically the system involves the pushing or thrusting of a drivage machine through

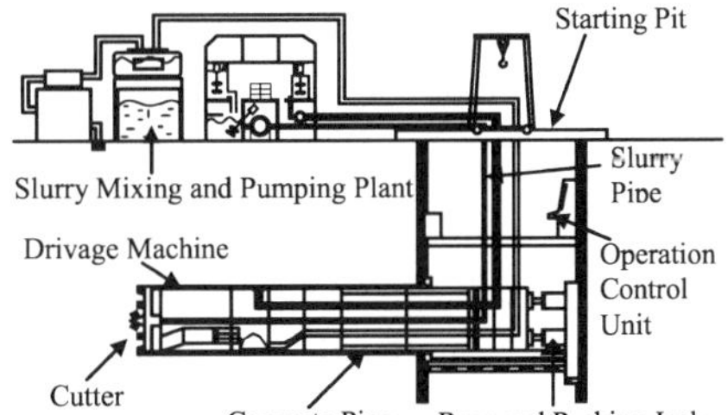

Figure 1. Slurry pipe-jacking system scheme.

concrete pipes ahead of jacks. This method utilizes mud slurry that is formed around the pipes in order to stabilize the surrounding soil. However, the behavior of the soil and mud slurry around the pipes is not clearly understood[1,2].

From these perspectives, this paper discusses the performance of the mud slurry around the drivage pipes and deformation behavior of the soil by means of Finite Element modeling.

2 Construction Outline

Slurry pipe-jacking, like many other methods of below ground construction, is best employed in stable, water free soil conditions. Unfortunately, with the demands on available space and the need to provide more services, it is not always possible to select stable strata, which means that contractors have to contend with difficult ground below the water table[3,4]. The slurry pipe-jacking system is applicable to the above situations.

Figure 1 shows the slurry pipe-jacking system scheme. This system is particularly suited to both cohesive and sandy soil, and can be used to construct pipe tunnels up to 2,000mm in diameter.

The features of this system are as follows:

1) It can be applied to sand with high-pressure or gravel-rich soil by using an excavating method.
2) It can construct both long (maximum length: 1000m) and curved tunnels (minimum curvature radius: 25m).

Recent technological developments have led to successful methods for stabilizing unstable strata by excluding water from excavations by means of the mud slurry around the pipes. This system meets modern environmental standards. Water, decomposed granite soil, carboxyl methyl cellulose, lost circulation control material, bentonite, fine aggregate, and other materials form this type of mud slurry. The proportional mixes vary the viscosity and the seepage capacity of the mud slurry and will be determined by whether mud slurry is required to strengthen the soil.

3 Finite Element Analysis

In an effort to identify the deformation behavior of the soil and the performance of the mud slurry around the pipes, finite element model was performed.

Figure 2 shows a two-dimensional finite element model used in this study. The model of the overburden depth and the pipe diameter were 5m and 1m respectively.

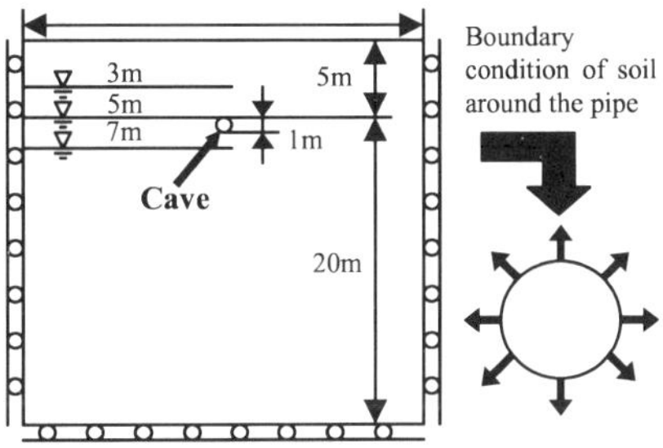

Figure 2. Two-dimensional finite element

Table 1. Finite element analysis parameters.

	Coarse sand	Clay
Dry density (g/cm^3)	1600	1550
Wet density (g/cm^3)	2300	2300
Void ratio	0.69	0.74
Shear modulus (Pa)	5880	196
Bulk modulus (Pa)	27400	915
Cohesion (Pa)	0	7000
Internal friction angle (deg)	32	20
Tensile Strength (Pa)	0	19200
Dilatancy angle (deg)	15	0

The depths of the underground water head from the surface were considered as 3, 5 and 7m. The stress release load and slurry pressure were applied on the soil around the pipes. The slurry pressure and characteristics of soils were simulated in the model as listed in Table 1. These variables are to be adjusted in situ. Materials such as soil used in this analysis were assumed to behave in a linearly elastic manner until their failure. Mohr's theory of yielding was adopted as the failure criterion.

Figure 3 (a)-(b) shows that the relationship between the surface deformation and the distance from the intersection point at the vertical line from the top of crown. Those figures show the cross sections of a circle and a rectangular respectively. This result indicates that a circle cross section in using slurry pipe-jacking is more stable for the surface deformation than that of a rectangular one.

Figure 4 (a)-(b) shows the failure zone around the concrete pipes of circle and rectangular cross sections. The failure zone in the case of the rectangular cross section developed more widely in the soil around the pipe than that of the circle cross

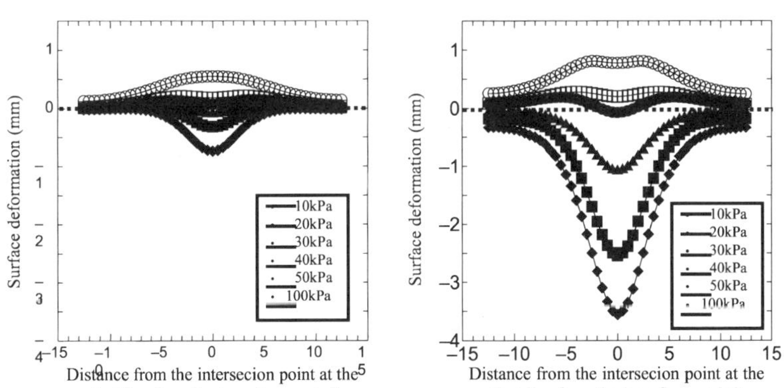

(a) Cross section : circle (b) Cross section : rectangular

Figure 3. Relationship between the surface deformation and the distance from the intersection point at the vertical line from the top of crown (soil:coarse sand, underground water head:3m).

section. It must be noted that ground load in the excavation of the tunnel was to be supported by the slurry pressure. Therefore, it is easy to support the circle cross section due to the own shape. Accordingly, if a rectangular shaped pipe is adopted, it is necessary to consider the stability of the surrounding soils and how to inject the mud slurry and so on.

4 Conclusion

This paper discusses the effect of shapes on stability of surrounding soil by means of Finite Element modeling.

Consequently, it is clear that if a rectangular shaped pipe is adopted, it is necessary to consider the stability of the surrounding soils and how to inject the mud slurry and so on.

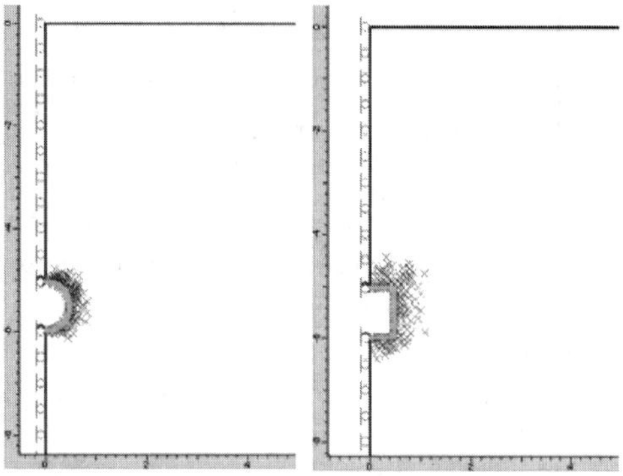

(a) Cross section : circle (b) Cross section : rectangular

Figure 4. Failure zone around the pipe (soil:coarse sand, underground water head:3m, ×:shear failure).

References

1. J.M.Cole, Tunnels and Tunneling, July, 91-93 (1977).
2. M. Hunt, Tunnels and Tunneling. July, 19-21(1968).
3. H.Shimada and K.Matsui, Proc. of Mining Science and Technology, 287-292 (1996).
4. H.Shimada, S.Kubota and K.Matsui, Proc. of International Conference on Geomechanics/Ground Control in Mining and Underground Construction, 2 , 803-811 (1998).

FINITE ELEMENT MODELING OF FULLY GROUTED ROCKBOLTS

ZIPING HUANG
Rogaland Research, N-4068, Stavanger, Norway
EINAR BROCH
Norwegian University of Science and Technology, Norway
MING LU
SINTEF Civil and Environmental Engineering, Norway

To simulate the behaviour of the fully grouted rockbolts in reinforcement of rock masses, a FE rockbolt model is proposed based on Aydan's theory. The element is a coupled truss and interface element facilitating bound slip and elasto-plastic behavior of steel bar and grout annulus. The FE formulation of 2-D and 3-D rockbolt element is given. The 2D rockbolt element is implemented into the DIANA code. Numerical pull-out tests of the rockbolt element have been carried out, which indicate that the smear of the grout annulus into the interface may overestimate the stiffness of the interfaces.

1 Introduction

In simulation of the rock mass reinforcement by fully grouted rockbolts, distinct modeling and equivalent approaches are often used. In distinct modeling, the rockbolt is specifically modeled by a rockbolt element (John and Van Dillen, 1983; Lorig, 1985; Aydan, 1989). And Lorig (1985) has implemented the models into the UDEC code. In the equivalent material approach, uniformly spaced rockbolts are smeared into the rock mass and their effects are taken into account through modification of the constitutive law of the rock mass (Chen and Egger, 1999).

In distinct modeling, FEM have been developed from the simple equivalent load model, bar element, reinforcement element to bolt element, spring model and bolt crossing joint element. For fully grouted rock bolt, the slippage on the bar-grout interface was modeled by linkage element consisting of two springs to represent bolt behavior in tangential and normal directions to the bar. Slippage is also modeled by special interface elements with axisymmetrical joint element (Ghabossi, 1973). The dowel effect of rockbolt across rock joints was accounted for by Lorig (1985) using short bar element attached to both sides of the joint. The difficulty may lie in the determination of parameters with a physical meaning in the representations.

2 Mechanical Behavior of Grouted Rockbolts

Pull-out tests and shear tests are fundamental for rock bolting. The shear behavior of grout annulus and the interfaces along the bolt axis are dependent upon various factors such as shear strength of the interfaces, thickness of grout annulus, surface smoothness of steel bar and hole wall, and the normal stress of the surrounding rock mass. The shear resistance of the interfaces is less than that of the grouting material (Aydan, 1989). Bolts contribute to shear resistance of joint by increase of joint friction and by the dowel action, which varies with bolt inclination to the normal of the joint, and the dowel action component makes a significant contribution, of as high as 50%, to the total shear strength of the bolted joint.

In general there are three typical failure modes of fully grouted rock bolts: axial failure of steel bars in tension; shear failure parallel to bar axis, in the grout annulus or at one of the interfaces of bolt-grout and grout-rock; shear failure of grouted bolt crossing rock joint. For shearing crossing joints, a simple constitutive model based on the assumption that the bolt deforms linearly in shear until yielding is used. The shear behavior of the grout annulus and interfaces parallel to bolt axis can be described with the classical Coulomb friction model. The shear failure is assumed to occur at the interface of bolt-grout or grout-rock. In FE formulation of rockbolt element, the grout annulus is smeared to represent the interfaces assuming that the shear stress is constant over the relatively thin grout annulus. However, the geometry and mechanical properties of the grout annulus are counted in the stiffness of the rockbolt element. The performance of the fully grouted rock bolt is governed by the mechanical interaction between bolt, grout annulus and rock mass(Figure 1)

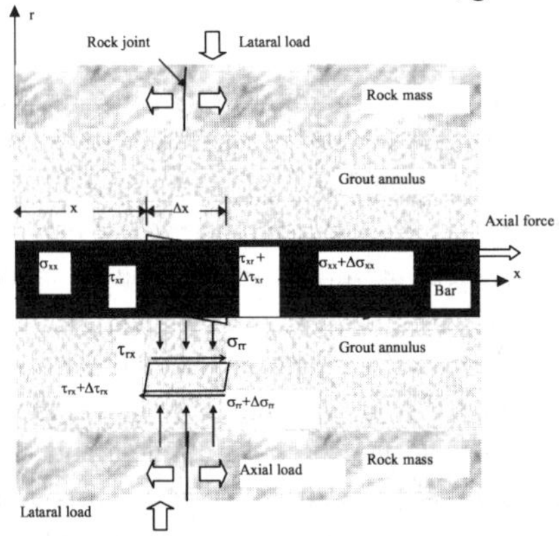

Figure 1. Stresses and displacements in a fully grouted rockbolt

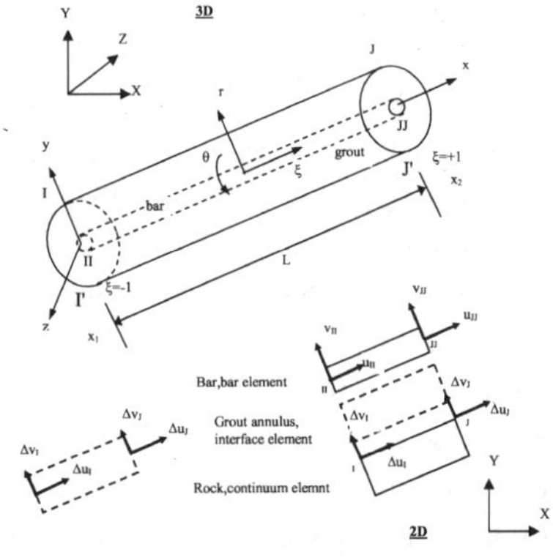

Figure 2. 2D and 3D rockbolt element

In the following, equilibrium equations of grout annulus are given and serve as strain displacement relationship in the proposed FE rock bolt model.

$$\gamma_{rx}^{g} = \frac{u_b - u_h}{r\ln(r_h / r_b)} = R\Delta u, \quad R = \frac{1}{r\ln(r_h / r_b)} \tag{1}$$

$$\varepsilon_{rr}^{g} = \frac{v_b - v_h}{r\ln(r_h / r_b)} = R\Delta v, \quad R = \frac{1}{r\ln(r_h / r_b)} \tag{2}$$

where u, r, v = axial displacement, radial distance and displacement of the considered point in r-direction, respectively; γ_{rx}^{g}, ε_{rr}^{g} = shear and radial strain; r_h, r_b = radius of bolt-grout interface and grout-rock interface.

3 Finite Element Formulation and Implementation of Rockbolt Element

The proposed rockbolt element couples the truss element with the interface element of Ghaboussi-type. The truss element with 4 DOFs represents the steel bar, the interface element represents the grout annulus. The final stiffness matrix of the rockbolt element is built up in terms of nodal displacements. Figure 2 shows the coordinate systems and the 2-D and 3-D rockbolt elements. The rockbolt element has 4 nodes for 2-D and 6 nodes for 3-D. Two of them are attached to the truss by *II, JJ*, others two (four) nodes are attached to outer surface of the grout annulus by *I, J* (*I',J'*). Assumptions have been made to simplify the formulation of the element.

From the displacements of the nodal points in the bar, the strain vector ε at any points within the bar can be obtained. Making use of Eq. (1)-(2), strains in grout annulus can be also established in terms of nodal displacement. Using linear elastic constitutive relation the desired 2-D rockbolt element stiffness matrix has the following form in the local coordinate system (*x,y,z*). The stiffness matrix in 3-D is straightforward as the element is axisymmetric.

$$\boldsymbol{K}^e=\begin{bmatrix} 2k_g^r & k_g^r & -2k_g^r & -k_g^r & & & & \\ k_g^r & 2k_g^r & -k_g^r & -2k_g^r & & \Theta & & \\ -2k_g^r & -k_g^r & k_b^x+2k_g^r & -k_b^x+k_g^r & & & & \\ -k_g^r & -2k_g^r & -k_b^x+k_g^r & k_b^x+2k_g^r & & & & \\ & & & & 2k_g^x & k_g^x & -2k_g^x & -k_g^x \\ & & & & k_g^x & 2k_g^x & -k_g^x & -2k_g^x \\ & \Theta & & & -2k_g^x & -k_g^x & k_b^r+2k_g^x & -k_b^r+k_g^x \\ & & & & -k_g^x & -2k_g^x & -k_b^r+k_g^x & k_b^r+2k_g^x \end{bmatrix} \quad (3)$$

Where $k_b^r=\dfrac{E_bA_b}{L}$, $k_b^x=\dfrac{G_bA_b}{L}$, $k_g^r=\dfrac{\pi LE_g}{3\ln(r_h/r_b)}$, $k_g^x=\dfrac{\pi LG_g}{3\ln(r_h/r_b)}$, $A_b=\pi r_b^2$, E_g, E_b are the Young's modulus of the grout annulus and the bar, G_g, G_b are the shear modulus of the grout annulus and the bar, L is the length of the element.

The proposed rock bolt element (RB8) is implemented into DIANA (TNO, Delft, The Netherlands) , coupling interface element of type L8IF as the smeared grout annulus, with the truss element of type L4TRU in DIANA as the steel bar. It superposes over the surrounding rock mass element of type Q8EPS (Figure 3a). The improved procedure includes the construction of the initial linear stiffness matrix and internal nodal force estimation of the rockbolt element. The Linear Stiffness Iteration scheme is adopted and updating of stiffness matrix is consequently avoided.

The rock bolt element, RB8, has been tested by numerical pull-out tests. Further shear tests are necessary. The numerical pull-out test has been done by comparing the rockbolt model (RB8) with other models available in DIANA, truss element combining with two Interface elements (L4TRU+2L8IF) and with one Interface element (L4TRU+L8IF), and with UDEC (Itasca, Minneapolis, US) modeling.

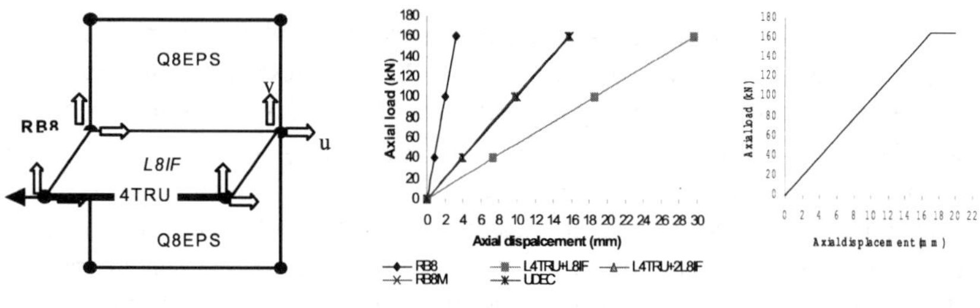

(a) (b) FE tests (c) UDEC test

Figure 3 (a) The rockbolt element (RB8); Axial load vs. displacement curves of FE test results (b) and UDEC test results (c)

Axial load vs. axial displacement curves are given for all models in Figures 3 (b) and (c). It is noted that the proposed rockbolt model is stiffer than other models. The stiffness contribution of the grout annulus to the rockbolt system is smeared at the interface in the formulation in the rockbolt model, which may overestimate the actual behavior of the bar-grout/grout-rock interfaces.

4 Conclusions

Numerical pull-out tests indicate that the proposed rockbolt element is capable of modeling the grouted rockbolt. However, it may overestimate the bond stiffness. Further verification by comparing to the field pull-out tests is needed.

References

1. Aydan, O. 1989. The stabilisation of rock engineering structures by rock bolts. Ph.D. thesis. Nagoya Univeristy, Japan.
2. Chen, S. H. et al 1999. 3D elasto-viscoplastic finite element anal. of reinforced rock masses and its application. Int. J. Numer. Anal. Meth. Geomech. 23, 61-78.
3. Ghabossi, J., Wilson, E. L., Isenberg, J. 1973. Finite element for rock joints and interfaces. ASCE J. Soil Mech. Found. Div. 99 (SM10), 833-848.
4. John C. M. and Van Dillen, D. E. 1983, Rockbolts: a new reprentaiton and its application in tunnel design, Proc. 24th U.S. Symp. On Rock Mech. Pp.13-25.
5. Lorig, L. J. 1985. A simple numerical representation of fully bonded passive rock reinforcement for hard rocks. Computers and Geotechnics, 1, pp79-97.

PREDICTING LOADS ON THE SUPPORT OF UNDERGROUND CONSTRUCTION

BORIS AMUSIN
40 Ocean PKWY # 2D, Brooklyn, NY 11218, USA
E-mail: boris_amusin@hotmail.com

One of the most disputed problems of rock mechanics is predicting of loads on the support of underground construction. There is no universal approach to the problem. Reliability of predicted parameters based on known empirical and conventional methods, as well as analytical and numerical methods, are very limited. The author strongly believes that the only Theoretical-Experimental Methods (TEM) could increase the reliability of design parameters of support of underground construction. The TEM developed by the author is based on a simple theoretical model of post-failure deformation of the rock mass, which is corrected by introduction of system coefficients and correlation functions in initial mechanical properties and stress state of the rock mass.

1 Introduction

Low reliability of known methods, used in practice design of the support of underground construction, is determined by:
1. Uncertainty of the initial rock mass properties and stress state in-situ;
2. Shortage of statistics for the load measurements in real rock mass;
3. Complexity of the interaction process between the support and the rock mass is determined by a variety of factors:

- depths, from shallow to the deepest mine more than 3000m;
- types of rock mass, from strongest intact rock to fractured or plastic shale;
- sizes of openings, from a drill hole to an underground power station and stadium;
- types of openings, from a separated (2D) to junction of openings (3D).

Reasons 1 and 3 lead to lack of theoretical methods. Shortage of the statistics compels authors of empirical methods to introduce geomechanical hypothesis, which in some cases have been proved to be wrong. Thus engineers have to be concerned about the danger of using empirical methods.

2 Empirical hypothesis of strata control

One of the most widespread defects of empirical methods is its approach for calculating loads on the support. For example, according to the Q-method[1], the load on the support is determined depend on Criteria Q (rock mass quality). If the Criteria's Q and a ratio of the rock strength Sc to the stress state q are the same for different depths (with equal other geometrical and technological parameters) then, the loads on support will be same for small depths and for large depths.

The Russian experimental U method [2] uses the same empirical hypothesis to determine load on the support. As a Criteria of tunnel stability is used long-term contour displacements (U_o) for conditions yielding support with minimum capacity (P_{min}).

Empirical hypothesis adopts equal loads on the rigid support for the different depths if values of U_o or Q factors are equal:

$$U_o \text{ (great depth)} = U_o \text{ (small depth)} \text{ or } Q \text{ (great depth)} = Q \text{ (small depth)};$$
$$\text{and } U_p \text{ (great depth)} = U_p \text{ (small depth)}$$
$$P_o \text{ (great depth)} = P_o \text{ (small depth)} \qquad (1)$$

Where U_p is displacement of tunnel contour with rigid support.

For many practical cases of using yielding support, the U-method gives a reliable result. However for the rigid structural support hypothesis (1) is not confirmed by the results of field data [3] and theoretical methods. [4, 5]

3 Theoretical hypothesis of strata control

As base for predicting the rock deformations, failure and load on support in the Theoretical-Experimental Method (TEM) the author used an axis-symmetrical visco-elasto-plastic solution. This simple model is corrected by introduction a system coefficients and correlation functions in initial mechanical properties and stress state of the rock mass.[4, 6, 7] It takes into account the mechanism of post failure deformation, creep of rock in time and the interaction between the support and rock mass. Empirical coefficients are taken into account a scale factor, degree of fracture and type of rock mass, technological and geometrical parameters (3D – problem: distance support installation from the face of the tunnel, junction of tunnels and chamber).

In setting up of the problem, the following boundary conditions are adopted. An extended circular shaped opening is located at a depth of H. Pressure P_o corresponding to stabilized resistance of the support is applied to the internal contour of the opening in time t_1 after driving the tunnel. The stress state of the rock mass in-situ is assumed to be hydrostatic: $Q_h = Q_v = q = gH$ (where g = the unit weight of rock). It is assumed that three zones of non-elastic deformations are formed around the tunnel. Their size depends on time, stress levels and mechanical properties of rocks mass:

A_1: zone of failed rocks close to the contour. Displacements here take place without dilatation;

A_2: zone of intensive fracturing (post-failure) with dilatation;

A_3: zone of plastic non-loosening (limit) deformations.

The boundaries of those zones (r_1, r_2, r_3 respectively) are determined by conditions of the continuity of radial stress (σ_3) and tangential strain (ε_θ) from one zone to the neighboring one. The zones A_j correspond with the curves of rock deformations on a rigid press.

The theoretical solution permits one to obtain the deformations for different types of rock. For example, in salts, A_1 and A_2 zones are not formed even for great depths because of considerable limited deformations without failures, which are typical for salt. The openings in this case can be exploited without supports. In brittle rocks (for example, coal) which are not subject to plastic deformations, A_1 and A_2 zones do get developed, and support is necessary for the stability of the opening. Finally the method was calibrated based on field measurements of displacements and loads on the support of underground construction.

The theoretical solution permits one to design support without a priory approaches and assumptions. For example, according to the theoretical solution, value U_o does not alone determine the value P_o as in empirical hypothesis (1). Assuming that for a given rock, any change in depth or stress will change both the strength (S_m) and shear modules (G_m) of the rock mass as follows:

$$q / S_m = \text{constant}; \quad G_m / S_m = \text{constant}; \quad \phi = \text{constant}; \qquad (2)$$

where: ϕ = the angle of internal friction of the rock mass.

In this case, the final displacement U_o remains the same independent on the depth, for the given zero resistance of support $P_o = 0$ (no support or $P = P_{min}$).

If P_o / q = constant, then the rock mass reaction curve for the different depths also remains constant. For the conditions (2) and equal values U_o / U_p (where U_p is the displacement of the opening contour under the support reaction P_o) the ratio loads is equal ratio the depths.

$$P_{great} / P_{small} = H_{great} / H_{small} \qquad (3)$$

This peculiarity is very important to the practice of design support of the underground constructions.

For example, though for the certain geomechanics conditions a concrete support could be recommended for the small depth, it could be wrong for the great depth with similar conditions:

$$[U_o / U_p] \text{ great depth} = [U_o / U_p] \text{ small depth.}$$

Similar results will be for the rigid rock bolt supports. [8]

The author found a proof of this peculiarity from field measurements of loads on concrete supports. [3] For the same geomechanical condition, the permanent value U_o was obtained when the support was yielding (with $P_o = P_{min}$) and the relation H/S_c was constant (where S_c = unilateral compressive strength in lab). However results of measurement of loads on the concrete supports are shown that for the same "geomechanical conditions" loads are increased with increase the depth even more than according to theoretical hypothesis (3).

Thus, the value U_o does not determine the value P_o uniquely and the empirical assumption is not correct.

To check theoretical hypothesis author set up series of experiments on model of equivalent materials. The experiments were conducted by Dr. Shkliarsky with participation of the author. Total number of four models were conducted in scale 1:50 (for two types model of rock mass with strength: Sm=70kPa and Sm=200kPa and two types of support: yielding support, k=0.1kPa/mm and k=60kPa. There were used following steps for loading model of the rock mass during the experiments. Symmetric equal steps of loading (stress state) Qv =Qh =q were applied after the support was installed in an opening. Radial displacements and loads on model support were measured for each step of loading. A limit of paper is not allowed to discuss results in details. In shortly, for example, according to theoretical hypothesis (3) relation the loads should be 200/70=2.86. Based on the experiment for $q / S_m = 0.8$ the relation of the loads was 2.6. The ratios can be considered close especially because in the experiments not all requirements of the hypothesis (3) were satisfied exactly (G_m / S_m and ϕ were not constants).

In the TEM the hypothesis (3) is not used directly. It shows only the tendency and peculiarity of the load relations for the similar geomechanics conditions. Using of a theoretical solution as base of design methods allows eliminate different a priori guesses, which sometimes could be wrong as for example discussed above experimental hypothesis (1).

References

1. Barton N., Lien R., Lunde J. (1974). "Engineering Classification of Rock mass for the Design of Tunnel Support," Rock Mech., V.6, No 4, 189-236.
2. Manual for Designing of Underground Mine Workings and Calculation of Support (1983). Moscow, Stroyizdat, p.273.
3. Basinski Y.M., Ivanov E.A. (1972). "Dependence of Rock Pressure Phenomena in Permanent Workings Deep Shaft.", Tr. VNIMI, Leningrad, No 85, 79-84 (in Russian).
4. Amusin B.Z. (1977). "Prediction of the Stability of Main Workings with Allowance for Gradual Fracture of the Rock in the Zone of Inelastic Deformation." Soviet Mining Science.V.13, No 5, 460-466.
5. Amusin B.Z., Ardashev K.A., Basinski Y.M. (1984)."Geomechanics Principles of Planning Mine Workings." *ISRM Symposium on Design and Performance of Undergr. Exc., Cambridge, 3-6 Sept. 1984, Publ. London British Geomech. Society, 81-87.*
6. Amusin B.(1994). Method for design support of Underground Constructions. *Intern. Conf. Glovik, Norway 14-17 June, 1994, 65-77.*
7. Amusin B. (2000). Eval. of Angle of Int. Friction of Fractured Rock Mass. *Proc. 4th NARMS. Seatle/Washington/USA/31July-3 Aug. 2000, 723-728.*
8. Lama R.D., Amusin B.Z. (1992). "Rigid or Yielding Roof Bolts: At the Face or away from the Face". *X1 Intern. Conf. on Ground Control in Mining, Wollongong, NSW, July 1992, 193-203.*

A NEW METHOD OF VISUALISING INTEGRATED KINEMATIC AND MECHANICAL STABILITY ANALYSES OF ROCK SLOPES

J.P. HARRISON and J.A. HUDSON
Department of Earth Science and Engineering, Imperial College, London, UK
j.p.harrison@ic.ac.uk and jah@rockeng.co.uk

A technique for incorporating both kinematic and mechanical stability indices on a single hemispherical projection plot is introduced. This allows rapid composite assessments of rock slope instability.

1 Introduction

One of the most common preliminary methods for the design and analysis of slopes in fractured rock is the use of kinematics in conjunction with hemispherical projection methods (Goodman[1]; Priest[2]; Hudson and Harrison[3]). This method allows assessment of potential instability, such that those slope orientations leading to instability can be identified and subjected to a more thorough mechanical analysis.

However, if a grid of points is placed over the hemispherical projection, each grid point can be taken to represent the orientation of a potential slope face. For each grid point some characteristic of slope instability can be computed (e.g. the factor of safety), and as a result the entire projection can be covered with such values. This paper introduces a presentational method based on this approach that allows the overall effect of slope orientation on instability to be assessed.

2 Example of the approach

As an example of its use, an analysis has been performed for the case of a large pre-split rock slope adjacent to the major A82 highway at Rubha Mor, Loch Lomond, in the West of Scotland (Fig. 1a). The highway follows a curve at this location, which means that the slope possesses a range of dip directions. After construction, wedge instability became apparent at the southern end of the slope (Fig. 1b), and it is instructive to see how this instability is related to the local orientation of the slope.

Fig. 2a shows the pole plot of a fracture survey conducted after construction (Saini[4]). The range of slope orientations (plotted in terms of the downward directed normal to the slope) is also shown on the Figure. This data set is shown in simplified form in Fig. 2b, where only those fractures deemed to be significant in terms of slope instability (e.g. natural fractures of substantial extent) are included. It is this

(a) View of slope, looking North (b) Typical wedge instability

Figure 1. Rubha Mor rock slope, west bank of Loch Lomond, Scotland.

simplified data set, comprising 79 and 39 fractures in each of sets 1 and 2, respectively, that has been used in the analysis of wedge instability.

In the plots shown in Figs. 3a and 3b, the results for mean factor of safety and mean wedge volume have been grouped into 6 categories, as shown by the legend to the right of the hemispherical projections. The legend also indicates how tinting has been used on the projections to highlight the most critical slope orientations through the use of the darker shades. In this way, we can see immediately from Fig. 3a how slopes that are almost vertical and with dip directions in the range of north-east through to east have the lowest factors of safety. More strikingly, Fig. 3b shows how wedges of large volume will be generated in slopes that cover a wide range of orientations — both dip direction and dip angle.

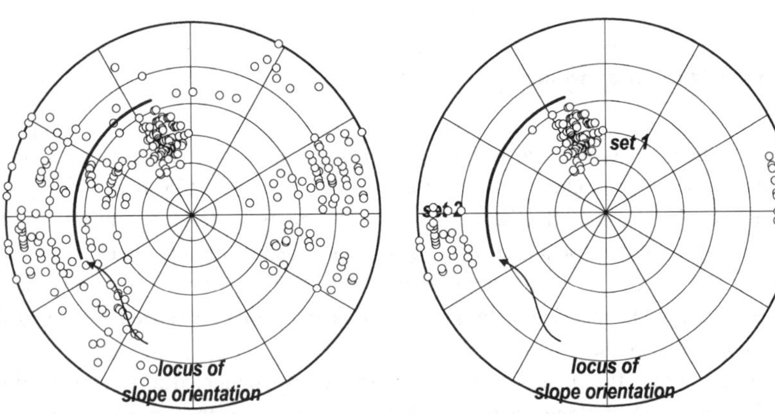

(a) all orientation data obtained (b) critical fractures

Figure 2. Fracture orientation data for Rubha Mor rock slope.

For the case of a friction-only wedge, stability is controlled by just the orientations of the slope face and the two fractures defining the wedge, together with the friction angle of the fractures. To compute the maximum volume of a wedge, the orientation of the slope top and the height of the slope face are also required. In this example, the friction angle has been set as 35° for both fractures, the slope top has been considered horizontal, and a nominal slope height of 10 m has been adopted. All combinations of fracture pairs have been used in the analysis (i.e. a total of 3081 combinations, resulting from 79 fractures in set 1 and 39 fractures in set 2).

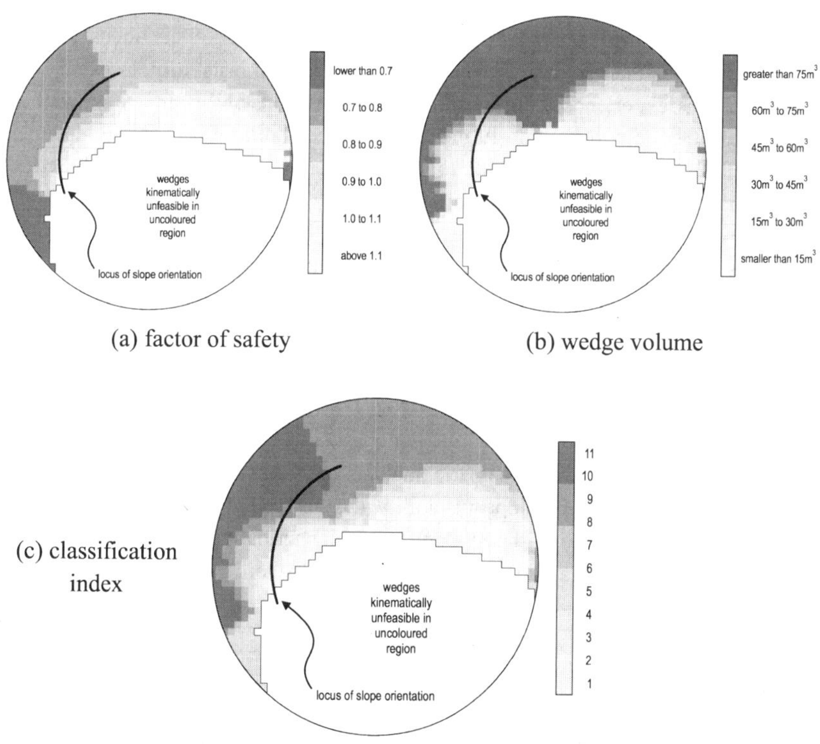

Figure 3. Rubha Mor stability plots.

Although it is of critical interest for designers to be aware of both the factor of safety and volume of potentially unstable blocks, one could argue that it is these two features in combination that is most important (see Table 1). Consequently, we can bring together the six categories from each of the factor of safety and wedge volume analyses, to form a combined classification index as shown on Table 2. In this index, which is plotted in Fig. 3c, high values indicate large wedges of low factor of safety, and low values indicate small wedges of high factor of safety. As Fig. 3c clearly

Table 1: Interaction of wedge volume and factor of safety.

		wedge volume	
		small (good)	large (bad)
Factor of safety	low (bad)	requires assessment	critical
	high (good)	not critical	requires assessment

Table 2: Classification index based on factor of safety and wedge volume.

			wedge volume					
			1	*2*	*3*	*4*	*5*	*6*
Factor of safety	low	*6*	**6**	**7**	**8**	**9**	**10**	**11**
		5	**5**	**6**	**7**	**8**	**9**	**10**
		4	**4**	**5**	**6**	**7**	**8**	**9**
		3	**3**	**4**	**5**	**6**	**7**	**8**
		2	**2**	**3**	**4**	**5**	**6**	**7**
	high	*1*	**1**	**2**	**3**	**4**	**5**	**6**

shows, there is a critical region towards the southern end of the locus of slope orientation: this is the very region where post-construction instability, as exemplified in Fig. 1b, was experienced. Furthermore, this plot of instability index shows how, by reducing the slope dip angle further towards the southern end of the slope (i.e. moving the locus of slope orientation towards the centre of the projection) the effects of the instability might have been ameliorated.

This analysis is presented only as an example. We recommend that appropriate categories and stability indices could be adopted on a project-by-project basis, so that this form of analysis and presentation can be used to highlight critical conditions.

3 Conclusion

A combined mechanical and kinematic analysis has been developed and applied to an example examination of slope instability due to wedge sliding. When used in conjunction with quantitative shading on the hemispherical projection, this produces a powerful technique for rapid assimilation of slope instability.

References

1. Goodman R.E. (1989). Introduction to Rock Mechanics. 2nd edn., Wiley, New York, 562pp.
2. Priest S.D (1985). Hemispherical projection methods in rock mechanics. George Allen & Unwin, London.
3. Hudson J.A. and Harrison J.P. (1997). Engineering Rock Mechanics: An introduction to the principles. Elsevier, Oxford, 444pp.
4. Saini G.S. (1990). Evaluation of a presplit slope cutting at Rubha Mor, Loch Lomond, Scotland. MSc dissertation, Imperial College, London.

APPLICATION OF DAMAGE INDICES FOR SEISMIC SAFETY ASSEEMENT OF ARCH DAMS AGAINST SLIDING AND OPENING FAILURE IN JOINTED ROCK MASS AT ABUTMENT

MAHMOUD YAZDANI[*] HIDEYUKI HORII[**]
[*] *Civil Engineering Department, The University of Tokyo, Japan*
[**] *Civil Engineering Department, The University of Tokyo, Japan*

One of the serious concerns of the engineers involved in the dam design for the areas of highly seismicity is the safety evaluation of dams against large earthquakes. In the case of concrete arch dams, which are often unavoidable to place them on a rock foundation consisting of joints, sliding and opening of rock joints at abutment are two possible failure mechanisms, which may lead to collapse of the dam or release of the reservoir under earthquake excitation. Based on these mechanisms, a method for the safety assessment of concrete arch dams against large earthquakes is presented. In this method using the results of three-dimensional dynamic analysis of the dam and foundation, a two-dimensional local analysis is carried out on critical regions of abutment. In the local analysis, which covers a large 2-D domain around the dam body at a critical elevation of abutment, the rock joints are modeled by interface elements. By this analysis, sliding and opening of rock joints can be obtained as a function of time. A shear softening constitutive law with a hyperbolic form is applied on the behavior of joints in sliding and also an artificial hyperbolic tensile softening has been used for the joints behavior in opening.

In this paper, two kinds of damage indices, which are applied for both failure mechanisms of sliding, or opening, are introduced to indicate the failure progress of abutment at each time of earthquake excitation. The maximum value of each damage index during the earthquake loading is plotted against peak ground acceleration after doing several dynamic analysis of the 2-D model under earthquake motions with the same waveform and different amplification. Using these curves, the dominant failure mechanism of abutment as well as the magnitude of critical earthquake in which the abutment ruptures will be obtained and from that the seismic safety of arch dam against a specified earthquake can be estimated.

1 Introduction

The safety assessment of dams against large earthquakes is of serious concern in many countries located in highly seismicity areas. The importance of the matter for concrete dams was highly demonstrated after the failure accidents caused by large earthquakes in Hsinfengkiang dam in China, Koyna dam in India and Sefidrud dam in Iran. In the all three cases, the damage was severe enough to require major repair and strengthening, although the reservoirs were not released. For a concrete arch dam, the failure mechanisms, which may lead to collapse of the dam or release of the reservoir under a large earthquake excitation are formation and propagation of shear and tensile cracks in the concrete body as well as sliding and opening of pre-existing joints in the dam foundation.

In literature, there are enough number of research and studies, which have been carried out on the failure mechanisms of dam body, while a few numbers can be found to study on sliding and opening in the jointed rock mass at abutment (Ahmadi et al, 1992; Hall, 1998; Lau et al, 1998; Singhal et al 1998; Alonso et al 1996). The authors have proposed a method for the seismic safety evaluation of arch dams based on the shear failure mechanism in the jointed rock mass at abutment (Yazdani and Horii, 2002). In this paper, using the damage indices, which are applied to both sliding and opening, it is shown that the opening failure mechanism can also govern the safety of abutment under a large earthquake.

2 The method for seismic safety evaluation of arch dams against sliding and opening failure in the jointed rock foundation

In the proposed method, at first a dynamic linear elastic analysis is carried out on the three-dimensional finite element model of the dam and foundation under a time dependent earthquake loading. Then using the results of this analysis, a local study is performed on a two-dimensional horizontal domain of abutment located at a critical elevation. The 2-D domain should contain a large area around the dam-abutment interface to cover all possible fractured points. In this domain, a refined mesh, in which the rock joints are modeled by interface elements, is generated. It is assumed that the joints in abutment are located on two major joint sets with different orientation, spacing and mechanical properties. After the analysis of this 2-D model, distribution of the sliding and opening of rock joints can be obtained as a function of time. To indicate the failure progress of abutment at each time of earthquake motion, it is very useful to employ an appropriate *Damage Index* based on the progressive rate of sliding and opening. In this paper two kinds of damage index, which can be applied to both failure mechanisms of abutment, are introduced:

1. ***Maximum Sliding (or Opening) Index**:* the maximum slippage (or opening) occurred on each set of joints.

2. ***Sliding (or Opening) Intensity Index***: the multiplication of slippage (or opening) at each point by its affected area, which is integrated on whole area of the 2-D domain.

Therefore, for each joint set two sliding and two opening damage indices can be obtained. After the dynamic analysis of the 2-D refined model under a seismic loading with specified peak ground acceleration (PGA), the time history of damage indices can be calculated and from that the maximum value of each index during the earthquake motion is obtained. By several dynamic analyses with different peak ground acceleration, the variation of maximum damage index can be plotted against PGA. If a failure mechanism (sliding or opening) leads to a critical state in which the abutment ruptures under a specified earthquake with PGA = A_{cr}, then the variation of corresponding maximum damage index against PGA becomes asymptotic at PGA = A_{cr} (Figure 1).

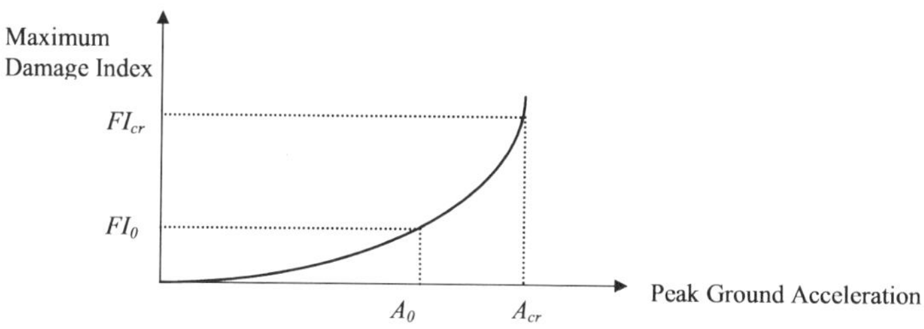

Figure 1. Variation of maximum damage index against peak ground acceleration for a failure mechanism leading to the rupture of abutment

From this figure it could be possible to estimate the safety factor of abutment at that special elevation against an earthquake with PGA = A_0:

$$SF = \frac{A_{cr}}{A_0} \tag{1}$$

Where FI_{cr} and FI_0 indicate respectively, the values of damage index at the critical earthquake and the earthquake with PGA = A_0. The value of A_0 can be the peak ground acceleration of Maximum Credible Earthquake (MCE), which is the largest conceivable earthquake possible to occur along a recognized fault.

3 Application of the method

In this section, application of the proposed method is illustrated by seismic safety evaluation of an arch dam. Figure 2 shows a three-dimensional FEM model of the arch dam and foundation and the critical elevation of abutment used in this study, which is located at elevation 90 m. A refined mesh generated in a closed area at above critical elevation along with the characteristics of rock joints is represented in Figure 3. For the behavior of joints in sliding, a shear softening constitutive law with a hyperbolic form is assumed. In this study, the 2-D model is analyzed for 7 earthquake motions with different amplification factors 1.0, 2.0, 2.75, 3.0, 3.15, 3.25, 3.3 which correspond to the peak ground accelerations of 438 ~ 1445 cm/s^2 (see Yazdani and Horii, 2002 for characteristics of the dam and earthquake loading). It should be noted that the other analyses with the PGA larger than 1445 cm/s^2 failed because of the convergence problem. The maximum value of damage indices during the earthquake motions with different PGA is obtained and then plotted against peak ground acceleration (Fig.4). From this figure it is obvious that the opening mechanism is dominant in this study and their damage indices become asymptotic at PGA ≈ 1445 cm/s^2.

According to the relationship (1), for example the safety factor of abutment at that critical elevation against a Maximum Credible Earthquake with PGA = 800 cm/s^2 is about 1.8.

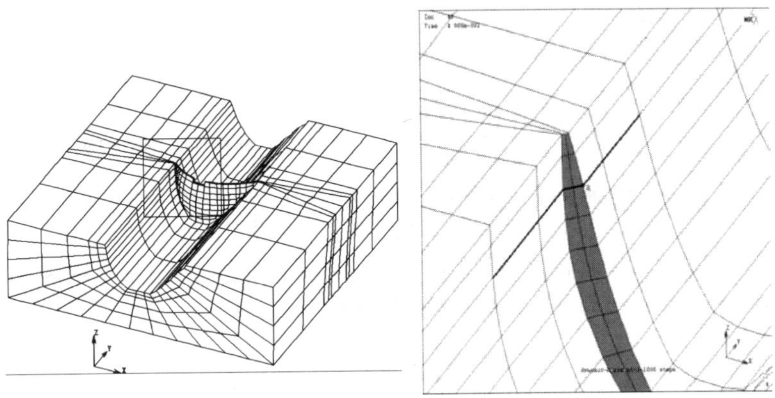

(a) (b)

Figure 2 (a). 3-D FEM model of the dam and foundation. (b) A local view of the abutment and the critical elevation used for this study

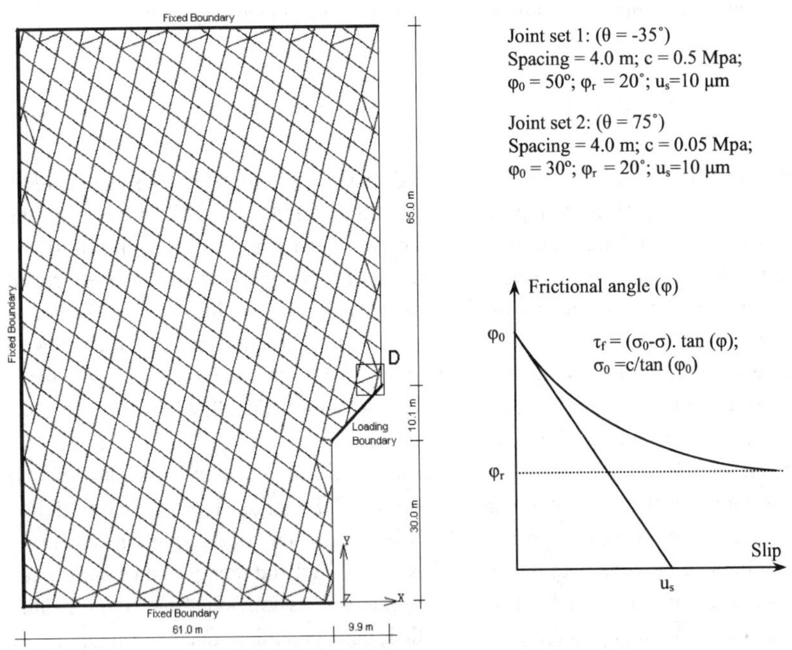

Figure 3. A 2D refined model at elevation 90 m and characteristics of the rock joints

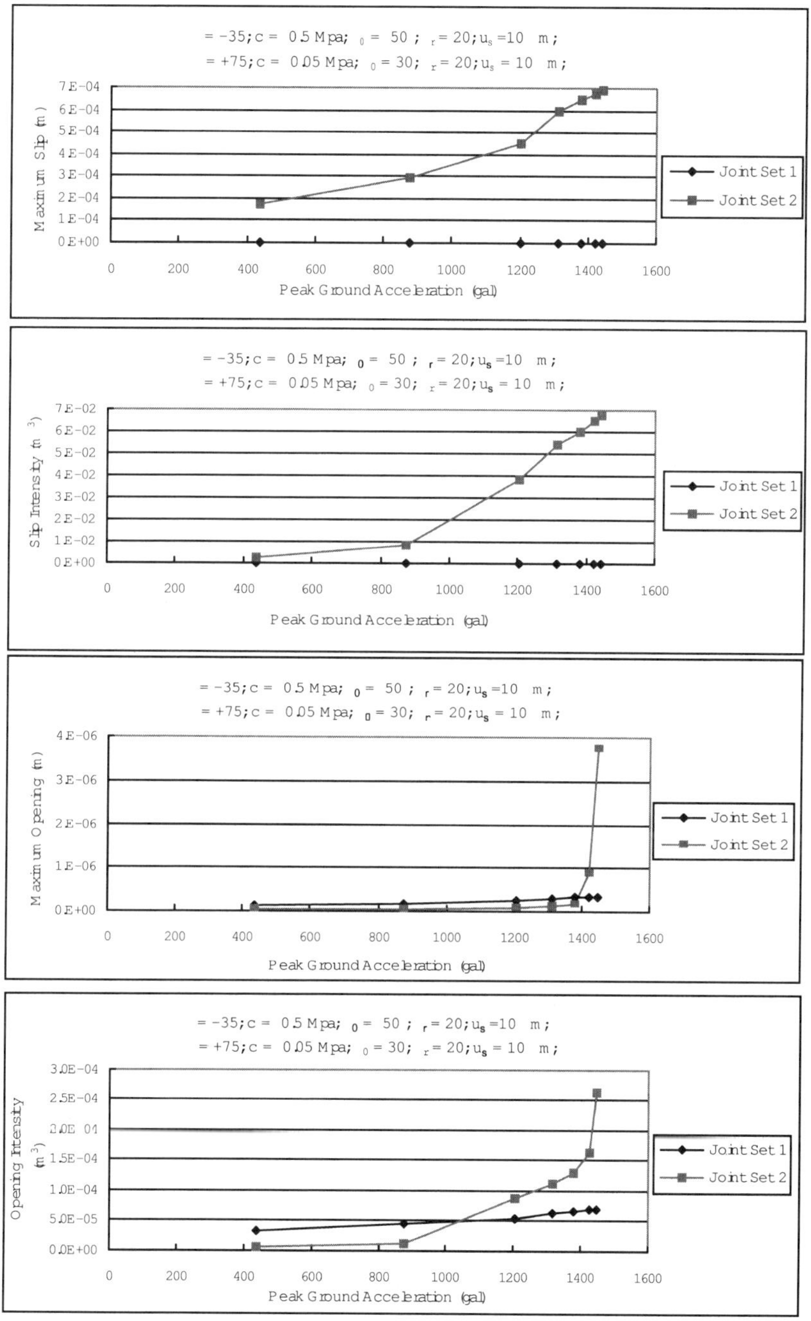

Figure 4. Variation of maximum value of each Damage Index against peak ground acceleration

4 Conclusion

Safety assessment of dams against large earthquakes for either existing or new dams to be built is of serious concern in many countries situated in intensive earthquake areas. In the case of concrete arch dams, the failure mechanisms, which may occur in the concrete body or rock foundation, can govern the safety of dam under a large earthquake. Using the proposed method it could be possible to estimate quantitatively the safety of a concrete arch dam against the sliding and opening failure in the jointed rock mass at abutment under its Maximum Credible Earthquake. The introduced damage indices are also useful to distinguish the dominant failure mechanism, which may occur in the abutment of arch dam.

References

1. Ahmadi, MT. & Razavi S. 1992. A 3-Dimensional joint opening analysis of an arch dam. *Computers & Structures.* 44: (1-2) 187-192.
2. Hall, JF. 1998. Efficient non-linear seismic analysis of arch dams. *Earthquake Engineering & Structural Dynamics.* 27: (12) 1425-1444.
3. Lau, DT., Noruziaan, B., Razaqpur, AG. 1998. Modeling of contraction joint and shear sliding effects on earthquake response of arch dams. *Earthquake Engineering & Structural Dynamics.* 27: (10) 1013-1029.
4. Singhal, AC., Zuroff, MS. 1998. Dynamic analysis of dams with nonlinear slip-joints. *Soil Dynamics and Earthquake Engineering.* 17: (3) 185-196.
5. Alonso, E., Carol, I., Delahaye, C., Gens, A., Prat, P. 1996. Evaluation of safety factors in discontinuous rock. *International Journal of Rock Mechanics and Mining Science & Geomechanics Abstracts.* 33: (5) 513-&.
6. Yazdani, M., Horii, H. 2002. Seismic safety evaluation of concrete arch dam based on the shear failure of jointed rock mass at abutment. *3rd International Conference on Dam Engineering March 20-22, 2002 Singapore.* 341-348.
7. Yazdani, M., Horii, H. 2002. The effects of mechanical properties of the rock joints at abutment on seismic safety of concrete arch dam. *ISRM Regional Symposium and 3rd Korea-Japan Joint Symposium on Rock Engineering-Problems and Approaches in Underground Construction July 22-24, 2002 Korea.*
8. Research Report on Ultimate Bearing Capacity of Dams during Earthquake, *Ministry of Economic, Trade and Industry of Japan, National Resources and Energy Agency, Electric Power Civil Eng. Association*, 1997 (in Japanese).

A MOMITORING AND WARNING MODEL FOR THE SAFETY OF STRATIFIED ROCK SLOPE

ZHIJIAN CHEN[1] QIANG LI[1,2] XIAOYAN LI[1] YINGXUE SUN[1]
[1] *Hohai University, Nanjing, Jiangsu Province, 210098;*
[2] *Xiamen United development company, Xiamen, Fujian Province, 361006*

This paper firstly presents the engineering characteristics of stratified rock slope and the load characteristics of tower base and rock mass to resist anchorage of big span suspension bridge to slide. On the base of the observed data on south bank slope of Jiangyin Bridge Project, using 3D nonlinear FEM, the monitoring and warning models for vertical load of the tower and residual horizontal load of the anchorage are developed, and their essentialities and methods to model are presented. In addition, the monitoring and warning model for shear displacement of possible sliding surface is developed by back analysis. These models are necessary to ensure the safety and stability of south base in Jiangyin Bridge Project.

1 Introduction

Stratified rock slope is composed of rock mass of stratified structure. And there are many stratified rock slope in nature. Because there is a group of predominant primary structure plane in stratified rock mass, and the characteristic of sedimentary rock is interbedded of soft and hard rock, stiffness of stratified rock mass to resist bend is poor. So fold and fault were grown. So the stability of these slopes is important and complex.

The base of main tower in the Jiangyin Bridge (approximately 190m high) must bear 1.2×10^{6}kN vertical load, and the anchorage must bear 6.4×10^{5}kN drawing force of main cable (5.5×10^{5}kN horizontal force). They are transferred to rock mass of foundation. The south tower, anchorage and approach of the bridge are all situated on the Xishan hill. The Xishan hill is free on three directions, and the strata of the hill are stratified sandstone and shale dipping to the river gently. South tower is based on a unsafe gently consequent slope, and the south anchorage must take the consequent slope as resistance body. So the stability of them is important and safety monitoring is the essential measure to ensure the bridge running safely.

2 Warning model for the vertical load of tower

As stated before, the stability problem of south tower base of Jiangyin Bridge Project is important. The effects of south tower to its base slope include: huge and concentrated engineering load, boundary condition worsening by excavation and blasting destroy. Calculation results by several common methods indicate that, the tower base side slope would slip, if all the load of south tower acts on the top of the tower base slope. So, the acting surface must be declined 26m from the top of slope. To ensure the slope stability of tower base, open-cut is a good method to decline the acting surface of the tower load. The deeper the acting surface is, the better is the stability of the tower base side slope. But heavy cut would endanger the safety of south tower and approach by worsening the stability of south foundation pit slope and weakening the resistance of anchorage. In addition, water proofing and discharging in deep foundation pit would endanger the stability of south foundation pit slope. Obviously, to ensure the safety and stability of south tower, it is important to optimize the foundation project of south tower to ensure the stability of tower base slope and south foundation pit slope. A new pile foundation project, socketed pile by removing friction on the upper 20m, is introduced. In this project, between the pile cap and foundation rock mass, cork and fine sand cushion are used to separate, so that the top of the tower base side slope does not bear the tower load. The foundation piles are cast-in-place piles, and their diameters are 2.8m, and their lengths are 35m. There are 12 piles on each cap. On the upper of each pile, double steel sleeve are used to separate rock mass with the pile to remove friction around pile, so the load of tower is transferred to the depth of the base, i.e., 26m under the top of the tower base side slope. As for bank engineering, this is a foundation method worthy of developing. It not only can decline the acting surface of load, but also avoid heavy cut affecting the slope stability. So, abundant monitoring works about the axial force of pile are doing to investigate the running characteristic and to test its design intention. On the base of these, the monitoring model for axial force of pile is developed. Although it is a monitoring model for stress, to the tower base side slope, it is a monitoring and warning model for the distribution of causation factor. Theoretic calculation, numerical analysis and qualitative evaluation are combined in this model.

2.1 Monitoring model for the load of the top of tower base side slope

As stated before, to ensure the slope stability of tower base, the acting surface of load must be declined 26m from the top. Because cork and fine sand cushion can't completely separate pile cap and foundation rock mass, the warning model for the axial force on the top of pile should be developed to monitor the engineering load on the top of the tower base side slope, as given by following expression:

$$4\bar{p}_{1i} + 12\bar{p}_{2i} + 8\bar{p}_{3i} \geq \lambda_1 p_i \tag{1}$$

where $\bar{p}_{1i}$ is the average axial force of all piles under the tower (i.e., the four piles in the center of the cap)observed at certain time. $\bar{p}_{2i}$ is the average axial force of piles in circumference (twelve piles) observed at certain time. $\bar{p}_{3i}$ is the average axial force of piles on every corner (eight piles) observed at certain times. p_i is the total engineering load observed at certain times. The coefficient λ_1 can be obtained by FEM. The 1200MN vertical load of tower can be formed 10 load combination (a, b): (0,1200),(120,1080),...,(1200,0). Taking the second as an example, it exhibits that there is 120MN vertical load on the top of slope, and 1080MN vertical load on the embedding bottom plane of pile. Using interference-displacement method[1], the most possible sliding plane of tower base side slope on each load combination can be obtained. Then, according to the stress response, using static method, the safety factor of the most possible sliding plane corresponding to each load combination can be calculated. When a safety factor less than 1 is obtained, λ_1 can be decided by the last load combination as following:

$$\lambda_1 = \frac{b_{i-1}}{a_{i-1} + b_{i-1}} \qquad (2)$$

If the friction removing around pile is complete, when λ_1 is bigger than or equal to 0.5, it means that the tower base side slope is stable.

Observed data exhibits that, after pouring of pile cap, the top axial force increases with its concretion. In about 40 days after the completion of pouring, ninety-five percents of the cap's weight is shared by piles. After pouring of across girder, their weight is transferred gradually to piles by cap in 30 days. The top axial force conforms to the monitoring model given by the firth expression. In the following observation, the top axial force always confirms to the firth expression, and λ_1 is approximately 0.95~0.98. It means that, the separate effect of cork cushion is good, and the top of tower base side slope does not bear load nearly.

2.2 Monitoring model for tension sliding surface of tower base side slope

Considering technical problem, it is difficult to remove friction completely around piles. Whether the axis of internal and external steel sleeve is parallel and fitted, and whether there are granule, concrete and the like enter the interspace between internal and external steel sleeve, would effect friction removing, and effect the stability of tower base directly. In order to ensure the stability of tower base side slope, while • $_1$obtain 0.9, it is still necessary to make monitoring model for single pile axial force as following:

$$p_i' \geq \lambda_2 p_i + G \qquad (3)$$

where p_i is the top axial force observed at certain time, p_i' is the axial force on the bottom of the friction removing section, G is the weight of the pile in friction removing section, and it is 3244kN for a single pile. λ_2 is a coefficient between 0 and 1, and can be obtained as λ_1. The bigger the λ_2 is, the better is the affect of friction removing. The difference is that a_i distribute uniformly on the tension sliding surface, The calculated results indicate that, to the first four load combination, the interlayer 227 is the most dangerous sliding surface, but its safety factor is bigger than 1.1, and there is a tendency to increase. When the firth load combination is acted, the interlayer 230 becomes the most dangerous sliding surface, and its safety factor decline to 1.19. So λ_2 can be decided by this load combination:

$$\lambda_2 = \frac{b_5}{a_5 + b_5} = \frac{720}{1200} = 0.6 \qquad (4)$$

Observed data indicates that, before 30 percent of pile cap and tower were poured, single pile axial force didn't satisfy the seventh expression. The effect of friction removing around piles is not obvious, and the reason is possibly that the across girder between two pile caps is very heavy. In following observation, the increment of axial force satisfy the seventh expression, and •$_2$ increase with the increasing of t. It means that the friction removing around piles have acted gradually. During the period of spanning the main cable and bridge running, the axial force and its increment all satisfy the seventh expression. And•$_2$ increase gradually. At the time of bridge running, •$_2$of each piles reach 0.7~0.9. So, the intention to decline the acting surface of engineering load to 26m under the top of the side slope had realized.

2.3 monitoring model for the plane distributing of tower load

Although the south tower is situated on rock foundation, there is much soft rock and weak intercalated layer, and fissure and bedding plane were grown. So the strength of rock mass is poor. Because the load of tower act centrally, and the load of single pile reaches 50MN, it is possible that certain pile would be overloaded and the structure plane of rock mass would yield. It would results to stress transfer and fail of local tower base or whole tower base side slope. Observed data indicates that, foundation piles are overloaded. During construction, especially spanning main cable and installing case and girder, the foundation pile under tower is overloaded all the time (the maximal overloading reach 35 percent of its load capacity). To avoid fail of tower due to local overloading, it is necessary to make monitoring model for axial force of single pile as following:

$$P_{1i} \le \lambda_3 \times \bar{p}^{*}_{1i} \qquad (5)$$

Where $\bar{p}^{*}_{1i}$ is average load of all piles observed at certain time. The •$_3$coefficient can be obtained by FEM. While calculating, the overloading rate of central two piles is increased gradually, until the foundation rock mass yield locally. The • $_3$can be obtained by adding 1 to the last overloading rate. According to the calculated result, •$_3$ takes 1.4.

3 To decide shear displacement warning value of possible sliding surface

To ensure the stability of tower side slope, besides monitoring tower load, monitoring system for internal and external deformation of side slope rock mass are made. There are two monitoring hole for internal horizontal displacement. The main purpose is to monitor shear displacement of possible sliding surface. The warning value for the shear displacement of possible sliding surface can be decided as following:

(1) To decide the possible sliding boundary. According to observation data by clinometer, combined the column chart of clinometer hole with the analysis result interference-displacement method, the position of possible sliding surface can be decided. Generally, several possible sliding surface can be obtained according to shear displacement along weak intercalated layer and its boundary condition. In the Jiangyin Bridge Project, interlayer 301,230, 227 and 222 are possible sliding surface. The interlayer 230 is the most dangerous one.

(2) To decide the sensitivity factor of slope stability. Analysis shows that, one of the factors to decide the south tower side slope is dynamic and static water pressure due to seepage of rainfall. Hereby, the hydraulic reduction factor of possible tension sliding surface can be taken as the sensitivity factor [2] of slope stability. So, according to shear displacement of possible sliding surface, the hydraulic reduction factor β of possible tension sliding surface can be obtained by back analysis. To simplify it, the β of each layer change in same proportion α.

(3) Back analysis to hydraulic reduction factor. In order to monitor on real time, forecast model for slope stability should be made ahead. After a period of observation, position of possible sliding surface is known. Hydraulic reduction factor β could be obtained by back analysis, according to shear displacement value, such as 0.4, 0.8, 1.2, etc.

(4). Forecast model for slope stability. On the base of hydraulic reduction, safety factor, corresponding to each possible sliding surface and each hydraulic reduction factor, can be calculated and given in table1.

Table 1. The relationship between shear displacement and safety factor of possible sliding surface

Shear displacement /mm	0.4	0.8	1.2	1.6	2.0	2.4	2.8	3.2	3.6	4.0
Coefficient α	0.4	0.5	0.6	0.7	0.8	0.9	1.0	1.1	1.2	1.3
Safety factor k	1.44	1.35	1.28	1.23	1.21	1.13	1.08	1.03	0.99	0.97

$$K = 1.477\, e^{-0.11s} \qquad R=0.995 \tag{6}$$

(5) Warning value for shear displacement of possible sliding surface. As indicated in the tenth expression, if observed displacement on interlayer reach 3.5mm, the

slope stability can't be ensured while the most disadvantageous condition. In addition, when calculating hydraulic reduction factor, it must based on the condition corresponding to observed shear displacement. However, when calculating safety factor, it must consider the most dangerous condition.

4 Conclusion

The stability problem of stratified rock slope is important and complex. At the present time, it is difficult to evaluate exactly. So, to important engineering slope, safety monitoring is necessary. Obviously, it is worthy of deep study to safety monitoring and warning model. These study results had been awarded the highest encouragement in science and technology advancement of Jiangsu province in China in 2001, as the main part of the project "study of key technique in Jiangyin Highway Bridge".

1 References

1. Z.J. Chen, J.S. Zhuo, S.J. Liu. Interference-displacement method and determination of slide boundary of layered rock slope[J]. Chinese Journal of Rock Mechanics and Engineering, 2000, 19(5): 630~633
2. Z. J. Chen, L. Ji. Distribution character and controlling stability effect of underground water in layered rock slope[J]. Journal of Hohai University, 2001, 29(1): 107~110

STARTING TUNNELLING SCHEME OF WANWU TUNNEL ON RIGHT LINE

WEIXING FENG [1], KAIZHI YANG [2]

[1]*School of Civil Engineering, Shijiazhuang Railway Institute, Shijiazhuang 050043, China; E-mail:fengwx@sjzri.edu.cn*

[2]*The No.5 Ltd. Company, The 15th China Railway Construction Bureau Group, Jiexiu 031200, China*

Safe starting tunneling is the premise of realizing smooth construction to a tunnel. In general, starting tunneling is difficult for bad surrounding rock tunnel. This paper discusses the starting tunneling scheme of Wanwu Tunnel on right line of Fuding to Ningde Expressway based on tunnel support mechanism, including such contents as tunnel support mechanism, general starting tunneling scheme, specific starting tunneling scheme, starting tunneling construction procedure, technical measures and technical parameters. By practice the scheme described in the paper is very successful, which provides a good reference to similar works.

1 Introduction

Wanwu tunnel on the Fuan connection line of Fuding to Ningde Expressway that is part of the main state road from Tongjiang to Sanya is located in northeast part of Guogailiang Mountain in Wanwu village, Wanwu town, Fuan city. The tunnel passes through Guogailiang Mountain in the direction from southeast to northwest, it is a up and down line 4-lane Grade 1 highway tunnel with maximum depth about 220m. The mileage of left tunnel is ZK0+233～ZK1+380, and the total length of left tunnel is 1,147m. The mileage of right tunnel is YK0+250～YK1+400, and the total length of right tunnel is 1,148m. The geological state is seaside structure to intrusive corrosion rock lower hill, it belongs to lower hill landform, is of strong land cutting and growthed ditches. The surrounding rock in right tunnel inlet section has weathered into fully weathered tuff in hard soil block state and fracture zone, it has poor stability, is easy to collapse and can be classified as Scale II according to Chinese Railway Tunnel Surrounding Rock Classification. There are plentiful water in rock creak belt and its influenced area. The water is simply supplied by rain.

Serious deformation of surrounding rock and support structure happened at the inlet section with mileage of YK0+278～YK0+300 on the right line at 15:00 on 28th Oct. Completed primary support appeared the phenomena of shotcrete creak and steel arc twisted, and serious settlement and creak also appeared on ground surface. Due to the over tunnel deformation and serious ground surface settlement, it had to cut openly the section of deformation and settlement (Fig.1). Because of the bad soil and easy collapse, how to realize normal start tunneling in collapse-cut openly area become an urgent problem to be solved to ensure tunnel time limit.

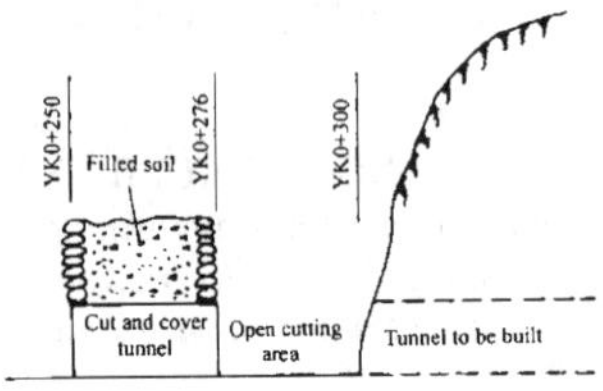

Figure1. State of inlet portal of Wanwu tunnel on right line

2 Start tunneling scheme

2.1 Tunnel support mechanism

It is known that the soil is in the stable and balanced state before tunnel excavation and tunnel excavation destroyed its stable and balanced state, and causes surrounding rock deformation into the tunnel. In order to keep the space after tunnel excavation and maintain surrounding rock stability, it is necessary to construct tunnel support. The function of the support is to replace the excavated soil. So, what needs us to do is to maintain surrounding rock keeping original stable state, to reduce interference and damage to the soil as possible as we can during tunneling. Effectiveness can be obtained only based on above tunnel support mechanism to draw start tunneling scheme.

2.2 General scheme

According to condition of project and similar construction experiences (for example, the works of inlet section of Qiling Tunnel on Beijing to Jiulong Railway Line), the general construction of start tunneling is drawn up as follows: first carry out groundwater decline, then execute start tunneling procedure ---- pipe-roof, upper bench core soil reserved arc excavation, steel arch —mesh shotcrete primary support enclosed into ring quickly , and pour second concrete lining as early as possible.

2.3 Detailed scheme

1. Due to plentiful groundwater, a running water ditch on the right side of the tunnel and tunnel bottom elevation 10m lower than the ditch bottom elevation, the whole face of the tunnel is in water condition. In order to benefit start tunneling, pipe

well groundwater decline is carried out on the right side of the tunnel.

2. The construction procedure of start tunneling is shown in Fig.2.

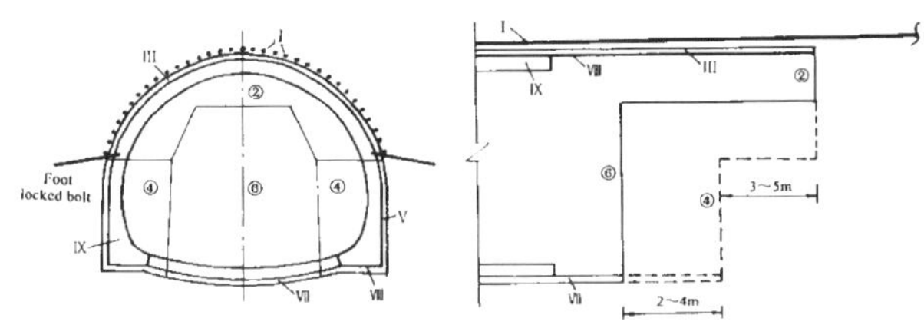

Figure2. Construction procedure of starting tunneling

(1) Pipe-roof Ⅰ

Large diameter pipe-roof Ⅰ is set along the outer side of tunnel arc excavation outline. The scope of the pipe-roof is 2×70° in arc part. Φ108×6mm hot rolling seamless bloom is adopted as pipe-roof (3m and 6m section length, 15cm thread connection) with 28m row length, 30cm ring spacing and 3° outward angle. The pipe is made into having interleaving small holes on its wall at front part. The pipe is not only pipe-roof pipe but also grouting pipe. After pipe-roof setting, double—slurry of cement and sodium silicate is grouted into the pipe both to increase the stiffness of the pipe and to enhance self—stability of surrounding rock. The volume ratio between cement and sodium silicate is 1:1.5, W/C=1:1. The modulus of sodium silicate is 2.4 and its concentration is 35 Be'. The first grouting pressure is about 0.5~1.0MPa and the final pressure is 2.0MPa.

(2) Upper bench core soil reserved arc excavation ②

Because of poor stability and easy collapse of surrounding rock, the upper bench core soil reserved arc excavation method is used to stabilize working face soil. The advance length of arc excavation is controlled within 0.5m.

(3) Arch primary support Ⅲ

After the completion of upper section arc excavation, I18 formed steel arch is set with 0.5m longitudinal spacing on time. Steel arch and pipe-roof are connected with electric welding. Φ22 deformed steel bars with 1.0m ring spacing are used for the connection between two steel arches to form integral steel support structure. Steel arch feet are matted with [-type steel, and Φ42 foot locked pipe-bolt in 4.0m long is set beneath the [-type steel in outward and downward direction. The longitudinal spacing of foot locked pipe-bolt is 25cm. The foot locked pipe-bolt can be made into having interleaving small holes on its wall at front 2.5m, and can properly grouted cement slurry to supply stable foundation for upper section primary support. Φ6 double-layer steel mesh (20×20cm) is hung and 26cm thickness C20 shotcrete is done to form arch primary support III.

(4) Lower bench side wall excavation ④

Lower bench side wall excavation ④ is started behind upper bench about 3～5m. The advance length of excavation is 0.5m.

(5) Side wall primary support V

Side wall steel arch is quickly set after side wall excavation and connected with upper section arch properly and fixedly. Then, Φ6 double-layer steel mesh (20×20cm) is hung and 26cm thickness C20 shotcrete is done to form side wall and wall foot primary support V.

(6) Pull central trough ⑥

This procedure can be carried out behind side wall excavation about 2～4m.

(7) Invert primary support VII

Invert steel arch is set up after the completion of pull trough. Two ends of invert arch are properly connected with side wall arches to form closed steel arch ring.

(8) Waterproof layer VIII

When the primary support become enclosed ring, compound BFP-A-1 waterproof layer is set foe the whole tunnel. The laying and welding of waterproof layer will be carried out accordance with relevant construction requirements.

(9) Second concrete lining IX

45cm thickness C25 concrete lining will be poured after the completion of waterproof layer setting.

3 Conclusion remarks

The starting tunneling scheme of Wanwu Tunnel described in this paper was evaluated and approved by the Owner, supervisor and design institute on December 19 of 2001, started on December 25 of 2001, and obtained success after about one month construction. The practice is a good reference to similar works.

References

1 Chief Bureau of Capital Construction of China Railway Ministry, Railway Tunnel NATM Guideline, Beijing: China Railway Press, 1988

2 Feng Weixing, The construction method of starting tunneling in special geological condition, Computer Methods and Advances in Geomechanics, Yuan(ed.), Balkema, Rotterdam, 1977,1413～1417

MIXED H_2/H_∞ OPTIMAL DESIGN OF ANTI-SEISMIC ACTIVE CONTROL OF CIVIL ENGINEERING STRUCTURES CONSIDERING DYNAMIC PROPERTIES OF ROCK OR SOIL OF SITE

CHENGMAN SHA ENDE WANG CHUNAN TANG
Northeastern University, 110004, Shenyang, China
E-mail: neuress@mail.sy.ln.cn

The influence of soil-structure interaction (SSI) is considered. On the basis of analysis of the norm of structural system and the merits and defects of H_2 and H_∞ control schemes, the mixed optimal control index J is cited and its physical implication is given. Therefore mixed H_2/H_∞ control schemes are worked out by solving a set of Linear Matrix Inequation (LMI) using convex optimal approach. Finally, A simulative example is given.

1 Introduction

In general, with respect to anti-seismic active optimal control schemes of large civil engineering structures, H_2 and H_∞ schemes are adopted[1,2]. For H_∞ scheme the condition in the smallest basic frequency (SBF) is considered, and the condition in other frequency scope is ignored. For H_2 scheme the condition in whole frequency-domain is considered but the condition in SBF that result in resonance isn't emphasized. So based on the norm analysis of the structure system, the index J is cited and mixed H_2/H_∞ control schemes are designed.

2 The system equation of structural model

The optimal design of anti-seismic active control of large civil engineering structure is considered .The equation of motion is derived by the substructure method, in which the system is divided into two substructures, the superstructure and the rigid foundation-soil system. The equation of whole structure subjected to a ground acceleration and a controller, is given by

$$\begin{cases} \dot{x} = Ax + B_1\xi + B_2U \\ z = C_1x + D_1U \\ y = C_2x + D_2U \end{cases} \tag{1}$$

where, $x = \begin{bmatrix} \dot{X} \\ X \end{bmatrix}$, $X = \begin{bmatrix} X_{SUP} \\ X_{CF} \end{bmatrix}$, $A = \begin{bmatrix} 0 & M \\ M & C \end{bmatrix}^{-1} \begin{bmatrix} M & 0 \\ 0 & -K \end{bmatrix}$, $M = \begin{bmatrix} M_{\sup} & M_{\sup}\Gamma \\ \Gamma^T M_{\sup} & \overline{M}_{CF} \end{bmatrix}$

$$K=\begin{bmatrix} K_{\sup} & 0 \\ 0 & K_{CF} \end{bmatrix},\ C=\begin{bmatrix} C_{\sup} & 0 \\ 0 & C_{CF} \end{bmatrix},\ B_1=-\begin{bmatrix} 0 & M \\ M & C \end{bmatrix}^{-1}\begin{bmatrix} 0 \\ I_N \end{bmatrix}\begin{bmatrix} M_{\sup}\Gamma \\ \overline{M}_{CF} \end{bmatrix},\ B_2=-\begin{bmatrix} 0 & M \\ M & C \end{bmatrix}^{-1}\begin{bmatrix} 0 \\ V \end{bmatrix}$$

$$V=\begin{bmatrix} I_n \\ \Gamma^T \end{bmatrix}W_u,\quad F=-\begin{bmatrix} M_{\sup}\Gamma \\ \overline{M}_{CF} \end{bmatrix}\xi,\quad \xi=\ddot{X}_g \quad \overline{M}_{CF}=M_{CF}+\Gamma^T M_{\sup}\Gamma$$

$$C_2=\begin{bmatrix} -M^{-1}C & -M^{-1}K \end{bmatrix},\ D_2=M^{-1}\begin{pmatrix} I_n \\ \Gamma^T \end{pmatrix}W_u,\quad \Gamma=\begin{bmatrix} \cdots \frac{1}{h_i} & \frac{1}{h_{i+1}} \cdots \end{bmatrix}^T$$

where, M_{SUP}, K_{SUP} and C_{SUP} are mass, stiffness and damping matrices of the superstructure obtained by finite-element method(FEM), U is control force vector; where $Wu \in R^{n\times r}$ shows the direction and presence of the control force. X_{SUP} is relative displacement vector of superstructure with respect to the foundation; $\ddot{X}_g$ is the acceleration vector of the ground measured at the foundation centroid level; and X_{CF} is linear and angel displacement vector of the foundation centroid with respect to the ground. z is linear quadratic optimal output target. K_{CF}, C_{CF} form the impedance matrix of the soil, which is relative with the frequency of seismic signals, but the term including the varying frequency can be replaced approximately by the basic frequency of the structure system [2]. M_{CF} is a matrix of mass and mass moment of inertia of the foundation. C_1, D_1 are weight matrices of U and X separately.

3 Mixed H_2/H_∞ optimal control scheme

The system described by Eq. (1) is considered. And the definition of the H_2 norm and H_∞ norm of the system in frequency-domain is described by

$$\|G\|_2=\sqrt{\frac{1}{2\pi}\int_{-\infty}^{\infty} trace(G^T * G)d\xi}=\sqrt{trace(C_1 L C_1^T)},\ \|G\|_\infty = SUP\sigma_{\max}[G]$$

where, σ_{MAX} is maximum singular value, which shows the gain of the system in frequency-domain , " *trace* " means computing the trace of the matrix, *SUP* means supper limit, L is a solution of the Lyapunov equation as follows

$$AL + LA + B_1 B_1^T = 0 \tag{2}$$

The optimal design of anti-seismic active control of large civil engineering structure is divided into two types, which are H_2 and H_∞control schemes. Because of such hypothesis that seismic signal is gaussian white noise process, it is easy to deduce that the physical meaning of H_2 norm express gain of the system as a whole when probability that earthquake will happen is identical at each frequency value. Its geometric meaning is the area enclosed by σ_{MAX} curve and frequency-axis. So H_2 control scheme that minimize H_2 norm of the system is an overall optimal scheme when general seismic signal is considered. But the worst condition like resonance is not considered although it has small probability, so the safety of the structure system can't be sure. From the definition we known that the physical meaning of the H_∞ norm is maximum gain of the system, and its geometric meaning is maximum value of σ_{MAX} curve ,which is so-called worst condition. Considering the worst condition of seismic signal, of which frequency is equal to structural basic frequency and result in resonance, H_∞ control scheme minimize the system gain at the structural basic frequency in frequency-domain. Therefore it ensures the safety of the system and makes system have robust performance. But seismic signals in other frequency scope cannot be considered, so that it is a conservative and waste scheme. For this reason , a mixed H_2/H_∞ control index J *is* cited as: J=trace (AYA^T), where Y>0 and satisfies the following Riccati equation

$$R(Y) = AY + YA^T + \frac{1}{\gamma^2} YC_1^T C_1 Y + B_1 B_1^T = 0 \tag{3}$$

where index γ is given and $\|G\|_\infty < \gamma$. The Riccati equation (3) is Considered, when $\gamma \to \infty$, it becomes Lyapunov equation (2) and implies Y=L. Furthermore, $\|G\|_2^2 = \lim J(\gamma \to \infty)$.That means the physical meaning of the J is the mean gain of the system on condition that γ is restrained. The target of mixed H_2/H_∞ control is to minimize $\|G\|_2$ of the system subject to $\|G\|_\infty < \gamma$.

The mixed H_2/H_∞ optimal control scheme can be computed by solving a set of Linear Matrix Inequality(LMI) using convex optimal method in metlab toolbox.

A simulative example is given. The structural model for example is a high steel tower, which has three floors and is based on a rigid reinforced concrete foundation. With regard to the super structure, the masses of three floors of the tower are 7.2,5.4 and 3.6(e05kg) separately, of which the stiffness of supporting structure are 2.3,1.9

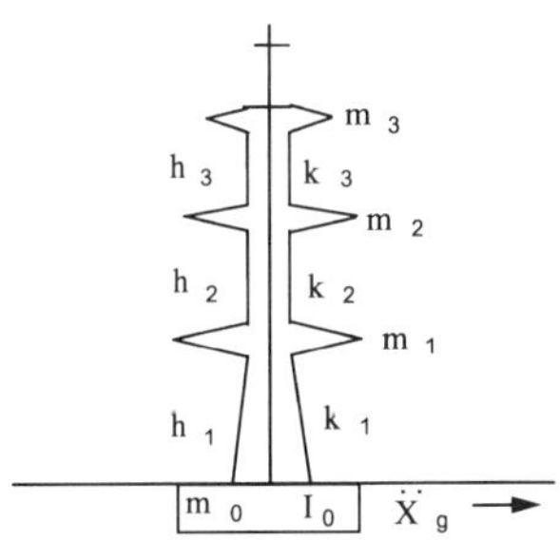

Fig.1 The simulative model

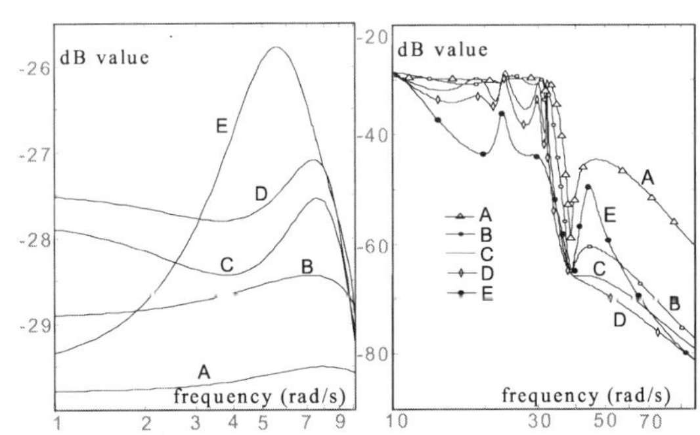

Fig 2. The maximum singular curves of close-loop system on 5 kinds of control scheme in frequency domain

Table 1. Close loop system parameters of 5 kinds of control scheme

Control Schemes	A	B	C	D	E
H_2 norm($\times 10^2$)	11.7	10.6	9.99	9.32	8.46
H_∞ norm($\times 10^2$)	3.35	3.78	4.19	4.42	5.14
The max displace of the top(cm)	3.15	3.40	3.82	3.98	4.68
The max force($\times 10^6$ N)	1.64	1.52	1.27	1.22	1.11

And 1.5(e08N/m) separately. And their masses are ignored. The heights among the floors are 80,65 and 50m separately; damping ratio is 1%. With regard to foundation, the radius is 12m, height is 5m, mass is 5.43e6 Kg, density is 2.4e3 Kgm^{-3}, and mass moment of inertia is 2.07e8Kgm^{-2}. For the site soil, the modulus of elasticity is 9.86e8 Nm^{-2}, Poisson ratio is 0.4, the density is 2.2e3Kgm^{-2}, and velocity of the seismic shear wave is 400ms^{-1}. The weight ratio of control force with displace is specified as 1.0e7N/m*.

Five different control schemes are designed using the method as above, which have different γ and J values. The maximum singular value curves were plotted (Fig. 2). The Close loop system parameters of 5 kinds of control scheme are computed and cited in table 1. It can be seen from the Fig.2 and the Table 1 that the A scheme (H_∞ scheme) is to minimize γ and not to restrain J, which is a best scheme at the smallest basic frequency of the close loop system , and a worst scheme in higher frequency-domain. Conversely, E scheme (H_2 scheme) is to minimize index J and not restrain γ ,which is a worst scheme at the smallest basic frequency of the close loop system , and a best scheme in higher frequency domain .In turn, B,C,D schemes are mixed schemes to minimize J on condition that γ is given, which compromise the defects and merits of scheme both A and E and are the real best schemes in whole frequency-domain. They assure the safety of the structure and save the control energy as much as possible simultaneously.

References

1. B.F.Spencer et al., Benchmark problems in structural control: part I --active mass driver system, Earthquake engrg. and struct. Dyn. 27, 1127(1998).
2. Shahid Alam S.M. etc. Robust active optimal control scheme including soil-structure interaction, J. of Struct. Engrg. 119, 2533(1993).
3. Kemin Zhou and Doyle J.C., Essentials of Robust Control, New Jersey, Pretice Hall (1998.)
4. Joln Dolye, Kemin Zhou etc. Mixed H_2 and H_∞ performance objective II: optimal control, IEEE Tran. on Automatic Control, 39, 1575(1994).

* The work was supported by the National Ministry of Education, P.R. China.

APPLICATION AND DEVELOPMENT OF SHIELD IN GUANGZHOU METRO

ZHI NAN LI
Design & Research Institute of Guangzhou Metro

Basic on engineering of Guangzhou metro line 1, 2 and 3, the paper summarizes the application and development of structure analysis of shield tunnel, profile design of segment, waterproof design of shield tunnel and technology of tunneling by shield in Guangzhou metro. Finally, the paper studies the prospect of the new technology and theory application in Guangzhou metro and point out what we should pay great attention, referring to the new development of technology and theory of shield. We hope it can give some useful experiences for design and construction of shield tunnel.

1 Introduction

In order to reduce the traffic difficulties, connect the urban and the suburbs, as well as accelerate the whole development of GuangZhou city, an underground railway system which consists seven lines has been designed. And now one has been constructed, two are under construction and other two will be constructed in the near future, which will form the basic traffic framework of GuangZhou subway. Line 1 of the subway is mainly from east to west, Line 2 is from south to north and then turn east to west, and Line 3 is mainly from south to west.

The geology is very complex in GuangZhou. The stratums that underground railway will go through include the Sinian system, Jurassic system, Cretaceous and the forth system of the Cenozoic Erathem. Meanwhile, GuangZhou is close to the sea and there are Zhujiang River and its branches through the city. At the same time, the irrigation network is developed almost everywhere in the city. The water level is very high, and in general, the water level is about 1.5m~2.0m under the ground surface. The strata that Line 1 goes through is mainly red stratum, basically with soil of red stratum remnant on it and some times with sandy soil above it while it is near the river. The strata that Line 2 goes through are very complex which include granite, red stratum and etc. with red stratum or soil of red stratum remnant. The variety of the lithology and soil structure makes the selection of the type of shield, also the design and construction of shield tunnel more difficult.

Through comparison in technology and economy, six sections are constructed by shield in Line 1. Three shields are used in the construction, two of which are slurry shield and the other is EBP shield. The period of construction was 36 months and the distance is 8.8km, which composed 26% of the whole length. Six sections which include the length going under through ZhuJiang river whose name is HaiZhuGuangChang station • JiangNanXi station are also constructed by EPB shield in Line 2. 11 sections will be constructed by shield in Line 3. The distance is 38km, which is 60% of the whole length. The sections are divided into 7 parts and will be constructed by 13 shields. The three times of construction while go under through the ZhuJiang river were completed by using shield.

2 The Shield Tunnel Design

1). Design of the Segment

Ordinary the thickness of the whole ring of the shield is 300mm in GuangZhou subway. The segments of a ring adopted the 5+1 model. The segments are jointed together by alternative model except the special district. The bolt is the mature bending bolt. The structure used single layer ring. At the outer side of a ring, a waterproof groove is set up. In the inner side, a wedge joint groove is preformed.

The segment width is 1.2m in Line 1. So in Line 2, in the section of going under through river, the used mould and the shield are same as the mode of Line 1. But in the last two sections of shield construction, the segment width is increased to 1.5m. Such improvement greatly increase the construction speed, reduce the number of joints and improve the waterproof effects.

Different models of standard ring, left-turn ring and right- turn ring are designed based on the practical construction and manage technique.

Combined with the choice of waterproof material, the outer construction of the segment is amended. The new construction can reduce the failure of angles of segments happened in the process of transportation or due to the concentration of stress in the period of construction. Because of the same reason, the inner wedge joint groove of segment is also reformed (Figure1).

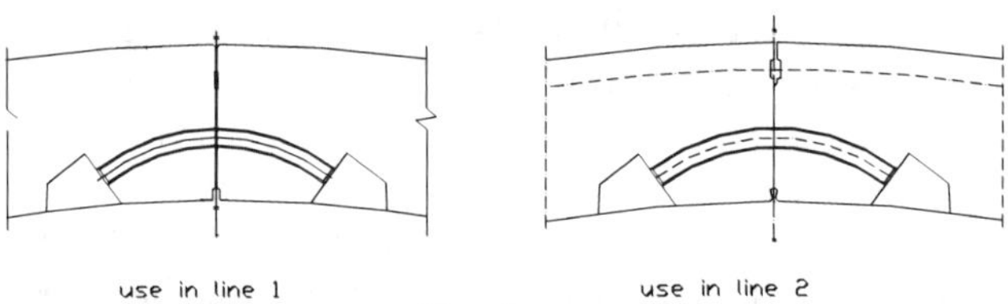

Figure1. segment joint

2). The Analysis of the Segment Construction

In construction of line 1,the analysis of the shield segment has mainly considered the effects of permanent load, which might be acted on the segment after construction, and neglected to the various factors, which shall also effect the construction, as well as the factors in the manufacture production of segment. In construction of line 2, we have not only analyzed the combination of load on segments, but also took into account the different factors occurred in the process of segment manufacturing, transporting, jointing together and etc. Based on the result of FEM, the amount and arrangement of the reinforcing steel rods were reduced and adjusted. Meanwhile, the ground pressure which acts to the segments has been measured (Figure2).

3). Design of waterproof

Hydrophilic is used to be waterproof material in line 1. The EPDM is used to be waterproof material in line 2. According to the practical construction effects, the EPDM is better than the hydrophilic. The hydrophilic used to waterproof due to it's expanding under the condition of absorbing water. So the waterproof effect is not very good.

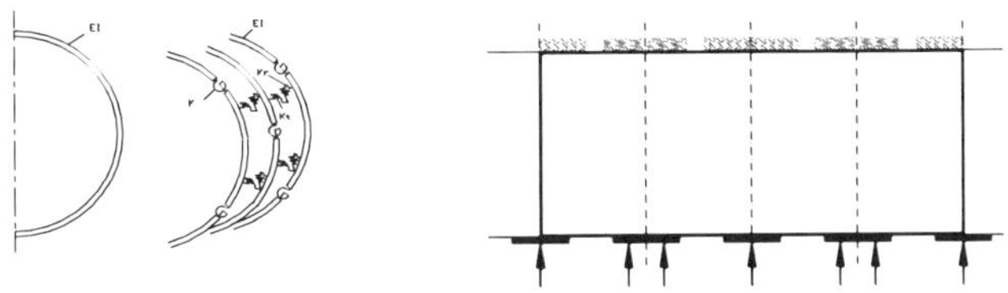

Figure 2. segment analysis model
1.Japanese correction consistently use method • 2. Load-Construction method considering the influence of the joints • 3. Segment internal stress analysis considering the jack functionary.

In the primary stage, the soil and sand sometimes nipped in and therefore influence the waterproof effect later. EPDM relied on the compressive stress coming from segments, Because of above reason, EPDM should be used in the line 3 (Figure3).

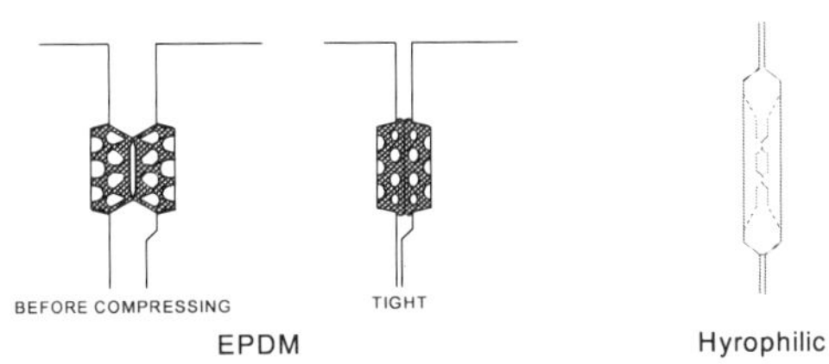

Figure3. profile of seal

3 Shield Tunnel Construction

1). Shield Choose

The earth shield and muddy shield were chosen in the first line construction. According the experiences in the construction, people all agreed that the earth shield is better in technique and economy, so in the second and third lines we only use the earth shields, even in the four times going under through ZhuJiang river.

2). Building Protect

Pile foundations of four apartments were located above the section of ChangShou-Lu • ZhongSan-QiLu in Line 1. The RC, pre-stress concrete, bore perfusion stake are used for the hold-exchange construction to the four apartments. The number of hold-exchange stakes is 143 and the hold-exchange area is 9700m^2. This method saved the payment needed for removing apartments.

4 New Technique and conclusion

Using of the shield expanding construction method: The application of shield construction method is limited because of the different size of subway and the existence of double lines.

Therefore the use of shield expanding construction method has very important meaning, especially in the development of the subway. Now the managers already recognize this serious problem and pay great attention to this problem combined with construction of line 3. The research is concentrated on theory of calculation, special style of segment and joints, as well as the construction technique.

Application of steel fibre segment: The steel fibre segment uses the steel fibre which can replace the steel rod system in the segment. By using the steel fibre segment, the speed of production will increased and the quality of segment will be improved. There is a few studies in China currently. If the design theory will be studied and reached to a new level, we believe the steel fibre segment will be used widely here.

Application of multi-circle shield: There are a lot of limitations of present shield construction such as: the least distance of two single tunnels, the arrangement of underground stations, the increasing of the subway protect region and so on. The use of multi-circle shield permits us to build two tunnels with minimum distance between them and realize the maximum utilization of finite underground space. So the study of multi-circle shield has important meaning in practice.

The advanced construction method such as ELC method, MSD method, segment automatic put together should be also researched furthermore.

No only the cost of the tunnel that constructed by shield is near to or lower than the cut-and-cover method and blasting method, but also the tunnel that constructed by shield has higher quality, lower influence to circumstance and higher level of mechanization, so shield construction will be adopted wider in GuangZhou Metro.

References

1. Zhang Yuan, Hold-exchange construct of metro line 1, construct. tech., 1998, vol 27
2. Li Zhi Nan, Discussion of the tolerance of segment, 14th Conference on Tunnel and underground engineering, BeiJing, China, November, 2001
3. Shi Zhong Heng, Design and construction of metro, ShanXi, 1991
4. Code of the design of metro (GB50157-92), BeiJing, 1993
5. Code for construction and acceptance of metro engineering (GB50299-1999), BeiJing, 1999
6. Design document of GuangZhou metro line 1, 2 and 3; GuangZhou , 1994 - 2002

STUDY FOR TOPOLOGICAL STRUCTURE OF A KIND OF NONLINEAR SYSTEM OF MACHINE FOUNDATIONS

W.B.BAO[1,2] Y.X.LIU[2] B. CH. WEN[2].

[1] *Shenyang University of Technology, Shenyang 110023, China;* [2]*School of Mechanical Engineering and Automation, Northeastern University, Shenyang 110006, China*

The article first starts from the strongly nonlinear dynamical model of a kind of machine foundation system, and analyses multifarious things about topological structure that the model shows in the operation of the machine foundation system: singular points and their changes, periodic oscillation and chaotic motion etc. The types lying on dynamical parameter and the changes with them varying of topological structure in the whole system have been studied. By using theoretic analysis and numerical simulation, the conditions on which the complicated dynamical behaviors occur in the machine foundation system and the influence on dynamical characteristics and stability of this system with the adjustment of the dynamical parameters have been discussed. This study might be instructive to dynamical design of machine foundations.

1 Dynamical Model

The dynamical system of machine foundation generally takes on strongly nonlinear characteristic, and the simplified dynamical model of the system is shown in figure 1(a), where longitudinal vibration is merely researched. The machine foundations are usually subjected to cyclic loading and a hysteretic loop will appear as shown in Fig.ure1(b). This backbone curve of the hysteretic loop can describe stress-strain law due to it's special stress-strain relation[1]. Namely hyperbolic model is introduced as shown in Figure.1(c). It is convenient to substitute below power function for hyperbola when displacement x is small [2,3]

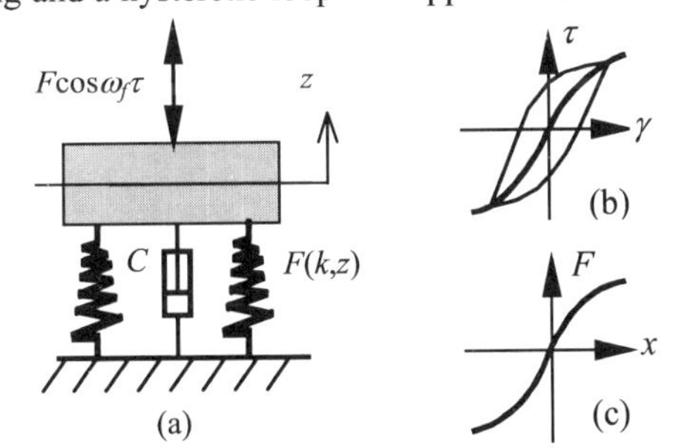

(a) The mechanical model of the machine foundation
(b) The hyperbolic model (c) The power function model
Figure1. The dynamical model of the machine foundation

$$F(k,z) = k[1-(\frac{z}{\alpha})^2]z \tag{1}$$

where k is the stiffness constant, α is coefficient of the nonlinear term of restoring force, which represents extent of nonlinearity. From mechanical model, and noting the equation (1), the following governing differential equation is obtained

$$M\frac{d^2z}{d\tau^2} + C\frac{dz}{d\tau} + k[1-(\frac{z}{\alpha})^2] = F\cos\omega_f\tau \tag{2}$$

where M is the sum of the mass of the machine, the foundation and join-vibrant soil, C is the coefficient of damping that is simplified to viscous one because the resistance is small as compared with the restoring force, and ω_f is the frequency of harmonic excitation force. Introducing following notations

$$\sqrt{\frac{K}{M}} = \omega_0 , \qquad \frac{C}{M\omega_0} = \mu,$$

$$\frac{F}{\alpha M {\omega_0}^2} = e , \qquad \frac{\omega_f}{\omega_0} = \Omega$$

and using the displacement and time transformation

$$\frac{z}{\alpha} = x , \qquad \tau\omega_0 = t$$

Substituting above into equation (2), got normalized governing differential equation

$$\ddot{x} + x - x^3 = -\mu\dot{x} + e\cos\Omega t \tag{3}$$

2 Topological Structure of the Autonomous System

The autonomous system from equation (3)

$$\frac{dx}{dt} = y = X(x, y) \qquad \frac{dy}{dt} = -x - \mu y + x^3 = Y(x, y) \tag{4}$$

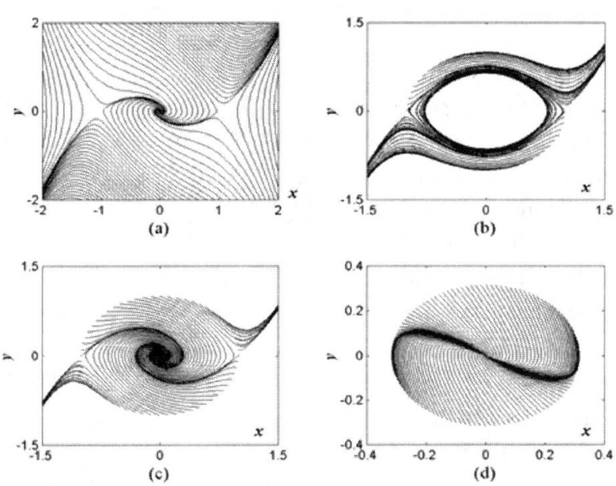

(a) The global trajectories near three equilibria; (b), (c), (d) The global trajectories near focus with $\mu = 0$, 0.5 and 2 respectively

Figure 2. Phase portraits of the autonomous system

can be viewed in a small region of the equilibrium points as equivalent in structure to their linear part only. From the characteristic equation of the linear parts of equations (4), three singular points have been found[3]: a stable focus at point (0,0), tow saddle points at (±1,0). The global trajectories at the local of these singularities are shown in Figure 2(a), which are obtained using numerical emulation method.

The topological structure of the autonomous system is governed by dynamical parameter. Different topological structure will appear with the adjustment of damping. Analysis indicates that origin will be a center with μ=0, a degenerate node with μ=2 and a stable focus with 0<μ<2. These conclusions have been proved by numerical emulation as shown in Figure 2, which accord with theoretical analysis.

3 Topological Structure of the Nonautonomous System

The topological structure of the autonomous system will be changed in a wide rage owing to the effect of the exciting force. First the method of rate of trajectories is used to research the change of topological structure near the focus at origin.[4,5,6] The energy integral of the nonlinear system

$$E(x,y)=\frac{1}{2}x^2-\frac{1}{4}x^4+\frac{1}{2}y^2=C$$

The amount of trajectories passing through a close curve S(i.e. the curve of E=C) which contains equilibrium (0, 0) is[5]

$$I=\oint_S \rho(Ydx-Xdy)=\frac{1}{C}\oint_S(-\mu y+e\cos\Omega t)dx \tag{5}$$

In order to calculate conveniently, a new reference frame is introduced

$$x=a\cos\theta+b(a)$$

It is proved that $b(a)$=0 and $\dot{x}=y$ is deducted as below[3]

$$y(a,\theta)=\frac{a\sin\theta}{2}\sqrt{(4-3a^2)(1-H\cos 2\theta)}$$

$$\approx\frac{a}{2}\sqrt{4-3a^2}\sin\theta[(1-\frac{H^2}{16})-\frac{H}{2}\cos 2\theta-\frac{H}{16}\cos 4\theta]$$

$$H=\frac{a^2}{4-3a^2}$$

Substituting above formula into equation (5) and just taking into account the primary resonance, that is

$$\Omega t=\theta-\theta_0$$

The amount of trajectories passing through a close S is obtained

$$I=\frac{\pi}{C}[\frac{\mu}{2}a\sqrt{4-3a^2}(1-\frac{H}{4}-\frac{H^2}{16})+e\sin\theta_0] \tag{6}$$

There is $a=a^*$ when $I=0$ and $I<0$ with $a=a_2>a^*$, $I>0$ with $a=a_1<a^*$. It means that a stable limit cycle exist at least one between curve $S(a_2)$ and $S(a_1)$.

In the same time, roughly identical conclusion is also obtained by using the method of numerical emulation. Figure 3 (a) shows the global trajectories near these singular points in the nonautonomous system (3), in which the change of the topological structure at the local of three singular points can be viewed obviously. The topological structure of the saddle points at $(\pm 1,0)$, merely some change in the form of trajectories, has unaltered in nature, but the focus at point (0,0) has occurred alteration essentially. Fig.3 (b), (c) and (d) display a set of trajectories near the focus at origin of nonautonomous system in the phase plane respectively with $\mu=0$, 2 and $0<\mu<2$, and these numerical calculations indicate clearly that stable focus at point (0,0) has turned into unstable and a 'broad loop' near it which is similar to limit cycle has presented. Numerical calculation accords generally with theoretical explanation as above.

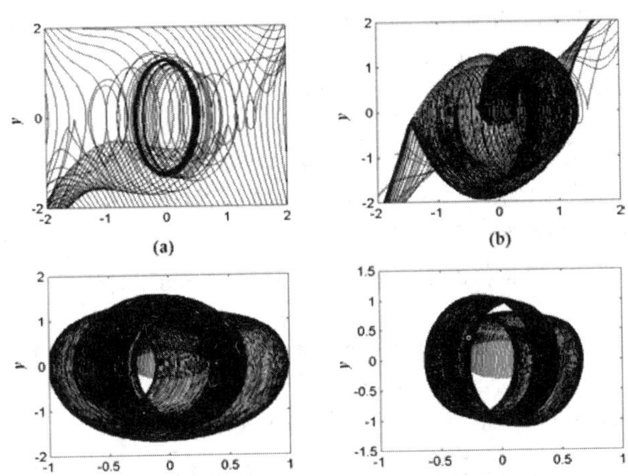

(a) The global trajectories near three equilibria; (b), (c), (d) The global trajectories near focus with $\mu=0$, 0.5 and 2 respectively

Figure 3. Phase portraits of the nonautonomous system

Figure 4 gives a set of phase portraits with parameters to take different magnitude. It shows that the change of topological structure is correlative with parameters e and μ.

The nonautonomous system (3) of machine foundation can be considered to coupling of the autonomous system (4) and the harmonic excitation $e\cos\Omega t$ that correspond to a linear system of harmonious oscillation. In the system, the disturbing energy from harmonious oscillation lies on e. The oscillation of linear harmonious system is weak when e is small and its effect to the nonlinear system is infirm. With e rising, the effect well be strong and it makes system to appear cyclic oscillation. The radius of the loop near origin is approximatvely proportional to the amplitude of disturbing force e as shown in Figure 4(a), (b) and (c), in which μ is identically equal to 0.1. Another parameter, damping ratio μ, has mighty influence on the topological structure of the system. Figure 4(d), (e) and (f) show a set of phase portraits with μ of different magnitude. When μ is bigger, a limit cycle appear in the phase plane as shown in Figure 4(f). Along with μ lessening, a broad loop appears near the focus, it seems that a lot of successive cycle has been collocated. It means that quasi-periodic oscillation or chaos motion occurs in this condition. It can be proved according to Poincare map and detail drawing of phase diagram as shown in Figure 5, in which μ equals 0.1.

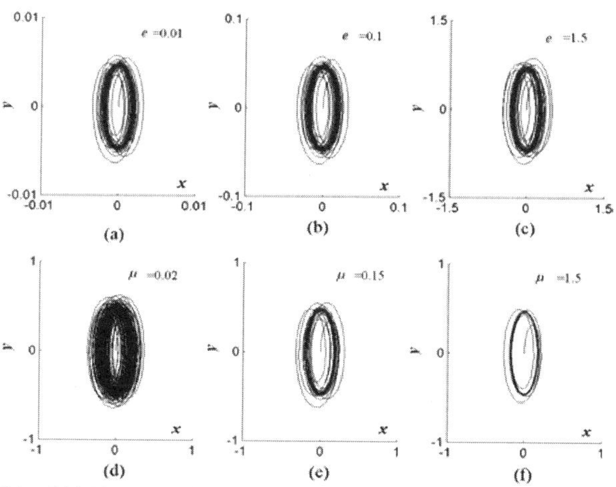

(a), (b) and (c) Phase portraits with μ=0.1; (d), (e) and (f) Phase portraits with e=1.
Figure 4. Phase portraits of the nonautonomous system with different parameter

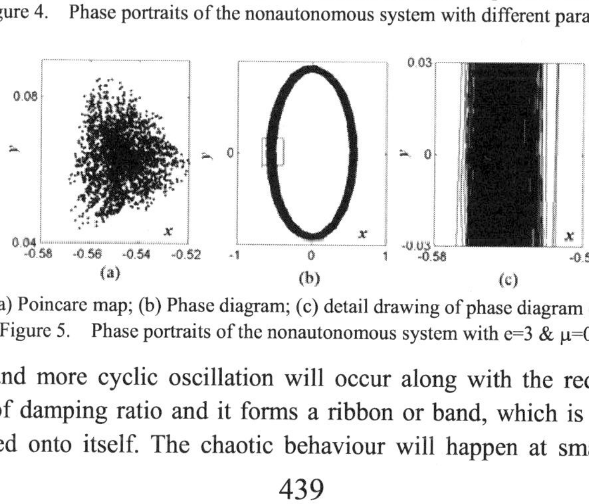

(a) Poincare map; (b) Phase diagram; (c) detail drawing of phase diagram (b).
Figure 5. Phase portraits of the nonautonomous system with e=3 & μ=0.1

More and more cyclic oscillation will occur along with the reduction of the coefficient of damping ratio and it forms a ribbon or band, which is smoothly and simply folded onto itself. The chaotic behaviour will happen at smaller damping ratio μ. This chaotic response occurs in a rather small region of control space, because the softening spring effect in system. Even then the stability of the system becomes worse and worse.

4 Conclusions

There are three equilibrium points in the autonomous system of machine foundation: a focus at origin and two saddle points at ($\pm$1,0). And the stable focus will be turned into unstable one in the nonautonomous system and unlimited unclosed limit cycles appear, which oscillate in a rather small region near the origin.

The topological structure depends on dynamical parameters: The radius of oscillatory loop increases with e enhancing, and the breadth of it thickens with μ decreasing. It indicates that the magnitude of disturbing force influences the amplitude of vibration but the stability of the system, on the contrary damping ratio mainly works on the stability of machine foundation system. With μ decreased, the range of the cyclic oscillation becomes bigger and the stability worse, and chaos motion occurs in this condition. This effect should be paid great attention to in the design of machine foundation despite the fluctuation is limited in a not too bound.

References

1. Yang Guitong. Soil Dynamics[M]. Beijing: Building Materials Industry Press, 2000: 46-51.
2. Chen longzhu, Ye Guiru, Chen Shengli. Applied analysis of non-linear vibration of foundation [J]. Journal of Vibration Engineering, 1996, 9(3): 292-295se).
3. W. B. Bao et al, Asymptotic Behavior and Energy Solution of a strongly Nonlinear Dynamical System[A]. Bangchun WEN ed. Proc.of APVC 2001[C], Hangzhou: Jinlin Science and Technology Press, 2001, Vol(I): 72-76.
4. J. M. T. Thompson, H. B. Stewart. Nonlinear Dynamiccs and Chaos: Geometrical Methods for Engineers and Scientists[M]. New York: John Wiley & Sons Ltd., 1986:200-207.
5. Li Li. Qualitative theoretics and quantificational method of strongly nonlinear oscillations system [M]. Beijing: Science Press, 1997: 25-34.
6. Wen Bangchen, et al Analytical method and engineering application in nonlinear vibration theory[M]. Shenyang: NEU Press,2001: 154-159.
7. XU Yu-xiu, BAO Wen-bo, W.Schiehlen and HU Hai-yan. A 1/3 pure subharmonic solution and transient process for the Duffing's equation [J]. Applied mathematics and mechanics (English Edition). 2001, 22(5): 586-592.
8. Peter Hagedorn. Non-linear oscillations [M]. New York: Oxford University Press, 1981.

NEW APPROACH FOR THE SAFETY ANALYSIS OF STRUCTURE SUBJECTED TO BOUNDED BUT UNCERTAIN ROCK MATERIAL PARAMETERS

XIANG JI-LAI LAI GUO-WEI
School of Water Resources and Hydropower, Wuhan University, Wuhan 430072, China
E-mail: xjl96@sina.com

A new approach is presented for the safety analysis of structures subjected to bounded but uncertain rock mechanics material parameters based on convex model theory. First, It defines the uncertain material parameters with a multidimensional box convex model when the amount of information available on the uncertain material parameters is not enough to accurately define. The worst cases are searched over this multidimensional box by using the successive quadratic programming (SQP) algorithms. The safety parameter's domain of is then got. The approach is reasonable, and the whole process is feasible.

1 Introduction

Most structural design problems include uncertainties. The most frequently used approach for studying uncertainties employs probabilistic methods. These methods are exceedingly useful, however, they require a wide pool of information on the uncertain parameters to descript the distribution of the events. The convex model does not require a probability distribution description of the uncertain parameters and may be used to specify uncertainties in the absence of detailed probability information. The convex model requires only the bounds of uncertainties. This makes the convex model an alternative way to analyze uncertainty when a limited amount of information is available.

Recently, the use of convex model of uncertainty has received considerable attention. Several studies have been performed using convex model theory in applied mechanics. Ben-haim. Y. [1] based on the theory of convex model described a deferent formulation about the reliability of a system. Marco Lombardi and Raphael T.Haftka [2] used convex models to improve an anti-optimization technique for structural design under load uncertainties. C.P. Pantelides[3] et al incorporated the convex model into an optimization computer algorithm. Sara Ganzerli [4] et al presented a convex model for the optimal design of structures subjected to uncertain loads. Zhiping Qiu[5] et al based on convex model to concerned with the problem of robust stability of linear dynamic systems. Nii O. Attoh-Okine [6] used convex models as a tool for addressing the variability and uncertainty in the structural number determination in pavement design.

The objective of this study is to use the convex model into rock mechanics and engineering, presented a new approach for the safety analysis of structures subjected to bounded but uncertain rock mechanics material parameters. It defines uncertain rock mechanics material parameters with convex model. The second objective is to search over this convex model to find the safety parameter's domain.

2 New approach for the safety analysis of structure based on convex model

Assume the safety of structure may be influenced by n numbers of uncertain material parameters:$x_1 \cdot x_2 \ldots\ldots x_n$, each of them is represent by a uniform bound convex set [1]:

$$x_i^l \le x_i \le x_i^u \quad (1 \le i \le n) \tag{1}$$

where x_i^l and x_i^u represent the lower and the upper bounds of the magnitude of x_i. This convex set is an N-degree multidimensional box. That means uncertain parameters may be locate everywhere in the box. Thus, the safety parameter $K(x)$ of engineering is not one value ,but one domain. Obviously, it is important to find the domain for the safety analysis of engineering.

The general formulation becomes the below form:

$$\min_{x \in R^n} \; K(x) \le K(x) \le \max_{x \in R^n} \; K(x) \tag{2}$$

subject to $G_i(x) = 0$, $i = 1, \cdots, m_e$;
$G_i(x) \le 0$, $i = m_e + 1, \cdots, m$;
$x^l \le x \le x^u$

Where x is the vector of design parameters, G(x) is the equality or inequality constraints vector function, m_e is the equation number, m is the constraints vector function number.

Actually, the solutions are two nonlinear optimizations. We can use some optional programming, such as the successive quadratic programming (SQP) algorithms, which have developed maturely to solve one side of above formulation.

3 Successive quadratic programming algorithms

The SQP algorithms for solving one half of above generates, at an iterate x_k, a search direction d_k by solving

$$\min_{d \in R^n} \frac{1}{2} d^T H_c d + \nabla K(x_k)^T d \; ; \tag{3}$$

s.t. $\nabla g_i(x_k)^T d + g_i(x_k) = 0$;
$\nabla g_i(x_k)^T d + g_i(x_k) \le 0$;
$x^l \le x_k \le x^u$

A new iterate is then computed as

$$x_{k+1} = x_k + \alpha_k d_k \tag{4}$$

where α_k is a step length parameter, H_c is a Hessian matrix.

4 A gravity dam example

To use this new approach analyze the sliding stability safety factor of gravity dam on concrete/rock interface, a relatively simple concrete solid gravity dam as follows (Fig. 1). Dam height is 100m,the top width is 15m,and the downstream slope is 1:0.7.A thickness of 1 meter is selected for computation. It is supposed that gravity dam is built and water is fully stored, and there is no tail water. In the global coordinate system, the origin of the coordinate is located at the dam heel, the axis x is downstream, the axis Y is upward. There are some uncertain material parameters :2350 • γ_c •2400;1.10• f'_R •1.30;1.10• c'_R •1.30;0.20• α •0.30,Where γ_c is unit weight of the concrete solid gravity dam(kg/m^3), f'_R is shearing friction strength factor of the concrete/rock interface; c'_R is Unit cohesion of the concrete/rock interface(Mpa); α is the reduction coefficient of uplift .

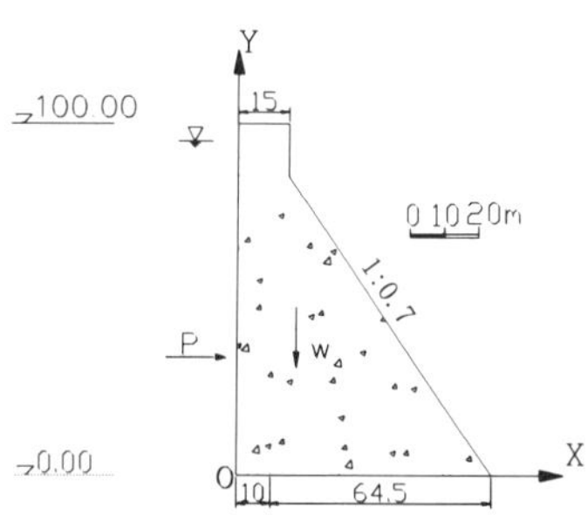

Fig.1 the computation section

Use the SQP algorithm to solve the nonlinear equations, the result domain of k(x) is: $3.4 \leq k \leq 4.2$.

Actually, if consider the uncertainty of loads, that means in this example the upstream water level is change from 95m to 100m,the downstream water level is from 0m to 3m. We can also calculate the domain of k(x) is: $4.2 \leq k \leq 5.2$

5 Conclusions

A new approach is presented for the safety analysis of structures subjected to uncertain rock mechanics material parameters based on convex model theory. This new approach can apply to the situations when the uncertain material parameters are not enough to accurately define, or there is a high level of uncertainty and too little information is available to use probabilistic methods.

References

1. Ben-Haim.Y.Robust reliability of structures. Advances in applied mechanics, volume.33, 1997
2. Marco Lombardi, Raphael T.Haftka. Anti-optimization technique for structural design under load uncertaimties. Comput. Methods Appl. Mech. Engrg. 157 (1998) 19-31
3. C.P. Pantelides, Berkeley C. Booth. Computer-aided design of optimal structures with uncertainty. Computers and Structures 74 (2000) 293-307.
4. Sara Ganzerli, Chris P. Pantelides. Optimum structural design via convex model superposition. Computers and Structures 74 (2000) 639-647.
5. Zhiping Qiu ; Peter C. Müller , Andreas Frommer. Stability robustness bounds for linear state–space models with structured uncertainty based on ellipsoidal set-theoretic approach. Mathematics and Computers in Simulation 56 (2001) 35–53
6. Nii O. Attoh-Okine. Uncertainty analysis in structural number determination in flexible pavement design – a convex model approach. Construction and Building Materials 16 (2002) 67–71.
7. David J. Ternet and Lorenz T. Biegler.Recent improvements to a multiplier-free reduced Hessian successive quadratic programming algorithm. Computers Chem. Engng Vol. 22, No. 7–8, 963–978, 1998

AN IMPROVED HIERARCHY DECISION MAKING MODEL FOR THE TREATMENT OF ROCK MASS SLOPE

QUANMIN XIE YUANYOU XIA
Wuhan University ofTechnology, Hubei Wuhan 430070, P.R.China

In this paper, the traditional analysis hierarchy process is improved by means of optimal transfer matrix. An improved hierarchy decision making model for the treatment of rock mass slope set up. The judgement matrix obtained by the method can naturally meet the requirement of identity, and its application is discussed in detail.

1 Introduction

The analysis hierarchy process has a good effect in solving multiobjective and multihierarchy optimization problem of treatment schemes of rock mass slope. Its basic principle is to compare relative importance each other in every two evaluating factors, and then ordered them. Because of the artificial one-side judgement, the comparative results aren't to be objective identity, then the identity test is often needed if it can't pass the test, and the traditional method is that the judgement matrix would be adjusted by approximate estimation. Though it is effective, it is blind and subjective. Sometimes, it need adjust many times in order to pass the identity test. In this article the traditional analytic hierarchy process is improved by the optimal transfer matrix, which meet the requirement of identity and needn't take the identity test. The application of the method is discussed in detail.

2 Improved analysis for hierarchy process of treatment decision making of rock mass slope

2.1 Define weighing matrix $W^{(p)}$ of evaluation indexes

If there are m evaluation indexes in the decision problem of slope treatment schemes, and there are r experts to give their evaluation, the evaluation index set of slope treatment question is $X=\{x_1, x_2, \cdots, x_m\}$, and the expert set is $S=\{s_1, s_2, \cdots, s_r\}$ $(m, r \geq 1)$. The expert who is included in set S assesses every weighing index, adopting the two-two comparative method, x_i and x_j are compared each other by any expert S_k in expert set S, forming a comparative matrix:

$$D^{(k)}=[d_{ij}^{(k)}]_{m\times m} \tag{1}$$

where $d_{ij}^{(k)} = \begin{cases} 2 & \text{the index } x_i \text{ is more important than the index } x_j \\ 1 & \text{the index } x_i \text{ is as important as the index } x_j \\ 0 & \text{the index } x_j \text{ is more important than the index } x_i \end{cases}$

and there $d_{ii}^{(k)}=1$, that is the objective X_i compares with itself, having the same importance.

According to the order of its importance degree, the judgement matrix can be formed as:

$$A^{(k)} = \left[a_{ij}^{(k)}\right]_{m\times m} \tag{2}$$

where $a_{ij}^{(k)} = \begin{cases} r_i^{(k)} - r_j^{(k)} & r_i > r_j \\ 1 & r_i = r_j \\ [r_j^{(k)} - r_i^{(k)}]^{-1} & r_i < r_j \end{cases}$

$$r_i^{(k)} = \sum_{l=1}^{m} d_{il}^{(k)} \quad (i=1, 2, \cdots, m; k=1, 2, \cdots, r). \tag{3}$$

Obviously, $A^{(k)}$ σ_{ij} is a mutual antimatrix, $B^{(k)}=\lg A^{(k)}=[\lg a_{ij}^{(k)}]_{m\times m}$ is an antisymmetry matrix. If σ_{ij} is the total standard error of expert's evaluation, there is:

$$\sigma_{ij} = \sqrt{\frac{1}{r-1}\sum_{l=1}^{r}[b_{ij}^{(l)} - \frac{1}{r}\sum_{t=1}^{r} b_{ij}^{(t)}]^2} \tag{4}$$

From σ_{ij} calculation results, there are two following situation:

(1)if $\sigma_{ij}<1$, it shows that the opinion of the experts is comparatively unified, and the arithmetic mean of the expert judicial value is looked as the judicial result of the group set. We can gain:

$B=[b_{ij}]_{m\times m}$, where $b_{ij} = \frac{1}{r}\sum_{l=1}^{r} b_{ij}^{(l)}$ (i, j=1, 2, ⋯, m) It doesn't surely have the identity , so applying the improved analysis hierarchy process, the matrix is formed as: $A^*=[a^*]_{m\times m}$, where $a^*=10^{c_{ij}} = 10^{\frac{1}{m}\sum_{l=1}^{m}(b_{il}-b_{jl})}$

From the theorem 2, it is known that A^* is A's optimal transfer matrix and it is identical. The correspondent vector of the A^*'s maximum character value got by root-square method is the index's weighing $\overline{w} = (\overline{w}, \overline{w}_2, \cdots, \overline{w}_m)$, and it is normalized, then it changes to be the standard weighing $w = (w_1, w_2, \cdots w_m)$, where $w_i = \overline{w}_i \Big/ \sum_{k=1}^{m} \overline{w}_k$, it is the standard weighing of every index.

(2) if $\sigma_{ij}<1$, (i, j=1, 2, ⋯, m), it shows the opinion of the expert group is very different, the arithmetic mean of every expert's judicial value isn't looked as the judicial result of the group set. But the optimal transfer matrix can be adopted to

calculate, that is: exist an optimal transfer matrix $B=[b_{ij}]_{m\times m}$, make $J=\sum_{i=1}^{m}\sum_{j=1}^{m}\sum_{l=1}^{r}[b_{ij}-b_{ij}^{(l)}]^2$ minimum, where $b_{ij}=\frac{1}{m}\sum_{t=1}^{m}\sum_{l=1}^{r}[b_{it}^{(l)}-b_{jt}^{(l)}]^2$

Let $C=[a^{b_{ij}}]_{m\times m}=[c_{ij}]_{m\times m}$, where a is objectively important ratio of close grade assessment, and there a=1.1-1.3. The matrix C is the comparative judgement matrix of the group set gained, and it is identical. The index's weighing can be obtained by consulting the description in above paragraphs

If some slope has p probable treatment schemes, forming the set $F=\{F_1, F_2, \cdots, F_p\}$. The standard weighing of the h scheme's evaluating index can be got by the above method:

$$w^{(h)}=(w_1^{(h)},w_2^{(h)},\cdots,w_m^{(h)}) \tag{5}$$

The index-weighing matrix made up of p schemes is:

$$W^{(p)}=[w_i^{(h)}]_{p\times m},\ (i=1,2,\ldots\ldots m;h=1,2,\ldots\ldots p) \tag{6}$$

2.2 *Define character matrix $T^{(p)}$ of evaluation index*

As to P treatment schemes having m evaluation indexes, the character value of individual evaluation index is $X^i=[x_{ih}^i]_{m\times p}$. The slope treatment scheme's evaluation involves the problems of the qualitative and quantitative indexes, which can be settled by the following method:

(1) Standardization of quantitative indexes

As to the index that it is better if it is bigger, let

$$\gamma_{ih}^i=x_{ih}^i\Big/\max_{1\le h\le p}\{x_{ih}^i\} \tag{7}$$

As to the index that it is better if it is smaller, let

$$\gamma_{ih}^i=(\max_{1\le h\le p}x_{ih}^i-x_{ih}^i)\Big/(\max_{1\le h\le p}x_{ih}^i-\min x_{ih}^i) \tag{8}$$

(2) Standardization of qualitative indexes

The character value of some index in the treatment scheme can be evaluated according to the different grade "excellent ,good, ordinary, bad and very bad", If the remark set of some index in treatment scheme may show $R_i=(r_1, r_2, \cdots, r_q)$(having q grades) and the value is given according to linear equal difference, the quantitative grade matrix of comment set is $C_i=[1,\frac{n-2}{n-1},\frac{n-3}{n-1},\cdots,0]$. Therefore, when individual expert gives the remark of the qualitative index x_{ih}^i is $r_i\in R_i$, the normalized character value of this index is $\gamma_{ih}^i=\frac{n-i}{n-1}$, (n=1,2,⋯,q;i=1,2,⋯,m).

(3) Character matrix of assessment indexes

The character matrix $T^{(p)}$ of the evaluation index after index normalization is

$$T^{(p)} = [\gamma_{ih}^{i}]_{m\times p} \quad (i=1, 2, \cdots, m; h=1, 2, \cdots, p), (i=1,2,......m; h=1, 2,\cdots,p). \tag{9}$$

2.3 Evaluation matrix Z of slope treatment schemes

Decision making evaluation matrix Z of P treatment schemes having m evaluation indexes is:

$$Z = W^{(p)} \times T^{(p)} = [z_h]_{1\times p} \tag{10}$$

The slope optimal treatment scheme can be defined conveniently according to the evaluation matrix Z.

3 Engineering example

The side facing the sky of the critical rock slope at the Slate Mountain is steep slope or cliff. There are landslides, factories, schools and inhabitant below the scar. The environment is very complex. On the base of geologic exploration and stability analysis, there five possible treatment schemes are put forwards as: blasting unloading method, drainage works method, flexible reinforcement method, rock-control drain method and anti-sliding concrete key method. In order to define the better treatment scheme to suit for the critical rock mass, these following indexes are chosen to be the standards of weighing the treatment schemes after analyzing and studying: total investment(x_1), annual maintenance cost(x_2), control effect(x_3), construction difficulty degree(x_4), construction period(x_5), risk(x_6), engineering validity term(x_7), environment affection (x_8).

The decision evaluation matrix of the critical rock-mass slope treatment scheme is:

$$Z = W^{(p)} \times T^{(p)} = (\ 0.65\ \ 0.52\ \ 0.62\ \ 0.49\ \ 0.58\ ) \tag{11}$$

According to evaluation matrix Z, the blasting unloading method should be mainly adopted in the treatment of the critical rock slope of the Slate Mountain.

Reference

1. XIA Yuan-you, ZHU Rui-geng. Multiperson multilayer fuzzy comprehensive decision making for control plans of unstable slopes[J].Chinese Journal of Natural Disasters,1998,8(1):88-91

THE SNS FLEXIBLE SYSTEMS FOR HOLDING ROCKFALL

YONGMEI HE YOUKUI YANG YINGQING ZHOU RUIQI JIANG
Geobrugg Chengdu Co, Ltd. Chengdu,China 611731
E-mail: yh-sns-a@chinagccl.com

As a kind of geological hazard, rockfall is difficult to be control due to its complexity. Based on the analysis of the current technology of rockfall protection, this paper introduces the functions, principles, basic characteristics, main structure and applicable fields of SNS which mainly made up of wire rope nets and mesh nets, and clearly know that SNS has technical and economic advantages comparing with traditional ways in protection and operations.

1 Introduction

Rockfall is a familiar geological hazard of single or clustered stone detached from matrix under the effect of the gravity and various energy, move suddenly down steep slope and reach the foot of slope finally due to falling, slipping, bouncing, rolling or the combination of above, which may cause traffic block, damage to the constructions and injury or death of people while occurring nearby the transits and buildings. Objectively, the scale of single rockfall and its loss caused is not evident to be compared with the typical landslip, so it hasn't call much attention from people to investigate and research an advanced and effective method for the use of holding it up to date. However, the rockfall may also cause great damage because of its high falling speed, great impact power, randomness, unpredictability and complicated movement in certain areas. What's more, the atrocious environment in which it happens is very hard to deal with. So such problem has great meaning for the harmonic development of economic construction and environmental protection through exploit and make good use of new technology to protect the slope.

In recent years, a new protection way-SNS flexible protection system has been developed rapidly to hold rockfall in various construction such as hydropower stations, mines, railways, transportation etc. This paper illustrated how the system can hold the rockfall hazard better under various circumstances.

2 The Main Measure of Rockfall Protection and SNS

The rockfall protection can be considered in two aspects that is to prevent from its occurring and avoid its damage. The appropriate methods depend mainly on the history of events, latent rockfall characteristics and risk level, locale condition, investment of protection construction and maintenance charge etc.. Figure1 shows the main measures of protection.

SNS-Safety Netting System is a flexible safety protection system mainly made up of wire rope net or steel wire mesh to prevent rockfall. It can be divided into two kinds: active and passive. Active system works through covering flexible nets on the slope surface with anchor and support rope, avoids the danger of rockfall by stopping rockfall or restricting falling of rocks (Figure2, Figure3). Passive one is a kind of flexible metal barrier, whose principal parts are wire rope nets with highly strength and flexibility, which can catch or retain the falling stones (Figure4).

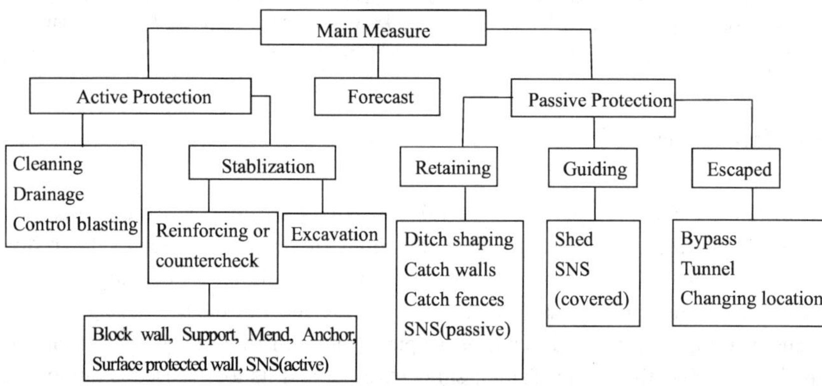

Figure1. Main Measure of Rockfall Protection

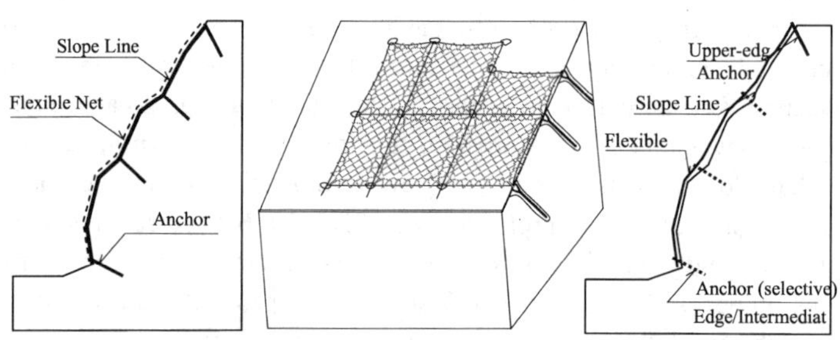

a. Standard Active System b. Active-Passive System

Figure2. Wire Rope Net/Mash Net System -SNS Active Protection System Sketch Map

According to the main components, the active protection system can be categorized into wire rope net system, Tecco® system (special type steel wire mesh) and mesh system (ordinary steel mesh). Wire rope net system and mesh system work through the way that fix wire rope anchors and support ropes on the slope (Figure2). Figure 2a is the standard active protection system usually applied to reinforce the slope, and restrain the rockfall falling, soil erosion, slip collapse, slope-slipping, by fixing wire rope net and/or mesh which have been pre-tensioned with anchor and support rope to support the whole slope continuously. With the aid of the pre-tension, the systems are firmly pressed onto the slope surface to resist the movement of local rock or soil and realize active protection (reinforcing) function thereby. Though the function of this system is similar to the traditional surface slope protection means such as the shotcrete, its flexible character make it possible to transfer the local dense load around equably, so this system can bear bigger load and reduce the anchor length. With flexible and on-limits character, this system can be installed with minimal alteration of the landscape and provide a permanently safe and secure environment for the surrounding population. As groundwater can be drained freely, it also avoids slope unstable caused by drainage block and hydraulic pressure raising after management. In some situation, we can plant some vegetation at the open space to fortify the soil combining with slope protection system.

Figure2b is active-passive system, namely SNS covering system. It is similar to standard active protection system in form, but its main function is passive protection. This system needn't systemic anchoring and pre-tensioning, it just need to cover slope under the deadweight of net or keep the certain space between system and slope surface. The potential rock can fall through its covering, but keeping stones in the close space between nets and slope surface, i.e. the system can make stones rolled down the slope to the base or flat of slope under control.

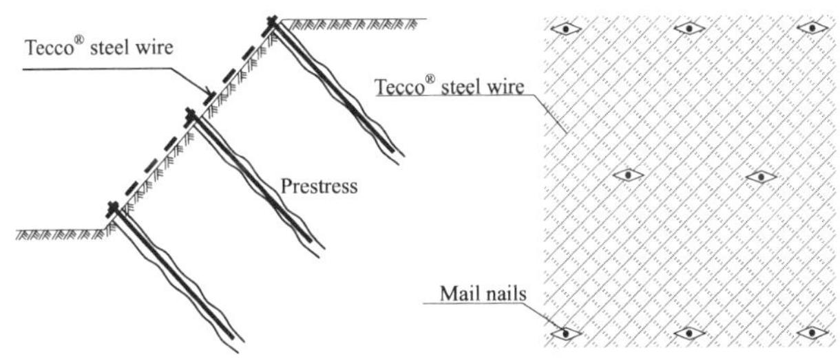

Figure3. Tecco® System Sketch Map

Figure 3 shows a new protection system- Tecco® system mainly made up of special steel wire mesh (with high-tensile and corrosion protection, generally called Tecco® steel wire mesh). It is the result of specific development work. Tecco® system provide a kind of slope stabilization means which is more economical, high corrosion protection (100 years of life expectancy) and its protecting ability is between the common mesh and wire rope net(up to 80% of wire rope net) . By means of the pre-stress ground or rock nail, the Tecco® steel wire mesh is anchored onto the sliding layer or the slope surface tightly to reinforce the slope based on the common anchor principle, which is the main difference compared with the standard active system , besides use the special Tecco® mesh.

The primary design parameter of the SNS passive system is the impact energy of the rockfall, and various standard types have been developed to adjust different forms and scales of the rockfall, including RX, CX, AX and CAN model. Generally, its protective energy range is about 40～2350kJ, and it also can give special protection to rockfall with energy up to 5000kJ. Fig.4 shows RX-500 system, its protective energy is 500kJ. The system can absorb and scatter the energy of fallen rocks by transforming it into deformation energy of the system within the designed energy limit, no matter where in the net the rocks fall. Compared with traditional construction such as gravity catch wall, the SNS passive system has enough flexibility and bearing ability to absorb and scatter the predicted rockfall energy and ensures minimum breach of the system, and we considered how to install the SNS system easily and instantly with least anchors and diggings in the very harsh environment like cliff in course of design. The system is mainly composed of wire rope net, brake rings, support ropes, steel posts and anchor rope, and the flexibility of the system mainly depends on the wire rope nets, support ropes and brake rings, and the movable joint between the steel posts and base.

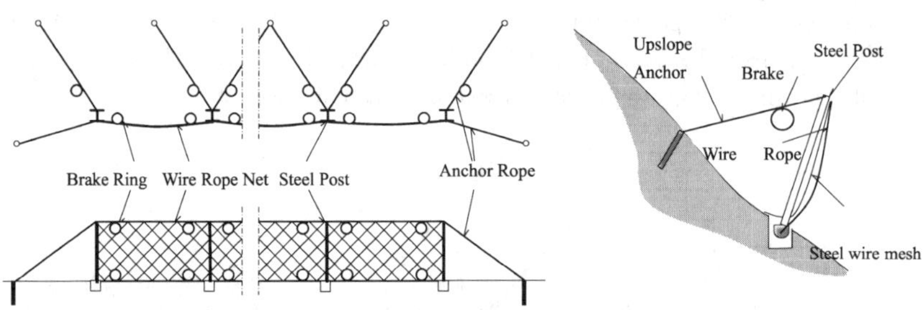

Figure4. Rx-050 Model SNS Passive System Sketch Map

When the falling stone impact the barrier , the impact energy is scattered by the flexibility of the net at first and transferred and reduced along the support ropes, seam ropes, and nearby wire rope nets gradually, finally reach the anchor rope and stable stratum, here the remaining energy has been reduced to the minimum. When the energy of the falling stones exceed the absorptive ability, the distortion of brake rings improve the flexibility of the system to reduce the maintenance work. As the energy transfer course is performed by various components which have different energy absorptive ability , we must make it clear that how much load each part of the system holds. A lot of field tests had been done to keep the balance and check the standard design, besides theoretical calculation under ideal conditions, which apparently ensures the standardization, rapidness of installation and optimization of design. That's why it is scientific, economical, credible and safe.

3 How To Design And Select SNS Model

According to the structure and function of the SNS, both active and passive protection system can achieve the purpose of preventing rockfall, so it is necessary to think how to make the system economical, practical, installed easily and minimal maintenance through reasonable type selection and arrangement. Design and type selection include two aspects, the one is selecting active or passive system, the other is selecting different model of two types. Following is the element which influence design and type selection: field landform, character of rockfall, construction condition, difficulty and cycle of construction, the interrelation of slope and area to be protected, maintenance requirement, economic status, view, etc..

Landform Generally, it is suitable to adopt active protection system for the slope which is lower than 20m. Otherwise, we should consider passive system even if placing several layer of protection system from upside to underside. Of course, according to the landform and character that rockfall appear, we may consider the combination of active and passive system either.

Character of Rockfall The character of rockfall includes its distribution, size, accumulating form and moving path. To some slope where a lot of rock distribute separately, it is appropriate to be inset passive system. To the edge slope where rock are concentrative, active system usually be used to cover it. If both of the two conditions exist , we should place active coverage on concentrative rock, at the same time, build passive system to hold separate rock.

Objective of Protection When the slope must be reinforced, we should adopt active system. Especially to some loose soil and rock slope, even rockfall is still the primary form of destroying, we must control it and reinforce slope by active system because the local falling of rock may drag backside of fallen ground gradually.

Installation, Maintenance and Rockfall Clearing SNS system is a light system that mainly be assembled by anchor and metal component with setting type of toy brick. Active system would consume a great deal material such as nets and anchors. For passive system, people would be in trouble with the transportation and installation of steel post. So we should adopt passive system if anchor condition is unfavorable on some slope, and adopt active system on some steep slope where steel post is difficult to be installed while the anchor condition is favorable.

Both active-passive system and passive system permit the rock falling. In the process of holding up rock, parts of system may be distorted and destroyed,even invalid, so it should be maintained or replaced; Besides, if rock held by system not be cleared immediately, the system would be tensioned under statically effect of rock, and this effect will debase flexility and virtual resistance of system, so it's very necessary for the normal works of system to clear fallen rocks in time and maintain the whole system. According to the two aspects above-mentioned, if field is unfavorable to maintain system and clean rock, the better choice is active system which requires few or no maintenance and rock clearing work.

Risk Estimate There is no protection way of geology hazard is safe absolutely, and the more safe the protective construction is, the higher it cost, vice versa. So, selecting a protective system, we should consider its reliability and investment roundly.

4 Discussion and Epilogue

In recent years, with the maturation of design principle and standard system exploitation of SNS, it has been widely applied in various fields oversea such as transportation, railway, mining, hydropower and city environment protection. Since 1995-Safety Netting System for slope protection was introduced to China market, it has made a real breakthrough in business with large success, totally finishing nearly 160 projects for rockfall protection. Its advantages are safety, rapid installation and economy etc. All in all, compared with traditionally protective construction, it takes advantages over: (1) Due to its high flexibility and rigidity, it can hold the large stones with high impact energy while the traditional method can not.(2)It is so adaptable to the original surface of the protective areas that no extra digging is needed, in this way ,the stability and green plant of the slope won't be damaged. (3) With the detailed and credible design principle, standard and balance design is developed.(4) The standard component ensures it's easy to install. Instant work just need little labor intensity and won't interfere other operation. (5) The system is easy to be maintained. (6) Generally 50 years' time limit. It will be more by replacing few new components on time.

References

1.Yingqing Zhou & Tianbin Li. Assessment of risks, hazard and mitigation of rockfall if Dahekou Hydropower Station. Proc. Of the 7th Int. Symposium on Landslides, De Terrain, July 17-21, 1996, pp.435-440

2.Duffy John D. & Smith Duane D. (1989). Field Tests and Evaluation of Rockfall Restraining Nets. Research Report No. CA/TL-90/05, California Department of Transportation, Office of Transportation Material and Research

3.Houtian Hu, Falling and Rockfall, Beijing : China Railway Pub. Company, 1989

4.Yongmei He, Youkui Yang, Design , Selecting Type and Arrangement of SNS Flexible Protection, HIGHWAY, 2001/11

RESEARCH ON CALCULATING FORMULAS OF CROSS SECTION BEARING CAPACITY OF STEEL REINFORCED LIGHTWEIGHT AGGREGATE CONCRETE (SRLAC) MEMBER

CHUNDONG JIANG
Construction engineering head office of Qitaihe coal group, Qitaihe, China

LI CUI, YUFEN DONG
LiaoNing Technical University, Fuxin, China

In this paper, we adopt ultimate-balance theory to establish the calculating expression of cross section bearing capacity of SRLAC according to section static-force balance condition and to obtain the experimental parameters in tests, considering concurrent work behavior of steel reinforcement (SR) and lightweight aggregate concrete (LAC). The formulas agree closely with the form of the current code of steel concrete structure design. They are easy to be accepted in engineering shield. Meanwhile, they are simple and convenient for calculation, the concepts in the formulas are clear. Either computer or handwork can finish the calculation using the formulas. The applicability is better.

1 Calculating assumption and value of corresponding parameters

Calculating assumption

(1) Calculated cross section coincides with plane section supposition;

(2) Don't consider tensile strength of concrete in tensile zone;

(3) The actual stress contribution diagram of compressive zone is a parabola, but in true calculation, it is supposed to an equivalent rectangle block according to the principle that the size and application point location of composite force doesn't vary. By analysis of test data and relative information, in this paper the corresponding parameters are given by α =0.6; β =0.37; $\gamma = \alpha/2\beta = 0.6/(2\times0.37) = 0.81$; $X = 2\beta X_0 = 0.74X_0$; Balanced-depth of relative compressive zone

$$\zeta_b = \varepsilon_{cu}\times2\beta/(\varepsilon_{cu} + f_y/E_s) = 0.003\times0.37/(0.003+499/(2\times105)) = 0.404$$

In the formula: ultimate compression strength of LAC, ε_{cu}; yield strength of longitudinal bar, f_y; elastic modulus of longitudinal bar, E_s; location parameter of composite force application point in equivalent rectangle block, β; balanced-depth of relative compressive zone, X_b; distance from the composite force application point of tensile bar to the external surface edge of compressive concrete, h_0. *Ultimate strain of axial compression member is 0.002, corresponding* maximal compression stress is f_c; ultimate compressive strain of eccentrically compression column is 0.003, corresponding maximal compressive stress is f_{cm}.

The stress diagram of shape steel in eccentrically loaded column with compression failure is simplified into trapezoid distribution. The stress diagram in eccentrically loaded column with tension failure is simplified into trapezoid-triangle contribution, the stress of shape steel and bar equals strain multiplied by corresponding modulus of elasticity:

$$\sigma_s = \varepsilon_s / E_s \le f_y$$

Modifying factor of eccentricity η,

$$\eta = (e_0 + f) / e_0 = (1 + 1/e_0) f = (1 + 1/e_0)(\phi l_0^2 / \beta)$$

$$\phi = ((\varepsilon_{cu} + f_y/E_s) / h_0) \zeta_1 \zeta_2, \qquad \eta = 1 + [0.0055(l_0/h)^2 \zeta_1 \zeta_2] / (l_0/h_0)$$

$$\zeta_1 = (N_0 - N) / (N_0 - N_b) \le 1.0, \qquad \zeta_2 = 1.15 - 0.01 \times l_0 / h \le 1.0$$

Tress of tensile steel and compressive bar

$$\sigma_s = \varepsilon_s E_s = 0.003 E_s (2\beta h_0 - X) / X \le f_y \qquad \sigma_s' = \varepsilon_s' E_s = 0.003 E_s (X - 2\beta a') / X \le f_y'$$

2 Calculating of cross-section bearing capacity of SRLAC compression-flexure member

According to the judge condition whether the longitudinal bar far away from axial load yields or not, classify steel SRLAC compression-members into two members: eccentrically loaded member with compression failure and eccentrically loaded member with tension failure. For the former, its stress contribution diagram is simplified into trapezoid contribution, that is, rectangle-triangle contribution (shown in Fig.1). It has been proved that mid-span web of SR doesn't yield when member fails in theory and test, so the mid-span stress contribution diagram of member may be approximately simplified into triangle contribution to coincide math model with actual structure. For the latter, we should choose true values according to actual conditions because the longitudinal bar far away from the application point of axial load doesn't yield yet.

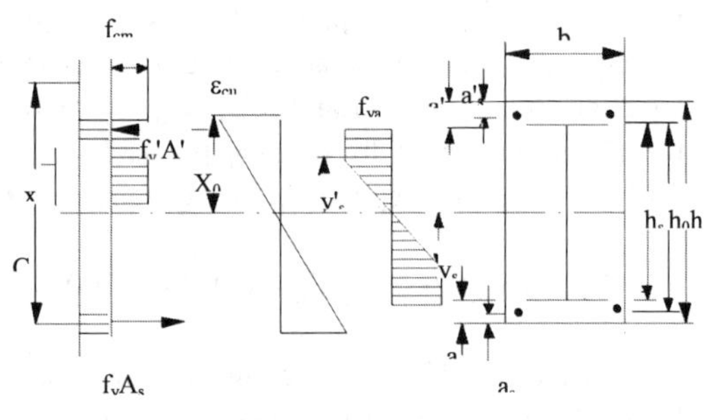

Fig.1 calculating sketch of eccentrically loaded member with compression

2.1 Eccentrically loaded member with compression failure

1). Calculating sketch is in fig.1; 2). Calculating formulas:

$$N_e \le f_{cm}bX + f_y'A'_s - f_yA_s + f_{ya}'A'_{ay} - f_{ya}A_{ay} + f_{yw}\delta_w (X/\beta - h_s - 2a') \qquad (1)$$

In formula (1): yield strength of compressive flange and tensile flange of shape steel f_{ya}', f_{ya}; cross-section area of compressive flange and tensile flange of shape steel A'_{ay}, A_{ay}; yield strength of shape steel web f_{yw}; thickness of shape steel web δ_w; $\beta = 0.37$; depth of shape steel h_s; distance from the center of gravity of compressive flange to external surface edge of compressive concrete a'. The meanings of other signs are the same as that of normal steel concrete.

Derive depth of compressive zone from formula (1):

$$X=(N-f_y'A'_s+f_yA_s-f_{ya}'A'_{ay}+f_{ya}A_{ay}+f_{yw}\delta_wh_s+f_{yw}\delta_w2a')/(f_{cm}+f_{yw}\delta_w/\beta) \qquad (2)$$

Obtain the calculating formula of cross- section bearing capacity by taking moment to tensile bar:

$$N_e\le f_{cm}bX(h_0-X/2)+f_y'A'_s(h_0-a'_s)+f_{ya}'A'_{ay}(h_0-a')-f_{ya}A_{ay}(a-a_s)+f_{yw}\delta_w(X_0-a')[h_0-(a'+X)/2]$$
$$-y_s'^2f_{yw}\delta_w/3- f_{yw}\delta_w [h_s- (X_o-a')]\times\{h_0-X_0- [h_s- (X_0+a')]/2\} \qquad (3)$$

In the formula, $y'_s = \varepsilon_{ya}X_0 / \varepsilon_{cu}$; $\varepsilon_{ya}= f_{ya} / E_{ya}$; $X_0= X/2\beta= 1.35X$

After rearranging formula (3), Obtain:

$$N_e\le f_{cm}bXh_0-f_{cm}bX^2/2+f_y'A'_s(h_0-a'_s)+f_{ya}'A'_{ay}(h_0-a')-f_{ya}A_{ay}(a-a_s)\ f_{yw}\delta_w[(1.35X-a')(h_0-a')-$$
$$(1.35X-a')^2/2-(h_s-1.35X+a')(h_0-1.35X)-(h_0-1.35X+a')^2/2-(f_{ya}X/(E_{ya}\varepsilon_{cu}\beta))^2/12] \qquad (4)$$

3). Applicable condition $X\le\zeta_bh_0=0.404h_0$ $X\ge 2a'_s$

2.2 Eccentrically loaded member with tensile

In Fig.2, ultimate compression strain of LAC ε_{cu}, ε_{cu}=0.003; modulus of shape steel E_{ya}; modulus of longitudinal tensile bar E_s; tension stress of tensile flange of shape steel σ_a; tension stress of longitudinal tensile steel σ_s; depth of triangle stress contribution diagram of elastic compressive web y'_s; depth of triangle stress contribution diagram of elastic

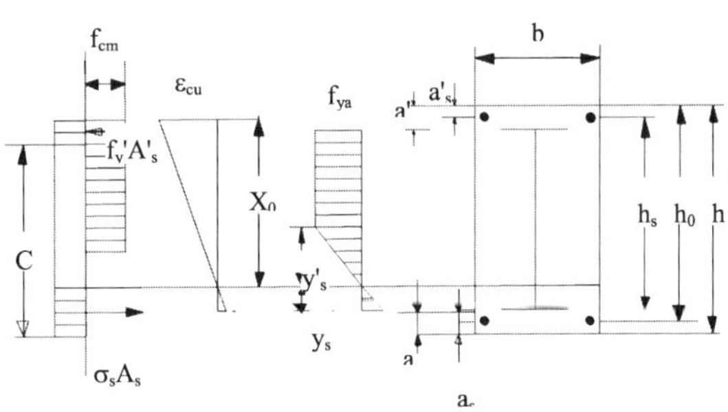

Fig.2 calculating sketch of eccentrically loaded member with tensile.

tensile web y_s; respective distance from the center of gravity of tensile flange and compressive flange to external surface of concrete a, a'.

1) Calculating sketch is in fig.2; 2) Calculating formulas:

$$N_e \le f_{cm}bX + f_y'A'_s + f_{ya}'A'_{ay} - \sigma_s A_s + f_{ya}'(X_0 - a')\delta_w - f_{ya}'y'_s\delta_w/2 - \sigma_a y_s\delta_w/2 - \sigma_a A_{ay} \quad (5)$$

where, $y'_s = f_{ya}'X_0/(\varepsilon_{cu}E_{ya})$; $y_s = h - X_0 - a$; $\sigma_a = \varepsilon_{cu}E_{ya}y_s/X_0 = \varepsilon_{cu}E_{ya}(h - X_0 - a)/X_0$; $\sigma_s = \varepsilon_s E_s = \varepsilon_{cu}E_s(h_0 - X_0)/X_0$

After putting the above data into formula (5), we obtain:

$$N_e \le 2\beta f_{cm}bX + f_y'A'_s + f_{ya}'A'_{ay} - \varepsilon_{cu}E_s(h_0 - X_0)A_s/X_0 f_{ya}'X_0\delta_w - f_{ya}'\delta_w a' - f_{ya}'\delta_w(f_{ya}'X_0/(\varepsilon_{cu}E_{ya}))^2 \varepsilon_{cu}E_{ya}\delta_w(h - X_0 - a)^2/2X_0 - \varepsilon_{cu}E_{ya}A_{ay}(h - X_0 - a)/X_0$$

After rearranging, calculate depth of compressive zone using formula (6):

$$X = (-B \pm \sqrt{B^2 - 4AC})/2.7A \quad (6)$$

where $A = 2\beta f_{cm}b + f_{ya}'\delta_w - (f_{ya}')^2\delta_w/(2\varepsilon_{cu}E_{ya}) - \varepsilon_{cu}E_{ya}\delta_w/2$

$B = f_y'A'_s + f_{ya}'A'_{ay} + \varepsilon_{cu}E_sA_s - f_{ya}'\delta_w a' + \varepsilon_{cu}E_{ya}\delta_w(h-a) + \varepsilon_{cu}E_{ya}A_{ay} - N$

$C = -\varepsilon_{cu}E_sA_s + \varepsilon_{cu}E_{ya}\delta_w(h-a)^2/2 + \varepsilon_{cu}E_{ya}A_{ay}(h-a)$

After taking moment to tensile bar, obtain:

$$N_e \le f_{cm}bX(h_0 - X/2) + f_y'A'_s(h_0 - a'_s) + f_{ya}'A'_{ay}(h_0 - a') + f_{yw}\delta_w(X_0 - a')[h_0 - (a' + X)/2 - a'] - y'_s f_{yw}\delta_w(h_0 - X_0 + y'_s/3)/2 - \sigma_a y_s(y_s/3 + a - a_s) - \sigma_a A_{ay}(a - a_s) \quad (7)$$

3) Applicable condition

(1) $\zeta > \zeta_b = 0.404$ (2) $\sigma_s \le f_y$; $\sigma_a \le f_{ya}$

From the above, the complete formulas that are analogous to that of normal steel concrete compression-flexure member is established to calculate bearing capacity of SRLAC compression-flexure member. The formulas are single and applicable, considering the norm in engineering shield. The conceptions and meanings of the parameters are clear in the formulas.

3 Conclusions

In this paper, the authors put forward calculating formulas about bearing capacity of SRLAC compression-flexure member on the base of traditional design method. Either computer or handwork can finish the calculation using the formulas. Through calculating, the maximal error of the ratio of calculated values to measured value is within 10%.

ANALYTICAL MODEL FOR PULLOUT RESPONSE OF INCLUSION REINFORCEMENT

YUE CAI[1)], YUJING JIANG[2)], TETSURO ESAKI [1)]

1) *Environmental Geotechnology Division, Institute of Environmental Systems, Kyushu University*
6-10-1 Fukuoka, Higashiku, 812-8581, Japan
e-mail: cai@ies.kyushu-u.ac.jp

2) *Department of Civil Engineering, Nagasaki University*
Nagasaki 852-8521, Japan
e-mail: jiang@civil.nagasaki-u.ac.jp

Although inclusion reinforcement systems are widely used in civil engineering, the coupling mechanism between reinforcement and rock/soil mass is not clear at present. Properties of rock/soil mass plays an important role in reinforcement system whereas it is neglected by many models. A normal analytical model for reinforcement has been developed based on refined Shear-Lag model. The model describes the mechanical coupling at the interface between reinforcement and rock/soil mass. On the basis of proposed model, coupling and decoupling behavior of reinforcement in pullout test is analyzed, and a new method is suggested to evaluate actual shear strength of interface between reinforcement and rock/soil mass. According to the model, average shear force, which is influenced by the anchor length in matrix, is not the same as actual shear strength. The difference is controlled by not only the embed length of reinforcement but also the deformation modulus of matrix material. Interaction influence of anchor length and matrix's deformation modulus is often ignored in standard pullout test, which may leads to big deviation. Rational anchor length is suggested for matrix with different deformation modulus in pullout test. Parametric studies are carried out for a range of deformation stiffness values. Suitable deformation of matrix material in laboratory pullout test is suggested for reinforcement installed in soft rock. The values calculated with proposed model are compared with other models and test data.

1 Introduction

Inclusion reinforcements such as rock bolts, rock anchors, soil nails, grids metal straps, geosynthetics and polymers etc., have been widely used in civil and mining engineering. The performance of the reinforcement system depends on the efficiency of shear stress transfer mechanism. Pullout test is often used to check the strength, integrity and effectiveness of the reinforcement system. Despite of extensive researches both experimentally and theoretically on rock/soil, only little has been understood on the coupling behaviors of reinforced system. According to past studies, interface properties between bolt and matrix are focused on, and relationship of shear force τ and displacement u are often simplified as $\tau = ku$. In order to satisfy the pullout experiment result, k is idealized by bilinear or hyperbolic idealization [1,2]. In most cases, tested shear stiffness k is not only the properties of interface, but also the properties of reinforcement and the surrounding matrix especially for extensible ones. At the same time, it is hard to define interface because the distance between reinforcement and matrix is infinitesimal. Shear-lag model (SLM) or fiber-loading theory is originally proposed by Cox [2], it has been widely used by material scientists and structural geologists as a powerful analytical method[4,5,6,7]. Abramento and Whittle proposed a model based on SLM for analysis pullout behavior of planar geosynthetic reinforcement assuming that shear stress decayed linearly [5]. However, his assumption cannot stand since the actual shear stress decays non-linely in the matrix according to the study of Mandal et al.[6]. Shear stress distribution in matrix is often assumed as Eq. (1) in order to get the parameters in the model [7,8].

$$\tau(r,x) = r_b \tau(r_b,x)/r \qquad (r_b \le r \le R) \tag{1}$$

However, anti-pressure from reinforcement to matrix tends to be zero according to Eq. (1). The equilibrium of reinforcement is usually focused on while that of matrix is often neglected in some models. Traditional SLM is improved and brief theoretical summary is presented in this paper. Coupling and decoupling behavior of rock reinforcement system in pullout test is discussed on basis of the model. Comparison and evaluation of some experimental pullout test are presented

2 interaction model for bolt and matrix

Before slippage, reinforcement and surrounding matrix work compatibly. Coupling behavior of pullout test is sketched in Fig.1.

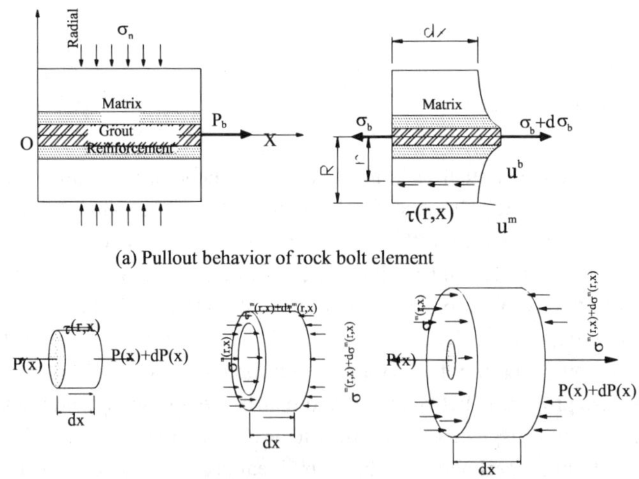

(a) Pullout behavior of rock bolt element

(b) Element balance of rock bolt, rock mass and composite

Figure1. Coupling behavior of reinforcement and Matrix

where, P_o is pullout load at the end of reinforcement. P(x) is pullout force at the position of x; R is the influence radial; σ_n is the normal stress perpendicular to reinforcement; $\tau(r,x)$ represents the shear stress at (r,x). According to the balance of infinitesimal element of reinforcement, surrounding matrix and matrix together with reinforcement in cylinder coordinate system, Eq. (2) is established.

$$\frac{dP(x)}{dx} = -2\pi r_b \tau_b \qquad (2a)$$

$$\frac{\partial \sigma_m(r,x)}{\partial x} + \frac{\partial \tau(r,x)}{\partial r} + \frac{\tau(r,x)}{r} = 0 \qquad (2b)$$

$$\frac{dP(x)}{dx} + 2\pi \int_{r_b}^{R} r \frac{d}{dx} \sigma_m(r,x) dr = 0 \qquad (2c)$$

It is assumed that

$$\frac{dP(x)}{dx} = H(u_b - u_m) \qquad (3)$$

where, u_b is displacement of reinforcement; where u_m and ε_m are the displacement and strain of Matrix at the edge of influence radial R; H is material parameter. Different stress distribution results in different model and parameters. In order to simplify the analysis and satisfy boundary condition, uniform stress distribution is assumed in the influence area of reinforcement. Properties of grout is also assume as the same with that of matrix. If R>>rb, then

$$H = 2\pi G_m / (\ln(R/r_b) - 1/2) \qquad (4a)$$

If the properties of grout is not the same as rock mass, neglecting the normal stress on grout paralleled to bolt, parameter H is expressed as followings:

$$H = \frac{2\pi G_g G_m}{[\ln(R/r_b) - 1/2]G_g + \ln(r_g/r_b)G_m} \qquad (4b)$$

where, r_g is the radial of borehole, Gg is the shear modulus of grout mortar. Decoupling behavior of the system depends on the type of reinforcement. According to Windsor's classification [9], reinforcement system is comprised four principle components such as matrix, reinforcing element, internal fixture and external fixture. Reinforcing element refers to reinforcement and external fixture refers to face plate and nut. The internal fixture is either a medium, such as cement or resin mortar for grouted reinforcement, or a mechanical action like friction at the reinforcement interface for frictionally coupled bolts. Therefore, system can be divided into three groups such as: (1) Continuously Mechanically Coupled (CMC) System; (2) Continuously Frictionally Coupled (CFC) System; (3) Discretely Mechanically or Frictionally Coupled (DMFC) System. The effect of external fixture will not be discussed here. Shear strength at interface is made up of three parts such as adhesion, inter lock and friction in axial direction. They are lost in sequence as the compatible of deformation is lost along the interface. For CMC system, the three components all exist while for CFC system, adhesion part is not considered. After decoupling, shear stress at interface becomes residual strength at slipping part. As a result, reinforcement behaviors differently for the difference of residual shear stress.

3 Reinforcements in pullout test

Theoretical analysis with improved SLM

Based on the suggested model, pullout behavior of reinforcement in laboratory is discussed. If uniform stress is assumed, matrix strain at the edge of specimen $\varepsilon_m = -P/(A_m E_m)$. According to Eq. (2), the relationship between pullout force and displacement at the end before decoupling is express as

$$P_o = k \cdot u_o \tag{5}$$
$$k = \alpha E_b A_b \tanh(\alpha L)$$
$$\alpha = \sqrt{H\left(1/(A_b E_b) + 1/(E_b A_b)\right)}$$

where, Po is axial load at loading end respectively. E_b and A_b are Young's modulus and cross section area of reinforcement. If the length is long enough, pullout force along reinforcement can be simplified as the formulas suggested by Farmer IW [8]. The total pull out force may not be max if the residual stress exists at debonding section. The ultimate pull out force after debonding is expressed as:

$$P_{\max} = 2\tau_{mo}\pi y r_b + 2\tau_m \tanh(\alpha(L-y))A_b/(\alpha r_b) \tag{6}$$

For friction types such as Swellex bolt, the residual stress equals the shear strength of interface. Therefore, debonding length y=L when it reaches it ultimate pullout load. Shear strength of friction bolt is calculated with Eq. (7).

$$\tau_{mo} = P_{\max}/2\pi r_b L \tag{7}$$

For reinforcement in situ, since it is loaded by the deformation of matrix, debonding reinforcements' behavior differs from those in laboratory.

Experiment verification

The pullout test performed in Holland Saka Tunnel in Nakasaki is taken as an example for analysis. It is known that: ultimate load P_{max}=100kN, length of rock bolt L=3.0m, radial of rock bolt r_b=10mm, Young's modulus of rock bolt E_b=210GPa, deformation modulus of rock mass E_m is assumed between the arrangement from 0.1GPa~0.5GPa. Residual shear strength is not considered and influence radial R=1.5m. Result of pullout force and displacement at pulling end is shown in Fig.5.

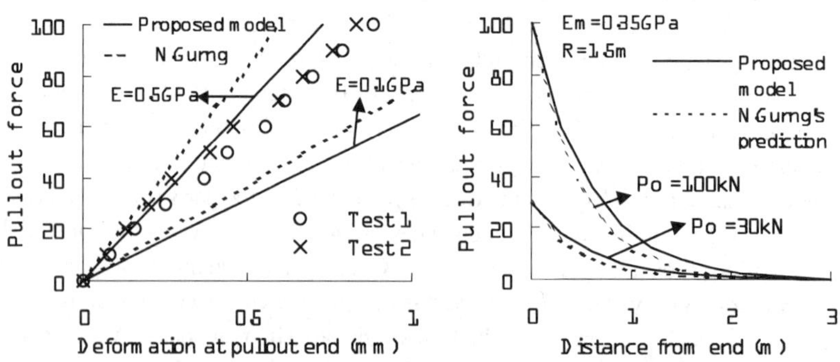

Figure 2. Comparison of pullout response (Holland Saka slope Tunnel in Nakasaki city, Japan)

When E_m=0.35GPa, theoretical result agrees with test data according to this model. Actual shear strength is calculated with Eq. (7) by back-analysis method and it equals 2.77MPa. Theoretical predictions are compared with that of N.Gurng's model. The comparisons clearly illustrate the application of the model. Deformations of rock bolts in pullout tests are observed to be linear with simplified equation. Predicted displacement is a little larger than that of N.Gurng's model according to the same parameter. Theoretical pullout force along reinforcement is almost the same as N.Gurng's prediction when pullout force is 30kN, while it is a little larger when pullout force is 100kN.

4 Average shear strength and actual shear strength in pullout test

Average ultimate shear stress in pullout test is often used to represent the shear strength of interface. However, the shear stress is not uniform along reinforcement especial when reinforcement is relatively long. Fig. 3 shows the shear stress distribution along reinforcement before and after decoupling. Obviously, it is suitable to use average stress as interface strength for frictional reinforcement but unsuitable for grouted reinforcement. Fig.4 shows the difference for different deformation modulus of specimen with different anchor length.

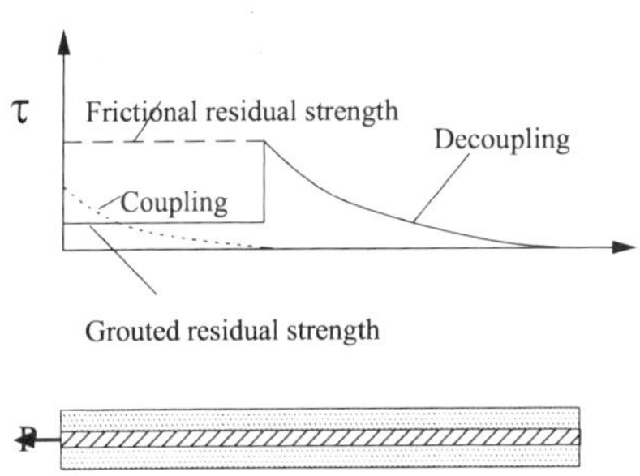

Figure 3. Shear stress distribution

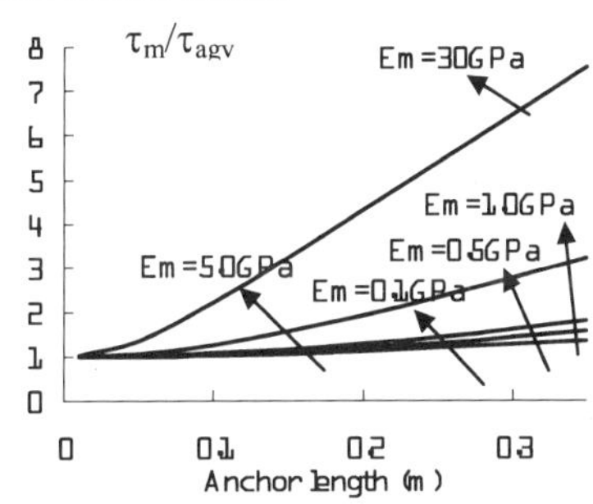

Figure4. Difference of average and actual shear

Actual shear strength is larger than average strength. When anchor length is relatively short, it is near to average strength. The difference increases together with the increasing of anchor length. At the same time, larger deformation modulus of specimen leads to more difference. According to JHS (Japanese Highway Standard), the anchor length of rock bolt is 0.35m and concrete is often used to simulate rock mass. Deformation modulus of common concrete is about 30GPa. Hence, actual shear strength is 7.5 times as average strength in pullout test. Since deformation modulus of soft rock mass is under 2.0GPa, specimen with low deformation modulus is recommended for pullout test in laboratory. Otherwise, back-analysis method should be adopted in order to get the exact shear strength of interface.

5 Conclusion

An analytical model for reinforcement system has been proposed on the basis of improved SLM. Coupling and decoupling behaviors of reinforcement system are discussed. Physical properties of matrix influence the distribution of axial load and shear stress at interface. Simplified formula shows that shear stress decreases exponentially along reinforcement before decoupling. Pullout load may not be max if residual stress exists. Theoretical prediction of rock bolt in soft rock mass agrees well with test result with certain parameters of rock mass. Predicted displacement is a little larger than that of N.Gurng's model according to the same parameters. Generally, average shear stress at interface is lower than actual strength for grouted reinforcement. The difference is not significant for fully frictionally coupled reinforcement while it is large for grouted types especially when the deformation modulus of specimen is relatively large. Specimen with large deformation modulus such as concrete is not recommended for pullout test of soft rock mass because it will leads to large difference between actual strength and average strength.

Reference

1. N. Gurung, 1-D analytical solution for extensible and inextensible soil/rock reinforcement in pull-out tests, Geotextiles and Geomembranes 19 (2001); pp195-212.
2. Cox, H.L., 1952, The elasticity and strength of paper and other fibrous materials. Br. J. Appl. Phys. 3, 72-79.
3. Holister, G.S. et al , 1966, Fiber Reinforced Materials, Elsevier, Amsterdam, 154 pp.
4. Clyne, T.W., Withers, P.J., 1993, An introduction to metal matrix composites, Cambridge University Press, 509 pp.
5. Abramento, M. and Whittle, J.A., Analysis of pullout tests for planar reinforcements in soil, Journal of Geotechnical Engineering, 1995,Vol. 121, No.6, pp.476-485.
6. Mandal, N., Deb, S.K., Khan, D., 1994, Evidence for a non-linear relation ship between fracture spacing and layer thickness. J. Struct. Geol. 16, 1275-1281.
7. Ji, S., Zhao, P., 1994, Strength of two-phase rocks: a model based on fiber-loading theory. J. Struct, Geol. 16, 253-262.
8. Farmer, I.W. 1975, Stress distribution along a resin grouted anchor, Int.J. Rock Mech. Min. Sci. & Geomech. Abstr. 12, 347-52.

6. Mining Practice, Rock Blasting and Rock Burst

THE FRACTAL RESEARCH ON FROZEN SAND FRAGMENTATION OF INSTANTANEOUS AND MILLISECOND IN HOLE BLASTING *

Qinyong Ma[1,2] Meifeng Cai[2]
1 Anhui University of Science and Technology, Huainan 232001, Anhui, CHINA
2 University of Science and Technology Beijing, Beijing 100083, CHINA

Processing an instantaneous and millisecond blasting crater test, collecting the blasting fragment to statistical analysis by griddle, according to the fractal theory, calculating the fractal dimension of the frozen sand blasting fragment. The fractal dimension of instantaneous blasting is 0.67~2.54, that of the millisecond is 1.68~2.37. The result shows that the fractal dimension indicates the distributing condition of the blasting fragment in quantity. At last, discussing the relationship between the fractal dimension and the least fragment ratio, the charging depth in the blasting crater, they have very good relativity. The more the least fragment is, the greater the fractal dimension is. Contrarily, the fractal dimensions become small. The less fractal dimension is corresponding to the best charging depth and the greatest blasting crater volume.

1 Introduction

The fractal theory is becoming an important mathematics way to study some things that is the ubiquitous and anomaly and fragmented, but having comparability in the statistical sense. Now, the fractal theory acts important action in studying the rock damage, fracture and the crack bifurcating. The scholars inside and outside of the nation gained the delectable progress in the studying of the rock granule and the frozen sand granule.

The blasting fragment is an important index to evaluate the blasting effect. According to the need of the engineering, some request to reduce the big fragment, some request to reduce the little fragment possibly, the others request to increase the output ratio of a certain average size. For the frozen sand blasting engineering, the bigger fragment should be decreased possibly to improve the loading effect. The aim of controlling the blasting fragment is to get the good blasting effect. The direct measure way--the griddle analysis way is used to measure the blasting fragment. The distributing condition of the blasting fragment is un-order character. In despite of the shape and the size is different, it satisfies the resemble disciplinarian in the statistical meaning. The relationship between the fractal dimension and the frozen sand blasting fragment ratio, the charging depth and the volume of the blasting crater is analyzed.

* The paper is supported by Outstanding Youth Foundation for Scientific and Technological Research of Anhui Province of China (№ 2001-28).

2 The fundamental of the fractal dimension calculation

According to the fractal theory:

$$N_i = C\, r_i^{-D} \tag{1}$$

Where N_i = the fragment number of the character size r_i , *C*= the constant of the material, *D*= the fractal dimension of the fragment distribution.
Seeking the derivative of the formula (1),

$$dN \propto r^{-D-1} dr \tag{2}$$

The increasing amount of the fragment is correlative to the mass of itself, that is,

$$dN \propto r^{-3} dM \tag{3}$$

Supposing, M(r) being the cumulating fragment mass of the diameter less than r, M_T being the overall mass, R being a quantum which is correlative to the average measure, then the size of the fragment and the frequency distributing is,

$$M(r)/M_T = 1 - \exp[-(r/R)^b] \tag{4}$$

In fact, $M(r)/M_T$ is the accumulating percent when the diameter of the fragment is less than *r*. If $(r/R) \ll 1$ exists, the formula (4) can change into,

$$M(r)/M_T = (r/R)^b \tag{5}$$

Seeking the derivative of the formula (5), then,

$$dM \propto r^{b-1} dr \tag{6}$$

Joining the formula (2) and (6) into (3), then,

$$r^{-D-1} \propto r^{-3} r^{b-1} = r^{b-4} \tag{7}$$

So, we can obtain,

$$D = 3 - b \tag{8}$$

Where *b*=the slope value of the logarithm coordinates of *r* and $M(r)/M_T$. So the formula (8) is convenient to calculate the fractal dimension of the blasting fragment distribution.

3 Test of frozen sand blasting crater

3.1 Model material and similar explosive

This test selects the frozen sand of Huinan xieqiao mine of China, its density is 2050kg/m^3, containing water ratio is 16%. The temperature of the test is confirmed at -17℃. According to the investigation in the local, the №2 rock ammonium nitrate explosive usually is used. The explosive can work under the condition that the temperature of the frozen wall is -18℃ and the atmosphere temperature at the work face inside the shaft is -5℃. According to the similar theory, DDNP is selected as the test explosive. Detonator is manufactured especially which is explosive package, and it is the component to rise blasting. Charging quantity is 4.8g. Charging density is 1050kg/m^3.

3.2 Model test of blasting crater

According to the Liveston blasting crater theory, the method of changing the charging depth and not changing the explosive quantity is applied to a series of crater test instantaneous blasting in single hole. The result of the experiment shows as Table 1.

Table 1. The blasting crater result of instantaneous blasting

Depth of L/mm	Parameter of crater			Piece composing /g		
	Depth H	Radius l	Volume	25-40mm	10-25mm	0-10mm
60	16	30	22	0	0	32
65	21	36	28	7	11	20
70	28	42	48	13	52	43
73	28	48	65	47	40	24
75	34	52	87	24	108	52
80	32	55	66	16	77	44
85	28	40	40	47	8	2
90	14	16	20	0	42	0
96	Sample break into four pieces					
100	A creak between the sample					

Table 2. The test result of the millisecond blasting in frozen blasting hole

Depth o L/m	Subsectic	Parameter of crater			Piece composing /g		
		Depth	Radius	Volume	25-40m	10-25m	0-10m
70	1.3	2	42	38	0	46	27
80	1.3	2	44	43	10	48	27
85	1.3	3	48	66	42	51	33
90	1.3	3	51	88	62	67	30
95	1.3	4	60	13	128	71	37
10	1.3	3	55	11	83	79	42
10	1.3	2	26	27	5	28	25
11	1.3	Sample break down while no crater generated					

Two manufactured especially detonators of millisecond blasting crater test encased in the hole. The quantity of charging explosive is 0.24g. The upper is first segment and the lower is departed into second segments, stemming mud between them. According to the experience formula and experiment, the millisecond time is confirmed 20ms. The result is shown at Table 2.

4 Calculating and analyzing the fractal dimension of the frozen sand blasting fragment

According to the method of least two multiplication's, regressing and calculating the relation between Log(r) and Log[$M(r)/M_T$]. The slope ratio of the line is the distributing index b of the size and the frequency. According to formula (8), calculating the fractal dimension D, which is shown as Table 3, Table 4.

Table 3. Fractal dimension of the fragment distribution of the instantaneous blasting

Depth of hole L/mm	Percent of the least fr r/%	Fractal nun	Correlation coef	Checkout
80	52.6	2.534	0.999	1569.93
85	39.8	2.308	0.978	21.710
90	20.5	1.899	0.999	2025.45
95	28.3	2.046	0.972	16.831
100	32.1	2.141	0.971	16.722
105	3.5	0.672	0.979	22.729

Table 4. Fractal dimension of the fragment distribution of the millisecond blasting in hole

Depth of L/mm	Percent of the le fragment r/%	Fractal nun	Correlation coef	Checkout v
60	37.0	2.232	0.943	8.001
65	31.8	2.132	0.971	16.692
70	26.2	2.026	0.999	605.384
73	18.9	1.788	0.999	456.628
75	15.7	1.687	0.996	113.312
80	20.6	1.858	0.999	12378.08
85	43.1	2.364	0.971	16.358

According to Table 3, 4 knowing that the relationship between Log(r) and Log[$M(r)/M_T$] is close to the linearity and the correlation coefficients are above 0.94. The fractal dimension of the instantaneous blasting is 0.67~2.54 and that of the millisecond blasting is 1.68~2.37. The greater the percent of the least fragment is, the bigger the fractal dimension is. Otherwise, the fractal dimension is smaller. The less fractal dimension is usually corresponding to the superior charging depth and the largest volume of the blasting crater. So the fractal dimension can indicate the distribution condition of blasting fragment in quantity and it can be used as a parameter to depict the blasting fragment of frozen sand.

According to the data of the fractal dimension D and the percent of the least fragment r in Table 3, 4, regressing the relation equation.

The instantaneous blasting:

$$D=0.942\text{Ln}(r)+0.901, \quad R=0.952 \tag{9}$$

The millisecond blasting in hole:

$$D=0.114r+1.557, \quad R=0.997 \tag{10}$$

From the data we can know that the relativity between the fractal dimension and the percent of the least fragment of the millisecond is better than that of the instantaneous blasting.

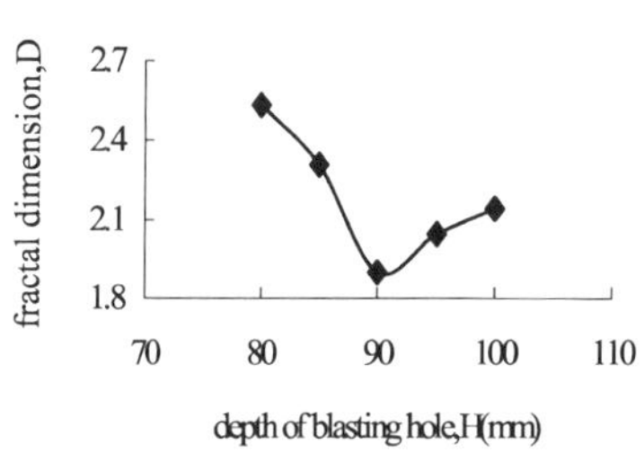

Figure 1. Relationship between depth of the blasting hole and fractal dimension of instantaneous blasting

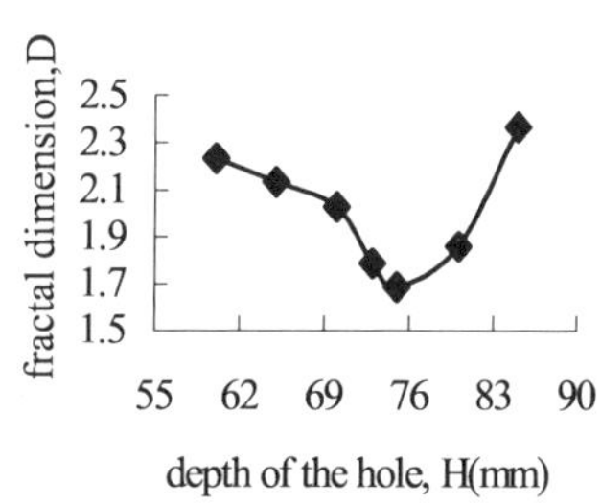

Figure 2. Relationship between depth of the blasting hole and fractal dimension of millisecond blasting

Figure 1,2 are the connection charts of the fractal dimension and the depth of the blasting hole of the instantaneous and millisecond blasting. From the charts knowing that the least value of the fractal dimension is corresponding to the best depth of the hole, which is the depth of the largest crater volume.

5 Conclusions

According to the fractal theory, the fractal dimensions of the frozen sand blasting fragment are calculated which of the instantaneous blasting is 0.67 ~2.54 and the millisecond blasting is 1.68 ~2.37. The relation equation between the fractal dimension and the least fragment percent is regressed. We can know that the relativity between the fractal dimension and the percent of the least fragment of the millisecond is better than that of the instantaneous blasting. The greater the percent of the least fragment is, the bigger the fractal dimension is. On the contrary the fractal dimension is smaller. The charts of the relation between the fractal dimension and the depth of hole of the instantaneous and the millisecond blasting are drawn. We can know that the least fractal dimension is usually corresponding to the best depth of the hole, the depth of the largest volume of the blasting crater namely. So the fractal dimension indicated the condition of the fragment of the frozen sand blasting in quantity and it can be used as a parameter to depict the blasting fragment. From the above mentioned, it can be shown that the blasting fragment of the frozen sand have good fractal structure. The fractal dimension figured the distributing condition of the blasting fragment. So involving the fractal dimension into the frozen sand blasting fragment will act an important action to forecast the fragment and optimize the blasting parameters further.

References

1. Jizhong Zhang, *The fractal, Tsinghua University Press*, 113(2001).
2. Heping Xie, *Fractals in Rock Mechanics, Science Press*, 240-241(1996).
3. Qinyong Ma, *Model test investigations of blasting crater and blastability of frozen soils. Coal Journal of science and technology*, 22(3), 288-293(1997).
4. Qinyong Ma, *Strength and Wave Velocity Test on Artificially Frozen Soils, Journal of Coal Science & Engineering*, 4(2), 58-63(1998).

STUDY ON THE STABILITY OF HIGHWALL IN PUNCH LONGWALL MINING

HAIXIAO CAO, HIDEKI SHIMADA, SHIRO KUBOTA, KIKUO MATSUI
Dept. of Earth Resources and Mining Engineering, Graduate School of Engineering, Kyushu University, 6-10-1 Hakozaki, Higashi-ku, Fukuoka, Japan
E-mail: cao@mine.kyushu-u.ac.jp

Punch longwall mining is a safe, economical and efficient Highwall mining system in which Longwall mining is employed. The stability of Highwall is a extremely important subject, while Punch longwall mining is used in practical. In this study,use 2D finite element numerical analysis to assess the stability of Highwall in Punch longwall mining; with the parameter is the distance from face area to Highwall, the angle of the coal seam and the stress ratio.

1 Introduction

Final highwalls of opencut coal mines can form the starting point for other mining methods, such as highwall or underground mining. In its basic application, highwall mining is a technique utilized after the opencut portion of a reserve has been mined, sometimes prior to the introduction of underground mining.

However, a large amount of coal tends to remain isolated and undeveloped as pillars due to previous mining operations performed by the use of the auger machine or the continuous miner. Major issues of the highwall mining system are less coal recovery and highwall instability due to pillar and roof failures. To overcome these issues, an introduction of the Punch longwall mining system is one of solutions [1].

Punch longwall mining is a mining system that longwall mining was introduced into Highwall mining so as to safely, economically and efficiently extract the coal, which cannot receive large amount be left as pillar in normal highwall mining. Its basic concept is shown in Figure 1. [2].

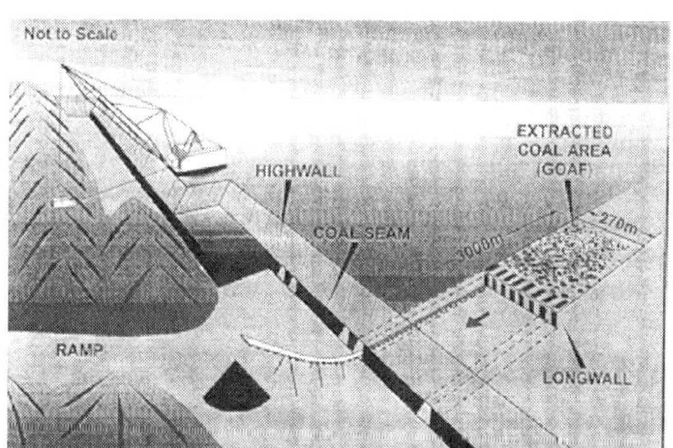

Figure 1: PUNCH LONGWALL MINING

When the punch longwall mining system is applied, the problem of the stability of highwall becomes a very important subject. It is essential to design an optimal highwall, which ensures the stability of highwall and increased to the biggest coal extraction income ratio in practical.

2 Analytic Condition

The analytic model is shown in the fig.2, 3. In model 1, siltstone seam of the thickness 3m exists under the coal seam with the thickness 3m,and siltstone seam of the thickness 4m exists above the coal seam, moreover it supposed sandstone to exist in siltstone seam up-down part. The analytic range is 200m on the right and left sides of the highwall bottom, under the entry floor downward 200m, and upward depth 100m from the surface. Moreover, highwall angle was 80, the stress ratio was 1.0, and it extracted 5m in the step from the back part 100m of highwall to 20m.

Furthermore, only the stress ratio was made to change in 1.0, 1.5, 2.0, 2.5 under this analytic condition for examine the stress ratio. The analytic model 2 is different to the analytic model 1 that the angle of the coal seam was made to change 5°,10°,15°, 20°.

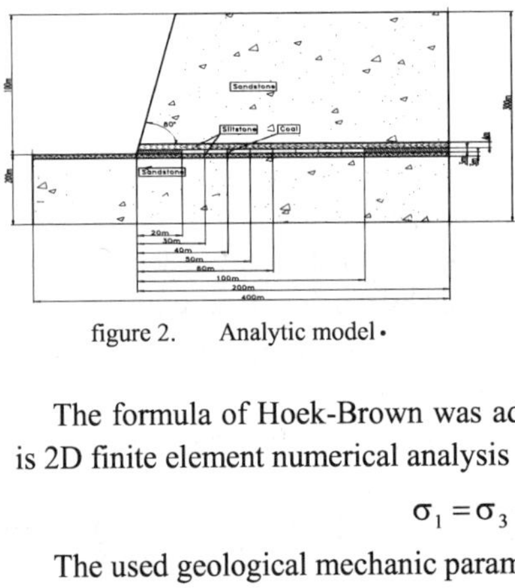

figure 2. Analytic model •

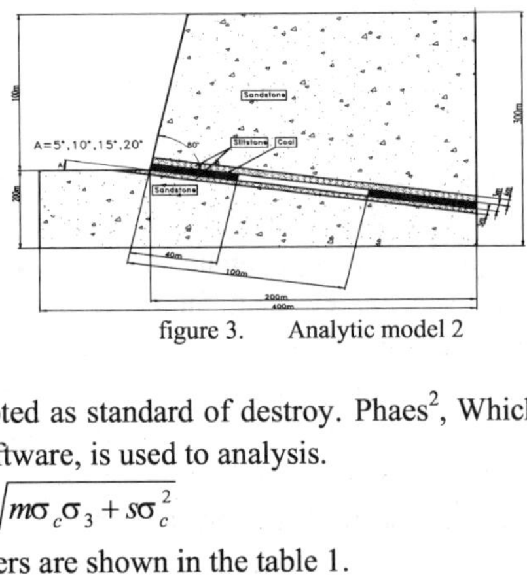

figure 3. Analytic model 2

The formula of Hoek-Brown was adopted as standard of destroy. Phaes[2], Which is 2D finite element numerical analysis software, is used to analysis.

$$\sigma_1 = \sigma_3 + \sqrt{m\sigma_c\sigma_3 + s\sigma_c^2}$$

The used geological mechanic parameters are shown in the table 1.

Table1. The parameter of geological mechanic

	Young's Modulus (MPa)	Poisson's Ratio	Compressive Strength (MPa)	m	s	Unit Weight (MN/ •)
Sandstone	5700	0.28	26.6	7.5	0.1	0.0243
Siltstone	4650	0.27	20.9	2.0	0.02	0.0247
Coal	1530	0.30	11.9	2.08	0.07	0.0134

3 Analysis Result and Consideration

First, it is examined about the influence of the distance from the longwall face area to highwall on the stability of highwall. The change of the destroy area by the difference in the distance from face area to highwall is shown in the figure 4. It is understood from this figure that destroy progresses in the roof and floor of the entry while face area approaches highwall. Especially, the tensile failure is remarkable in roof. In other words, In the case of the distance from highwall to face area is more than 40m, since destroy of the highwall bottom and the roof doesn't reach to highwall; it hardly influences the stability of highwall due to the existence of face area. On the other hand, when the distance from highwall to face area is within 40m, destroy of the highwall bottom and the roof increases rapidly, and for that it has the possibility that the stability of highwall is spoilt.

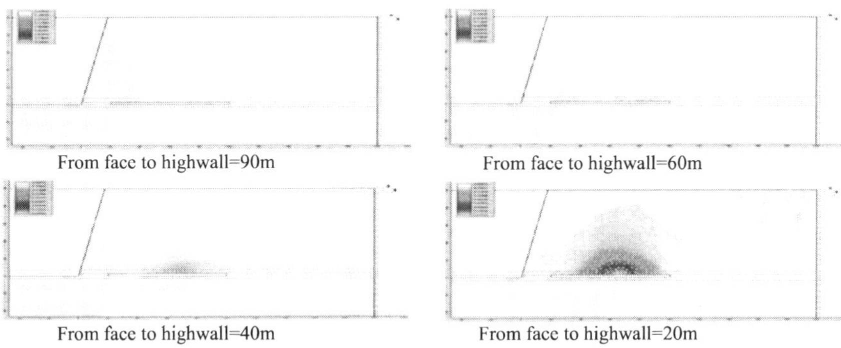

Figure 4. The change of destroy area with the difference in a distance from face area to Highwall

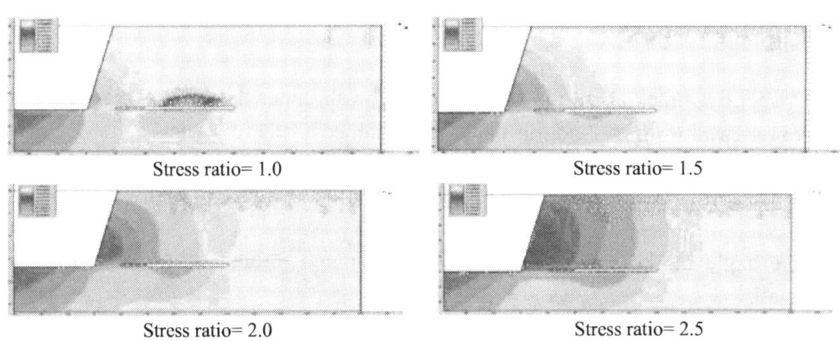

Figure 5. The change of the destroy area with the difference angle of coal sea

Next, it is examined about the influence of the stress ratio. The distance from face area to highwall is 40m, and the changes of destroy area that affected by the stress ratio is shown in the figure 5. From this figure we think that destroys areas decrease in the roof and floor of the entry, while the stress ratio increases. Nevertheless, destroy of the highwall bottom and highwall surface increases, and destroy of highwall increases too. As the stress ratio increase, destroy of the surface of the earth expands more by exploitation, and it can think the possibility of the highwall collapse rises.

Finally, we examine about the influence with angle of the coal seam on the stability of highwall. The change of destroy area by the difference in the angle of coal seam is shown in the figure 6. It is understood from this figure, as the angles of the coal seam increase, destroy decreases greatly in the root and floor of the entry. It is considered that the angle of the coal seam doesn't influence the stability of highwall.

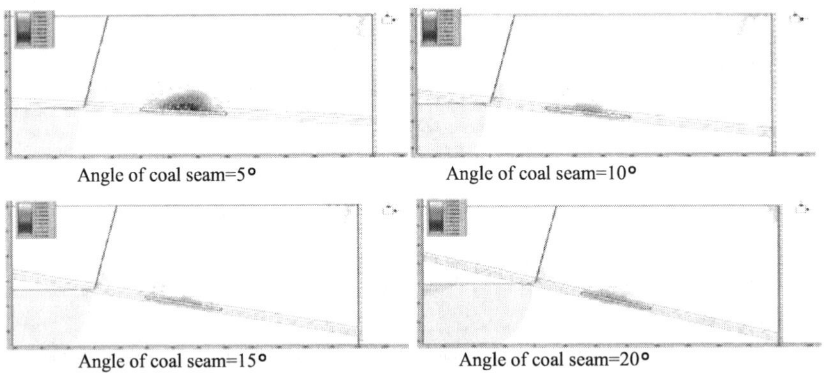

Figure 6. The change of the destroy area with the difference angle of coal

Conclusions

This paper examined about the stability of highwall under the various conditions of Punch longwall Mining. And this paper is concluded as the following.

1) Highwall becomes unstable when the face area of longwall approaches highwall.

2) As the stress ratio increase, the possibility that the collapse of highwall is caused rises.

3) The angle of the coal seam doesn't make much influence in the stability of highwall.

References

1) Matsui et.al.: Proc. of 17th International Congress and Exhibition of Turkey , pp.269-276, 2001
2) Matsui et.al.: Proc. of 10th International Symposium on Mine Planning and Equipment Selection, pp.325-330, 2001
3) Matsui et.al.:11th Japan Symposium on Rock Mechanics, K02(CD-ROM), 2002

EXPLOSIVE MODELING IN SINGLE BOREHOLE BY USING AUTODYN™

XIANHONG MENG
314 WBB Building, 135S 1460E,SLC, UT 84112,USA
E-mail: mengxh@hotmail.com

WILLIAM A. HUSTRULID
318 WBB Building, 135S 1460E,SLC, UT 84112,USA
E-mail: hustrulid@mines.utah.edu

AUTODYN™ is used to simulate field blasting. Decoupled charges has a good blasting results than full coupled charges because of a long time pressure action time and second increase of pressure after the first peak. Both the first peak pressure and peak particle velocities, under the decoupled situation, decrease almost two times compared with these conditions without air decoupled charge. These definitely decrease the damage to the zones necessary for protection without injuring the blasting result by the increase of action time of pressure and particle velocity on the rock.

1 Introduction

In blasting field, knowing all of the blasting parameters is very important to study the explosive basic theory, evaluate blasting result and finally to aim at the option blasting design. AUTODYN™ provides the ideal method to model the blasting processing and predict all these parameters in a convenient and energy-saving way.

This paper simulates the blasting process of explosive in signal blasting hole underground. The diameter of the hole is 300mm. To short the calculating time of the computer, we use 4m as the borehole length, 3m as the borehole effective charge length. Load diameter of the hole is 150mm, the left 75mm thick from the charge to the wall of the hole is used as decoupled area. The figure 1 shows the structure

The linear question of state is chosen for the model by using Hooke's Law in which p=K μ, where $\mu = (\rho/\rho_0) - 1$, and K is the material bulk modulus. We choose Mohr-Coulomb mode as rock strength model by a piecewise linear variation of yield stress with pressure up to a value of maximum yield to define the material strength curve. Damage factor is used to reduce the elastic moduli and yield strength of the material as the calculation proceeds. Explosive ANFO is used and Jones - Wilkins - Lee form is chosen to be the EOS of the explosive.

Ideal gas equation of state is chosen for air and only the value of γ need be supplied by the user.

We design the explosive area and decoupled air, when air decoupled charge is used, as Euler sub-grid with flow-out boundary condition allowing the constant flow of material and the surrounding rock area as the Lagrange sub-grid with transmit boundary condition. Symmetric scheme is used. Node detonation is specified and put on the bottom of the borehole. Different monitor targets, which are shown on figure 2, are place on the two sub-grids to record all kinds of blasting parameters.

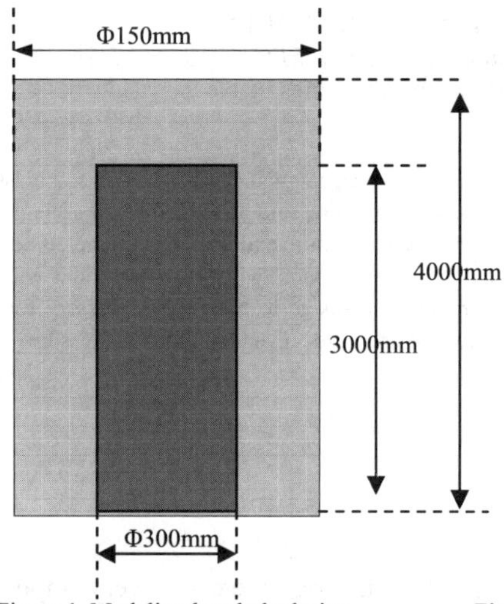

Figure 1. Modeling borehole design

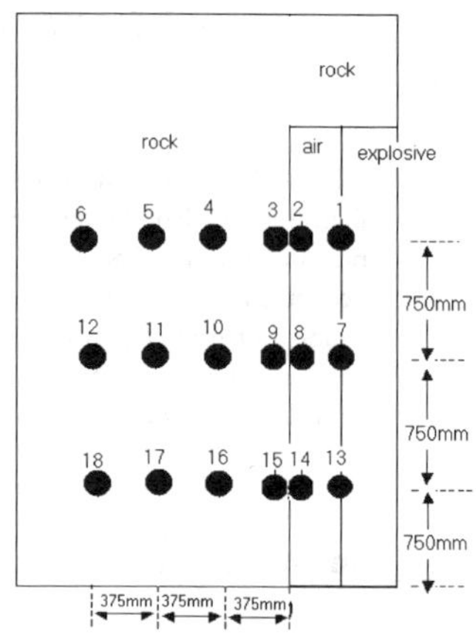

Figure 2. Targets map on the two sub-grids

2 Wave Propagation While Blasting

Different parameter field, such as pressure contour, absolute velocity contour, X velocity contour, Y velocity contour, stress contour, strain contour, temperature contour, have been inspected. The pressure contour and absolute velocity contour are shown in figure 3, from which we can find the Mach stem produced by the reflection of the shock wave, traveling in the annular space between the explosive and the rock, on the wall of the borehole. The modeling result shows the Mach stem propagates ahead of the detonation wave and compress the charge.

2 **Pressure-Time Curves**

The pressure-time curves of each targets are investigated and show that, with the increase of the distance from the bore hole, the pressure and its peak value decrease. On the same target point, the pressure peak under the air coupled charge condition is nearly half the value of that while there is no air de-coupled charge. This is advantage to decrease the blasting damage to blasting surround environment. For the target under air decoupled charge situation, such as target15 and 16, there are the second pressure peak, which is even bigger than the same target under the full charge situation, just after the pressure arrive the lowest point. The second pressure peak not only compensate the energy loss because of the air couple but also keep a long time for the rock to be compressed until broken. The pressure-time curves of the same position targets 15 under different charge type are show on figure 4. Pressure-time curve of target 16 are also shown on the same figure. The pressure values with air decoupled situation are consider reasonable and comparable to the measured and theoretical predicting formula: $\sigma = P_{max} * (R/b)^{-3.6}$, where P_{max} is the pressure peak value, b is the radius of the charge and R is the distance from the center of the charge (Hustrulid,W.).

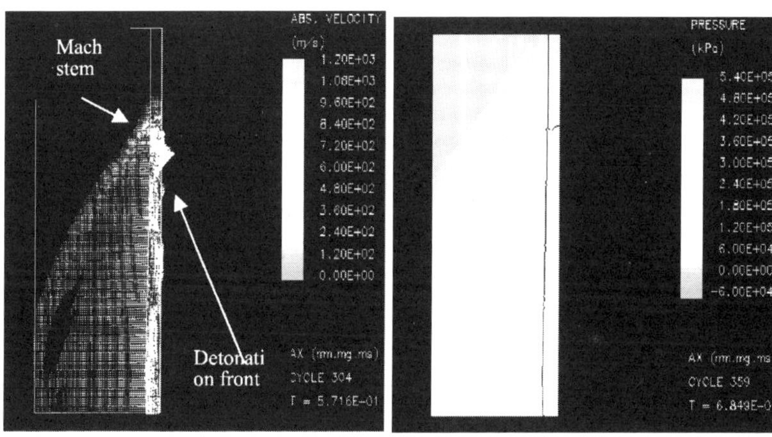

Figure 3. pressure contour and absolute velocity contour

4 Particle Velocity-Time Curves.

The velocity-time curves of each target are investigated and show that, with the increase of the distance from the borehole, the velocity and its peak value decrease. On the same target point, the velocity peak under the air decoupled charge condition is nearly half the value of that while there is no air de-coupled charge. For the target under air decoupled charge situation, such as target15 and 16, there are the second velocity peak-value just after the pressure arrive the lowest point. The velocity-time curves show the rock broken mechanics under the action of the air-coupled charge, which imply that the rock, while acted under air coupled charge, will break under a different velocity field.

The velocity-time curves of the same position targets 15 under different charge type are show on figure 5. Velocity-time curve of target 16 are also shown on the same figure.

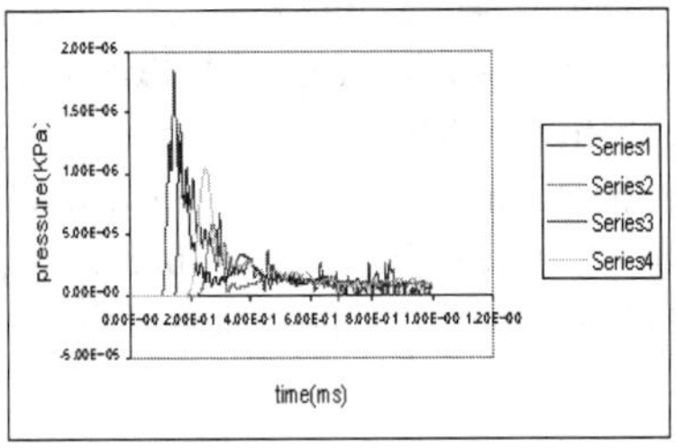

Figure 4. Pressure-time relationship.

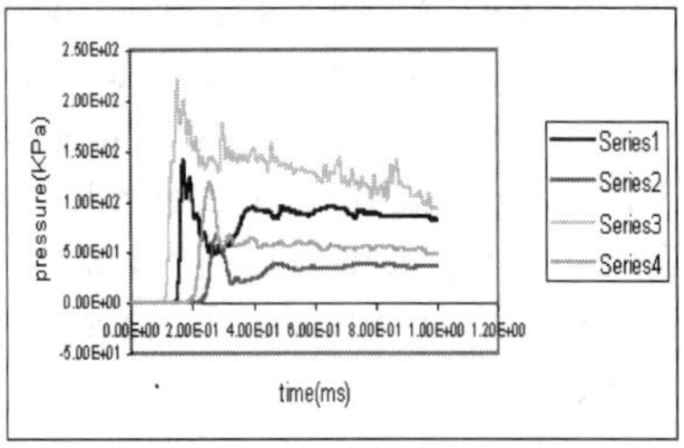

Figure 5. Velocity-time relationship.

In figure 4, series1 for target 15 with air decoupled charge, Series3 for target 15 without air decoupled charge, Series2 for target 16 with air decoupled charge, Series4 for target 15 without air decoupled charge.

In figure 5, series1 for target 15 with air decoupled charge, Series3 for target 15 without air decoupled charge, Series2 for target 16 with air decoupled charge, Series4 for target 15 without air decoupled charge.

5 Conclusion

The results of the numerical method proves the blasting theory that air decoupled charge has advantage by decreasing the blasting damage to surround environment through lowing the pressure peak value, improving blasting result through increasing action time of pressure. Blasting pressure and particle velocity are calculated directly and prove to be comparable to other papers concerning blasting.

References

1. AUTODYN™ 2001. Users's Manual. Century Dynamics Inc., San Ramon, California.
2. Hustrulid, W. 1999. Blasting Principles for Open Pit Mining. Volume 2. Theoretically Calculations. A.A. Balkema, Netherlands, pp. 703-716.

NUMERICAL SIMULATION STUDY ON THE PROCESS OF THE ROCKBURST IN DEEP ROADWAY

HAIHE JING[1,2] MANCHAO HE[2] SHUREN WANG[2]
[1]*Hei longjiang Institute of Science & Technology, Hei longjiang 158105, China*;
[2]*Beijing Campus, China University of Mining & Technology, Beijing 100083, China*

Rockburst is often encountered in the underground engineering of high geostress area, it is a kind of dynamic geodisaster. According to the phenomena of rockburst in deep roadway, dynamic calculation analysis of the rockburst in deep roadway are proceeded by program FLAC-Dynamic, depending on the problems of the rockburst problems in mining of deep coal mine of HUAFENG in ShanDong province. The calculation results indicate that the velocity curve and acceleration represent discontinuous change when the occurrence of the rockburst. The process of rockburst is the one that take place in the roof of the roadway first, then the both sides and the floor.

1 Introdution

Along with the depletion of resources in shallow coal bed, more and more coal mines in the world come to mining in deep[1-3]. The most deep coal mining level has reached to 1070m in china. In South Africa, the mining deep of gold mine had been to 3800m(D. F. Malam, 1998). According to the prediction[4], thirty percent of the gold mines in South Africa would be 3000m below the surface in 2010. With the increase in mining depth, geological environments are become more and more bad, underground stress level is much more high, and dynamic hazard, for example rockburst, is commonly takes place, all these would becomes a critical problem of the mine.

2 Geological Model

This study on the rockburst process is based on the problems of rockbursts in deep coal mining at HUAFENG coal mine in ShanDong province. In the exploitation of coal seam 4,The rockburst sometimes occurs in the roadway. Coal seam 4 is about 6.5m thickness in average. The bed is with tilt angle of 29°, substantial coefficient (*f*) of 1.5~2.0, density of 14.9kN/m^3, and uniaxial compression strength of 21.5MPa. The roof of the coal bed is siltstone with the thickness of 4~9m, the density of 27.5 kN/m^3, the uniaxial compression strength of 86.83MPa, and the tensile strength of 7.78MPa. The floor of the seam is moderate sand stone with the thickness of 40m, the density of 27.6 kN/m^3, the uniaxial compression strength of 78.4MPa, and the tensile strength of 7.08MPa. The parameter of a kind of the seam is shown in table 1.

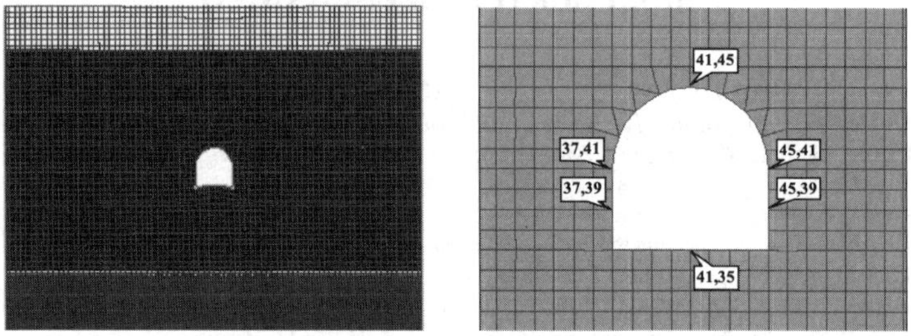

(a)coal and rock seam (b)identification points for analysis

Figure1. the model for calculation of rockburst process

Table 1 physical and mechanical parameters of the coal and rock

kindred	σ_t /MPa	R_c /MPa	E /GPa	μ	C /,MPa	φ/(°)	σ_b /,MPa
Coal	1.46	25.1	7.31	0.31	5.65	44.8	
siltstone	7.78	86.83	36.02	0.30	12.45	46.3	9.99
sandstone	7.08	78.4	30.02	0.24	16.42	45.0	22.25

3 Calculation Model

In order to make calculation simplify, the gravity of the rock seam is only in consideration. The roadway for calculation is 1000m below the surface. The model is with the area of 40m×40m, as shown in figure 1. The boundary condition of the model are that the bottom edges are fixed, the horizontal direction of both sides of the model are fixed, and the top of the model is free and subjected to the gravity of 25MPa. There are two steps for simulation, one is load to the model, and the other is the excavation of the roadway. In the numerical simulation for rockburst process, the program FLAC-Dynamic with Mohr Coulomb criteria to rock material strength is used. The parameters of the material strength can be seen in table 2.

4 Results and Analysis

Before the roadway excavation , the rock mass is in the state of high underground stress and high confining pressure. After the excavation of the roadway, however, the stress of the roadway perimeter comes about new distribution, which results in

Table 2. table of the material parameter for calculation

seam	C /MPa	φ /(°)	Bulk modulus /GPa	Shear modulus /GPa	Dilation angle /(°)	Elastic modulus /GPa	Poisson ratio
roof	12.45	46.3	30	13.85	30	36	0.3
coal	5.65	44.75	6.41	2.79	30	7.31	0.31
floor	16.42	46.7	18.87	12.15	30	30	0.235

the stress concentration at the perimeter of the roadway. At same times, a large number of energy is accumulated in the surrounding rock mass of the roadway, the confining pressure of the rock mass deduced quickly ,and the perimeter of the excavation comes on a high stress difference. Which may results in occurrence of rockbursts. As soon as rockbursts occur, the curve of the velocity and acceleration of the points(figure1) in the rockmass change into jump discontinuity, which can be graphically presented, as in Figure 2 and Figure3.

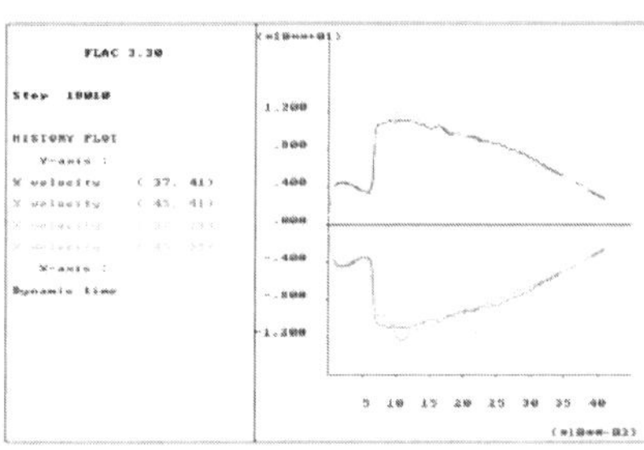

Figure2. horizontal velocity of the points

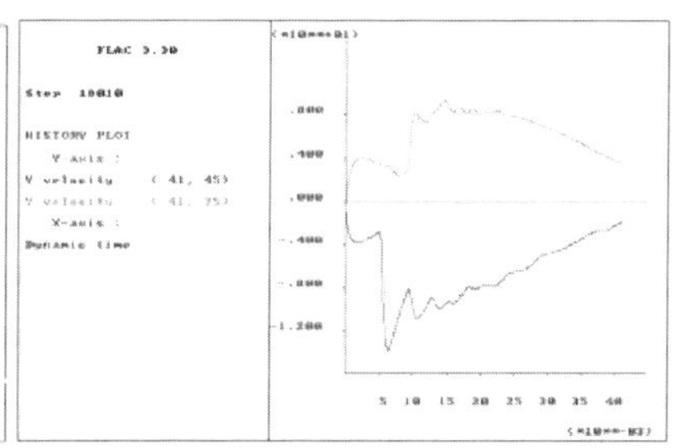

Figure3. vertical velocity of the points

From Figure2and Figure3, it can be seen that the velocity of identification point in Fig.1 bring about changes in sharply and inconsecutively when rockburst occurs. There are very big differences between the velocity of point (41,45) and point (41,35). The value of velocity of roof (point (41,45)) is much bigger than that of floor(point(41,35)), and the time of which the velocity curve of roof changes sharply and inconsecutively is a little earlier than that of the floor. There are also a little difference between both sides of the roadway in their velocity diagram. Although the identification point (37,41) and point (45,41) on the areas of base line of arch of the roadway appear suddenly change in velocity as same time as point (37,39) and (45,39) in the middle parts of the roadway, the velocity at the points of the middle parts of the roadway change wavily at a little behind the points of the base line of the arch. All these indicate that the rockburst probably begins from the base line of the arch, and then the sides and the floor.

From the diagram of the acceleration in Figure4 and Figure5, it can be seen that the acceleration of the point (41,45) on the roof occurs suddenly change at first, then the point (41,35) on the floor. The amplitude variation of acceleration at the roof is 4 times as large as that of the floor. These once more indicate that the rockburst probably begins from the base line of the arch, and then the sides and the floor.

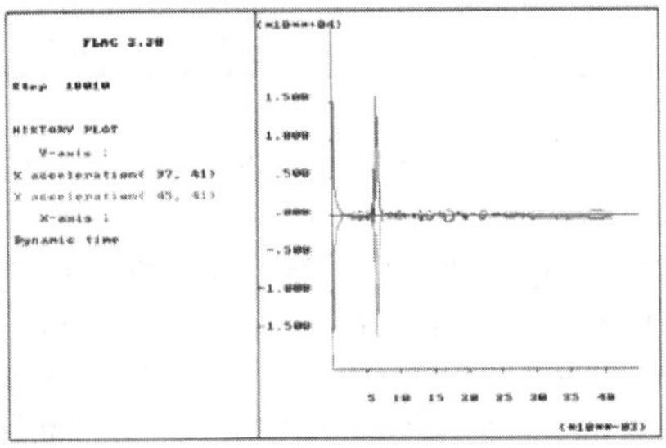

Figure4. acceleration in horizontal direction

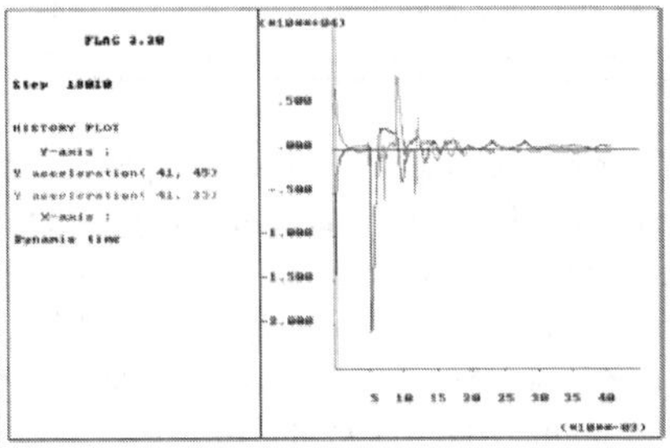

Figure5. acceleration in vertical direction

5 Conclusion

Through mechanical calculation analysis for the process of rockburst in the deep roadway, it can be concluded that the rockburst is resulted from that a large number of energy accumulated in the surrounding rock mass of the roadway after the excavation of the roadway, and the high stress difference perimeter of the excavation because the confining pressure of the rock mass deduced quickly as soon as the excavation of the roadway. The process of rockburst is probably the process that it begins from the base line of the arch, and then the sides and the floor. The process of rockburst, of course, is very complicated due to the complicated stress condition and the structure of the rock mass. It should be studied much deeply in the future.

References

1. Diering D.H., Mining At Ultra Depths In The 21st Century, CIM Bulletin 93 1036 2000 CIM 141~145
2. Malan D. F. And Basson F. R. P., Ultra-Deep Mining: The Increased Potential For Squeezing Conditions, The Journal Of The South African Institute Of Mining And Matallurgy, November/December, 1998,353~362
3. Fangjun Xu,Debing Mao. Rockburst Mechanism of the floor of roadway in HuaFeng Coal Mine, coal science and technology, 2001,Vol.29(4):41~43
4. Gurtunca. R. G, Keynote Lecture, Mining Below 3000m And Challenges For The South African Gold Mining Industry, Proc. Mech. Of Jointed And Fractured Rock, Bulkema, 1998, 3~10

PREDICTION OF POTENTIAL SUBSIDENCE OVER ABANDONED LONGWALL MINES

GAILING ZHANG, XIAOBING ZHAO, QINGHONG DONG
School of resources and earth sciences, China University of Mining and Technology, Xuzhou, Jiangsu Province, P.R.C, 221008

The ground surface which might have been stable for long periods of time after mining has been ceased may subside again due to the deformation of the caving and fissuring rocks under the influence of underground water erosion, overburden load and adjacent mining and etc. On the basis of engineering geological investigation, the relation between the potential subsidence and the modulus of deformation of caving materials was analysed in this paper by using elestoplastic finite element method. The results show that the potential settlement above abandoned longwall mine is mainly induced by the compression of the caving rocks. As an example, the potential movements and deformations at a power plant site in Xuzhou city, Jiangsu province are predicted and the results can be used to the subsidence resistant.

1 Introduction

More and more buildings have been constructed in the ground surface above mined-out area with the economic development and population growth in China [1-3]. These abandoned mines can be subdivided into two types according to the mining method: that under shallow cover using irregular room and pillar mining and that using longwall mining. Gray and Bruhn [4] provided a detail review about subsidence above abandoned coalmines in the United States. As far as to the second one, the ground surface above abandoned longwall mine which might have been measured stable after mining has been ceased may subside again under the influence of underground water flow, overburden load and adjacent mining and etc. The subsidence may not take place concurrent with the construction of buildings, but take place a period of time after the construction completed. The possibility of this subsidence and its risk must be anticipated in the design and construction of buildings.

2 Calculation Model and Parameters

An engineering geological model –the compression model to calculate the potential subsidence based on geological investigation is established in this paper. The potential subsidence is recognized in the model as the process of the compression of caving rocks when the existing balance is upset and reaches a new equilibrium. The investigation in an intended power plant in Xuzhou city indicated that the primary mined-out coal seams are the coal seam No.1 and No.3 of Permian System. The thickness of the coal seam No.1 is approximately 1.5m and the cover varies from 65m to 67m. The coal seam No.3 is around 2m thick and its bottom is generally 12m to 15m below the coal seam No.1. The mine roof consists of sandy shales, claystones and fine sandstones, and the floor is sandy shales. Different extracting methods were used. The tunnel and pillar mining was generally employed in the 30's of the 20th century. The method used narrow tunnel (around 2m wide) to extract coals and wide pillar (15m to 20m wide) left to support the overlying strata. The safety pillars for preventing the power plant and the industry square were left in 1949 (Area A1and A2 in Fig.1). The coals were mined from the pillars of the seam No.3 also using the original tunnel and pillar mining in the period 1950 to 1952 (Area B2). In the period 1957 to 1964, longwall mining was utilized to extract the left pillars of the seam No.3 (Area B1). Fig.1 provided the distribution of different kinds of mined-out area. Five exploratory boreholes were drilled to prospect the deformation and failure of the overlying strata above the mined-out area. A calculation model (Fig.2) was established for plane strain analysis with a length of 280m and a height of 117m, consisting of 1900 elements and 2000 nodes.

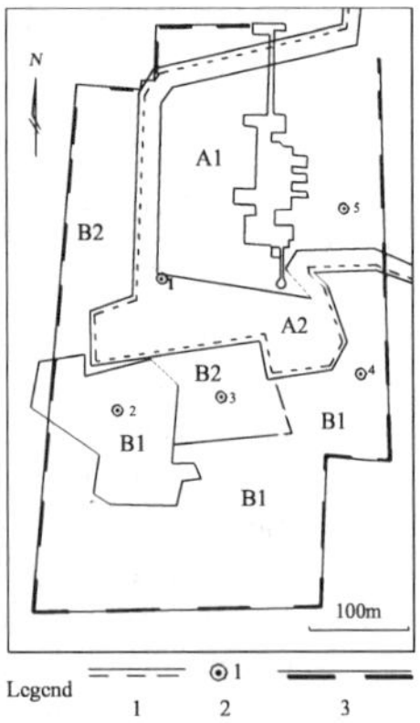

Figure1. Distribution of All Types of Mined-out area [2]
1--line of safety pillars;
2--exploratory boreholes;
3--extent of research area

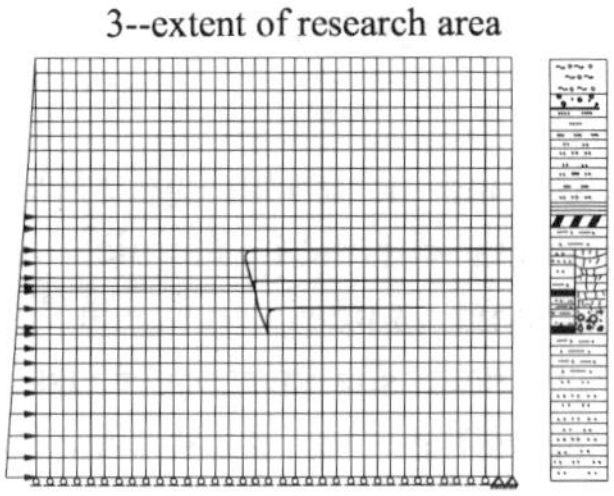

Figure2. Calculation Model (a half is presented)
①--the first slice of caving zone;
②--the second slice of caving zone;
③--the fissuring zone

The compression properties of fractured rock mass was studied by the Bureau of Mines and et al [5] and the results show that the Yong's modulus of fractured rocks is 7.1-32.3Mpa, being about 1/100-1/1000 of that of the original rocks. Dell and et al [5] deduced that the Yong's modulus of caving rocks was 1/4 of that of original rocks with anti-analysis using Coulomb and Von Mises' criteria. Of course, the deformation modulus of caving rocks does not only relate to that of original rock mass but also have closer relation with the degree of filling and consolidation of caving rocks. In calculations, a group of Yong's modulus was selected, taking the possibility of the maximum and the minimum into consideration to simulated the potential subsidence under different conditions of caving and filling. E_1, E_2 and E_r are the deformation modulus of the first slice caving, the second slice caving and original rocks, respectively.

3 Relations between Subsidence and the Deformation Modulus of the Caving Zone

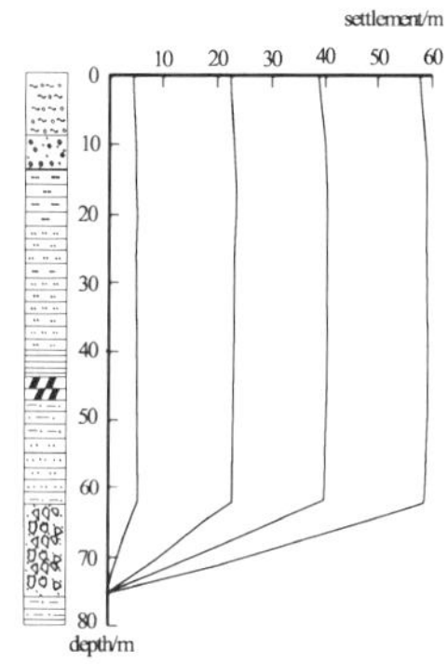

Figure3. Distribution of the Maximum Potential Settlement of Rock Layers with Depth

Calculations for the above-mentioned model were carried out using elastoplastic finite element method and some remarkable results were obtained.

The variation of the maximum potential settlement in rock mass with the depth is shown in Fig.3. The relation between the maximum potential settlement and the depth is a straight line from the bottom to the roof of caving zone. The smaller the Yong's modulus of caving rocks, the larger the slope of the line. The settlements from the fissuring zone to bedrock surface are approximately the same. It can be concluded that the potential settlement is mainly induced by the compression of caving rocks.

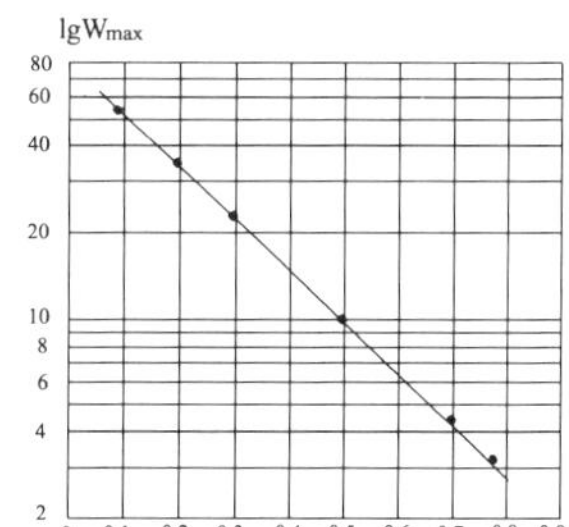

Figure4. Relation between the Maximum Subsidence and the Modulus of Deformation of Caving Rocks

The relationship between the maximum potential settlement and the deformation modulus of caving rocks is shown in Fig.4. Set $E_1=E_2$, $H_1+H_2=14.58$m, $E_1/E_r=n$ in calculations. The following equation can be easily gained:

$$\lg W=-1.92n+1.935 \quad (1)$$

The simulation shows that the maximum slope, curvature and strain in ground surface also depend on the deformation modulus of caving rocks. The following relations were obtained from an quantitative model with the parameters of $E_1=E_2$, $H_1+H_2=14.58$m and $n=E_1/E_r$.

The maximum slope i (mm/m)

$$\lg i=-1.811n+0.975 \quad (2)$$

The inverse curvature K (10^{-3}/m)

$$\lg K=-2.468n-0.615 \quad (3)$$

The compressive strain ε (mm/m)

$$\lg \varepsilon=-2.01n+0.834 \quad (4)$$

4 A Case Study

The primary buildings of the power plant can be located in the tunnel and pillar mining area (A1) in Fig.1 because the measurements and calculation show that the pillars, the roof and the floor in this area are stable. The construction should be avoided to be constructed in the boundary area (A2) of the safety pillars because its poor stability. Exploratory boreholes show that the thickness of the caving zone is 14.95m and the overlying strata fractured remarkable. The borehole No.2 and No.4 in the area B1 show that the goaf was basically filled by caving materials which has been slightly consolidated. The assistant buildings can be located in this area and the potential subsidence (in Table 1) calculated by the method this paper recommended must be taken into consideration. The caving materials in area B2 are very loose and tools struck and falling occurred sometimes during drilling. This area is not suitable for construction.

Table 1 The Maximum Potential Subsidence in the Area B1

Settlement (cm)	Horizontal movement (cm)	Slope (mm/m)	Curvature (10^{-3}/m)	Strain (mm/m)
55.30	19.61	6.55	-0.1375 +0.0675	-4.55 +1.18

5 Conclusions

The risk of the potential subsidence above abandoned mines must be taken into consideration in building design and construction. This paper indicates that the potential settlement over abandoned longwall mines is mainly induced by the compression of caving zone, the maximum settlement decreases logarithmically with the modulus of deformation and the potential subsidence at an site above abandoned longwall mines have been predicted and can be used for the subsidence resistant design.

References

1. Q. Di et al, Engineering geological study on mining induced subsidence. China building industry press. (1992)
2. W. Sui et al, Engineering geological analysis and finite element simulation of stress-strain failure of rock mass above the mined-out area at shallow depth. Journal of China coal society. 15(3): 72-82 (1990)
3. G. Guo, The deformation mechanism and control of construction basement above abandoned mines. CUMT Press. (2001)

STUDY ON THE SAFETY PILLARS UNDER WATER-BEARING STRATA USING FULLY MECHANIZED SUB-LEVEL CAVING METHOD IN TAIPING COAL MINE, SHANDONG PROVINCE, CHINA

WANGHUA SUI, SIGUANG YANG, QIANSHENG DI
School of resources and earth sciences, China University of Mining and Technology, Xuzhou, Jiangsu Province, P.R.C, 221008

DEJUN CHEN, QINGJIE ZHAO, ZUNXIN LIU
Taipin coal mine, Jining mining group, Shandong province, Shandong province, P.R.C, 272100

The engineering geological model and numerical simulation models for fully mechanized sub-level caving under thick unconsolidated water-bearing strata are constructed to predict and simulate the stress distribution and the extent of failure zones in the overlying rock mass. The distribution of the principal stresses in overburden rock mass is divided into three types. The failure conditions of overburden are classified into tensile cracked area, tensile-fissured area, and plastic deformation or shear fractured area. The extent and height of water flowing fractured zone can be determined on the stress distributions and failure conditions. Based on the calculation results and comprehensive engineering geological and hydrogeological evaluation, the height of safety pillars under water body of mining thick coal seam No. 3 in Taiping coal mine is then defined to be 60m, which is the minimum size of pillars in similar geological and mining conditions in Yanzhou coalfield.

1 Introduction

The overlying strata of the coal seam No. 3 in Permian Coal Measures in Taiping coal mine, Shandong province is approximately 30m to 80m. The strata dip on a regional scale at 3 to 5 and the geological structure is simple. The Quaternary cover varies from 150m to 190m. The safety pillars of more than 78m high under water-bearing strata should be left according to the fully mechanized sub-level caving mining experience in Yanzhou coalfield, and it would be impossible to mine the coal seam herein. So rational design of the height of the safety pillars under water-bearing strata relates directly to the development prospect of the mine.

2 Engineering Geological and Hydrogeological Conditions

The coal seam No. 3 with the average thickness of 8.85m in Shanxi Formation of Permian System is economic significant and mined over the entire coalfield. The overburden rock mass of the coal seam No. 3 can be classified into nine engineering geological types (ET in abbr. Table 1). The red sandstones belong to Jurassic System and the other types belong to Permian System.

The Quaternary System above the coal measures varies from 150m to 190m, which can be subdivided into three groups namely a top set, middle set and bottom

Table 1 Engineering Geological Types of the Overlying Strata

Engineering geological types (ET)	Average thickness (m)
Red sandstones (ET1)	29.43
Claystones interbeded fine sandstones (ET2)	10.51
Silt and fine sandstones (ET3)	17.00
Claystones (ET4)	7.00
Medium and fine sandstones（ ET5 ）	10.00
Claystones interbeded the coal seam No. 2(ET6)	5.35
Sandstones (ET7)	29.33
The coal seam No 3 and claystones (ET8)	10.58
Medium sandstones (ET9)	>10.63

set in accordance with lithology and water-bearing capacity. The middle set is s stable aquifuge all over the coalfield and isolates effectively the hydraulic relationship between the top and the bottom set. The bottom set is the main water supplying strata when overburden failure reaches the upper rock mass. The hydrogeological prospecting show that the bottom set belongs to the medium water-bearing capacity (Table 2). The depth of water level has fallen down approximately 90m since 1960's when the detailed exploration conducted, and approximately 10m since 1996. Its annual settlement is about 2.5m recently and these indicate that the discharge is greater than the recharge in the bottom set .

Table 2 Results of Pumping Tests for the Bottom Quaternary Aquifer

Bore hole number	Depth for water pumping (m)	Depth of water level (m)	Specific discharge (l/s m)	Types of water quality	Degree of mineralization (g/l)	Time of pumping tests
K 1	118.60-151.20	78.90	0.380-0.472	HCO_3-CaNa	0.400	1996.12
K 2	119.70-153.5	80.43	0.130-0.198	HCO_3-CaNa	0.420	1996.6
K 3	117.50-161.30	79.10	0.0152-0.090	HCO_3-CaNa	0.420	1996.10
K 4	115.23-149.25	81.11	0.197-0.248	HCO_3-CaNa	0.428	1996.5

3 Calculation Model and Parameters

On the basis of the engineering geological model constructed using the method recommended by Sui [1] the calculation model along the dipping direction of working face No. 6301 is established (Fig. 1). The lateral length of the model is 460m and vertical size ranges from 92.48m to 158.63m. The width of the working face is 120m. The model consists of 403 nodes

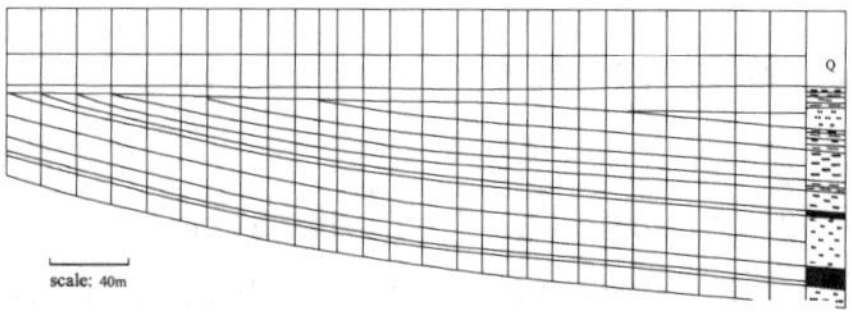

Figure1. Finite Element Model to Simulate the Full Mechanized Sub-level Caving

and 414 elements. The load of unconsolidated soils on the upper boundary is 2.3 MPa. The fully mechanized sub-level caving method is employed, mining the 2.6m thick bottom slice first and then caving top coal seam of about 5.8m thick.

The Drucker-Prager-cap with tensile crack calculation material model is selected and the calculation parameters are listed in Table 3.

Table 3 Calculation Parameters

Engineering geological types	Density (g/cm^3)	Yong's modulus (MPa)	Poisson's ratio	Parameters of yielding function		Tensile strength (MPa)
				α	K (MPa)	
Soil layer	2.08	11.0	0.35	0.0637	0.45	0.4
ET1	2.40	2670	0.24	0.1223	1.24	0.9
ET2	2.40	8230	0.21	0.1157	2.25	1.67
ET3-1	2.40	7660	0.24	0.1160	2.50	0.96
ET3-2	2.40	11500	0.17	0.1160	2.16	1.99
ET4	2.40	3950	0.24	0.1092	1.13	1.19
ET5	2.38	11300	0.22	0.1232	2.58	3.16
ET6	2.53	5000	0.25	0.1178	1.08	1.23
ET7-1	2.58	8300	0.21	0.1178	3.70	5.42
ET7-2	2.58	14500	0.21	0.1160	3.26	3.84
ET8 (coal seam No.3)	1.50	2830	0.25	0.1136	1.37	1.3
ET9	1.83	12500	0.19	0.1087	2.26	3.2

4 Distribution of Stress in the Overlying Rock Mass

The distribution of the principal stresses in overburden rock mass is divided into three types according to their magnitude, directions, tensile or compressive behaviour, namely a tensile-tensile stress zone, a tensile-compressive stress zone and a compressive-compressive stress zone [2]. As illustrated in Fig. 2, the extent of the tensile-tensile stress zone has extended to the upper part of the ET7 sandstones and the tensile-compressive zone to the ET3 silt and fine sandstones.

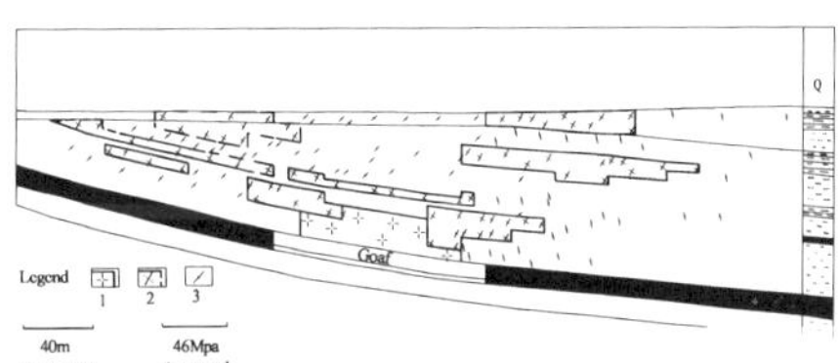

Figure2. Distribution of main stresses in rock mass after mining
1—tensile-tensile zone; 2—tensile-compressive zone; 3-compressive-compressive zone

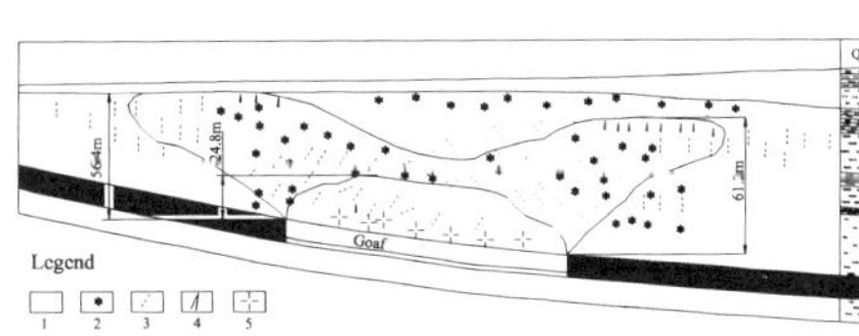

Figure3. Failure zone in rock mass after mining
1—elastic; 2—plastic; 3,4—tensile fissured; 5—tensile cracked area

5 Failure Zone in the Overlying Rock Mass

The failure conditions of overburden are classified into tensile cracked area, tensile-fissured area, and plastic deformation or shear fractured area [2]. The plastic deformation zones are located in ET1, ET2, ET4, and ET6 clayed rock mass, the weathered mantle over the mined-out area and above and beneath the coal pillars. The continuous tensile-tensile stress zones above the mined-out area where the tensile cracked areas are forming and the tensile and shear failure area are developing together constitute the caving zone in the mine roof. The caving zone in the cross section along dipping direction reaches ET7 sandstones and has the maximum height of 24.8m. When the tensile stress in the tensile-tensile and tensile-compressive stress zone reaches the tensile strength of rock mass the water flowing fractured zone are forming and enters the top of ET3 sandstones in the positive curvature region above the edge of the mined-out area and pillar at the direction of uphill. Maximum height of water flowing fractured zone is 56.4m as shown in Fig. 3.

6 Conclusions

The bottom set in the Quaternary System is an aquifer of medium water-bearing capacity. In order to ensure the mining safety, the waterproof pillars of certain size is necessary in accordance with the regulations for mining under water body. The maximum height of the caving zone and the water flowing fractured zone for fully mechanized sub-level caving method are 24.8m and 56.4m, respectively. Taking the engineering geological and hydro geological conditions into consideration the height of safety pillars under water body of mining thick coal seam No. 3 in Taping coal mine is defined to be 60m, which is the minimum size of pillars in similar geological and mining conditions in the same coalfield. 7.8Mt of coal reservoirs has been released and some workings under layers with medium water-bearing capacity have been mined out using fully mechanized sub-level caving method in the coal mine and distinct economic benefits have been gained.

References

1. W. Sui et al, Engineering geological model and its application to the study of safety pillars under water body. J. of CUMT, 28(5): 417-420 (1999)
2. W. Sui, Theory and practice of engineering geological prediction of mining-induced overburden failure. Chinese J. of Engineering geology. 2(2): 29-37(1994)

RESEARCH AND PRACTICE OF EME MONITORING ROCK BURST

XIYONG LI
HuaFeng Mine, Shang Dong,.271413, China
E-mail: **lixiyong@sohu.com**

XIUFENG ZHANG
HuaFeng Mine, Shang Dong,.271413, China
E-mail: xiufengzhang-sd@sohu.com

LINMING DOU
College of Mineral and Energy Resources, China University of mining and Technology, Xuzhou, 221008, China
E-mail:lmdou@cumt.edu.cn

The Electromagnetic Emission(EME) is caused by the inhomogeneous variable speed and the deformation of various parts of the coal and the variable speeding movement of electrical particles during crack expansion. This article is about the EME rules before and after the coal sample failure in 4# coal seam of HuaFeng Mine by experiments. Deviation values methods to monitor and forecast the rock burst have been held. KBD5 apparatus has been used to monitor the 4# coal seam which has high danger of burst in HuaFeng Mine, many times of pressure rock were forecasted according to the EME rules.

1 EME Mechanism of Rock Burst Failure

1.1 EME Mechanism of Rock Burst Failure

The research shows that the EME is the result of the deformation and the failure of the loaded inhomogeneous materials, such as coal and rock, and is formed in the variable speeding movement of electrical particles during crack expansion caused by the inhomogeneous variable speed and the deformation of various parts of the coal.

Variable deformation of various parts of the variously loaded coal and rock makes the free and escaping electrons move from the high-stress zones to the low-stress zones. At the same time, a lot of charges are covering the coal sample's surface, thus Coulomb's field or low frequency EME comes into being. Before the crack expands, there are a great number of charges or electrons on it. The electrons escape, during the crack expansion at variable speed, EME and AE (acoustic emission) are radiated at the same time. After the crack expansion, parts of the coal around the crack contract, meantime the Coulomb's field is formed because of the high concentration of electrons at the two sides of crack point. With the influence of Coulomb's field , the escaping electrons are accelerated with EME radiated. EME are radiated when the mediator molecules and atoms around the moving charges, which are decelerated, are ionized[1,3,4].

1.2 EME Phenomena and Experiment Result of Rock Burst

The EME characteristic of HanFeng coal sample was described by the experiment in lab. Figure 1 represents the typical curves of the stress distribution, EME impulse distribution and EME amplitude distribution of HuaFeng coal sample, which has high danger of rock burst.

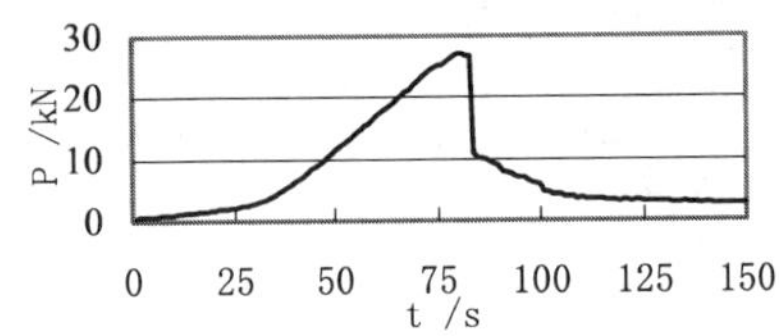

(a) The relationship between load and time

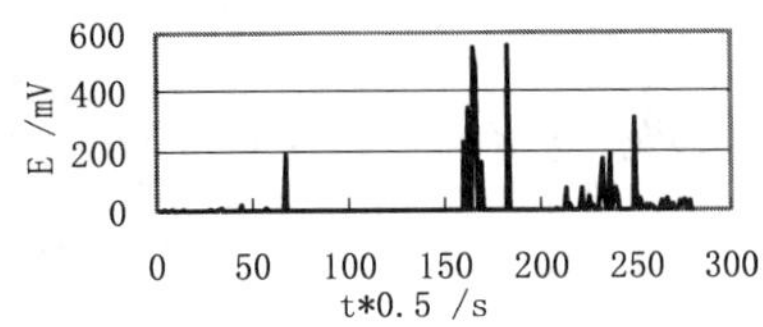

(b) The relationship between EME amplitude and time

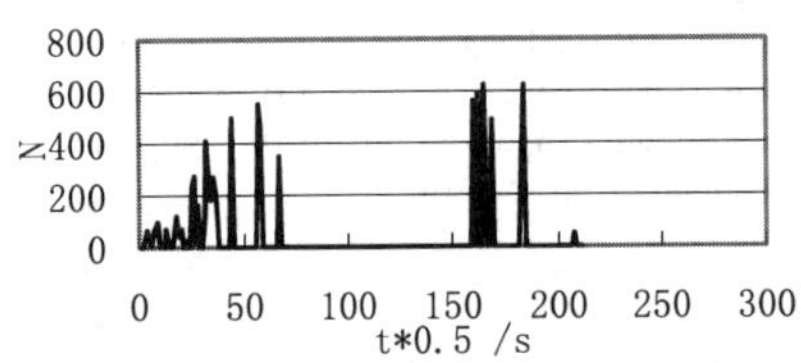

(c) The relationship between EAE impulse and time

Figure 1. The Result of Coal Sample Experiment of 5# coal sample

From the experiment results we know that EME signals appear in the process of deformation and failure under loading the various coal and rock. EME basically grows stronger with the increase of loading and the rate of deformation in coal. EME amplitude is below a certain value before the brittle failure, while it suddenly increases during the brittle failure of the loaded coal sample.

2 Forecasting of Rock Burst

The coal on the working face is in an unstable state when it's stress balance is broken after the heading and working space has been formed. There is EME radiated in the process of the rib coal's deformation and crack's forming to have a new stress balance, causing the coal to deform and to crack[2].

The higher the stress in the coal is, the stronger the EME value is. The higher the EME frequency is, the greater the danger of rock burst is. From the low-stress zone to the high-stress zone with the working face advancement, the EME signal is becoming stronger. In the abutment area, the stress is the highest, and the deformation of the coal is also great, therefore the EME signal is the strongest. In the low-stress zone, the EME amplitude decreases and intends to the balance. The EME signal received by isolated way is a collectivity of the ones in both low-stress zone and high-stress zone. When the stress in the monitored zone is high, the EME signal gives the just character of high stress zone, therefore we can use the EME signal to describe the high stress zone and then forecast the rock burst.

The EME deviation value[2] is a method of forecasting the rock burst by analysis of the difference between and the monitored value. Practice proves that the deviation value changes greatly before the rock burst. The formula is:

$$d = \frac{E - Eave}{Eave} \times 100\%$$

Where E is the monitored value and $Eave$ is the average value.

3 Application of the EME Technique

EME technique was used in Huafeng Mine on 3406(1) working face, which is located in District 2 of mining area 3 of layer 4 at level –750m.

KBD5 apparatus is used to monitor the EME on the working face, in the way of non-touched directional testing, accepting a wide range of frequency to the top of 500HZ,testing a distance from 7m to 22m, and the tested point is 10m far from each other.

The research show the change of the EME on the tested working face and in the roadway is usually small, and there is less EME impulse. The EME amplitude in different area of the working face is different. It is high where the pressure and the danger of rock burst is high. There are obvious signs of the great increasing of the EME amplitude and impulse before rock burst. The EME impulse changes greatly after the reliving shot. This proves that there is great deformation in the coal wall, it is very likely to have rock burst.

Figure 2 is about the regularity of the deviation values of the three features' parameters. According to these figures, the heir deviation values have exceeded the critical value before the rock burst. So using EME technique can greatly improve the accuracy of the forecasting rock burst.

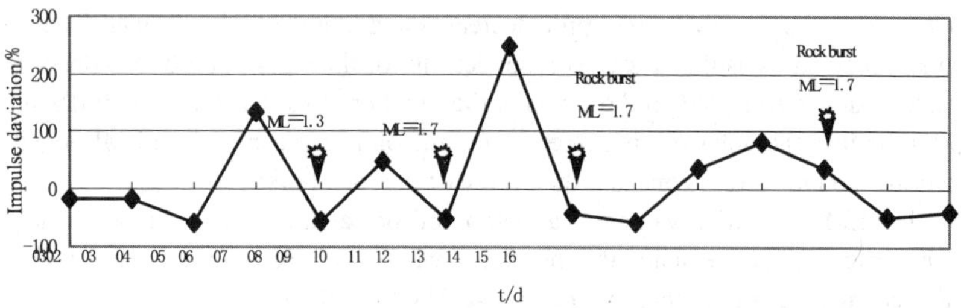

Figure 2. The changes of EME deviation of 130th prop of the working face

4 Conclusions

EME signals appear in the process of deformation and failure under loading the coal and rock. EME basically grows stronger with the increase of loading and the rate of deformation in coal. The on-site practice shows that the EME amplitude and the impulse increase with the increase of the danger of rock burst. The EME character of different state of working face can forecast many times of rock burst.

Acknowledgements

This study was jointly financially sponsored by NSFC (the National Science Foundation of China) with grant No. 50074030.

References

1. Dou Linming ,He Xueqiu, *Theory and Technology of Rock Burst Prevention.* Xu Zhou: CUMT Press,2001.(in Chinese)
2. Dou Linming, B.Drzezla, *Przeglad Gorniczy*. 54(11)(1998)20-25. (in Polish)
3. Dou Linming: *port-Ph. D. Thesis,* China Univ. of Mining & Tech.,2001.(in Chinese)
4. Dou Linming ,He Xueqiu. *Journal of TsingHua University*. 41(12)(2001)86-88. (in Chinese)

GEOSTATISTICAL APPLICABILITY IN SEISMIC INVERSION

CHENGYUAN ZHANG   DAOYING XI   XIAOYAN LIU
Dept. of Earth & Space Science, University of Sci.&Tech. of China
HeFei, 230026, China
E-mail: zhang@mail.ustc.edu.cn

Geostatistics can effectively incorporate well log with seismic data. In this paper, the geostatistical methods used in seismic prospecting are shown. Through case study we discuss what geostatistical applicability depends on and how to avoid negative factors. It is important to get data statistical distribution when applying geostatistics to inversion methods.

1 Introduction

There has been rapid growth in the use of attributes derived from 3-D seismic data to define reservoir physical properties. Statistics methods, including geostatistics, multiple regression, neural networks etc. are being explored to help correlate well log and seismic data and then to distribute reservoir properties throughout 3-D seismic image volumes. Geostatistics, such as co-kriging and collocated-kriging, is a kind of hopeful tool in rapid growth. It becomes a powerful analysor on spatio-temporal distributions of reservoir properties because of its description ability on regionalized variables. We can integrate seismic data, petrophysical property and geologic information into a more detailed reservoir model in less time by geostatistics (compared with ordinary methods). Moreover, it involves evaluation of uncertainty which most mapping tools cannot supply. Recently geostatisticians have overpassed the isolated use of kriging such as mapping and attempted to combine it tightly with seismic prospecting methods. Wook B.Lee[1] incorporated geologically feasible lateral and vertical velocity variations and all available well log into a 3-D velocity model that is the base of salt imaging. Hass & Dubrule (1994) attempted to combine kriging with post-stack seismic inversion by construct a geostatistical constrain. Carlos[2] moved one step further and applied trace-based and geostatistical inversion to San Jorge Basin. This seismic-guided and log-involved approach to predicting reservoir properties has been popularized. It bases on that log-derived reservoir properties sometimes correlated with seismic attributes and that the regression equations so derived could be used to predict reservoir properties away from existing wellbores. The attributes used in correlation process can include simple time-derived attributes (e.g., time structure, isochrons), amplitudes (e.g., instantaneous amplitude, average amplitude in an interval), complex trace attributes (e.g., reflection strength, instantaneous frequency), or others (e.g., impedance, coherence, velocity anisotropy). The map and volumes derived in this way can help engineers identify potential development drilling targets[3].

This charming predictive method make one wonder whether it can be applied to any given subsurface reality and what relevance grade should be. In this paper some fundamental characters are discussed through investigating real examples. Especially the applicability of kriging constrain in seismic inversion is shown.

2 Geostatistical analysis and Seismic inversion

Seismic data provide indirect, low-resolution image of rock velocity and density but good space coverage while well logs provide direct, high-resolution sampling of these and other rock parameters. Combination of seismic and log information can give us better reservoir models. An effective integrated method is to construct kriging constrain on acoustic impedance inversion which is basic tool in reservoir characterization.

The effect of inversion depends on initial model very much. Iteration process becomes unstable even failed when initial model is far from the true physical model in parameter space. Generally, fine fitness is not the sufficiency to reliable parameter estimation while low-frequency components control stability of iteration (Kolb, 1986). Kriging, which can provide good initial model, may be a suitable constrain on inversion.

Kriging is linear interpolation with existing log information. It provides reservoir estimation by utilizing not only wellbore petrologic physical data but also spatial-correlated model of reservoir properties in all distances and directions. As a weighted average method, ordinary-kriging (a basic form of kriging) shows that the weights λ_i derive from:

$$\sum_{i\ 1}^{n}\lambda_i C(x_i,x_j)-\mu = C(x_0,x_j), j=1,2,...,n \qquad (1)$$

$$C(h) \quad C(0)-\gamma(h)$$

Where C is the covariance, γ is the semivariogram, and h is the lag .

It can be seen that the weights λ_i depend on theoretical variogram $\gamma(h)$ completely. Therefore the initial model in constrained inversion is concerned with selection of theoretical variogram that is derived from fitting calculated variogram.

3 Examples and discussion

We make kriging on two wellbores sonic data and result is shown in figure 1. It can be seen that there are obvious variations between theoretical variograms for different optimum strategies. However the kriging interpolation with different variogram models yield similar results. Cross validation shows fine goodness of fit. High similarity indicates that there is little unstability on inversion process when in not-strong-nonlinear solution space. Kriging constrains can work by auto-selection of theoretical variogram. Inversion results come to the same conclusion.

The 'strange ' result derives from big enough sampling dataset which is sparsely shown in figure 1 for display convenience. In fact, current geostatistical practice in selecting a variogram model is rather subjective[4]. Due to lack of sampling, different variogram fittings yield different initial models that lead to obviously different convergence attributes especially in strong-nonlinear solution space.

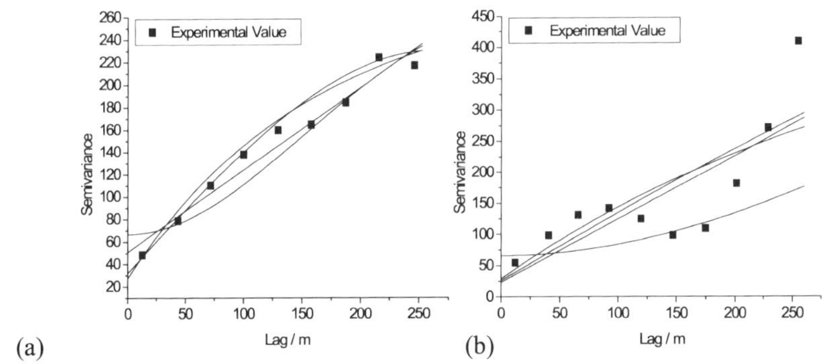

Figure 1. Experimental and theoretical variogram of Log 1 (a) and Log 2 (b, hole effect). Curves represent spherical, exponential, linear and Gaussian model respectively

In fact, kriging provides a unique estimation map with a characteristic uneven smoothing effect that increases further away from the sampling data location (A.G.Jouenel, 2000). Therefore it is reliable to obtain low-frequency information from seismic and well data. Excessive high frequency initial model may lead to unstable iteration process and unreliable results.

Essentially the reason kriging constrain fails is that target data do not follow the basic intrinsic and stationarity assumption of geostatistical methods[5]. Not all petrophysical property data sets follow these assumptions that ensure the applicability of kriging.

Figure 2 shows another analyzing results of porosity distribution of lithoherm reservoir[6]. Double-hump distribution of porosity disobeys the stationarity assumption of kriging. In actual estimation there is only one distribution of data that is taken into account. It directly leads to overrating reservoir average porosity.

Above geostatistical practices indicate that investigation at the beginning of project on target data statistical distribution is important.

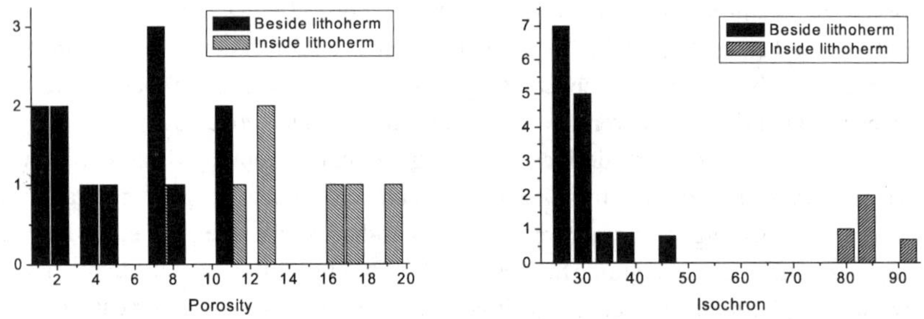

Figure 2. Double-hump distribution of porosity and isochron

4 Conclusion

When sampling is large enough and data follow the implied assumptions of geostatistics, it can be the powerful assistant tool of geophysics, geology and reservoir engineering. However, lack of sampling and multi-distribution of data are not rare in petrophysical field. Therefore it is necessary to take data statistical distribution into account before processing data from different sources with different uncertainty.

Acknowledgments

The National Natural Science Fund supported much of this work under contract number 40174050.

References

1. Wook B.Lee and Wenlong Xu et al, The Leading Eage, 19, 32, (2000).
2. Carlos T., et al,The Leading Eage, 18, 1070(1999).
3. Bruce S.Hart and Robet S.Balch, Geophysics, 65, 368 (2000).
4. David J.Gorsich and Marc G.Genton, Math.Geology, 32, 249 (2000).
5. C.V.Deutsch and A.G. Journel, New York,Oxford University Press, (1992).
6. Keith Hirsche et al, The Leading Eage, 16, 253, (1997).

REVIEW OF GEOLOGICAL-CALAMITY BY ROCK-BURST IN DEEP MINING PROCEDURE OF HONG -TOU-SHAN COPPER MINE

CHANGYAN SHI HAOCHUN MA

Fushun Hongtoushan Copper Mine is the deepest mining metal mine of China, its stresses characters and ground pressure activities are very typical, especially rock bursts occur frequently at stope. The strength of rock bursts has been beyond moderate strength, which is the only existence among the domestic homogeneous mines. This provides a very rare example for studying on inter-relative fields. In this paper, parts of rock bursts are described and the mechanism of rock bursts and tendency evaluating method adopted by the mine are also discussed. The author introduces support measures, meanwhile, makes a primary estimate to applying Acoustic Emission (AE) inspection and prediction methods. Therefore, it has some value to refer.

1 The engineering geology and characters of original stresses in mine-field

1.1 Brief description of engineering geology of mine-field

Hongtoushan copper deposit is of pyrite veins copper-zinc deposit and occurs in Archen metamorphic rocks. The rocks coming out in minefield all belong to Anshan Group of Archeozoic erathem and mainly consists of biotitize plagiogneiss. Geological structure is simple too. But rock or ore are very strong, ore's strength: f=8~10, rock's strength: f=10~12. What's more, support was not needed for drifts underground. Ore body is very complex and has many branche or compound phenomena. At present the depth of mining level is –767 meters, while development level is –827meters. Mining depth is up to 1197meters and development depth is 1337meters.

1.2 Properties of ore and rock underground and results of stresses test

To master the physical-mechanical parameters of ore and rock and the characters of original stresses, Beijing Institute of Non-ferrous Metallurgical Designing Research and Northeastern University have ever been given commissions of having a series of relative tests in order by mining company. The results of these tests are shown in table 1

Table 1. physical-mechanical parameters of ore and rock

Type of mine rock	Uniaxial compressive strength(Mpa)	Tensile strength (Mpa)		Shear strength (Mpa)		Modulus of elasticity (Mpa)	Poisson ration	Internal cohesion (Mpa)
Rubble plagiogneiss	103.7	vertical	8.2	60°	36.9	4.32×10^4	0.2	20.2
Biotite plagiogneiss	109.6	vertical	8.2	60°	33.0	6.25×10^4	0.2	16.3
Ore	102.4	5.24		60°	23.3	4.76×10^4	0.16	14.5
Diabase	91.3					8.08×10^4	0.25	25.0

2 Estimate of prime progeny on studying of ground pressure

In 1976, A ground pressure-study group was established. Later a technology contract of deep ground pressure's study was singed with NEU. In 1997, The Program was authenticated successfully, important progeny and prominent progresses were acquired.

2.1 Identifying the characters of original stresses

Based on the testing results before and many years' local investigation information, an important conclusion was drown, that is the height of –347 meters is a boundary of field stress in minefield. Gravity stress field is up the height, so the ground pressure activities effected by geological structure is static. Structure stress field lies under the height and the level stress whose direction is almost vertical to ore body trend main specificity of ground pressure activities is dynamic (rock bursts).

2.2 Systematic study on rock bursts

There were kinds of theories on the judgment of rock bursts. Several-index group was adopted to judge rock bursts by the mining company and NEU. The results is:
1). Conditions of rock property, see table 2

Table 2 the conditions of rock bursts' lithology

Tendency of rock bursts	Crisp parameters ($R=\sigma_c/\sigma_t$)	RQD	Integrity coefficient (Rv)
Strong	R>18	RQD>0.7	Rv>0.75
Moderate	R=10~18	RQD=0.5~0.7	Rv=0.5~0.75
None	R<10	RQD<0.5	Rv<0.5
Hongtoushan mine rock	R=10~19	RQD=0.78	Rv=0.88

Note: $б_C$ is uniaxial compression strength, and $б_t$ is uniaxial tensile strength.

See table 2,mine rocks of Hongtoushan fulfill generally the conditions of rock bursts whose strength are beyond moderate strength about the conditions of rock property.

2). Index of impacting tendency

There were kinds of theories to verdict the tendency of rock bursts. By reference many abroad and domestic dates and the index of elastic energy (We), the index of impact energy (Wb) and the time of dynamic destruction (ms) were adopted as the verdict indexes. The results may be seen from table 3.

We can approximately classify the rock bursts' strength of typical mine rock according to this. Rubble plagiogneiss, sillimanite gneiss and diabase have a tendency of strong rock bursts occurred. While, biotite gneiss and copper ore have a tendency of moderate rock bursts occurred.

Table 3. Estimating results of impacting tendency of Hongtoushan ore and rock

Rock property	Index of elastic energy (We)	Index of impact energy (Wb)	Time of dynamic destruction (ms)
Rubble plagiogneiss	15.6	4.8	50
Sillimanite gneiss	7.1	5.3	23
Biotite gneiss	5.3	3.1	70
Diabase	14.5	4.7	13
Copper	7.3	5.0	23
Grade of rock bursts			
Strong	>5.0	>2.0	<50
Moderate	2.0~5.0	1.0~2.0	50~500
None	<2.0	<1.0	>500

3 Introduction of deep ground pressure activities (rock bursts)

Since the middle of 1970,slight rock bursts that occurred in a form of bouncing small block have appeared during shallow mining. With the increase of mining depth, flaking and falling or rock shot off is frequent at level or stope. The peculiar questions of deep mining has slowly grown predominant, especially geological hazard in a typical form of rock bursts is the most difficult one.

As a serious geological hazard, the frequency of rock bursts increased quickly (it's about 20 times) after deep mining, especially recent tendency. The most serious rock burst happened twice continuously on May, 18, 1999, this is a sign that ground pressure activities are progressing towards a new phase when geological hazards of rock bursts are more predominant. Therefore the mining safety is facing a flintier cross.

4 Inspection and prediction practice of ground pressure activities

In order to study ground pressure ,some comprehensive measures such as glass tube, stopper and photic-stressometer have been ever adopted. Through cooperating with NEU and other institutes, the feasibility of applying AE technology based on the Kaiser principle were discussed. The credibility of using this method in local is decided by the value of n, n= $б_C$ / $б_k$ The value of $б_C$ expressed uniaxial compression strength, while $б_k$ is property stress of Kaiser.

The testing results of the value of **n** can be seen from table 4.

Table 4 Testing results of the value n about Kaiser

lithology	Copper ore	Rubble plagiogneiss	Biotite plagiogneisss
Value of **n**	1.9	1.3	1.7

According to the testing results above, the credibility of prediction for rock bursts using AE method is very high because of the greater value of **n** in the mine. Therefore, AE method was adopted as a main technologic way for inspection and prediction.

Then, we bought a DYF-2 type intellect-sonic inspection multiple function instruments from Changsha Institute of Mine of China. The instrument has some memorize function and can output original wave type for referring by connected with a computer .The instrument is more advanced about total energy of AE parameters, energy rate, rate of event and frequency than before.

The inspection technology about ground pressure was improved by using this instrument and there were more specific objects to ground pressure activities in form of rock bursts, because inspection has been transfer from ore body surface to internal-micro cracks. Finally we generalized long-term's local inspection and acquired some important regular recognitions by using AE technology. The following is: ① Explosions in next engineering impact on surrounding rocks evidently and cause partial stresses (AE energy) growing; ② There are different degree about ground pressure activities during every energy summit(25000~30000mv) declining; ③ The summit approaches 90 percent of limited stress of rock samples, the second obvious increasing point of AE and the obvious increasing point of tendency; ④ The summit point may be referring point of predicting rock bursts.

5 The recognition for mechanism of rock burst of mine

There are different expressions in different data for the conception of rock burst. In this article, it is defined as this: Rock burst is a special ground pressure phenomenon which is characterized by sudden resilience energy releasing and rock dynamic being shot off. The mechanism of rock burst denotes usually the physical procedure and principle of its occurrence. The means to study rock burst depends on the difference theory. a large number researches show that geological condition and mining technology is the main reason to cause rock burst happen. So the geological condition and mining technology of mine may be discussed.

Geological condition, the internal cause of rock burst, can be analyzed in the following two aspects:1) Rock Character --generally speaking, the harder and the more integral the rock is, the more easily that rock burst would happen.2) Stresses Condition--according to the mainstream viewpoint, the primary requirement is that the rock stresses must be over its intensity limitation, and at the same time, it should be able to amass lots of energy.

The research about the trend of rock burst engaged by the Corporation and Northeastern University is based on this technology route. On the viewpoint of author, the trend may be understood as the measurement of extends to geological requirement. As research showed, the deposits of Hongtoushan fulfill the geological requirements.

There's no doubt that mining technological condition should be taken as external cause of rock burst--it's just because of the mining work, formed the exposure face, and destroyed the balance of virgin rock, which make it possible for rock burst to happen.The factor's effects could be grouped as the following according their appearances.

1) Because of the redistribution of rock stresses, stresses concentration must happen at pillars and walls, which would satisfy the stress requirement.

2) Human's action such as explosion and so on, perhaps occurs rock burst, it has been proved as this: explosive shock (disturbance) may been taken as a kind of pulsation attached load. It makes the stress concentration of vicinity worse, the weak unstable limited balance status can easily be destroyed and so rock bursts happen.

The stress concentration caused by mining work is usually characterized slow, long-term and gradual, and is easy to monitor and measure, through technology means. On the contrary, the stimulation of explosion is often known short-term and sudden. It's difficult to measure and prevent. On the whole, rock burst can be looked as one of the characters of deep mining. Rock burst is the destiny of sequence of mining work (external cause) and some certain rock characters and stress condition (internal cause). In some how, human actions are usually the directed reason of rock burst.

6 Discussion about the prediction and prevention of rock burst

Because of the problems being singularity and complexity, the research about rock burst in metallic ferrous mine is still at rudiment phase in china. There aren't many typical examples as Hongtoushan. Most of the researchers based their theory on indoors experiences and trend mark. Therefore, there exists a great gap compared to the developed countries in rock burst's prediction, prevention and support, and not achieved obvious progress.

As to the deep mining of Hongtoushan, the following measure can be adopted to prevent rock burst.

1) Reduce the amount of excavation to avoid stresses concentration.
2) Find out practicable stress relief technology; relieve the existed stresses concentration to avoid rock burst.
3) Find out practicable and feasible supports technology; try to avoid the disaster of rock burst.
4) Control the explosive work strictly to avoid the destruct to work face.

7 Peroration

Nowadays, there're not many reports about rock burst of Metals Corporations in China. Along with the fact that most of the mining work will be done deeply underground in succession, and many of them have reached or will soon reach the depth of 1,000m, rock burst may be the most serious problem to be faced. It's very essential to use the developed countries' experience for reference, to study the need and reasonability to build model monitor-measure stations in some large deeply mining corporations.

In China, not only there are few corporations can afford the expensive equipments, but also the digestion and absorption of technology itself is a question. So the practicable choice is that some representative corporations and academies, which own specialty strength, should be organized together to do scientific research, and afforded special capitals by national scientific administrative parts. This may take the whole industry develop more quickly by the breakthrough in parts. It's strategic to the steady development of whole mining engineering.

STUDY ON THE APPLICATION OF DYNAMIC EFFECT OF UNLOADING ROCK MASS

YONGXIN ZHANG HAIJIA WEN
Faculty of Civil Engineering, Chongqing University, Chongqing, 400045, P.R.China

The subject matter of the thesis is to study the dynamic effect problems on unloading rock mass. Based on a simplified one-dimensional model about rock mass excavation, the dynamic effect equation is proposed, the unloading displacement is solved by analytical method, and some key points of dynamic effect of rock mass excavation are studied. Then, some application cases are analyzed in their unloading properties and displacement of rock mass. It concludes that, for the rock excavation engineering, the analysis results by means of *unloading rock mass mechanics* is more identical with the in-situ value than the results by means of *loading rock mechanics*.

1 Introduction

Up to now, the calculating or testing result of the problems on the rock excavation, based on the classic theories, is usually far from the actual condition in-situ. There are widespread cases to show that. The value of calculating and testing is only 3cm but the actual value is over 2m for the deformation of the Lianziya deformed rock mass at the Yangtze Gorges (Ha et al, 1995). And that is 20cm vs. 5m to the slope deformation for open pit Jinchuan Nickel mine (China, 1985). It shows there are some mistakes in studying on the mechanics dynamics and properties of the excavated rock mass by the classic theories.

In fact, the main dynamic condition, the value and scale of unloading and the released geostatic stress, are different from loading mechanics (Zhang et al, 1997). So, to analyze the rock mass unloading mechanics problem, it is necessary to study the dynamic effect of excavation unloading for the rock mass firstly.

2 Simplified Model

Here, a simplified one-dimensional model is put forward to simulate the rock slope unloading excavation. See the figure 1a. Suppose the left boundary of the model is fixed, and the right surface of the model is excavated. Here assume the initial bound force F_0, initial stress σ_0, the initial strain ε_0 (the corresponding initial displacement y_0), and the modulus of deformation E_0 before rock mass excavation.

Now, if the bound force F_0 is removed to simulate the model right rock excavated, the deformation to the right reproduced and the energy released in short time by rock mass excavation. See figure 1b.

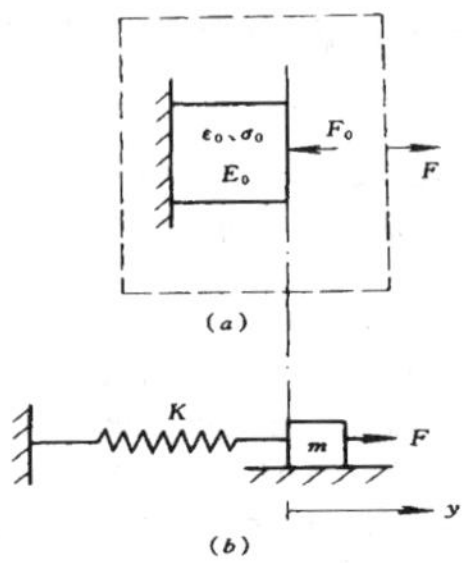

Figure 1. Simplified excavation model

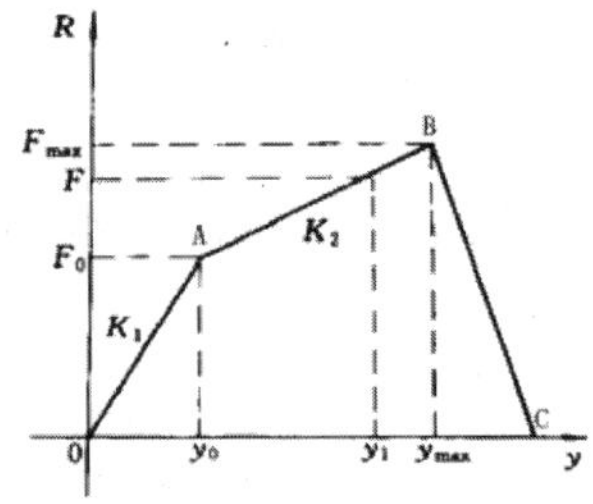

Figure 2. Varying properties of R

3 Dynamic Effect

3.1 Mathematics Model

From dynamic equilibrium, we can obtain the motion equation of the above system

$$m \cdot y'' + R - F = 0 \qquad (1)$$

where R is the spring force, see figure 2, its properties for the rock show below:

1) Before the system recover to static balance y_0 in figure 1a, the phase may be elastic, and the system stiffness K_1 (or E_1) is a bigger value. 2) Between y_0 and y_{max} which corresponds with the peak tensile strength of rock mass, the crack of rock mass spread out rapidly because rock can bear small tensile stress only, which shows the rock mass have large deformation properties. And the recoverable deformation may be negligible. The corresponding system stiffness is K_2 or E_2 in the phase. 3) After y over y_{max}, boundless deformation will come into being by rock mass failure.

Then the equation (1) may be show below

$$\begin{cases} m \cdot y'' + K_1 \cdot y - F = 0 & 0 < y < y_0 \\ m \cdot y'' + K_1 \cdot y_0 + K(y - y_0) - F = 0 & y_0 < y < y_{max} \end{cases} \qquad (2)$$

Moreover, we can assume F constant that does not vary by the time for rock excavation. And the other interference force to the system must be considered and put it into system in addition.

3.2 Model Solving

From dynamics, the equation (2) shows a single degree of freedom dampingless system. Assuming $\omega_1 = (K_1/m)^{1/2}$ and $\omega_2 = (K_2/m)^{1/2}$, the following solving expressions for the differential equation (2) are obtained

$$\begin{cases} y = C_1 \cdot \sin \omega_1 t + C_2 \cdot \cos \omega_1 t \\ y = C_3 \cdot \sin \omega_2 t + C_4 \cdot \cos \omega_2 t \end{cases} \quad (3)$$

And the two special solutions are as follows respectively

$$\begin{cases} y = F/K_1 \\ y = (F - F_e)/K_2 + y_0 \end{cases} \quad (4)$$

So the common solutions of the equation (2) can be obtained as

$$\begin{cases} y = C_1 \sin \omega_1 t + C_2 \cos \omega_1 t + F/K_1 \\ y = C_3 \sin \omega_2 t + C_4 \cos \omega_2 t + (F - F_e)/K_2 + y_0 \end{cases} \quad (5)$$

For $0<y<y_0$, when t=0, y=0 and y'=0, we can obtain $C_1=0$ and $C_2=-F/K_1$. Then the first item of equation (5) can be written as

$$y = -F/K_1 \cos\omega_1 t + F/K_1(1-\cos\omega_1 t) \quad (6)$$

Then the time t_0 corresponding y_0 can be shown as $\cos\omega_1 t_0 = 1 - y_0 K_1 / F$ or $\sin\omega_1 t_0 = (1-\cos 2\omega_1 t_0)^{1/2}$.

From the second phase starting, assuming $t_1=t-t_0$, and the initial conditions of the phase are: $y|_{t1=0}=y_0$ and $y'|_{t1=0}=F/K_1\,\omega_1\sin(\omega_1 t_0)$. Using initial conditions to the common solutions (5), the expressions for the other two constants are obtained

$$C_3 = F/K_1\omega_1/\omega_2 \sin \omega_1 t_0$$

$$C_4 = (F - F_e)/K_2$$

3.3 Problems Discussing

1) From the above hypothesis that the deformation is not recoverable when deformation over the equilibrium position, so the system maximum displacement means the in-situ test displacement. The corresponding time may be obtained by choosing the derivative zero for the second expression of equation (5), and written as

$$t_{\max} = [arctg\ (C_3/C_4)]/\omega_2 \quad (7)$$

Then using t_{max} to the second expression of equation (5), the value y_{max} (i.e. the calculation displacement of rock deformation in-situ) can be obtained. Only the dynamic inertia and the unrecoverable rock cracks' spreader but no other factor (e.g. disturbing force, rheological property) are thought over for the above value y_{max}.

2) The value y_{max} mainly depends on the properties of rock mass ω_1 and ω_2, i.e. the rock mass unloading modulus E_1 (or stiffness K_1) before zero stress state and the tensile modulus E_2 (or stiffness K_2) during rock cracks spreading out. And y_{max} also depends on the value of unloading and the value σ_0 of geostatic stress field. Assuming the results by numerical analysis correspond to the value of in-situ test, then the used stiffness must be given by $K=F/y_{max}$, or converts it into the dimensionless form $E = \sigma/\varepsilon_{\max}$. The value E is commonly between E_1 and E_2.

3) $E<E_1$ means the rock properties weaken due to dynamic effect and the cracks extension by the rock energy transition with the excavation and unloading rebound.

4 Application Cases

4.1 The Properties of Unloading Rock

For the weathering rock at the shiplock high slope of Yangtze Three-Gorge Project, σ_0 =10MPa, E_1=25GPa, and E_2=2.5GPa. So we can analyze the maximum strain and unloading deformation modulus E due to unloading excavation by above method.

Figure 3 shows the weathering rock mass' constitutive relationship of unloading stress vs. strain obtained by using above parameter. Then we can see that the value E reduces to 6*GPa* (only E_1/4.16) when the unloading value is 10MPa on above calculating conditions.

The value E will reduce larger if the rock quality weakens due to repeated unloading or bursting considered in addition. So here results are the most conservative because of only rock quality weaken due to unloading dynamic effect considered.

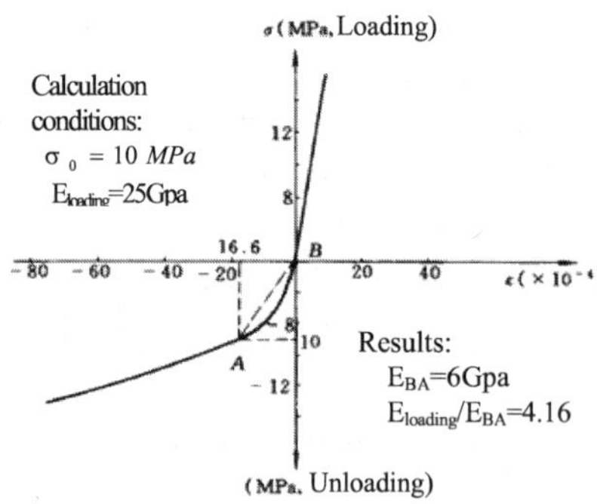

Figure3. The curve of stress vs. strain of weak weathered rock

4.2 Deformation Analysis

Case 1: The Lianziya deformed rock at Yangtze Xiling gorge is over 100m in high with

Table 1. The deformation value of analysis sections in deformed rock mass

Calculation parts	Study method	Analysis section		
		I—I	II—II	III—III
T8 Joint	Unloading mechanics	37.2	59.3	86.7
	Model test	40.0	62.3	95.7
	In-situ value	50.0	80.0	100.0
T9 Joint	Unloading mechanics	20.8	42.6	55.1
	Model test	15.0	56.5	70.4
	In-situ value	30.0	50.0	80.0
T8+T9 Joint	Unloading mechanics	58.0	101.9	141.8
	Model test	55.0	118.8	166.1
	In-situ value	80.0	130.0	180.0
Numerical analysis of loading rock mass mechanics		The maximum value of all parts displacement is about 3cm		

stick limestone at the mid-top and 1.6~4.0m coal stratum at the bottom. The rock mass is separated by tens of joints.

The table 1 shows the displacement value based on the above method, model experiment, in-situ test and numerical analysis of loading rock mechanics respectively.

From table 1, we can see that the displacement value based on the above method is accorded with the results by in-situ test and model experiment, but it is 50~60 times to the analysis result by loading rock mechanics.

Case 2: Jinchuan open mine pit is elliptical with about 1000m length, 600m width, and 180~310m height slope. It was constructed from 1965 to 1982.

The figure 4 shows the slope deformation process by in-situ test results, figure 5 shows the result of unloading deformation and damage deformation based on the above method. Then figure 6 and 7 show the displacement values by FEM based on the loading rock mass mechanics.

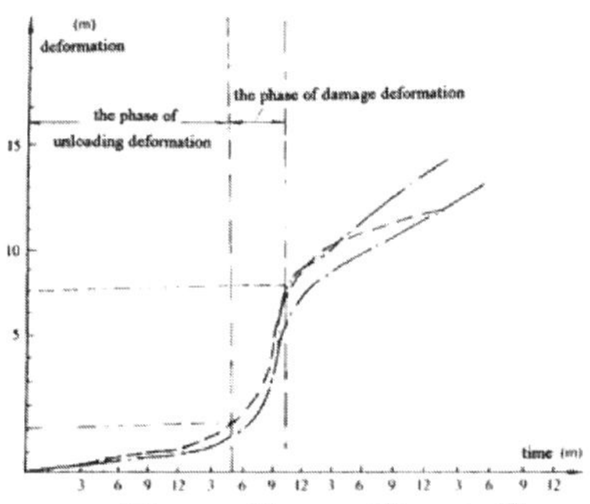

Figure4. The deformation process by in-situ test of Jinchuan slope (By Geol. Res. Chinese Sci. Ins.)

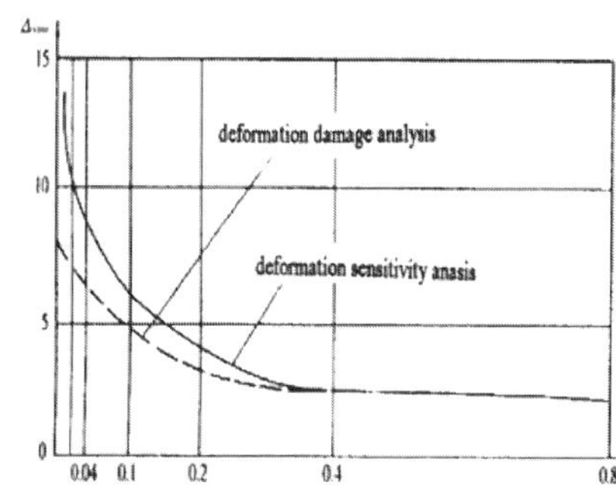

Fig.ure5. The unloading deformation of jinchuan slope

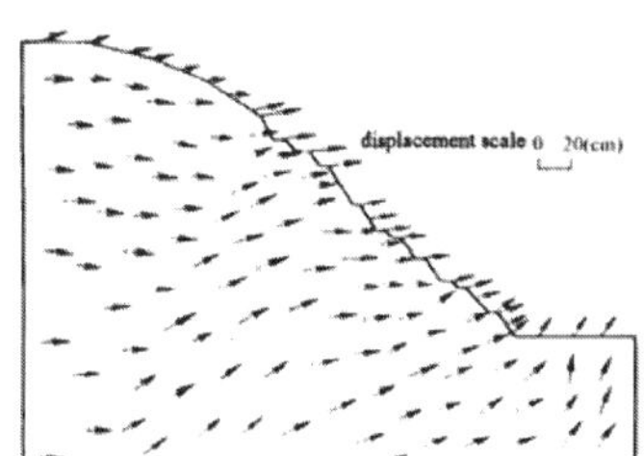

Figure6. The displacement vectors of the slope rock mass (By Geol. Res. Chinese Sci. Ins.)

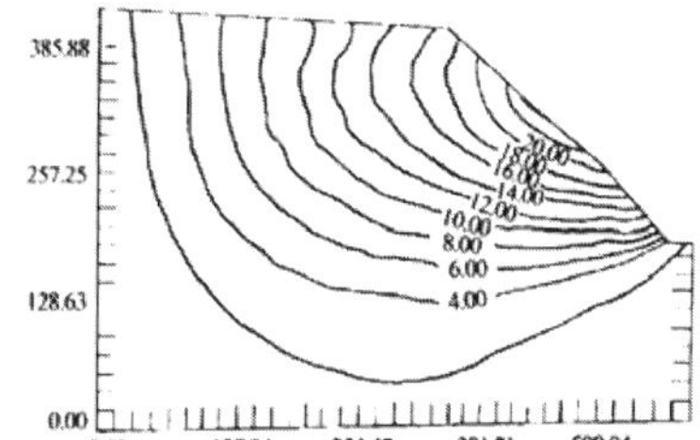

Figure7. The results of X-displacement by loading condition of the Jinchaun slope

From these figures, we can see that the results by the above method are accorded with the deformation values by in-situ test. But the results by the classical method, because of using the loading parameters and calculation model differing from the in-situ condition, are different about 40 times to the in-situ test data.

5 Conclusions

From above analysis and case studies, 1) the most conservative results of the rock mechanics properties by dynamic effect analysis are different from the results by loading rock mechanics. 2) The displacement values by unloading dynamic effect analysis are accorded with the values by in-situ test and model experiment. But the results by the classical method, because the mechanics parameters and calculating model based on the loading rock mechanics are used, are far more different from the in-situ test values. 3) The main reason of above difference is that many conditions of rock unloading excavation, including scale effect, numeral value of unloading, releasing level of geostatic stress and the mechanics dynamic effect, are all different from those of "*Loading Rock Mechanics*". The unloading dynamic effect acts as a crucial role in the rock excavation.

References

1. Q.L. Ha and Y.X. Zhang. Rock Slope Engineering. Chongqing University Press, Chongqing, P.R.China, 1995
2. Geology Laboratory of China Sci. Ins. Geology Mechanics Problems of Rock Mass Engineering. Science Press, Beijing, P.R.China, 1985
3. Y.X. Zhang et al, Nonlinear Analysis of Unloading Rock Mass Related to the Shiplock High Slope of Three-Gorge Project. Journal of Chongqing Jianzhu University. 1997, Vol. 19 No 2: 72~79
4. Q.L. Ha and Y.X. Zhang. The Rock Slope Anisotropic Unloading Nonlinear Rock Mass Mechanics of Yangtze Three-gorge Project—Rock Mass Mechanics Studies on the Shiplock High Slope (3). Chinese Construction Industry Press, Beijing, P.R.China, 1997
5. Q.L. Ha et al. Studies on the Macroscopic Mechanics Parameters of Slope Unloading Rock Mass of Yangtze Three-gorge Project—Rock Mass Mechanics Studies on the Shiplock High Slope (2). Chinese Construction Industry Press, Beijing, P.R.China, 1996
6. Q.L. Ha and Y.X. Zhang et al. Nonlinear Unloading Rock Mass Mechanics of Joints Rock Mass. Chinese Construction Industry Press, Beijing, P.R.China, 1998

RANDOM SIMULATION OF GROUTING IN ROCK MASS

ZHE HAO   SHUHONG WANG   BIN LIU
Resource and civil Engineering School, Northeastern University, Shenyang, China, 110006

In this paper, on basis of the practical grouting in Laiwu Iron Mine, Shandong Province, a program was developed to simulate the true flow situation of grout in rock mass.

1 Introduction

Rock mass is geological body consisted of rock and cross cracks. Through flowing and solidifying of grout in cracks, controlling underground water and enhancing strength are attained. Therefore, to study diffusion rule of grout in cracks is very important. To crack grouting in rock mass, only single crack and uniform distributing cracks can be simulated, and analysis of crack grouting are on the basis of engineering comparison and grouting theory of soil mass, have big error.

Aiming at grouting practice in site, this paper simulated crack grouting process, a 2-D crack grouting random simulation program was edited.

2 Theoretical analysis

Parameters of cracks were measured by method of measuring line. Through observing and analyzing typical cracks on outcrop, the probability distribution and correspond geometry parameters of cracks in rock mass can be obtained.

According to measured result of crack parameters in site, the random variables were grouped and distribution functions were fitted. Then the Monte-Carlo method was applied to simulate the distribution of cracks networks.

The iterative equation of grouting diffusing simulation in cracks is theoretical equation of diffusion in single crack by Newton liquid calculation model.

$$q = -\frac{\Delta P \cdot b^2}{12\mu L} \Rightarrow P_2 = -\frac{12\mu q L}{b^2} + P_1 \text{ ,} \qquad t = \frac{L}{\upsilon} \tag{1}$$

where, ΔP -pressure difference in crack, b -width of crack, L -length of crack, μ -viscosity, q -speed flux, t-diffusion time.

The model applied is dispersed medium model. To suppose that grout don't seep in rock, flow is only in cracks.

3 Measured result of cracks parameters in site

The Laiwu iron mine locates in the area of Laiwu basin of Shandong Province. It is a great magnetite. It consists of four strata. Calcareous rock (3rd stratum) contains rich water and seeps into the mining area. It leads to hard mining work. Therefore, we finished the grouting for calcareous rock by single grout.

By measured original data, the grouping of cracks was finished (Table 1). The results to fit parameters distribution are in table 2.

Table 1 Results of grouped cracks

Group	Strike (°)	Trend (°)	Dip (°)
I	71	161	72
II	32	122	33
III	214	304	62

Table 2 Plotted results of crack distribution parameters

Group	Quantity	Parameter / Eigenvalue	Strike (°)	Dip (°)	Spacing (m)	Length of Trace (m)	Width (mm)
I	106	Average	71.35	70.534	0.313	1.46	2.810
		Standard error	20.04	10.52	0.116	0.50	0.902
		Distribution	Normal	Normal	Exponent	Normal	Normal
II	48	Average	32.39	31.24	0.562	1.866	1.945
		Standard error	5.47	7.36	0.103	0.584	0.624
		Distribution	Lognormal	Normal	Normal	Lognormal	Normal
III	93	Average	214.12	62.27	0.353	1.682	2.649
		Standard error	16.51	10.0	0.110	0.50	0.750
		Distribution	Lognormal	Normal	Exponent	Normal	Normal

4 Random simulation of grouting diffusion

Developed program includes four modules. Its functions are drawing net distribution of cracks, drawing dynamic flow situation of grout on screen, calculating diffusion radius of grout, and drawing isobaric lines, etc. Simulation area is 12m×10m. Crack net distribution figure is as fig.1. Isobaric lines in diffusion area is as fig.2. Diffusion at the different moments is as fig.3.

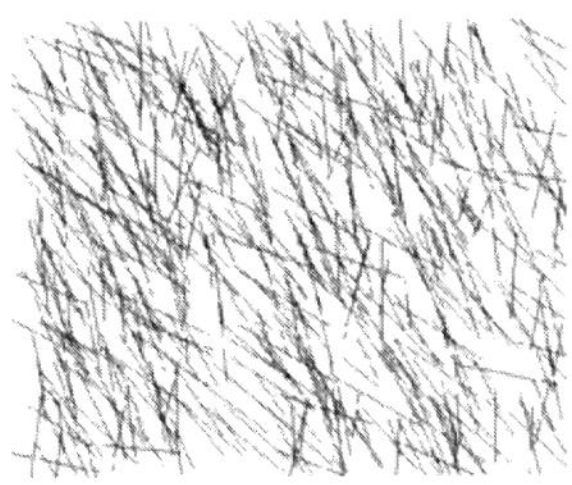

Figure 1. Crack net distribution

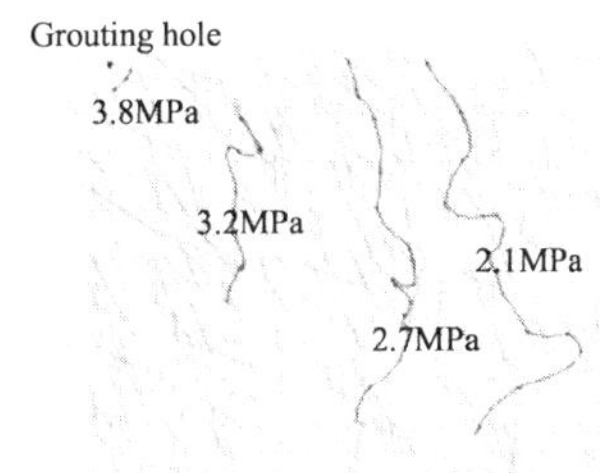

Figure 2. Isobaric lines (P=3.8，3.2，2.7，2.1(MPa))

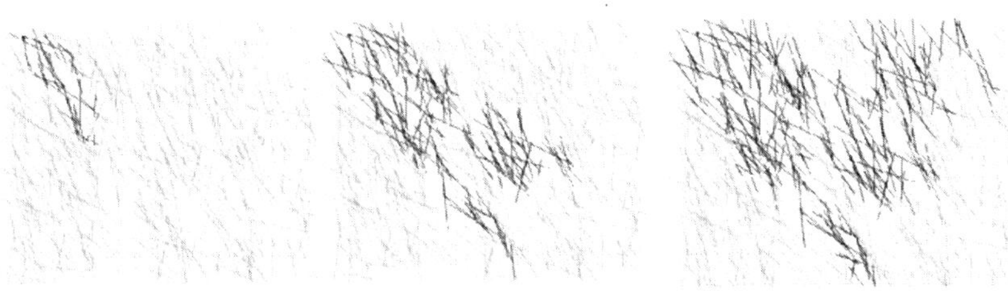

Figure 3-1. 500s Figure 3-2. 2000s Figure 3-3. 3000s

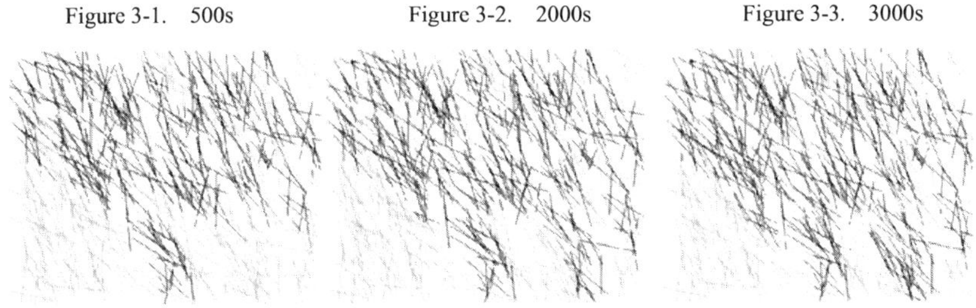

Figure 3-4. 5000s Figure3-5. 7000s Figure3-6. 10000s(final diffusion result)

Figure3. Diffusion of grout at different moments

5 Main conclusions

The simulation results of grouting indicated:

(1) Diffusion of grout has obvious directivity. The main direction of diffusion is identical with main direction of cracks growth. Diffusion speed of grout slower gradually, its reason is that pressure and flux in cracks decrease.

(2) Diffusion shape is fixed after some times (about 7000s). This reason is

solidifying and equilibrium of grouting pressure and hydraulic pressure.

(3) Distribution of isobaric lines is no regular. Generally, the shape of isobaric lines also is corresponding with growth characteristics of cracks.

(4) The calculation result of diffusion average radius is 8.763m. Design need of grouting in site was satisfied, but least diffusion distance of calculation is just 2.80m, some auxiliary hole must be increased to ensure need of grouting in site.

(5) This program doesn't consider cleavage effect. Cleavage will increase connectedness of cracks, therefore, some isolated cracks will be diffused, and the diffusion speed will be quickened, flux of grout will increase, and grout will diffuse even farther. Therefore, actual grouting effect will better than simulated.

(6) Non-uniformity of grout diffusion make the grouting in rock mass more complicated, and security is worse than in soil mass. We must observe cracks earnestly in construction, add repetitive grouting and auxiliary grouting.

This simulation program has wide application foreground in theoretical analysis, design and construction in site. But because the distribution of cracks in site is very complicated, a great deal of situation is difficult to be considered fully, and 2-D simulation results can't exactly accord with actual diffusion situation in all kinds of grouting site. These questions are awaited improving.

References

1 Zhu Wenbin. simulation of crack networks. Mine and metallurgy engineering, 1992,12(3).

2 Xu Zhongji. Monte-Carlo method. Shanghai science and technology press,1985.

3 Hao Zhe. Computerized simulation of crack grouting in rock mass. Chinese journal of geotechninal engineering,1999,21(6).

4 Nonveilley E. Grouting Theory and Practice. Elsevier Science Publisher B.V..The Netherlands,1989.

Appendix

Zhe Hao, 1972-, male, associate professor, doctor, post-doctor of engineering mechanics at post-doctoral station of NEU.

ANALYSIS AND TESTING OF CONTROLLED BLASTING VIBRATION UNDER COMPLEX CONDITIONS IN COLDER AREA

HAIWANG YE RUIGENG ZHU WENHUA ZHU
Wuhan University of Technology Wuhan 430070 P. R. China

Experiment of blasting vibration, controlled measures and tracking testing are discussed. The spreading rule of blasting shockwave in conglomerate is found by similar simulate experiments. Based on the results of testing, a conclusion can be drawn that the shock absorption is effective.

1 Introduction

Xinjiang Shimenzi reservoir lies in the southwest of Manasi County. The reservoir dam is built at the gorges of the upper reaches of Taxi River, about 1.2km far away from Taxi River hydrological station. The dam is made of ground and pressed concrete and the figure of the dam is double arch, therefore it is very important to make the dam shoulders stable. During the process of the dam building, there were evidences to show the existence of F5 fault in the left shoulder of the dam, and the connectivity rate of F5 is 100 percent. Considering the safety of the dam, measures should be taken to reinforce the left shoulder. That is to dig two laneways along the F5 fault at the altitudes of 1289m and 1350m. And below the 1350 laneway, at the altitude of 1289m a laneway is dug at the same time, which is used to drain the groundwater in the hill. After digging, the laneways must be reinforced with reinforced concrete.

In the rock entity of the left shoulder of the dam, there is a special concrete curtains that is used to prevent groundwater in the reservoir from filtering into the rock entity of the left shoulder.

Therefore, in order to ensure the security of the concrete curtains, it is necessary that the spot experiments be conducted before the formal blasting construction. As a result, a good blasting way arises. A good blasting way refers to the fact that the blast does no harm to the curtains. And every time the blasting can be tested by vibration instrument during the process of blasting.

2 Similar Simulate Experiments of Shock Absorption Blasting

2.1 Experiments on the Spreading Rule of Blasting Shockwave without Shock Absorption Measure

The similar experiments are performed with the absence of shock absorption hole and shock absorption slot. The blasting holes and the charges are shown in Figure 1. With different charges and in different detonating orders, we can get a series of data about the blasting vibration velocities of rock particles.

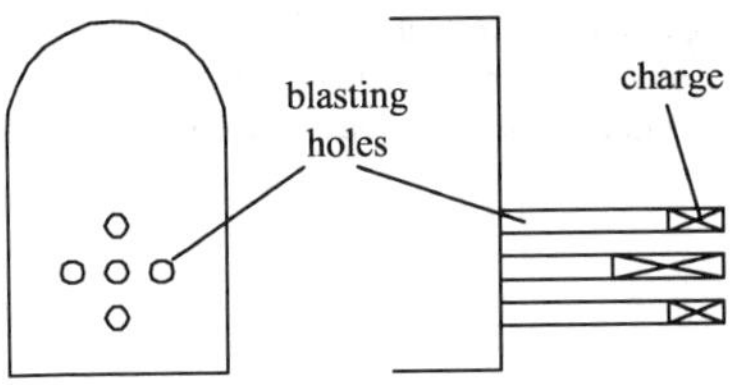

Figure 1. Disposal of blasting holes and charges under no shock absorption measure

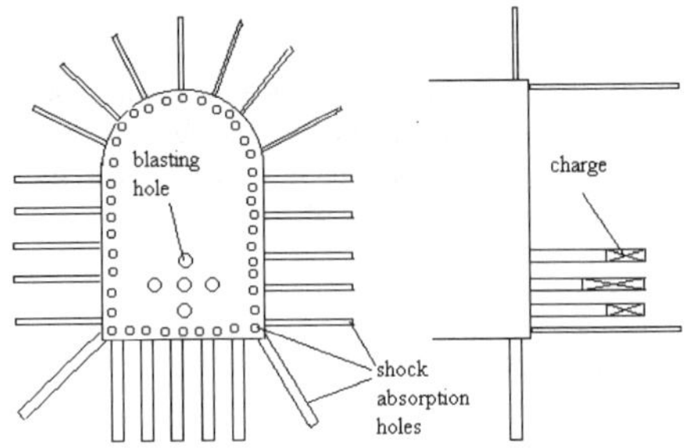

Figure 2. Disposal of blasting holes and charges under shock absorption measure

2.2 *Experiments on the Spreading Rule of Blasting Shockwave with Shock Absorption Measure*

The contrasting similar experiments are performed with the shock absorption hole or shock absorption slot. The blasting holes and the charges are shown in Figure 2. With different charges and in different detonating orders, we can also get a series of data of blasting vibration velocities of rock particles.

2.3 *Experiment Results and Analysis*

Based on the data from the experiments, a conclusion can be drawn that the vibration velocities under the condition with shock absorption measure are smaller than those under the condition without shock absorption measure.

According to the experiment data, a regression analysis can be made, and then a formula of computing the blasting vibration velocities can be drawn as follows

$$v = 176\left(\frac{\sqrt[3]{Q}}{R}\right)^{2.0} \tag{1}$$

Where v is the blasting vibration velocity, Q is the biggest charge at one time, R is the distance from testing spot to the heart of charge.

3 Testing of Blasting Vibration

Based on the contrasting experiments and the requests of blasting, controlled blasting with two shock absorption is adopted. All holes are not detonated at the same time during the blasting, between which there is a delay about 10~50 millisecond. With the augment of the distance between the blasting spot and the curtains, the charges get bigger.

Since the curtains below the altitude of 1289m is finished, the blasting will be held at the altitude of 1289m. Thus the influence of the blasting on the curtains is the worst and it is necessary to test the vibration of every blasting. The testing spots are laid out in Figure 3. The testing spots are designed on the top of the curtains in 1289 covered way.

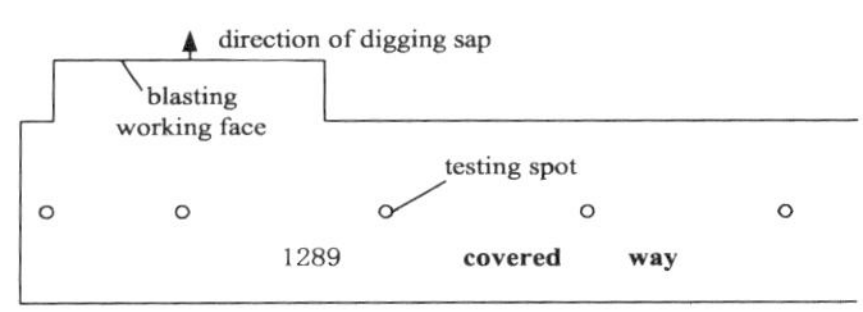

Figure 3. Layout of testing spots

4 Analysis of Testing Results

It took us eight months to carry out the tracking testing. Part of testing data is available(Table 1).

Table 1. Part of testing data

Serial number	Size of dynamite /kg	Distance /m	vibration velocities /cm/s
1	0.0375	3.0	1.2435
2	0.0375	3.5	0.9662
3	0.0375	4.0	1.1239
4	0.050	7.5	0.9255
5	0.075	9.5	0.6031
6	0.150	10.5	0.5482
7	0.300	12.0	0.6805
8	0.450	20.0	0.8439
9	0.600	22.0	0.9594
10	0.750	26.0	0.7622

According to the analysis of the testing data and the testing on site, another conclusion can be drawn that all blasting vibration velocities of particles are less than 1.5cm/s, which is the designed criterion. Typical blasting vibration wave and frequency charts are shown in Figure 4. During the testing, it is found that the vibration of the blasting holes near the floor is stronger than that of other parts.

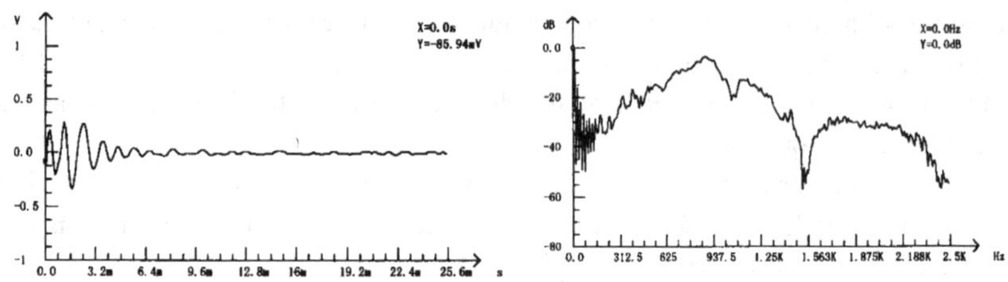

Figure 4. Typical blasting vibration wave and frequency chart

5 Conclusions

In the colder area, the integrality of conglomerate is very good, so the energy of blasting vibration shockwave attenuates slowly when the blasting vibration shockwave transmits in the rock. It is very necessary to adopt shock absorption measure. Judging from the experiment and the testing results the effect of the shock absorption slot is obvious. All vibration velocities of particles are less than 1.5cm/s.

During the course of security testing, information construction is adopted, which means adjusting the parameters of later blasting according to the testing results. In this way, not only the blasting vibration is effectively controlled but the schedule of construction is made quicker on condition that the safe of the curtains is ensured.

References

1 Jifu Meng, Hongbin Xi, Blasting Testing Techniques, Beijing: Metallurgy Technology Publishing Company, 1992, P.R.China.

ABUTMENT PRESSURE TRANSFERRING RULES IN THE FLOOR OF DEEP ISOLATED FULLY-MECHANIZED MINING FACE WITH SUBLEVEL CAVING

ZHONGCHENG QIN CHENGLUN LIU
Shandong University of Science and Technology, Taian, 271019, *China*

Aimed at the mining condition of No.4403 isolated fully-mechanized mining face with sublevel caving in Dongtan Coalmine, the abutment pressure transfer law in the floor is studied on the basis of analyzing the abutment pressure distribution rules of face No.4303. It can provide basis for controlling additional roadway deformation efficiently and ensuring the regular production of the mine.

1 Introduction

Dongtan coalmine is a super modernized mine founded in 1989, the mining depth is reached 700m. By using sublevel caving mining method, the technique of road driving along next goaf has been adopted in entries layout, thus appears several isolated fully-mechanized mining faces with sublevel caving. Face No.4303 is a typical example, it is using longwall mining to the dip with fully-mechanized mining face with sublevel caving method. The west is the gob of Face No.4304 (slice mining with fully-mechanized mining), while the 4302's gob is on other side (full-seam extraction with sublevel caving). Its influence range and transfer intensity in the floor of the abutment pressure are increasing evidently.

2 Abutment pressure distribution of Face No.4303

Stress increasing coefficient (K) is usually used to express the magnitude of abutment pressure, while the abutment pressure distribution is expressed as the fractured zone width (x_1), plastic zone width (x_0) and influence distance (L_1), etc. By in-situ observation and theoretical analysis, the abutment pressure distribution of face No.4303 owns the following main characteristics:

1. Before the face extraction

The roof activity of face No. 4302 and No. 4304's goaf has tended to stable, and a stable "*rock bridge structure*" has been formed at the common boundaries.

(1) The peak value coefficient of the fixed abutment pressure K=1.5~2.0.

(2) The abutment pressure distribution area on the solid coal:

On the side of face No.4303, K=1.5, x_1=3.75m, x_0=4.50m, L_1=47.3m;

On the side of face No.4302, K=2.0, x_1=4.10m, x_0=5.61m, L_1=66.8m;

The width of original stress zone in the middle of face, L_0=86m.

2. During the face extraction

During the extraction of face No.4303, the "*rock bridge structure*" has also been formed along the strike of the face. With the face advancing, the "*front moving abutment pressure*" (Ⅰ) is formed on the solid coal ahead of the middle part of the face; While a more complicated "*superposition abutment pressure* " (Ⅱ) is formed on the solid coal both ahead of the face ends; the "*rear moving abutment pressure*" (Ⅲ) is also formed in the back waste. The distribution rules are shown in table 1.

Table 1 Abutment pressure distribution of each region

Types		K	X_1 (m)	X_0 (m)	L_1 (m)
Ⅰ		2.0~3.0	4.0	7.1	86.8
Ⅱ	face No.4303 side	3.0~3.5	4.0	7.6	93.2
	face No.4302 side	3.5~4.0	4.0	8.1	98.5
Ⅲ		1.2~1.5	3.8	3.6	22.0

3 Transferring of abutment pressure in the floor

For the conditions of face No.4303, two sides has mined out, so a 3-D calculating model was used to calculate the stress at any point in the floor.

The vertical stress at arbitrary point $M(x,y,z)$ in the space for a concentrated force P acting on a 3-D semi-infinite space is

$$\sigma_z = -\frac{3Pz^3}{2\pi(x^2+y^2+z^2)^{5/2}} \tag{1}$$

Let y=0, it yields
$$\sigma_z = -\frac{3Pz^3}{2\pi(x^2+z^2)^{5/2}} \tag{2}$$

For plane problems
$$\sigma_z = -\frac{2P}{\pi}\frac{z^3}{(x^2+z^2)^2} \tag{3}$$

Obviously, Eq. (2) is different from (3), the equivalent conditions is that acting load is uniformly distributed in region $(-\infty, +\infty)$ along y-direction. This is just about the conditions for plane problem satisfying.

Then, Eq. (2) becomes
$$\sigma_z = -\frac{3Pz^3}{2\pi}\int_{-\eta_1}^{+\eta_2}\frac{d\eta}{[x^2+(y-\eta)^2+z^2]} \tag{4}$$

It can be proved that the integral result of Eq. (4) is Eq. (3).

However, the abutment pressure nearby the face line does not satisfy the above conditions, it must use Eq.(2). The additional stress (σ_z) caused by the face abutment pressure in the floor can be obtained by summing method.

$$\sigma_z(x,y,z) = \sum_{i=1}^{n}\sigma_z^{(i)}(x,y,z) \tag{5}$$

Where, $\sigma_z^{(i)}(x,y,z)$ is the additional stress caused by the face abutment pressure of ith zone at point $M(x,y,z)$ in the floor.

Suppose the ith zone in the abutment pressure distribution coordinate system is: $[\xi_{1i}, \xi_{2i}]$ in x-direction, $[\eta_{1i}, \eta_{2i}]$ in y-direction, the abutment pressure (use increasing coefficient k) distribution does not change with x, but with y linearly, i.e.:

$$k(\xi,\eta)=k_{1i}+\frac{k_{2i}-k_{1i}}{\eta_{2i}-\eta_{1i}}(\eta-\eta_{1i}) \tag{6}$$

Then

$$\sigma_z^{(i)}(x,y,z)=-\frac{3z^3}{2\pi}\int_{\xi_{1i},\eta_{1i}}^{\xi_{2i},\eta_{2i}}\int\frac{[k_{1i}+\frac{k_{2i}-k_{1i}}{\eta_{2i}-\eta_{1i}}(\eta-\eta_{1i})]d\xi d\eta}{[(x-\xi)^2+(y-\eta)^2+z^2]^{5/2}} \tag{7}$$

By integral,

$$\begin{aligned}\sigma_z^{(i)}(x,y,z)=&-\frac{3z^3}{2\pi}[k_{1i}+\frac{k_{2i}-k_{1i}}{\eta_{2i}-\eta_{1i}}(\eta-\eta_{1i})][f_{11}(u=y-\eta_{2i})-f_{11}(u=y-\eta_{1i})]\\&-\frac{3z^3}{2\pi}\frac{k_{2i}-k_{1i}}{\eta_{2i}-\eta_{1i}}[f_{21}(u=y-\eta_{2i})-f_{21}(u=y-\eta_{1i})]\end{aligned} \tag{8}$$

Where $f_{11}=\frac{u}{3}[g_1(t_2)-g_1(t_1)]+\frac{2}{3}u[g_2(t_2)-g_2(t_1)]$

$f_{21}=\frac{1}{3}[g_3(w=x-\xi_{2i})-g_3(w=x-\xi_{1i})]$

In which $g_1(t)=\frac{1}{(z^2+u^2)u^2}[\sin t-(1+\frac{z^2}{u^2})\frac{u}{z}arctg\frac{u\sin t}{z}]$

$g_2(t)=-\frac{\sin t}{z^2+u^2\sin^2 t}(\frac{1}{2u^2}+\frac{1}{2z^2})+\frac{1}{2zu}(\frac{1}{u^2}-\frac{1}{z^2})arctg\frac{u\sin t}{z}$

$t_1=arctg\frac{x-\xi_{1i}}{\sqrt{z^2+u^2}}$; $t_1=arctg\frac{x-\xi_{2i}}{\sqrt{z^2+u^2}}$; $g_3=\frac{w}{[u^2+z^2][w^2+u^2+z^2]^{1/2}}$

For the abutment pressure coefficient k of certain zone is changing linearly only along x-direction, the only thing is to replace η_{2i},η_{1i},y with ξ_{2i}、ξ_{1i}、x.

For the fixed abutment pressure far away from the faceline, the induced additional floor stress $\sigma_z^{(i)}$ can be calculated by using plane strain model. Suppose the ith zone with left boundary [ξ_{1i}, k_{1i}] and right boundary [ξ_{2i}, k_{2i}], then

$$\begin{aligned}\sigma_z^{(i)}=&-\frac{2z^3}{\pi}\int\frac{[k_{1i}+\frac{k_{2i}-k_{1i}}{\xi_{2i}-\xi_{1i}}(\xi-\xi_{1i})]d\xi}{[(x-\xi)^2+z^2]^2}=\frac{2z^3}{\pi}[k_{1i}+\frac{k_{2i}-k_{1i}}{\xi_{2i}-\xi_{1i}}(x-\xi_{1i})]\cdot[g_1(x-\xi_{2i})\\&-g_1(x-\xi_{1i})]-\frac{2z^3}{\pi}\frac{k_{2i}-k_{1i}}{\xi_{2i}-\xi_{1i}}[g_2(u=x-\xi_{2i})-g_2(u=x-\xi_{1i})]\end{aligned} \tag{9}$$

In which $g_1(u)=\frac{u}{2z^2(u^2+z^2)}+\frac{1}{2z^3}arctg\frac{u}{z}$; $g_2(u)=-\frac{1}{2(z^2+u^2)}$

4 Additional pressure distribution law in the floor of face No.4303

The condition of abutment pressure is constant, the magnitude of additional pressure is only related with the floor point position $M(x, y, z)$. According to these above analysis, we have programmed and calculated the additional pressure distribution rules before and during face No.4303 extraction respectively.

1. Before face No.4303 extraction

Only fixed abutment pressures caused by face No.4302 & 4304 were acting on floor, so plane strain model was used to calculate the floor additional pressure(Figure1)

2. During face No.4303 extraction

the calculated results by using the 3-D model are shown in Fig.2,3.

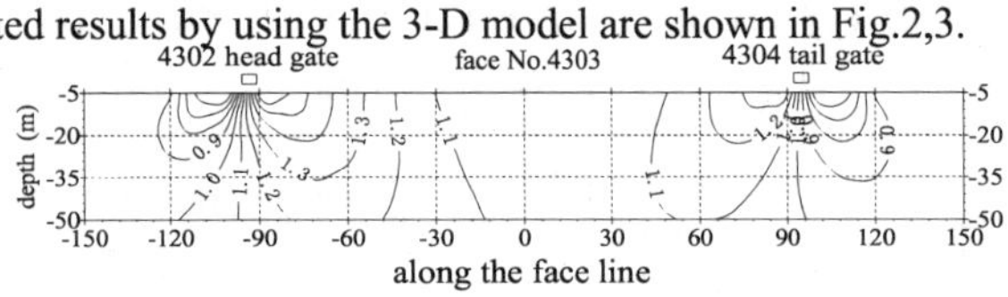

Figure1. Additional pressure distribution on the vertical section (y=10m)

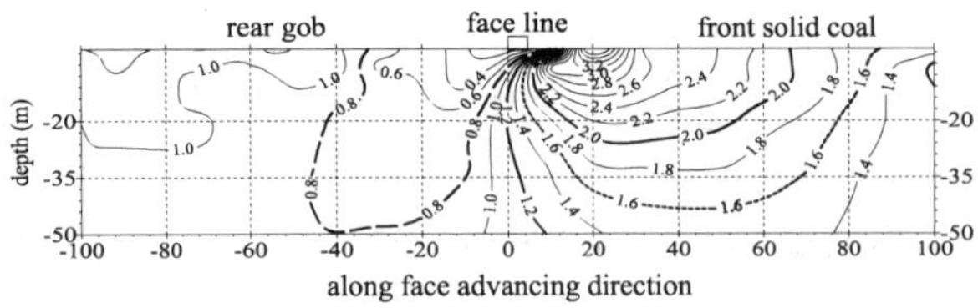

Figure2. Additional pressure distribution on the vertical section (x=10m)

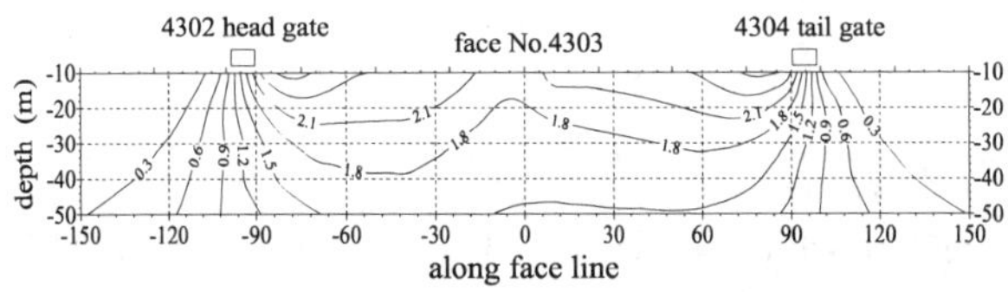

Figure3. Additional pressure distribution on the vertical section (y=10m)

5 Conclusions

(1) Before face No.4303 extraction, the additional pressure reduces gradually as the interval to the seam bottom increases; But the corresponded pressure distribution range increases gradually[2]. Due to the different conditions between face No.4302 and 4304, the floor additional pressure at the 4302 side is high than the 4304 side.

(2)Along with the face advancing, every floor point will suffering the dynamic changing process of additional pressure, i.e., increasing→decreasing→reverting. The magnitude and range of the additional pressure are related with the strike cross-section position; on the same cross-section, it relates to the vertical interval.

References

1 Yanguang Chen, Shiliang Lu. Roadway Surrounding Rock Control in China Coalmine. China University of Mining and Technology Press.1994

2 ZhongCheng Qin, Tongxu Wang. Additional Stress Analysis of Soft Rock Roadway Bestrided over Mining by Isolated Fully-Mechanized Mining Face with Sublevel Caving. Ground Pressure and Strata Control. 2002(2)

PRACTICAL APPLICATION OF THE GROUTING TECHNOLOGY IN SEPARATED STRATA FOR CONTROL SUBSIDENCE AT DONGTAN MINE

DESHEN ZHAO
Dalian University, Dalian 116622, P.R..China

HUIQUAN SUI
Dalian Institute For National Minorities, Dalian, 116600, P.R..China

LIEDONG WANG
Dalian University, Dalian 116622, P.R..China

The technology of grouting into overburden-separation for control subsidence is a patent technology. It has been used for several years at many mines in our country and acquired remarkable effects to control mining subsidence. One of the practical application at Dongtan Mine, face №14308(E), is introduced. The effects of ground subsidence reduced, benefits of economy and interests of public are analyzed in this paper.

1 Introduction

Yanzhou mine area located in Shandong province of China, where hypsography evenness, village denseness, soil fertility, not only is one of our important coal production bases, but also is main agriculture regions.

Dongtan is one of the main production mines in Yanzhou mine area, every year excavate more than 5 millions coal from underground. The №14308(E) was mostly working face in 1999, at that time the face was all ready to be mined underground. But at the surface there is a village.

If any measures had not been taken, the workface would not have been excavated in time. The village must be moved at first. For solving the problem between underground mining in gear and some buildings safety in the village, after Dongtan mine leaders and some experts demonstrating, the technique method to grout ash-water slurry into overburden-separation for control subsidence, was taken for the best way.

2 General Situation

№14308(E) is a full side working face used breaking coal technology, located at Dongtan mine, 14 area, section 8, east. The overburden like this: alluvium 131m, red sandstone and fine sandstone in Jurassic Period 269m. Carboniferous Period coal measures strata composed of the mixed mudstone and fine sandstone, even thickness is 162m. Here, geological structures are simple and not bigger faults. The ground average elevation is +45.5m. Even mine depth is 563m, coal strata inclination angle is 3° ,188m long and 914m wide of the underground working face, to be excavate coal thickness is 5.6m. The mining method of full side face and breaking roof technology would be taken. The face advance would be about 5 meter everyday.

The village named Tantaimu at that region, 90 thousands square meters, would be affected by the ground subsidence due to the underground mining. In the area there are some villager's houses, Yan-Zou road, Ni river and bridge, besides, are all good farmlands. Most farmer houses made up brick and rock, which could not resist any deformation.

3 Engineering Design of Separated Strata Grouting

The technology of grouting in overburden separation to control and reduce subsidence is developed by author, associate with Fushun coal mine department in the 1980s. During past 15 years, it has been used in some coal mines in our country, such as Fushun, Kailan, Datun, Xinwen, Yanzhou, Fengcheng and so on, and obtained better benefits of economy and society. The technology based on the theory of structure of "arch-plate balance" [2]. The ultimate principle is through the drill to breathe ash slurry into the overburden-separation zone to control the rock strata settling and realize the purpose of control ground subsidence. Comparing other ways, the technique has some special points: 1. It need not take any other measures at underground working face, do not affect the system of developing and working. And these are propitious to high-yield and high-effectiveness. 2. The technique is simple for operating. 3. That investment is smaller but the income greater. 4. To reduce subsidence, the effects are obvious (the subsidence modulus can be reduced at 0.2 to 0.4). 5. Using the ashes come from power station that could decrease the ashes pollution to the environment.

Considering the situation of the high rapidity underground mining, requiring more volumes slurry per unit time and some other reasons, this time we take some new experiments in our project as the position of bore holes, hole number, complete sets of equipments, grouting technique development and so on. Those experiments are all achievement.

Fix on the hole positions: At the ground surface set two bore holes for grouting that located the strike main section of subsidence basin. The two holes is apart from 120 meters, the first one is 150 meters from the start place of mining. According to the structure and the separated pattern of strata, by synthesis study all data, the two holes depth were decided, one was 438 meters, another was 432 meters.

Complete sets of equipment: Because the project to be demanded seriously, large numbers ash-slurry will be needed to grout in short time, so in the project five pressure pumps were fit out in the grouting station, one high pressure pump type

DSB-300, three middle high pressure pumps type TBW-850, one low pressure pump type NBB-250, which build up a large grouting system. The system can ensure in any times and any situations work well. The grouting techniques, system and some methods were developed. For the first time, the double pipeline technique was used to meet the needs of large flux grout. In the grouting pump station every pipes were set flux instruments to control and memorize the quantity of ash-slurry.

4 Effect Analysis

In 1999, from March 31, to July 12, the project had grouted ash-water total 291 thousand cubic meters, hereinto ash 69 thousand cubic meters, which were 6 thousand cubic meters less than plan's. At that time of the project over, the pressure at mouth of bore hole was 0~3 MPa at ground which did not reach the design value of 15 MPa.

Effect analysis of subsidence reduced: For measure movement and deformation of ground subsidence and buildings and estimate the effect of reduce subsidence, some survey points were set up at ground and built up a observation station. Now taking the points number 801~810 distributed the strike main section for example to analyze the effect.

In 1999, from March 4, to September 20, the observation line had been surveyed 23 times. The greatest settle value measured finally was 1.496 meters, it was point number 810, compared with the value of calculated, the reducing rate was 38.5 per cent.

To the actual result, separated strata grouting have controlled the ground subsidence effectually and achieved the engineering purpose, but to the value of reduce settle rate, it did not reach the design required (50%). The reasons are following: ① More than 6 per cent impurity in the coal ash get from local power station, made the pumps more trouble, the grouting system malfunction rate increase, so that it affected the time and lowered the quantity of grouting. ② At the end of the project, the pipes pressure had been lower and could keep on grouting yet, but it was interrupted for some reasons. ③ Near the working face there is face 14307(E) that had been excavated before that. So during the face 14308(E) was mining, extra subsidence would be brought. At the same time, it was possible that the ash and water liquid should run off other place.

Analysis of ground settling speed: It is presented in Fig. 2 that the correlation with position of working face and subsidence and settling speed of point number one. And also something is showed that: ① When the face advance part from start position had not reached 112 meters, ground to begin movement distance in Yanzhou, the point 1 had been measured settle 122mm.It is obvious that the result is old excavation reactivity affected. If these factors were taken into account that the reduce rate of point 810 settle should be 44 per cent. ② During the ground subsidence the points settle speed had not initial stag, which also had showed the

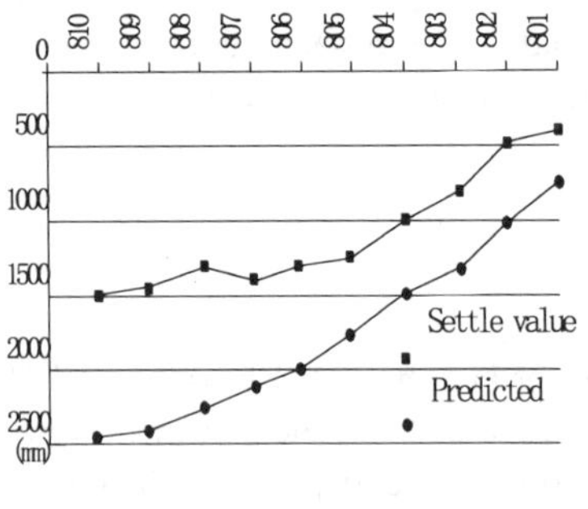

Figure1. contrast map between the measured values and the predicted values of the earth's surface subsidence

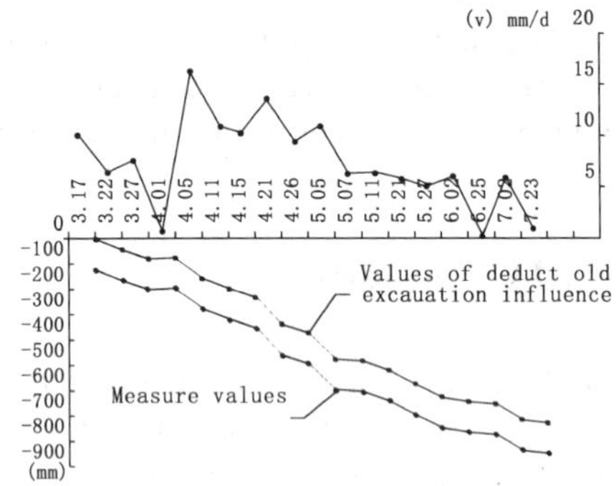

Figure 2. Graph of correlation with position of working face and point 1 subsidence and settling speed

fact that the ground subsidence was affected by old excavation. In the curve of point settling speed, it can be seen that the speed changed in the mode of waves with time. The maximum of peak value last a short time that illuminate the ground settle speed to be reduced obviously by the separated strata grouting project. ③ As mining advance, the settle speed of point 1 has been slowed down rapidly. During June 2 to July 23, the speed was only 1.06 mm per day. But after end grouting, at September 20 measure data showed that the point speed turn round increased, which indicated the effect of grouting.

Analysis of benefits of economy: According to the law of ground subsidence, if the ground buildings were not damaged when the coal face mining, more than 500 thousand tons coal should remain in underground as protecting pillar. After taken the technique of separated strata grouting, 19000 thousands Yuan had been received. That economized the cost, 946 thousands Yuan for highroad maintain and bridge rebuild, 2975 thousands Yuan pay for the farmland. The project invested 5400 thousands Yuan for operating and 1200 thousands Yuan for maintaining houses partly. So the total benefits of economy Q given by

$$Q=(1900+94.6+279.5)-(540+120)=1614.1 \quad \text{ten thousands Yuan}$$

HYPERSINGULAR BIE: METHODS AND APPLICATIONS IN RUSSIAN MINING PRACTICE

VICTOR V. ZOUBKOV

Dr., State Scientific Research Institute for Geomechanics and Mining Surveying, VNIMI, Saint-Petersburg, 199106, Russia. E-mail: vzoubkov@pop3.rcom.ru

Two advantages of using hypersingular forms of the elasticity BIE in geomechanics are emphasized: (i) these BIE are formulated in terms of the very values which characterize contact interaction (displacement discontinuities and tractions) and (ii) they have zero index what saves from additional conditions usual for singular equations. For 3-D problems, simple quadrature rules for evaluation of hypersingular integrals are available. For 2-D problems, real and complex variable (CV) formulations are compared and computational advantages of the CV-BEM are emphasized. In the case of a layered medium, the spectral BEM is recommended as the most efficient method *ad hoc* for solving hypersingular BIE. Special attention is paid to the Fast Fourier Transform and to accounting for asymptotics at high and low frequencies. It is noted that the programs FAULT3D, HYPERBLOCKS2D and LAYERS3D, based on these methods are developed and systematically used in Russia for solving problems of planning mining workings, pillar stability, dynamic phenomena in mines and also environmental problems caused by earth subsidence and mine closure.

1 Introduction

Boundary integral equations (BIE) proved to be a useful tool for solving problems arising in daily mining practice. Programs based on the BIE usually employ the boundary element method (BEM). Such are, for instance, the programs MULSIM [1], MINSIM [2], MAP-3D [3], SUIT3D [4-5], developed and used in a number of countries, including Russia. Immense literature, which covers this topic, is beyond the scope of the paper: it is impossible to review it here even briefly. Our aim is to present and to share with colleagues specific features of approaches, numerical techniques and codes developed in Russia[1] and to illustrate their applications in Russian mining practice. We hope that the gained experience may be of use for further enhancements of existing methods and codes.

2 Hyper-singular equations as the best choice among BIE

There exist various forms of BIE, obtained by the direct and indirect approaches, which may serve for numerical solution of boundary value problems. One needs to choose among them those forms which are most appropriate for a particular problem

[1] Some of the results were obtained in close collaboration with our French colleagues [8, 10].

under consideration. Our mining problems have two features influencing the choice:

(i) we are interested in extraction of tabular deposits and

(ii) we need to account for surfaces of discontinuities such as faults, boundaries of interacting blocks and cracks.

Both these features impel us to use *hyper-singular* equations (HSIE) because they are perfectly adjusted to account for displacement discontinuities (DD) and contain the vary values which characterize contact interaction: DD and tractions. For the same reason we prefer to solve the HSIE by employing *finite part* integrals [6-10] without reducing the HSIE to singular or regular equations: such reduction involves derivatives of the DD what strongly complicates boundary conditions, generates eigenfunctions or produces artificial singularities not present in an initial problem. The HSIE normally has zero index what saves from additional conditions usual for singular equations.

Our approach, based on finite-part integrals and used in 3D problems since 1980, resulted in developing the programs SIUT3D [4-5], and FAULT3D [8]. It has close relation to the 'seam' ('face') elements [11] that eventually resulted in the codes MULSIM and MINSIM. Note also that the DD method [12] although avoids using the finite-part integrals, is practically equivalent to our method when boundary elements are rectangles with constant DD. To the date, we developed simple quadrature rules for hyper-singular integrals over triangular elements [8] and implemented them into a code allowing us to account for 3D curvilinear surfaces of discontinuities.

3 Complex variable HSIE as the best choice for 2D problems

In 2D, further enhancements are gained when the HSIE is used in a *complex variable* (CV) form. The CV-HSIE, suggested by Linkov [13-14], has significant advantages over their real counterparts [9]. Not listing all the advantages, mention the main ones: they provide easy evaluation of hyper-singular and singular integrals over *arbitrary curvilinear elements*, simple quadrature rules for all regular integrals over straight and circular-arc elements, and also easy accounting for asymptotes at crack tips and common vertices of wedges.

These virtues of using the complex variable rather than pairs of real variables hold also for singular equations. For this reason, we had used CV *singular* equations for solving 2D mining problems [15] before the CV hyper-singular BIE became available. But now, accounting for the said above, we focus primarily on developing and employing codes based on the complex variable *hypersingular* equations. Such a code HYPERBLOCKS is used for 2D blocky systems with arbitrary contact conditions on interacting boundaries of blocks and cracks. The code also accounts for extraction of a tabular deposit: as usual in [12], an extraction opening is represented by elements with displacement discontinuity.

4 Spectral BEM as the best choice for multi-layered strata

Spectral BEM, when used with the fast Fourier transform (FFT), significantly reduces time expense [16] for solution of a system of algebraic equations even for a homogeneous medium. It becomes indispensable for 3D *multi-layered* systems. For such systems, its efficiency strongly increases if one uses reduction to three-point difference equations solved by the sweep method [17-18]. Additional reduction in memory and time expense is reached when we employ asymptotic formulae for low and high frequencies [10]. The program LAYERS3D, developed by combining these techniques, allows one to consider up to tens of openings on adjacent seams in a layered medium comprised of up to hundreds of layers with different elastic modules and arbitrary contact conditions.

Acknowledgements.

The author would like to thank to the friend and colleague Dr. A.M. Linkov for support and invaluable help in preparation of this manuscript. The author is also grateful to the Russian Fund of Fundamental Investigations (RFFI) for the support of this work (Grants № 00-05-64316).

References

1. Zipf, R.K. MULSIM/NL theoretical and programmer's manual. US Dept. InteriorBureau of Mines. Information Circular 9321, 1992.
2. Napier, J. A. L, Stephansen, S. J. "Analysis of deep-level mine design problems using the MINSIM-D boundary element program", Proc. Int. Symposium on the Application of Computational Mathematics in the Mineral Industries. Johannesburg, S.A. Inst. Mining and Metall., 1988, **1**, 3-19.
3. Plewman, R.P., Deist F.H., Ortlepp W.D. The development and application of a digital computer method for the solution of strata control problems. J. South African Inst. Mining Metallurgy, 1969, **70**, 33-44.
4. Zoubkov V.V., Zoubkova I.A. At calculate of the protected zones at improvement of the neighbouring seams of suite. Management of rocks pressure and struggle with rock bursts. Leningrad, Proc. of VNIMI, 1980, 99-101.

5. Zoubkov V.V. et al., Evaluation the stresses state of a rock bed near the breakage working of an arbitrary three-dimensional shape. Soviet Mining Science, 1986, **22**, 176-182.
6. Zoubkov V.V. et al., Calculation of the stress state near extraction openings of an arbitrary form in plane, Fiziko-tehnicheskie problemi razrabotki poleznih iskopaemih, 1986, **3**, 24-30 (In Russian).
7. Linkov A.M., Mogilevskaya S.G. Finite-part integrals in problems of three-dimensional cracks. J. Appl. Math. Mech., 1986, **50**, N 5, 652-658.
8. Linkov A.M., Zoubkov V.V., Al Heib M. Computer aided analysis of stressed state and rockburst hazard in veins and coal seams influenced by faults. Proc. 1st South African Rock Engineering Symposium. SARES'97. G. Gurtunca (ed), Johannesburg, 1997, 44-55.
9. Linkov A.M., Mogilevskaya S.G. Complex hypersingular BEM in plane elasticity problems. Singular Integrals in Boundary Element Methods. V. Sladek and J. Sladek (eds), Computational Mechanics Publications, Southampton, 1998, 299-364.
10. Linkov A.M., Zoubkov V.V., Sylla M., Al Heib M. Spectral BEM for multi-layered media with cracks or/and openings. Proc. Symp. IABEM-2000, Association for Boundary Element Methods, Brescia, 2000, 141-142.
11. Salamon M.D.G. Elastic analysis of displacements and stresses induced by mining of seem and reef deposits. J. South African Inst. Mining Metallurgy, 1963, **64**, 128-149; 197-218, 468-500.
12. Crouch S.L., Starfield A.M. Boundary Element Methods in Solid Mechanics. George Allen and Unwin, London. 1983.
13. Linkov A.M., Mogilevskaya S.G. Complex hypersingular integrals and integral equations in plane elasticity. Acta Mechanica, 1994, **105**, 189-205.
14. Linkov A.M. Complex Variables Boundary Integrals Method. St-Petersburg, Science, 1999 (In Russian).
15. Linkov A.M., Zubkov V.V. Boundary integral equations in problems for jointed rocks, Computer Methods and Advances in Geomechanics. Beer, Booker & Carter (eds), Balkema, Rotterdam, 1991, 1747-1750.
16. Peirce A.P., Spottiswoode S., Napier J.A.L. The spectral boundary element method: a new window on boundary elements in rock mechanics. Int. J. Rock Mech. Min. Sci. & Geomech. Abstr., 1992, **29**(4), 379-400.
17. Linkov A.M., Filippov N.A. Difference equations approach to the analysis of layered systems, Meccanica, **26**, 1991, 195-209.
18. Linkov A.M., Filippov N.A., Milova L.A., Zoubkov V.V. A method to calculate stresses and deformations in 3D layered media. Advances in Rock Mechanics. Yunmei Lin (ed), USA-World Scientific Publications Co, NEW YORK, 1998, 135-144.